Mathematics for Elementary Teachers via Problem Solving

Student Resource Handbook

Mathematics for Elementary Teachers via Problem Solving

Student Resource Handbook

Joanna O. Masingila
Syracuse University

Frank K. Lester
Indiana University

Anne M. Raymond
Bellarmine University

Prentice Hall
Upper Saddle River, New Jersey 07458

Acquisition Editor: Quincy McDonald
Editor in Chief: Sally Yagan
Executive Project Manager: Ann Heath
Vice President/Director of Production and Manufacturing: David W. Riccardi
Executive Managing Editor: Kathleen Schiaparelli
Senior Managing Editor: Linda Mihatov Behrens
Production Editor: Jami Darby, WestWords, Inc.
Manufacturing Buyer: Alan Fischer
Manufacturing Manager: Trudy Pisciotti
Marketing Manager: Patrice Lumumba Jones
Marketing Assistant: Rachel Beckman
Editor in Chief, Development: Carol Trueheart
Assistant Editor of Media: Vince Jansen
Editorial Assistant: Joanne Wendelken
Art Director: Maureen Eide
Assistant to the Art Director: John Christiana
Interior Designer: Jill Little
Cover Designer: Joseph Sengotta
Managing Editor, Audio/Video Assets: Grace Hazeldine
Creative Director: Carole Anson
Director of Creative Services: Paul Belfanti
Cover Photo: Bill Aron, Photo Edit
Art Studio: Network Graphics

Upper Saddle River, New Jersey 07458

Printed in the United States of America
10 9 8 7 6 5 4 3 2 1

ISBN 0-13-017879-9

Pearson Education LTD., *London*
Pearson Education Australia PTY, Limited, *Sydney*
Pearson Education Singapore, Pte. Ltd
Pearson Education North Asia Ltd, *Hong Kong*
Pearson Education Canada, Ltd., *Toronto*
Pearson Educación de Mexico, S.A. de C.V.
Pearson Education, Japan, *Tokyo*
Pearson Education Malaysia, Pte. Ltd
Pearson Education Upper Saddle River, *New Jersey*

Brief Contents

Chapter 1: Getting Started in Learning Mathematics via Problem Solving 1

Chapter 2: Numeration 29

Chapter 3: Operations on Natural Numbers, Whole Numbers, & Integers 40

Chapter 4: Number Theory 102

Chapter 5: Data & Chance 127

Chapter 6: Fraction Models & Operations 169

Chapter 7: Real Numbers: Rationals & Irrationals 181

Chapter 8: Patterns & Functions 231

Chapter 9: Geometry 257

Chapter 10: Measurement 328

Resources and Tools 385

Glossary 413

Index 421

Contents

Preface *xxi*

Chapter 1: Getting Started in Learning Mathematics via Problem Solving 1

Set Theory

Topic 1.1 ***Sets and Set Terminology 2***
1.1.1 Sets, Elements, and Set Notation 2
1.1.2 Empty Sets, Finite Sets, and Infinite Sets 2

Topic 1.2 ***Subsets and Set Relationships 4***
1.2.1 Subsets 4
1.2.2 Venn Diagrams, Sets, and Disjoint Sets 4
1.2.3 One-to-One Correspondence 5
1.2.4 Equivalent Sets 5
1.2.5 Equal Sets 6

Topic 1.3 ***Set Operations 7***
1.3.1 Union of Sets 7
1.3.2 Intersection of Sets 7
1.3.3 Complement of a Set 8
1.3.4 Relative Complement (or Difference Set) 8
1.3.5 Cartesian Product of Sets 9
1.3.6 Properties of Set Operations 9

Logic and Problem Solving

Topic 1.4 ***Logic 11***
1.4.1 Inductive Reasoning 11
1.4.2 Deductive Reasoning 12
1.4.3 Conjectures, Hypotheses, Conclusions 14
1.4.4 Statements and the Negation of Statements 15
1.4.5 Compound Statements 15
1.4.6 Truth Tables 15
1.4.7 Conditional Statements 16
1.4.8 Inverse, Converse, and Contrapositive Statements 16
1.4.9 If and Only If Statements 16
1.4.10 Direct and Indirect Reasoning 17

Topic 1.5 ***Problem-Solving Topics 19***
1.5.1 Polya's Process of Problem Solving 19
1.5.2 Problem-Solving Strategies 20

In the Classroom 26

Bibliography 27

Chapter 2: Numeration 29

Topic 2.1 ***Characteristics of Numeration Systems 30***
2.1.1 Characteristics of the Hindu-Arabic System 30
2.1.2 Unique Representation 30

Topic 2.2 *Base-Ten Introduction* 31
2.2.1 Characteristics of the Egyptian Numeration System 31
2.2.2 A Subtractive Characteristic Not Associated with the Hindu-Arabic System 32
2.2.3 Other Bases: The Example of Base Two 32

Topic 2.3 *Place Value and Zero* 34
2.3.1 Place Value 34
2.3.2 A Symbol for Zero 35

Topic 2.4 *Decimal System and More on Base Ten* 36
2.4.1 Decimal System and Expanded Notation 36
2.4.2 Multiplicative Characteristic 36

In the Classroom 38

Bibliography 39

Chapter 3: Operations on Natural Numbers, Whole Numbers & Integers 40

Topic 3.1 *Natural Numbers* 41

Whole Numbers

Topic 3.2 *Introduction to Whole Numbers* 42
3.2.1 Number Representations 42
3.2.2 One-to-One Correspondence 43
3.2.3 Whole Numbers and the Set of Whole Numbers 44
3.2.4 Whole Numbers and Counting 44

Topic 3.3 *Addition with Whole Numbers* 45
3.3.1 Definition of Addition: Joining Sets 45
3.3.2 Addition Terminology and Pictorial Representation 45
3.3.3 The Number-Line Model of Addition 46
3.3.4 Addition and Mental Mathematics 46

Topic 3.4 *Addition Properties and Patterns (Whole Numbers)* 48
3.4.1 Whole-Number Addition Table 48
3.4.2 General Patterns in the Addition Table 48
3.4.3 Additive Identity Property 49
3.4.4 Commutative Property 49
3.4.5 Closure Property 49
3.4.6 Associative Property 49

Topic 3.5 *Addition Algorithms (Whole Numbers)* 51
3.5.1 Addition Algorithms 52
3.5.2 Base-Ten Block Model 52
3.5.3 The Traditional Addition Algorithm 52
3.5.4 Partial Sums Algorithm 53
3.5.5 Scratch Algorithm 54
3.5.6 The Scratch Algorithm in Other Bases 55

Topic 3.6 *Subtraction with Whole Numbers* 56
3.6.1 Definition of Terms 56
3.6.2 Take-Away Subtraction 56
3.6.3 Missing-Addend Subtraction 56
3.6.4 Comparison Subtraction 57
3.6.5 A Pictorial Model 57
3.6.6 Number-Line Model 57
3.6.7 Subtraction and Mental Mathematics 57

Topic 3.7 *Subtraction Algorithms (Whole Numbers)* 59

3.7.1 Connections to Addition 59
3.7.2 Subtraction Using Base-Ten Blocks 59
3.7.3 Notation for the Subtraction Algorithm 60
3.7.4 Subtraction in Base Six 61

Topic 3.8 *Multiplication with Whole Numbers* 62

3.8.1 Definition and Terminology 62
3.8.2 Repeated Addition Multiplication 62
3.8.3 Array Multiplication 62
3.8.4 Cartesian Product 63
3.8.5 Tree Diagram 63
3.8.6 Array Model and Area Model 64
3.8.7 Number-Line Model 64

Topic 3.9 *Multiplication Properties and Patterns (Whole Numbers)* 66

3.9.1 Multiplication Table 66
3.9.2 Zero Property of Multiplication 67
3.9.3 Multiplicative Identity 67
3.9.4 Commutative Property 67
3.9.5 Closure Property 67
3.9.6 Associative Property 67
3.9.7 Distributive Property of Multiplication over Addition 68

Topic 3.10 *Multiplication Algorithms* 69

3.10.1 Multiplication Algorithm 69
3.10.2 Base-Ten Blocks 69
3.10.3 Partial-Products Multiplication Algorithm 70
3.10.4 Standard Multiplication Algorithm 71

Topic 3.11 *Whole-Number Division* 73

3.11.1 Definition and Terminology 73
3.11.2 Repeated Subtraction 73
3.11.3 Sharing 74
3.11.4 Repeated Subtraction versus Partitioning 74
3.11.5 Discrete Models of Division 75
3.11.6 The Array Model 75
3.11.7 The Number-Line Model 75
3.11.8 Division by Zero 76

Topic 3.12 *Division Algorithms* 77

3.12.1 Concrete Models of the Algorithm 77
3.12.2 Scaffolding 78
3.12.3 Another Transitional Algorithm 79
3.12.4 Links to Estimating Products 79

Integers

Topic 3.13 *Introduction to Integers* 80

3.13.1 Natural Numbers and Whole Numbers 80
3.13.2 Positive and Negative Numbers 80
3.13.3 Notation for Positive and Negative Numbers 80
3.13.4 Opposites 81
3.13.5 Absolute Value 81
3.13.6 Integers and the Set of Integers 81

Topic 3.14 *Integer Addition* 82

3.14.1 Properties of Integer Addition 82
3.14.2 The Additive Inverse Property 82
3.14.3 Addition Rules for Integers 82
3.14.4 Models of Integer Addition 83

Topic 3.15 *Integer Subtraction* **86**

3.15.1 Closure Property for Integer Subtraction 86
3.15.2 Rules for Integer Subtraction 86
3.15.3 Models of Integer Subtraction 87

Topic 3.16 *Integer Multiplication* **90**

3.16.1 Properties of Integer Multiplication 90
3.16.2 Multiplication Rules for Integers 90
3.16.3 Models of Integer Multiplication 91

Topic 3.17 *Integer Division* **94**

3.17.1 Rules for Dividing Integers 94
3.17.2 The Set Model for Illustrating Integer Division 94

Topic 3.18 *Order of Operations* **97**

3.18.1 Definition of the Term Order of Operations 97
3.18.2 The Rules of Order of Operations 97
3.18.3 The Calculator and Order of Operations 98

In the Classroom **99**

Bibliography **100**

Chapter 4: Number Theory 102

Topic 4.1 *Prime and Composite Numbers* **103**

4.1.1 Definitions of Prime and Composite 103
4.1.2 Sieve of Eratosthenes 103

Topic 4.2 *Prime Factorization and Tree Diagrams* **105**

4.2.1 Definition of Prime Factorization 105
4.2.2 Tree Diagrams 105
4.2.3 Notation of Prime Factorization 105

Topic 4.3 *Fundamental Theorem of Arithmetic* **107**

Topic 4.4 *Divisibility* **108**

4.4.1 Definition of Divisibility 108
4.4.2 Divisibility Notation 108
4.4.3 Divisibility Theorems 108
4.4.4 False, But Often Believed Divisibility Theorems 109
4.4.5 Related Terms 109
4.4.6 Divisibility Tests 109

Topic 4.5 *Greatest Common Divisor* **113**

4.5.1 Definition of Greatest Common Divisor 113
4.5.2 Relatively Prime 113
4.5.3 Notation for GCD 113
4.5.4 Methods for Finding the GCD 113

Topic 4.6 *Least Common Multiple* **116**

4.6.1 Definition of Least Common Multiple 116
4.6.2 Notation for LCM 116
4.6.3 Methods for Finding the LCM 116
4.6.4 Relationships Between the LCM and GCD 117

Topic 4.7 *Clock Arithmetic & Modular Arithmetic* **118**

4.7.1 Telling Time 118
4.7.2 Clock Arithmetic 118
4.7.3 Modular Arithmetic 119

Topic 4.8 *Patterns* 121
4.8.1 Definition of a Pattern 121
4.8.2 Arithmetic Sequences 121
4.8.3 Geometric Sequences 122
4.8.4 Growing Patterns 122
4.8.5 Fibonacci Sequences 122
4.8.6 Finding the *n*th Term 122
4.8.7 Figurate Numbers 123

In the Classroom 125

Bibliography 126

Chapter 5: Data & Chance 127

Probability

Topic 5.1 *Probability Notions* 128
5.1.1 Probability Terminology 128
5.1.2 Computing a Simple Probability 128
5.1.3 Simulation with a Spinner 129

Topic 5.2 *Equally Likely Outcomes* 130
5.2.1 Equally Likely 130
5.2.2 Certain and Impossible Events 130
5.2.3 Theoretical Probability 131
5.2.4 Empirical Probability 131
5.2.5 Relating Empirical and Theoretical Probabilities 131
5.2.6 Expected Value 132

Topic 5.3 *Mutually Exclusive and Complementary Events* 133
5.3.1 Mutually Exclusive Events 133
5.3.2 Non-Mutually Exclusive Events 133
5.3.3 Understanding "Or" Probability Situations 133
5.3.4 Using a Venn Diagram to Look at Mutually Exclusive Events 134
5.3.5 Interpreting an "Or" Situation 135
5.3.6 Table of Data Simulation 135
5.3.7 Complementary Events 136
5.3.8 Odds In Favor and Odds Against 136

Topic 5.4 *Multistep Experiments* 138
5.4.1 Independent versus Dependent Events 138
5.4.2 Conditional Probability 138
5.4.3 Understanding "And" Probability Situations 139
5.4.4 Tree Diagrams 140

Topic 5.5 *Counting Principles* 142
5.5.1 Factorial 142
5.5.2 Definition of Permutations 143
5.5.3 Computing Permutations 143
5.5.4 Permutation Notation and the Calculator 143
5.5.5 Definition of Combinations 143
5.5.6 Computing Combinations 143
5.5.7 Combination Notation and the Calculator 144
5.5.8 Fundamental Counting Principle 144

Statistics

Topic 5.6 *Statistics Notions* 145
5.6.1 Statistics and Statistical Data 145
5.6.2 Hypotheses 145
5.6.3 Populations, Samples and Types of Sampling 146

Topic 5.7 *Measures of Central Tendency* 147
5.7.1 Mean 147
5.7.2 Mode 148
5.7.3 Median 149
5.7.4 Range and Midrange 149
5.7.5 Weighted Average 149

Topic 5.8 *Data Dispersion* 151
5.8.1 Variance and Standard Deviation 151
5.8.2 Z-Scores 152
5.8.3 Normal Distribution of Data 152
5.8.4 Skewed and Bimodal Data Distributions 153
5.8.5 Quartiles 153
5.8.6 Percentiles 155
5.8.7 Box-and-Whisker Plots 156

Topic 5.9 *Representing Data* 157
5.9.1 Pictographs 157
5.9.2 Bar Graphs 158
5.9.3 Frequency Tables 159
5.9.4 Histograms 160
5.9.5 Line Graphs and Frequency Polygons 161
5.9.6 Line Graphs and Scatter Plots 161
5.9.7 Circle Graphs/Pie Charts 163
5.9.8 Stem-and-Leaf Plots 164

In the Classroom 167

Bibliography 168

Chapter 6: Fraction Models & Operations 169

Topic 6.1 *Fractions* 170
6.1.1 Fractions Representing Part of a Whole 170
6.1.2 Fractions Representing Ratios 171
6.1.3 Fractions as Representing a Division Problem 172

Topic 6.2 *Equivalent Fractions* 173
6.2.1 Finding Equivalent Fractions 173
6.2.2 Testing Whether Fractions are Equivalent 173
6.2.3 Ordering Fractions 174

Topic 6.3 *Simplifying Fractions* 175
6.3.1 Relatively Prime 175
6.3.2 Simplifying Fractions by Looking for Common Factors 175
6.3.3 Simplifying Fractions Using Prime Factorization 176

Topic 6.4 *Improper Fractions and Mixed Numbers* 177
6.4.1 Improper Fractions and Mixed Numbers 177
6.4.2 Changing Mixed Numbers to Improper Fractions 178
6.4.3 Changing Improper Fractions to Mixed Numbers 178

In the Classroom 179

Bibliography 180

Chapter 7: Real Numbers: Rationals & Irrationals 181

Topic 7.1 *Ratio* 182
7.1.1 Definition of Ratio 182
7.1.2 Ratio Notation 182
7.1.3 Types of Ratios 182
7.1.4 Rates 183
7.1.5 Comparing Ratios and Fractions 183

Topic 7.2 **Proportion 184**

7.2.1 Definition of Proportion 184
7.2.2 Solving Proportions Using Equivalent Ratios 184
7.2.3 Solving Proportions Using Cross Multiplication 184
7.2.4 Directly and Indirectly Proportional 185

Decimals

Topic 7.3 **Introduction to Decimals 186**

7.3.1 Definition of Decimal 186
7.3.2 Decimals, Base Ten, and Place Value 186
7.3.3 Expressing a Decimal Number 187
7.3.4 Zeros in Decimals 187
7.3.5 Base-Ten Blocks as a Model for Decimals 187
7.3.6 Rounding of Decimals 188

Topic 7.4 **Decimals and Fractions 189**

7.4.1 Decimals and Common Fractions 189
7.4.2 Converting Decimals to Common-Fraction Form 189
7.4.3 Converting Certain Common Fractions to Decimals 190
7.4.4 Fractions That Do Not Have Power-of-10 Denominators 190
7.4.5 Using a Calculator for Converting 190
7.4.6 Comparing and Ordering Decimals Using Base-Ten Blocks 190
7.4.7 Ordering Decimals by Lining Up Decimal Points 191
7.4.8 Ordering Decimals on a Number Line 191

Topic 7.5 **Terminating and Repeating Decimals 192**

7.5.1 Terminating Decimals 192
7.5.2 Repeating Decimals 192
7.5.3 Rational Numbers as Repeating Decimals 192
7.5.4 Converting Repeating Decimals to Rational-Number Form 193

Topic 7.6 **Decimal Addition and Subtraction 195**

7.6.1 Adding Decimals 195
7.6.2 Modeling the Addition of Decimals 195
7.6.3 Subtracting Decimals 196
7.6.4 Modeling the Subtraction of Decimals 197

Topic 7.7 **Decimal Multiplication 199**

7.7.1 Rule for Decimal Multiplication 199
7.7.2 Connection Between Decimal and Fraction Multiplication 199
7.7.3 Area Model of Decimal Multiplication 200

Topic 7.8 **Decimal Division 201**

7.8.1 Rule for Decimal Division 201
7.8.2 A Key Fact About Division 202
7.8.3 Connection Between Decimal and Fraction Division 202

Topic 7.9 **Percent 203**

7.9.1 Definition of Percent 203
7.9.2 Percents and Decimals 203
7.9.3 Solving Basic Percent Problems 203
7.9.4 Percent Change 204

Topic 7.10 **Interest 205**

7.10.1 Basic Terms in an Interest Problem 205
7.10.2 Computing Simple Interest 205
7.10.3 Computing Compound Interest 206

Rational Numbers

Topic 7.11 **Rational Numbers 207**

7.11.1 Definition of a Rational Number 207
7.11.2 Rational Numbers versus Fractions 207
7.11.3 Three Models of Rational Numbers 207

Topic 7.12 ***Looking at Rational Numbers as the Set of Rational Numbers 210***
7.12.1 Comparing Sets of Numbers 210
7.12.2 Infinite Sets 211
7.12.3 The Density of the Set of Rational Numbers 211
7.12.4 Other Properties of Rational Numbers 212

Topic 7.13 ***Adding Rational Numbers 214***
7.13.1 Addition of Rational Numbers with Common Denominators 214
7.13.2 Formal Definition of the Addition of Rational Numbers 214
7.13.3 Addition of Mixed Numbers 215

Topic 7.14 ***Subtracting Rational Numbers 217***
7.14.1 The Subtraction of Rational Numbers with Common Denominators 217
7.14.2 Formal Definition of the Subtraction of Rational Numbers 217
7.14.3 Subtraction of Mixed Numbers 218

Topic 7.15 ***Multiplying Rational Numbers 220***
7.15.1 Understanding Multiplication of Two Rational Numbers 220
7.15.2 Definition of Multiplication of Rational Numbers 221
7.15.3 Multiplication of Mixed Numbers 221

Topic 7.16 ***Dividing Rational Numbers 222***
7.16.1 Understanding Division of Rational Numbers 222
7.16.2 Definition of Reciprocal 223
7.16.3 Definition of Dividing Rational Numbers 223
7.16.4 Why Can We Multiply by the Reciprocal? 223
7.16.5 Dividing Mixed Numbers 224

Topic 7.17 ***Irrational Numbers 225***
7.17.1 Nonterminating, Nonrepeating Decimals 225
7.17.2 Definition of Irrational Numbers 225
7.17.3 Square Roots 225
7.17.4 Roots 226
7.17.5 Rational Exponents 226

Topic 7.18 ***Real Numbers 227***
7.18.1 Set of Real Numbers 227
7.18.2 Relationship Between the Reals and Other Number Systems 227
7.18.3 Properties of the Real Numbers 228

In the Classroom 229

Bibliography 230

Chapter 8: Patterns & Functions 231

Topic 8.1 ***Functions 232***
8.1.1 Definition of Functions and Function Notation 232
8.1.2 Domain and Range 232
8.1.3 Ways to Describe Functions 233
8.1.4 Vertical-Line Test 235
8.1.5 Composition of Functions 235

Topic 8.2 ***Equations and Inequalities 237***
8.2.1 Definition of Equation 237
8.2.2 Properties of Equations 237
8.2.3 Definition and Notation of Inequalities 238
8.2.4 Properties of Inequalities 238

Topic 8.3 ***Cartesian Coordinate System 239***
8.3.1 Definition of the Cartesian Coordinate System and Related Terms 239
8.3.2 Equations of Horizontal and Vertical Lines 240

Topic 8.4 *Graphs and Equations of Lines* 241
8.4.1 Slopes of Lines 241
8.4.2 Computing the Slope of a Line 242
8.4.3 Lines of the Form $y = mx$ 242
8.4.4 X- and Y-Intercepts 243
8.4.5 Slope-Intercept Form of a Line $y = mx + b$ 244
8.4.6 Point-Slope Formula for the Equation of a Line 244

Topic 8.5 *Systems of Linear Equations* 245
8.5.1 Simultaneous Equations 245
8.5.2 Geometrical Relationships of the Graphs of Two Lines 245
8.5.3 Solving Simultaneous Linear Equations 247

Topic 8.6 *Solving Systems of Linear Equations Symbolically* 249
8.6.1 Solving Systems of Equations by Substitution 249
8.6.2 Solving Systems of Equations by Elimination 249
8.6.3 Determining the Equation of a Line Through Two Points 250

Topic 8.7 *Exponents* 251
8.7.1 Definitions of Terms Associated with Exponents 251
8.7.2 Squares and Cubes 251
8.7.3 Zero as an Exponent 252
8.7.4 Negative Exponents 252
8.7.5 Properties of Exponents 252

Topic 8.8 *Scientific Notation* 253
8.8.1 Definition of Scientific Notation 253
8.8.2 Writing Numbers Expressed in Scientific Notation as Decimals 253
8.8.3 Writing Decimal Numbers in Scientific Notation 254

***In the Classroom* 255**

***Bibliography* 256**

Chapter 9: Geometry 257

Two-dimensional Geometry

Topic 9.1 *Two-dimensional Geometry Basics* 258
9.1.1 Points and Lines 258
9.1.2 Relationships Among Points and Lines 259

Topic 9.2 *Planes* 260
9.2.1 Planes 260
9.2.2 Relationships Among Points, Lines, and Planes 261

Topic 9.3 *Line Relationships* 262
9.3.1 Coplanar, Noncoplanar, and Skew Lines 262
9.3.2 Intersecting and Concurrent Lines 263
9.3.3 Parallel and Perpendicular Lines 263
9.3.4 Relationships Among Lines and Planes 264

Topic 9.4 *Line Segments* 266

Topic 9.5 *Rays and Angles* 267
9.5.1 Rays 267
9.5.2 Angles 267

Topic 9.6 *Angle Relationships* 269
9.6.1 Adjacent Angles, Linear Pair Angles, and Vertical Angles 269
9.6.2 Complementary and Supplementary Angles 270
9.6.3 Angles Formed by a Transversal and Two Lines 270
9.6.4 Angles Formed by a Transversal and Two Parallel Lines 271

Topic 9.7 *Polygons* 272
9.7.1 A Polygon and its Features 272
9.7.2 Convex and Concave Polygons 273
9.7.3 Names of Polygons 273

Topic 9.8 *Regular Polygons* 274
9.8.1 Regular Polygons 274
9.8.2 Sum of the Measures of Angles in a Convex Polygon 274
9.8.3 Sum of the Measures of the Exterior Angles in a Convex Polygon 275

Topic 9.9 *Triangles* 277
9.9.1 Features of a Triangle 277
9.9.2 Special Kinds of Triangles 277
9.9.3 Sum of the Measures of the Interior Angles in a Triangle 278

Topic 9.10 *Quadrilaterals* 279
9.10.1 Features of a Quadrilateral 279
9.10.2 Special Kinds of Quadrilaterals 279

Topic 9.11 *Circles* 281
9.11.1 Features of a Circle 281
9.11.2 The Compass 282

Topic 9.12 *Congruence* 283
9.12.1 Congruent Segments 283
9.12.2 Congruent Angles 283
9.12.3 Congruent Polygons 283

Topic 9.13 *Symmetry* 285
9.13.1 Reflection Symmetry 285
9.13.2 Rotational Symmetry 285
9.13.3 Plane Symmetry 286

Constructions

Topic 9.14 *Line and Angles Constructions* 287
9.14.1 Geometric Tools 287
9.14.2 Constructing a Line Segment Congruent to a Given Line Segment 287
9.14.3 Constructing an Angle Congruent to a Given Angle 288

Topic 9.15 *Angle-Bisector Constructions* 289
9.15.1 Angle Bisector 289
9.15.2 Constructing an Angle Bisector 289

Topic 9.16 *Perpendicular-Line-Through-a-Point Constructions* 291
9.16.1 Constructing a Line Perpendicular to a Line Through a Point on the Line 291
9.16.2 Constructing a Line Perpendicular to a Line Through a Point Not on the Line 291
9.16.3 Constructing a Perpendicular Bisector of a Line Segment 292

Topic 9.17 *Parallel-Line Construction* 293

Topic 9.18 *Circumscribed-Circle Constructions* 294
9.18.1 Circumscribed Circles 294
9.18.2 Constructing a Circumscribed Circle About a Triangle 294

Topic 9.19 *Inscribed-Circle Constructions* 296
9.19.1 Inscribed Circles 296
9.19.2 Constructing an Inscribed Circle Within a Triangle 296
9.19.3 Orthocenters and Centroids 297

Topic 9.20 *Triangle Congruence* 298
9.20.1 Congruent Triangles and Their Features 298
9.20.2 How Two Triangles May or May Not be Congruent 298

Topic 9.21 *Triangle-Congruence Properties* 300
9.21.1 Side-Side-Side Triangle-Congruence Property 300
9.21.2 Side-Angle-Side Triangle-Congruence Property 300
9.21.3 Angle-Side-Angle Triangle-Congruence Property 301
9.21.4 Angle-Angle-Side Triangle-Congruence Property 301
9.21.5 Isosceles-Triangle-Congruence Properties 302

Topic 9.22 *Quadrilateral Properties* 303
9.22.1 Properties of Trapezoids 303
9.22.2 Properties of Parallelograms 304
9.22.3 Properties of Rectangles 304
9.22.4 Properties of Kites 305
9.22.5 Properties of Rhombi 306
9.22.6 Properties of Squares 306

Topic 9.23 *Similarity* 307
9.23.1 Similarity and Similar Figures 307
9.23.2 Similar Triangles 307
9.23.3 Indirect Measurement 308

Topic 9.24 *Triangle Proportion Properties* 310

Topic 9.25 *Fractals and Rep-tiles* 312
9.25.1 Fractals 312
9.25.2 Rep-tiles 313

Three-dimensional Geometry

Topic 9.26 *Three-dimensional Geometry Basics* 314
9.26.1 Plane Relationships 314
9.26.2 Dihedral Angles 315

Topic 9.27 *Polyhedra, Prisms, and Pyramids* 316
9.27.1 A Simple Closed Surface 316
9.27.2 A Polyhedron and Its Features 316
9.27.3 Convex and Concave Polyhedra 317
9.27.4 Prisms 318
9.27.5 Pyramids 318

Topic 9.28 *Regular Polyhedra* 320
9.28.1 Regular Polyhedra 320
9.28.2 Semiregular Polyhedra 320

Topic 9.29 *Cylinders, Cones, and Spheres* 322
9.29.1 Cylinders 322
9.29.2 Cones 323
9.29.3 Spheres 323

Topic 9.30 *van Hiele Levels* 325

In the Classroom 326

Bibliography 327

Chapter 10: Measurement 328

Topic 10.1 *Angle Measurement* 329
10.1.1 Angle Measurement and Notation 329
10.1.2 How to Use a Protractor 329
10.1.3 Angle Classification 330
10.1.4 Dihedral Angle Measurement 330

Topic 10.2 **Linear Measurement 332**
10.2.1 Length and How to Measure It 332
10.2.2 The English System of Linear Measurement 332
10.2.3 The Metric System of Linear Measurement 332

Topic 10.3 **Perimeter 334**
10.3.1 Perimeter 334
10.3.2 Perimeters of Polygons 334

Topic 10.4 **Circumference 336**
10.4.1 Circumference and π 336
10.4.2 Arc Length 336

Topic 10.5 **Measuring Area 338**
10.5.1 Area and How to Measure It 338
10.5.2 The English System of Area Measurement 338
10.5.3 The Metric System of Area Measurement 339

Topic 10.6 **Area of Quadrilaterals 341**
10.6.1 Area of a Square 341
10.6.2 Area of a Rectangle 341
10.6.3 Area of a Parallelogram 342
10.6.4 Area of a Trapezoid 243

Topic 10.7 **Area of Triangles 343**

Topic 10.8 **Area of Regular Polygons and Circles 344**
10.8.1 Area of a Regular Polygon 344
10.8.2 Area of a Circle 344
10.8.3 Area of a Sector 345

Topic 10.9 **Surface Area of Prisms and Cylinders 346**
10.9.1 Surface Area 346
10.9.2 Surface Area of a Right Prism 346
10.9.3 Surface Area of a Right Circular Cylinder 346

Topic 10.10 **Surface Area of Pyramids, Cones, and Spheres 349**
10.10.1 Surface Area of a Right Regular Pyramid 349
10.10.2 Surface Area of a Right Circular Cone 350
10.10.3 Surface Area of a Sphere 351

Topic 10.11 **Measuring Volume 352**
10.11.1 Volume and How to Measure It 352
10.11.2 The English System of Volume Measurement 352
10.11.3 The Metric System of Volume Measurement 353

Topic 10.12 **Volume of Prisms and Cylinders 355**
10.12.1 Volume of a Right Prism 355
10.12.2 Volume of a Right Circular Cylinder 355

Topic 10.13 **Volume of Pyramids, Cones, and Spheres 357**
10.13.1 Volume of a Right Pyramid 357
10.13.2 Volume of a Right Circular Cone 357
10.13.3 Volume of a Sphere 358

Topic 10.14 **Cavalieri's Principle 359**

Topic 10.15 **Mass and Temperature 361**
10.15.1 Mass and Weight 361
10.15.2 Temperature 361

Topic 10.16 ***Pythagorean Theorem and Other Triangle Relationships 363***

10.16.1 The Pythagorean Theorem 363
10.16.2 The Converse of the Pythagorean Theorem 364
10.16.3 Special Right Triangles 364
10.16.4 The Triangle Inequality 365

Topic 10.17 ***Distance Formula 366***

Topic 10.18 ***Tessellations 368***

10.18.1 What is a Tessellation? 368
10.18.2 Any Triangle Can Tessellate the Plane 368
10.18.3 Any Quadrilateral Can Tessellate the Plane 369

Topic 10.19 ***Tessellations of Regular Polygons 370***

10.19.1 Which Regular Polygons Can Tessellate the Plane? 370
10.19.2 Semiregular Tessellations 371

Topic 10.20 ***Transformations 372***

Topic 10.21 ***Translations 374***

Topic 10.22 ***Rotations 376***

Topic 10.23 ***Reflections and Glide Reflections 378***

Topic 10.24 ***Size Transformations 380***

In the Classroom 382

Bibliography 383

Resources and Tools 385

Glossary 413

Index 421

Preface

We developed the *Student Resource Handbook* in response to requests from our students for a complete, easy to use reference book to support the *Student Activity Manual.* Because the activities in the *Manual* do not include discussions and explanations—to provide this information with the activities would be detrimental to helping you become a good problem solver and an independent learner of mathematics—we have developed the *Student Resource Handbook.* The purpose of this *Student Resource Handbook* is to provide you with useful explanations of the mathematical concepts and procedures that are explored in the activities, and to better prepare you to teach mathematics in a manner that is consistent with the new view of mathematics instruction promoted in the *Student Activity Manual.*

The *Student Resource Handbook* and *Student Activity Manual* are designed to work hand-in-glove. For example, when you work activities in the *Student Activity Manual* the FYI boxes will direct you to the necessary foundation content in the *Student Resource Handbook.* In addition, the following features are built into the *Handbook*:

Features

- **Chapter Overview and Topic Outline**

The handbook is organized into the same 10 chapters as is the *Student Activity Manual.* Each chapter begins with an overview of the mathematical ideas contained in the chapter and the corresponding chapter in the *Student Activity Manual* and includes a list of all of the topics included in the chapter. The content of each chapter is divided into numerous bite-sized sections that review the key ideas for each concept and provide worked examples. Dividing the content into many topics and sub-topics will allow you to find the exact material you need quickly and easily.

- **Cornerstone**

Each topic begins with the "Cornerstone" which provides a list of topics that should be understood or reviewed before the present topic is studied. The cornerstone identifies the mathematical building blocks that lay the foundation for the content reviewed in the topic. At the conclusion of each topic, a list of "Related Topics" is provided.

- **Did You Know**

Each topic includes a brief section, "Did You Know?", that provides relevant historical, cultural, and social contexts for some of the key mathematical ideas found in the chapter. Unfortunately, many students, especially those in grades 5 and higher, view mathematics as irrelevant, dull, and routine. Good teachers are people who not only have a firm understanding of the subject matter they are teaching, but also have a good grasp of what it takes to make mathematics interesting and relevant for their students. Further, they have knowledge of the social and cultural role mathematics plays in our everyday lives. Teachers who have this broader base of knowledge are more likely to excel at preparing their students for the real world outside of school.

- **In the Classroom**

An essential goal of any mathematics course for prospective teachers of young children should be to develop an understanding of how the mathematical concepts, procedures, and processes considered in the course relate to the mathematics that is taught in the elementary school. With this goal in mind, we have developed a special feature titled "In the Classroom" which falls at the end of each chapter. This feature relates the mathematics content in the chapter to the NCTM *Standards* and to what the *Standards* propose for mathematics in pre-K through grade 5.

- **Bibliography**

We believe that prospective teachers should begin immediately to collect ideas for classroom use, to deepen their understanding of important mathematical concepts and processes, and to increase their awareness of how to help children learn mathematics. Accordingly, the end of each chapter contains a bibliography of useful books and articles to augment the activities and explanations as well as the historical, social, and cultural facts. These readings have been selected from journals of the National Council of Teachers of Mathematics (especially *Teaching Children Mathematics* and *Mathematics Teaching in the Middle School*) and other professional references whose audience is the classroom teacher.

- **Tools Appendix**

A special appendix is included to provide prospective teachers with a set of tools to use as they embark upon their teaching careers. Included are:

  - a synopsis of the NCTM *Standards* by grade level
  - information about useful manipulatives and technological tools and a list of resources for finding them
  - a compendium of selected black-line masters for activities presented in the *Student Activities Manual*. These may be reproduced for use in class now and in the future.

- **Glossary**

A list of important words and terms that appear at various places in the *Student Activity Manual* and in the *Student Resource Handbook* are found in the Glossary at the end of this text. This list includes definitions or descriptions of these words and terms with examples whenever appropriate.

Acknowledgments

Although it is impossible to completely acknowledge all the people who have helped us in writing this textbook, we want to thank some people specifically. Since these materials were developed out of two projects funded by the National Science Foundation (one at Indiana University and one at Syracuse University), there have been a number of people who developed activities and ideas for activities that have contributed to this textbook set. We are grateful to these people and want their contributions recognized here: Jean-Marc Cenet, Rapti de Silva, K. Jamie King, Norman Krumpe, Diana V. Lambdin, Sue Tinsley Mau, Francisco Egger Moellwald, Preety Nigam, and Vânia Santos. There were also people who helped in the latter stages of writing the preliminary and first editions by generating exercises and solutions, checking for accuracy, and developing particular sections of the *Student Resource Handbook*. We thank them for their help: Fran Arbaugh, Zaur Berkaliev, Ernesto Colunga, Rapti de Silva, Priscilla Gathoni, Dasha Kinelovsky, Levi Molenje, Jean Palm, Sandra Reynolds, and Robert Wenta.

We also thank the following persons who reviewed our manuscript. Their comments and suggestions have helped us make this a better textbook set: Rita M. Basta (California State University at Northridge), George Csordas, (University of Hawaii at Manoa), Richard Friedlander (University of Missouri at St. Louis), Michael Hall (University of Mississippi at Oxford), Guershon Harel (University of California at San Diego), Theodore Hodgson (Montana State University), Donald Hooley (Bluffton College), Eric Milou (Rowan State University), Kathy Nickell (College of Dupage), Dale Oliver, (Humboldt State University). Finally, we thank our editors, Ann Heath, Quincy McDonald, and Sally Yagan, for their encouragement and desire to publish a reform textbook for prospective elementary teachers.

We have worked hard to make sure that this book is clean and accurate and will provide you with a useful reference as you embark on your career as a teacher. However, if you identify any errors in the text, please send us the information so that it may be corrected in subsequent printings of the text and posted on our web site.

Joanna O. Masingila
(jomasing@syr.edu)

Frank K. Lester
(lester@indiana.edu)

Anne M. Raymond
(araymond@bellarmine.edu)

CHAPTER 1 ONE

Getting Started in Learning Mathematics via Problem Solving

CHAPTER OVERVIEW

Most mathematics educators agree that problem solving is a very important, if not the most important, goal of mathematics instruction at every level. In this chapter, you can learn some interesting facts about problem solving and examine eight of the most commonly used problem-solving strategies. You will also learn about set-theory ideas and logic.

BIG MATHEMATICAL IDEAS

Problem-solving strategies, generalizing, verifying, using language and symbolism, multiple representations

NCTM PRINCIPLES & STANDARDS LINKS

Problem Solving; Reasoning; Communication; Connections; Representation

TOPIC **1.1** **Sets and Set Terminology**
1.2 **Subsets and Set Relationships**
1.3 **Set Operations**
1.4 **Logic**
1.5 **Problem-Solving Topics**
In the Classroom
Bibliography

TOPIC 1.1 Sets and Set Terminology

Introduction

Sets provide a means of categorizing and organizing objects into specific groups. Sets can be large, like the set of people living in the United States, and sets can be small, like the set of days in the week. In this section, we introduce the concept of sets and ways to describe sets. We also explain terminology of sets, including the notions of finite and infinite sets.

Renowned mathematician Georg Cantor (1845–1918) spent most of his childhood in Germany and completed his doctoral studies in mathematics at the University of Berlin. His research in mathematics centered on sets and comparing the size of infinite sets. He is known particularly for his work with the set of numbers in the closed interval [0,1] that are not in the open middle third of [0,1] and not in the open middle third of any interval obtained by separating [0,1] into three equal intervals, then dividing each of these intervals into three equal intervals, etc. This fascinating set has been named the *Cantor Set* after Cantor himself.

For More Information Bunt, L. N. H., Jones, P. S., & Bedient, J. D. (1988). *The Historical Roots of Elementary Mathematics*. New York: Dover Publications.

1.1.1 Sets, Elements, and Set Notation

A **set** is a collection of objects or ideas and is typically denoted by a list or description of what is in the set in brackets. For example,

The set of natural numbers could be described this way: $\{1, 2, 3, 4, 5, \dots\}$.

The set of vowels could be described this way: {a, e, i, o, u}.

The set of boys on the debate team is: {Jay, Tommy, Chris, Mike, John, Frank}.

An **element** of a set is a member of the set. The notation meaning "is an element of" is represented by the symbol, $\in$. The letter "a" is an element of the set A = {a, e, i, o, u}; we denote this by writing, $a \in A$. Likewise, because the letter "h" is not an element of set A, we can denote the phrase, "is not an element of," using the symbol $\notin$, saying $h \notin A$.

Rather than indicate a set by listing all the elements in a set within brackets, we can describe a set using **set-builder notation;** that is, we can summarize the members of a set using the notation $\{x|x \text{ is}| \dots\}$, which is read "$x$ such that x is" to describe the elements.

For example, instead of describing set A = {a, e, i, o, u} by listing all the elements, we could say $A = \{x|x \text{ is a vowel}\}$, which is read "the set of all x such that x is a vowel."

Another example of a set written in set-builder notation includes $B = \{x|x \text{ is an even whole number less than } 20\}$, which is read as "the set of all x such that x is an even whole number less than 20," meaning the set $\{0, 2, 4, 6, 8, 10, 12, 14, 16, 18\}$.

1.1.2 Empty Set, Finite Sets, and Infinite Sets

A set with no elements is called the **empty set** or the **null set.** The empty set is denoted either as { } or $\varnothing$.

Sets can be **finite,** that is, have a limited number of elements, or **infinite,** that is, have an unlimited number of elements. Examples of finite sets include:

$\{1, 2, 3, 4, 5, 6\}$

{x|x is a student in this class}.

{red, yellow, blue}

{x|x is a prime number between 1 and 100}.

Examples of infinite sets are:

$\{x|x \text{ is a whole number}\}$.
$\{1, 3, 5, 7, 9, \ldots\}$
$\{\ldots, -2, -1, 0, 1, 2, 3, \ldots\}$
$\{x|x \text{ is an idea}\}$.

Notice that infinite sets are often described using ellipses, which are the symbols "…", indicating that the pattern described by the previous elements in the set is continued in the same manner.

Subset and Set Relationships
Set Operations
Set of Whole Numbers

TOPIC 1.2 Subsets and Set Relationships

cornerstone **Sets and Set Terminology**

Introduction

Many relationships we observe involve noting how one set relates to another. For example, when we invite a few friends to go to the movies, we are identifying a subset of our total set of friends. In fact, we usually categorize our total set of friends into a variety of subsets such as our friends from high school, our friends from college, our friends from the neighborhood, our friends at work, and so on. In this section, we investigate relationships between sets, particularly focusing on the concepts of subsets, disjoint sets, and equivalent and equal sets.

Venn diagrams, which are used to illustrate logical relationships between sets, are named after John Venn (1834–1923), an Englishman who studied logic and logical symbols. Figures 1.1 and 1.2 are examples of Venn diagrams.

For More Information Bunt, L. N. H., Jones, P. S., & Bedient, J. D. (1988). *The Historical Roots of Elementary Mathematics.* New York: Dover Publications.

1.2.1 Subsets

A set A is a **subset** of another set B if every element in set A is also in set B. We denote this relationship symbolically as $A \subseteq B$. For example, the set $A = \{a, e, i, o, u\}$ is a subset of the set $B = \{x|x \text{ is a letter of the alphabet}\}$. It is important to note that the empty set is considered a subset of every set. Also, every set is considered a subset of itself. Using set notation, we say that $A \subseteq A$ to denote that A is a subset of itself and that $\varnothing \subseteq A$ to denote that the empty set is a subset of set A.

We say A is a **proper subset** of B, denoted $A \subset B$, if A is a subset of B, but there is at least one element of B that is not an element of A. For example, the set $\{1, 2\}$ is a proper subset of $\{1, 2, 3, 4\}$. However, although $\{1, 2, 3, 4\}$ is a subset of the set $\{1, 2, 3, 4\}$, it is NOT a proper subset. Using notation, we say, $\{1, 2, 3, 4\} \not\subset \{1, 2, 3, 4\}$; instead we would write $\{1, 2, 3, 4\} = \{1, 2, 3, 4\}$ or $\{1, 2, 3, 4\} \subseteq \{1, 2, 3, 4\}$.

1.2.2 Venn Diagrams, Sets, and Disjoint Sets

We can illustrate sets and relationships between sets using **Venn diagrams.** For example, given set $A = \{1, 3, 5, 7, 9, 11, 13, 15\}$ and set $B = \{1, 5, 9, 13\}$, we note that $B \subset A$. We can show this subset relationship with the Venn diagram shown in Figure 1.1.

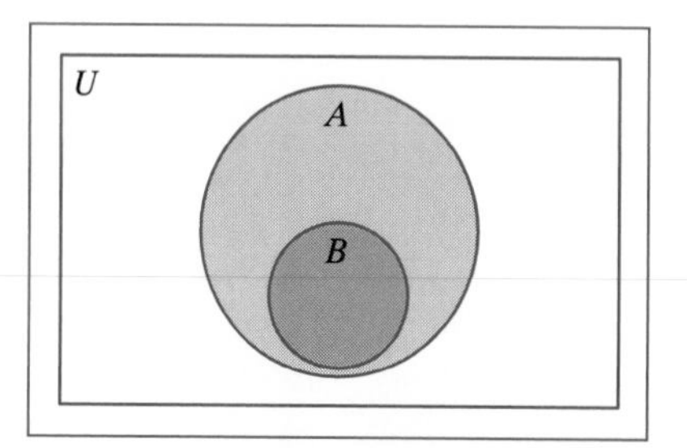

Figure 1.1

Note that the Venn diagram shows set B inside of set A, indicating that all elements in set B are also in set A. Also note that the sets A and B are shown within a rectangular box labeled *U*. We usually show sets in a Venn diagram situated within the **Universe,** U, of a larger set of elements. Sometimes, we are very specific about U. For example, in Figure 1.1, U could stand for the set of all counting numbers. Thus, A and B are shown as being sets within the larger universal set.

Sometimes, sets have no elements in common. If two sets share no elements, then we say that these sets are **disjoint.** For example, given set A = {Bill, Bob, Bart} and set B = {Jack, Jill}, we would say sets A and B are disjoint. A Venn diagram representation of the disjoint sets A and B can be seen in Figure 1.2.

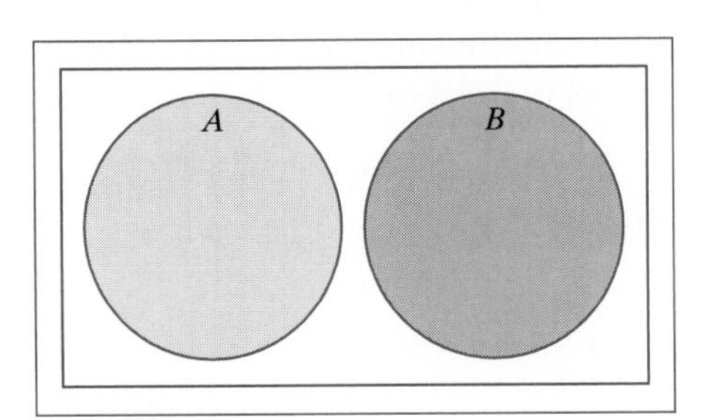

Figure 1.2

1.2.3 One-to-One Correspondence

We say two sets, A and B, have **one-to-one correspondence** if, and only if, each element of A can be paired with exactly one element of set B and each element of set B can be paired with exactly one element of A. For example, Figure 1.3 shows a one-to-one correspondence between two sets.

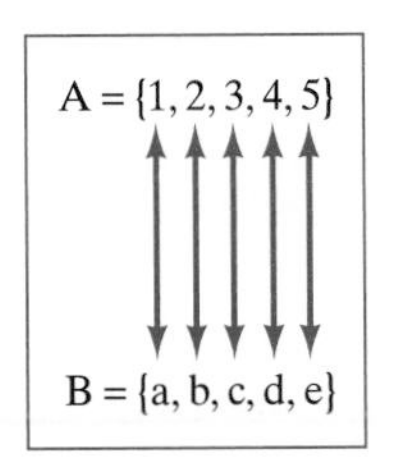

Figure 1.3

1.2.4 Equivalent Sets

We say two sets A and B are **equivalent,** denoted A ~ B, if we can show a one-to-one correspondence between them. For example, we say A = {Sue, Sally, Sara, Sam} is equivalent to B = {red, blue, green, yellow} because we can show a one-to-one correspondence between them (see Figure 1.4).

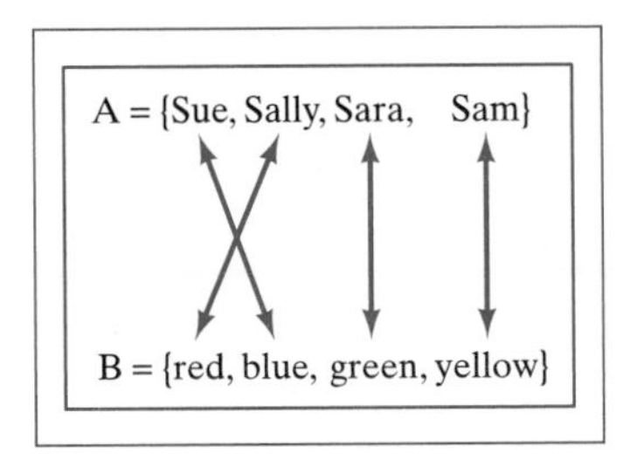

Figure 1.4

Another way of describing equivalence is by saying that two sets are equivalent if they have the same number of elements. In set language, we say the number of elements in set A, denoted $n(A)$, is called the **order** of set A. For example,

Given A = {a, b, c, d, e}, n(A) = 5.
Given B = $\{x|x$ is a letter in the alphabet$\}$, n(B) = 26.
Given C = {1, 2, 3, 4, ... }, n(C) is infinite.

1.2.5 Equal Sets

We say two sets A and B are **equal** if they contain the exact same elements. This is denoted A = B. For example, set A = {2, 4, 6, 8} is exactly the same as set B = $\{x|x$ is an even counting number less than 10$\}$ because they both describe exactly the same elements: 2, 4, 6, and 8.

Related Topics **Set Operations**

Set Operations

cornerstone

Sets and Set Terminology
Subsets and Set Relationships

Introduction

The thinking involved in set operations arises in everyday situations. For example, couples planning their wedding often talk about inviting friends on the groom's side of the family and friends on the bride's side of the family, but they acknowledge that there are some friends who fit into both categories. In this situation, we must deal with combining two sets, with an eye on noting where the two sets overlap. In this section, we examine the mathematical process of operations between sets, including finding the union of sets and determining the intersection of sets, and we explore a means of illustrating these operations through Venn diagrams.

Many people know Lewis Carroll (1832–1898) as the author of the children's story, *Alice's Adventures in Wonderland*. However, what many people do not know is that Lewis Carroll (his real name was Charles Lutwidge Dodgson) was a mathematics lecturer at Oxford University. In 1896, he published a book entitled, *Symbolic Logic: A Fascinating Mental Recreation for the Young*, which was filled with logical puzzle problems to be solved for fun.

For More Information Reimer, L., & Reimer W. (1995). *Mathematicians Are People Too: Stories from the Lives of Great Mathematicians, Volume 2*. Palo Alto, CA: Dale Seymour Publications.

1.3.1 Union of Sets

The **union** of sets A and B, denoted A ∪ B, is the set of all elements that are in either set A or set B or in both sets. For example, given $A = \{1, 2, 3, 4, 5\}$ and $B = \{2, 4, 6, 8, 10\}$, we find $A \cup B = \{1, 2, 3, 4, 5, 6, 8, 10\}$. Notice that even though the numbers 2 and 4 are in both sets, they are each presented only once in the union of the two sets.

Suppose we have sets $C = \{1, 2, 3\}$ and $D = \{4, 5, 6\}$ that have no elements in common. We can find the union of the sets by combining the two sets. Thus, $C \cup D = \{1, 2, 3, 4, 5, 6\}$.

Figure 1.5 illustrates a Venn-diagram representation of the union of sets A and B. The shaded portion shows the union.

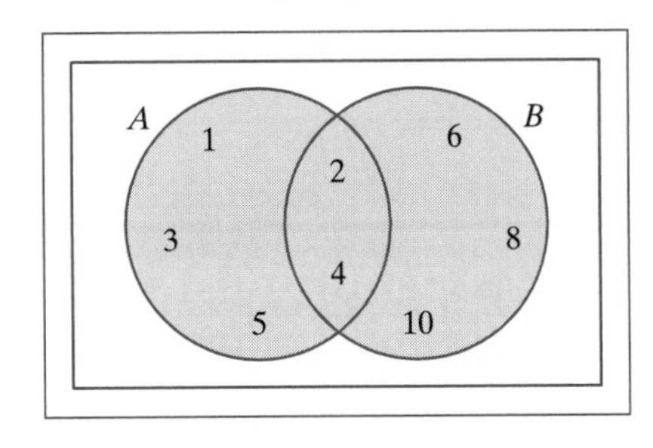

Figure 1.5

1.3.2 Intersection of Sets

The intersection of sets A and B, denoted A ∩ B, is the set of all elements that are in both sets A and B. For example, working with the sets $A = \{1, 2, 3, 4, 5\}$ and $B = \{2, 4, 6, 8, 10\}$, we find $A \cap B = \{2, 4\}$. Note that when two sets are disjoint, their intersection is the empty set. For example, given C = $\{a, b, c\}$ and D = {x, y, z}, we say, C $\cap$ D = $\varnothing$

The Venn diagram in Figure 1.6 illustrates the intersection of sets A and B. Notice that the shaded portion, the overlap, shows the intersection.

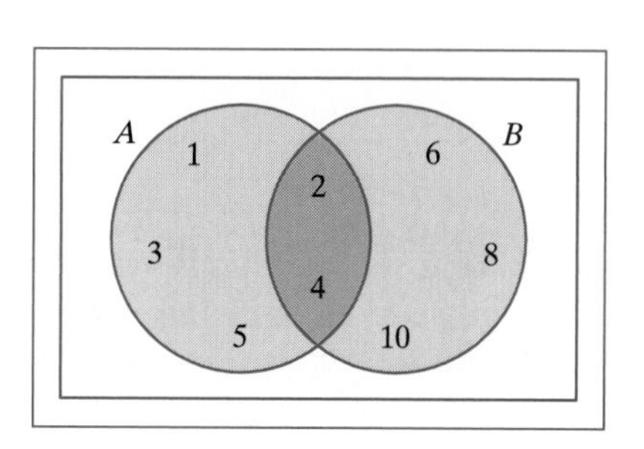

Figure 1.6

1.3.3 Complement of a Set

Given a set A within a universal set U, we say the **complement** of set A, denoted $\overline{A}$, is the set of all elements in the universe that are not in set A. We can write this as $\overline{A} = \{x \in U | x \notin A\}$. For example, if the universe we are working within is the set of whole numbers and set A = $\{x | x$ is an odd whole number$\}$, then $\overline{A}$ = $\{x | x$ is an even whole number$\}$.

To show the complement of set A in a Venn diagram, Figure 1.7 shows set A within the universe, U. The shaded portion represents the complement of set A.

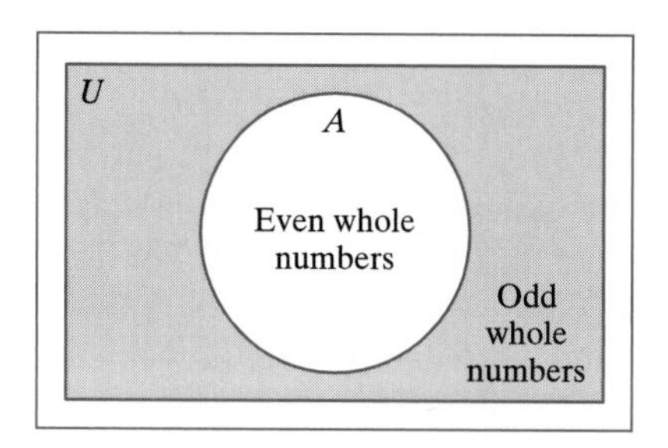

Figure 1.7

In another example, suppose

U = {x|x is a whole number less than 20}
A = {x|x is a multiple of 5 less than 20}
B = {x|x is an even number less than 20}.

Then we can determine that

$$A \cup B = \{0, 2, 4, 5, 6, 8, 10, 12, 14, 15, 16, 18\}$$
$$A \cap B = \{0, 10\}$$
$$\overline{A \cup B} = \{1, 3, 7, 9, 11, 13, 17, 19\}$$
$$\overline{A \cap B} = \{1, 2, 3, 4, 5, 6, 7, 8, 9, 11, 12, 13, 14, 15, 16, 17, 18, 19\}.$$

1.3.4 Relative Complement (or Difference Set)

Given sets A and B in a universal set U, we define the **complement of A relative to B** as the set of all elements in A that are not in B. This is denoted symbolically as $A - B = \{x \in U | x \in A \text{ and } x \notin B\}$. For example, suppose U = {x|x is a whole number less than 20}, A = {x|x is an odd number less than 20}, and B = {x|s is a multiple of 3 less than 20}. Then A − B = {1, 5, 7, 11, 13, 17, 19}. The Venn diagram in Figure 1.8 illustrates sets A and B, and the shaded portion represents A–B.

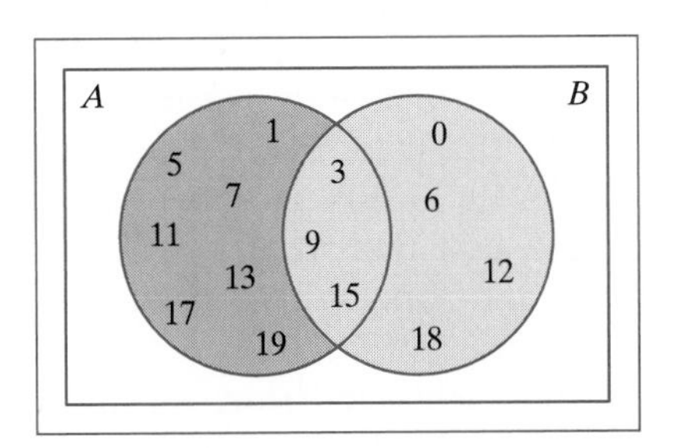

Figure 1.8

1.3.5 Cartesian Product of Sets

The **Cartesian product** of sets A and B, denoted A × B, is the set of ordered pairs (*x*,*y*) you can form where the first element in the ordered pair, *x*, is an element from set A and the second element in the ordered pair, *y*, is an element from set B. We can write the Cartesian product, $A \times B = \{(x, y)|x \in A, y \in B\}$. For example, given set A = {1, 2, 3} and set B = {a, b}, we find A × B = {(1, a), (1, b), (2, a), (2, b), (3, a), (3, b)}. Note that there are six elements in the set A × B and each element in the set is an ordered pair written in parenthetical notation.

In a second example, suppose we want to form singing duets, matching one soprano singer with one bass singer. The set of soprano singers include = {Janet, Dina, Jean} and the set of bass singers includes = {Paul, Ted, Dale}. The duets that can be formed are Soprano × Bass = {(Janet, Paul), (Janet, Ted),(Janet, Dale), (Dina, Paul), (Dina, Ted), (Dina, Dale), (Jean, Paul), (Jean, Ted), (Jean, Dale)}. Note that there are nine elements in the Cartesian product.

From both examples, we can confirm that the size or order of a set formed by a Cartesian product, given sets A and B, is n(A × B) = n(A) × n(B).

1.3.6 Properties of Set Operations

There are five properties associated with set operations. The first property is the **Commutative Property of the Union of Sets.** This property says, given sets A and B, A ∪ B = B ∪ A. Take, for example, sets A = {*x*|*x* is a vowel in the alphabet} and B = {*x*|*x* is a consonant in the alphabet}. A ∪ B is the set of all elements that are in A, in B, or in both A and B. Similarly, $B \cup A$ is the set of all elements that are in B, in A, or in both A and B. Therefore, A ∪ B = B ∪ A = {*x*|*x* is a letter in the alphabet}.

Another property of set operations is the **Commutative Property of the Intersection of Sets.** This property states that, given sets A and B, A ∩ B = B ∩ A. Given A = {red, yellow, blue} and B = {blue, red, green, black}, we would say that A ∩ B is the set of all elements that A and B have in common and that B ∩ A is also the set of all elements that A and B have in common. Therefore, A ∩ B = B ∩ A = {blue, red}.

Next we consider the **Associative Property of the Union of Sets,** that is, given sets A, B, and C, (A ∪ B) ∪ C = A ∪ (B ∪ C). Suppose we have set A = {a, b, c, d, e}, set B = {a, e, i, o, u}, and set C = {e, f, g, h, i}. We find (A ∪ B) ∪ C by first finding (A ∪ B) = {a, b, c, d, e, i, o, u}. Then (A ∪ B) ∪ C = {a, b, c, d, e, i, o, u} ∪ {e, f, g, h, i} = {a, b, c, d, e, i, o, u, f, g, h}. Now, to find A ∪ (B ∪ C), we first find (B ∪ C) = {a, e, i, o, u, f, g, h}. Then, we find A ∪ (B ∪ C) = {a, b, c, d, e} ∪ {a, e, i, o, u, f, g, h} = {a, b, c, d, e, i, o, u, f, g, h}. Thus we have confirmed that (A ∪ B) ∪ C = A ∪ (B ∪ C).

A fourth property of set operations is the **Associative Property of the Intersection of Sets.** This property states that, given sets A, B, and C, (A ∩ B) ∩

C = A ∩ (B ∩ C). To show this is true, consider again the sets A = {a, b, c, d, e}, B = {a, e, i, o, u}, and C = {e, f, g, h, i}. First, let us find (A ∩ B) ∩ C. We determine (A ∩ B) = {a, e}. Then (A ∩ B) ∩ C = {a, e} ∩ {e, f, g, h, i} = {e}. Now, to compute A ∩ (B ∩ C), we first find (B ∩ C) = {e, i}. Then A ∩ (B ∩ C) = {a, b, c, d, e} ∩ {e, i} = {e}. Therefore, (A ∩ B) ∩ C = A ∩ (B ∩ C).

A final property of set operations is the **Distributive Property of Intersection over Union.** This distributive property asserts that, given sets A, B, and C, A ∩ (B ∪ C) = (A ∩ B) ∪ (A ∩ C). Given A = {1, 3, 5, 7, 9, 11}, then B = {1, 5, 10, 20}, and C = {5, 10, 15, 20}. To show that the left- and right-hand sides of the distributive property equation are equal, we compute both sides and compare.

To compute the left-hand side, we find A ∩ (B ∪ C). First, we find (B ∪ C) = {1, 5, 10, 20, 15}. Then A ∩ (B ∪ C) = {1, 3, 5, 7, 9, 11} ∩ {1, 5, 10, 20, 15} = {1, 5}. Now, computing the right-hand side, we find (A ∩ B) ∪ (A ∩ C). First, we find (A ∩ B) = {1, 5}. We also find (A ∩ C) = {5}. Then, we determine (A ∩ B) ∪ (A ∩ C) = {1, 5} ∪ {5} = {1, 5}. Therefore, we have shown that A ∩ (B ∪ C) = (A ∩ B) ∪ (A ∩ C).

Related Topics

Functions

TOPIC 1.4 Logic

Introduction

Logical reasoning plays a role in many conversations that we have. When we try to convince someone to support a certain stance, we have to provide compelling and valid arguments. In mathematics, we use the same tools of deductive and inductive reasoning that are used in the courtroom in settling cases to solve mathematical problems and to prove mathematical theorems. In this section, we examine the rubrics of deductive and inductive reasoning and the types of statements involved in logical arguments.

Mathematician Thales of Miletus (636–546 b.c.) was the first known Greek philosopher and scientist. He is considered the first person to have made use of deductive methods in mathematics. Once, while traveling in the desert of Giza in Egypt, a guide commented to Thales that the base of the pyramid they were viewing measured 518 cubits on each side. Interested, Thales asked the guide how high the pyramid stood. The guide was embarrassed that he did not know. After a short period, Thales announced that he had figured out the height of the pyramid to be 329 cubits. Thales explained that he used the pyramid's shadow and his own shadow to find the height of the pyramid by solving the proportion:

$$\frac{\text{Height of Thales}}{\text{Height of Thales's shadow}} = \frac{\text{Height of pyramid}}{\text{Height of pyramid's shadow}}$$

Many people were impressed with Thales's ability to solve problems. His was the first name on the list of "Seven Wise Men" of Greece compiled by the Greeks themselves.

For More Information Reimer, L., & Reimer W. (1990). *Mathematicians Are People Too: Stories from the Lives of Great Mathematicians, Volume 1*. Palo Alto, CA: Dale Seymour Publications.

1.4.1 Inductive Reasoning

Inductive reasoning is reasoning from the particular to the general; that is, given a set of specific cases, we induce a generalization by observing patterns. There are a number of mathematical problems that can be solved using inductive reasoning. For example, consider the following problem:

> *At the class reunion in the conventional hall, there was a first knock on the door, and one person entered the room. At the second knock, two more people entered than on the first knock. On each successive knock, two more people entered than on the previous knock. After the 200th knock, how many people had entered the room for the reunion?*

It is not likely that we would want to work each step of the problem out to the 200th knock. Instead, we can generate a table that helps us create specific information about how many people were in the room after certain knocks. We can then use these specific pieces of data to look for a pattern and to induce a generalization of how many people will be in the room after the nth knock. We can then use this generalization to say how many people will be in the room after the 200th knock. See Table 1.1 for the table of information.

After observing a number of cases, we might notice the pattern that after knock number 1, the total number of people in the room is $1^2 = 1$, after knock number 2, the total number of people in the room is $2^2 = 4$, after knock number 3, the total number of people in the room is $3^2 = 9$, and so on. Thus, after knock number n, the total number of people in the room will be n^2. Thus, after the 200th knock, there will be $200^2 = 40,000$ people at the reunion.

Table 1.1

Knock #	How Many Entered	Total Number in the Room
1	1	1
2	3	4
3	5	9
4	7	16
5	9	25
6	11	36
7	13	49
8	15	64
9	17	81
10	19	100
⋮	⋮	⋮
n		

Note that there are limitations to inductive reasoning. For example, there are some mathematical problems that appear to generate a pattern that could lead to a generalizable solution, but at some point, the pattern fails. However, not all mathematical problems can be solved using inductive reasoning. Rather, sometimes other types of reasoning or problem-solving strategies are more likely to lead to a solution. Further, in some cases, a combination of strategies will lead to a solution, particularly the combination of inductive and deductive reasoning.

1.4.2 Deductive Reasoning

Deductive reasoning involves working from a set of axioms or given general statements and striving to derive or prove other specific statements or conclusions. Unlike inductive reasoning, which is reasoning from specific to general, deductive reasoning is reasoning from general to specific. A mathematical problem that can be solved using deductive reasoning is the following:

> *Four friends, Nancy, Molly, Erin, and Caroline, are going away to college together. They each are pursuing different majors, English, mathematics, biology, and education. Erin and Caroline usually receive help from tutors on their mathematics homework. Nancy was hired as a feature writer for the school newspaper. Caroline's volunteer work with children at a local elementary school will count for credit in her major. Match each girl with her major.*

To solve this problem deductively, we can create a grid that shows each girl and each major as shown in Figure 1.9. As we work through the given general informa-

	math	biology	English	education
Nancy				
Erin				
Molly				
Caroline				

Figure 1.9

tion provided about the girls, we can eliminate possibilities, noting these eliminations on the grid.

From the first clue, "Erin and Caroline usually receive help from tutors on their mathematics homework," we can eliminate Erin and Caroline as pursuing the math major. Thus, in Figure 1.10, we eliminate the possibility of Erin and Caroline as math majors by crossing an X through the boxes that indicate the connection between Erin and math, and Caroline and math. This leaves either Nancy or Molly as the math major.

Figure 1.10

	math	biology	English	education
Nancy				
Erin	X			
Molly				
Caroline	X			

The next clue, "Nancy was hired as a feature writer for the school newspaper," implies that Nancy is probably the English major. Thus, not only can we pinpoint Nancy with this major, but we can eliminate others from this major and eliminate Nancy from the other majors, leaving Molly as the math major. We can then eliminate Molly from the other majors. See Figure 1.11.

Figure 1.11

	math	biology	English	education
Nancy	X	X	Yes	X
Erin	X		X	
Molly	Yes	X	X	X
Caroline	X		X	

After the second clue, we only have two people left to match with their majors. The final clue, "Caroline's volunteer work with children at a local elementary school will count for credit in her major," leads us to believe that Caroline must be the education major. Therefore, we can eliminate the possibility that Erin is the education major, leaving Erin as the biology major. See Figure 1.12.

Figure 1.12

	math	biology	English	education
Nancy	X	X	Yes	X
Erin	X	Yes	X	X
Molly	Yes	X	X	X
Caroline	X	X	X	Yes

Thus, from a few general statements, we were able to draw a specific conclusion using the process of elimination in our deductive reasoning.

1.4.3 Conjectures, Hypotheses, Conclusions

A **conjecture** is a guess or opinion based on inconclusive evidence. Conjectures often lead to investigations that can show whether or not the guess or opinion has any merit. To test a conjecture, we can form an hypothesis. A **hypothesis** is a statement that is assumed to be true. We generally prove whether or not an hypothesis is actually true through deductive reasoning. A **conclusion** is a decision or judgment made regarding the truth of an hypothesis. The conclusion is drawn through a valid deductive process.

For example, suppose Gina notices that $3 \times 1 = 3, 5 \times 7 = 35, 3 \times 13 = 39$ makes the conjecture, "When I multiply whole numbers by an odd number, my answer is odd." We can form the hypothesis, "The product of an odd number times any whole number is an odd number." We can try to prove or disprove this hypothesis by forming an organized table, showing the multiplication of an odd number times a variety of whole numbers.

Table 1.2 shows a table of products. One column of factors includes representative whole numbers, starting with 0. The other column of factors includes several odd numbers.

Table 1.2

Whole Numbers	Odd Numbers	Product
0	3	0
1	3	3
2	3	6
3	3	9
4	5	20
5	5	25
6	5	30
7	7	49
8	7	56
9	7	63
⋮	⋮	⋮

We find from the list of products that not all the products are odd. Therefore, the hypothesis is shown to be false. However, while looking at the list of factors and products, Gina makes another hypothesis. She notices that all of the odd products came from two factors that were both odd. We form the hypothesis, "The product of two odd factors will always be odd." Testing this hypothesis, we can see that all of our examples indicate that this hypothesis is true.

To prove formally that the hypothesis is true, we can work with the following given statements. Any number of the form $2m$ where m is a whole number is an even number. So, a number of the form $2m + 1$ is odd. Suppose we have two odd numbers expressed generally as $2m + 1$, and $2n + 1$. The product of these two numbers is $(2m + 1) \times (2n + 1) = 4mn + 2n + 2m + 1$. This number can be rewritten as: $2(2mn + n + m) + 1$. This product is also an odd number. Thus, we can draw the conclusion that the hypothesis, that the product of any two odd numbers is an odd number, is true.

1.4.4 Statements and the Negation of Statements

In logical reasoning, we work with **statements** that are declarative sentences that are either true or false (not both). Examples of statements are: (a) It is raining; (b) The door is open; (c) $5 + 6 = 11$; (d) I won the game; and (e) Babies can drive. In symbolic logic, we usually denote a statement by a symbol, either p or q.

The **negation** of a statement p, which is denoted $\sim p$, is the statement "It is false that p." For example, if p is the statement "It is raining," then $\sim p$ is the statement "It is false that it is raining," or, rather, "It is not raining." If p is the statement, "No numbers between 10 and 20 are prime numbers," then $\sim p$ is the statement, "At least one number between 10 and 20 is a prime number."

1.4.5 Compound Statements

Compound statements are formed by combining two statements. We can combine two statements with the words (operations) "and" and "or." If p and q are each statements, then the compound statement, ***p* and *q*,** denoted $p \cap q$, is true when both p and q are true and is false when either p, q, or both p and q are false. Such a compound statement is called a **conjunction.**

For example, suppose p is the statement "5 is a prime number" and q is the statement "8 is an even number." The statement p is true, and the statement q is true. Therefore, the statement *p and q*, that is, "5 is a prime number, and 8 is an even number," is also true.

Consider the example where p is the statement "5 is a prime number" and q is the statement "10 is an odd number." The statement p is true, but the statement q is false. Therefore, the compound statement *p and q*, which says, "5 is a prime number, and 10 is an odd number," is false.

Given statements p and q, the compound statement ***p or q*,** denoted $p \cup q$, is true when either p or q or both p and q are true and is false when both p and q are false. For example, suppose the statement p is "$2 + 3 = 5$" and the statement q is "$5 > 8$." Because the statement p is true and the statement q is false, then the statement *p or q*, which reads "$2 + 3 = 5$ or $5 > 8$," is true. Such a compound statement is called a **disjunction.**

Suppose p is the statement "Dogs can fly" and q is the statement "Everyone speaks Spanish." The compound statement "Dogs can fly, or everyone speaks Spanish" is false because both individual statements are false.

1.4.6 Truth Tables

We can form logical tables, called **truth tables,** that list all the possible outcomes regarding the truth or falsity of statements and related compound statements. If we have statements p and q, the possible combinations of the truth or falsity of these statements are: (a) both statements are true, (b) p is true, and q is false, (c) p is false, and q is true, and (d) both statements are false. We can place these possibilities in a table as shown in Table 1.3.

If we expand the truth table in Table 1.3 to include the possible outcomes of truth or falsity of compound statements involving p and q, we create the truth table shown in Table 1.4. Note that the table shows the cases when the statement p or q is true and false and cases when the statement p or q is false.

Table 1.3

p	q
T	T
T	F
F	T
F	F

Table 1.4

p	q	$p \cup q$	$p \cap q$
T	T	T	T
T	F	T	F
F	T	T	F
F	F	F	F

Table 1.5

p	q	$\sim p$	$\sim p \cap q$
T	T	F	F
T	F	F	F
F	T	T	T
F	F	T	F

We can use truth tables to determine any combinations of statements. Suppose we want to determine outcomes associated with the compound statement $\sim p \cap q$. We can make the truth table shown in Table 1.5 to determine the answer.

Truth tables can help find the possible outcomes associated with even more-complex statements such as $\sim(p \cup \sim q) \cap r$, where p, q, and r are all simple statements. See Table 1.6 for the truth table.

Table 1.6

p	q	r	$\sim q$	$p \cup \sim q$	$\sim(p \cup \sim q)$	$\sim(p \cup \sim q) \cap r$
T	T	T	F	T	F	F
T	F	T	T	T	F	F
T	T	F	F	T	F	F
T	F	F	T	T	F	F
F	T	T	F	F	T	T
F	F	T	T	T	F	F
F	T	F	F	F	T	F
F	F	F	T	T	F	F

1.4.7 Conditional Statements

A **conditional statement** is a statement of the form "*if p then q*," denoted $p \rightarrow q$. The statement p is called the hypothesis, and the statement q is called the conclusion. We can determine the truth or the falsity of a conditional statement by the rule that $p \rightarrow q$ is true except in the case when p is true and q is false. The truth table associated with the conditional statement is shown in Table 1.7.

Table 1.7

p	q	$p \rightarrow q$
T	T	T
T	F	F
F	T	T
F	F	T

For example, suppose the statement p is "I weigh more than 150 pounds" and q is the statement "I will skip breakfast," then the statement. The conditional statement reads, "If I weigh more than 150 pounds, then I will skip breakfast." If the hypothesis is true and the conclusion is true, then the conditional statement is true. However, if the hypothesis is false and the conclusion is true and is read as "If I don't weigh more than 150 pounds, then I will skip breakfast," the conditional statement is still true. This makes sense because the conclusion that "I will skip breakfast" may still come about even, though the hypothesis is not true.

1.4.8 Inverse, Converse, and Contrapositive Statements

For every conditional statement $p \rightarrow q$, we have a related **contrapositive statement,** denoted $\sim q \rightarrow \sim p$. For example, given the conditional statement. "If it snows, then school will be canceled," the contrapositive of this statement is "If school is not canceled, then it did not snow."

Every conditional statement $p \rightarrow q$ also has a related **converse statement,** denoted $q \rightarrow p$. For example, given the conditional statement, "If I run fast then I win the race," the converse of this statement would read, "If I win the race, then I run fast."

Finally, for a conditional statement $p \rightarrow q$, there is an **inverse statement,** denoted $\sim p \rightarrow \sim q$. For example, given the conditional statement, "If $a + b$ is even, then a and b are both even," the inverse of this statement is "If $a + b$ is not even, then a and b are not both even."

1.4.9 If and Only If Statements

Given statements p and q, we make the statement **"p if and only if q,"** denoted p *iff* q, or $p \leftrightarrow q$, to describe the situation where both of the conditional statements

$p \to q$ and $q \to p$ hold true. For example, consider the *if and only if* statement, "A number is divisible by 2 if and only if the digit in the ones place is divisible by 2." The truth of this *if and only if* statement implies that both the conditional statements "If a number is divisible by 2, then the digit in the ones place is divisible by 2" and "If a digit in the ones place is divisible by 2, then the number is divisible by 2" are true.

1.4.10 Direct and Indirect Reasoning

Direct reasoning is a mode of deductive reasoning that follows through the logic of a conditional statement $p \to q$. Direct reasoning follows these steps:

$$\begin{array}{ll} \text{Hypothesis:} & p \to q \\ & \quad p \\ \hline \text{Conclusion:} & \quad q \end{array}$$

For example,

Hypothesis:	If it rains, then I bring my umbrella.
	It rains.
Conclusion:	I bring my umbrella.

Generally, with direct reasoning, if we want to prove the conditional statement is true, we assume the hypothesis is true and show how the conclusion follows from the hypothesis through deductive reasoning.

An example of direct reasoning in problem solving is this. Let us prove the statement, "If we add two even numbers, then our sum will be an even number." By definition, an even number is any number of the form 2 times a whole number. Given any two even numbers, which can be expressed as $2m$, and $2n$ (where m, n are whole numbers), we find the sum to be $2m + 2n = 2(m + n)$. Because $m + n$ is a whole number, the sum is expressed in the form of 2 times a whole number and, therefore, is an even number.

Indirect reasoning uses reasoning that makes use of the contrapositive. The steps of indirect reasoning are:

$$\begin{array}{ll} \text{Hypothesis:} & \sim q \to \sim p \\ & \quad\;\; p \\ \hline \text{Conclusion:} & \quad\;\; q \end{array}$$

For example, suppose we have the conditional statement, "If there is a traffic jam, then I will be late for work." We can use indirect reasoning with the contrapositive statement, "If I am not late for work, then there is not a traffic jam" to work through the logical possibilities of this statement.

Hypothesis:	If I am not late for work, then there is not a traffic jam.
	There is a traffic jam.
Conclusion:	I am late for work.

In general with indirect reasoning, in trying to prove that a certain conclusion follows from a hypothesis, we assume the conclusion is not true and try to show, then, that the hypothesis is not true. What happens is that we find that the hypothesis is in fact true (we find a contradiction to our false assumption), which means that the conclusion cannot be false. Therefore, the conclusion is true.

An example of indirect reasoning in problem solving is this. Let us prove that there are infinitely many natural numbers. To do so, we will assume there are finitely many natural numbers. Through deductive reasoning, we will show that this false assumption leads to a contradiction of a fundamental mathematical property, and, therefore, we prove that our original assumption is false. Therefore, we will conclude that there must be an infinite number of natural numbers.

Proof: Assume there is a finite number of natural numbers. Then there is a largest natural number, n. However, the set of natural numbers is closed under addition; that is, the sum of any two natural numbers is also a natural number. Therefore, because both n and 1 are natural numbers, the sum $n + 1$ will be a natural number. But $n + 1$ is greater than n, which contradicts the assumption that n is the largest natural number. Therefore, there must be an infinite number of natural numbers.

Related Topics

Patterns
Problem Solving

TOPIC 1.5 Problem-Solving Topics

Introduction

Of course, solving problems is not restricted only to mathematics. There are examples of problems being posed and solved in many ancient cultures. What is new about the idea of problem solving is the fact that more and more people are coming to believe that solving problems is at the very heart of learning mathematics. In this section, we examine the process of solving mathematics problems, giving special attention to four key phases of problem solving and the eight popular strategies for solving a variety of mathematical problems.

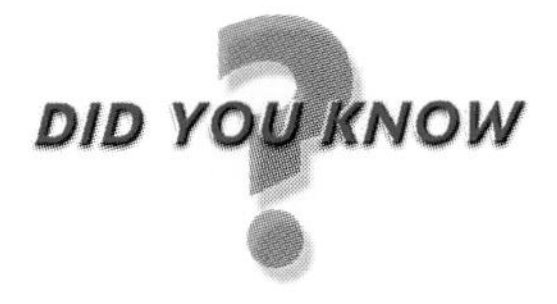

George Polya (1887–1985) was a Hungarian mathematician who came to the United States in 1940. He is often referred to as the "father of problem solving" due to his treatise on problem solving in his work entitled, *How to Solve It*, first published in 1945. This work was inspired through Polya's experience in trying to tutor a young man in mathematics and science who struggled with solving problems. Polya developed a model for the process of solving problems to help his young student solve mathematical problems, based on his self-examination of the processes he himself followed when solving problems.

For More Information Reimer, L., & Reimer W. (1995). *Mathematicians Are People Too: Stories from the Lives of Great Mathematicians, Volume 2*. Palo Alto, CA: Dale Seymour Publications.

1.5.1 Polya's Process of Problem Solving

George Polya is credited for developing a methodical process for finding solutions to mathematical problems. This **process of problem solving,** along with a description of what each step involves for the problem solver, includes:

Understand the Problem. To be successful in solving any problem, you must understand it well. To help yourself understand the problem, you can (a) restate the problem in language that is clear and sensible to you; (b) clarify the question being asked, including any hidden questions; (c) organize the information in the problem by identifying key information and getting rid of unnecessary information.

Devise a Plan. Analyze the problem with the goal of identifying a systematic method for solving the problem. At this step in the process, you can (a) ask yourself, "Have I ever solved a problem like this one before? In what ways was it like this one? To solve that problem, what did I do that might be helpful in solving this one?"; (b) consider the various problem-solving strategies that you have learned, and decide if any of them will help; and (c) determine what you want to do first, second, and so on.

Carry Out the Plan. Carry out the plan you have chosen, being sure to implement it correctly. When carrying out your plan, (a) check your work along the way as you proceed with your solution effort; (b) don't be content to check only your computations. Also, check to make sure that you are using all the important information and that you have not misinterpreted anything; (c) ask yourself, "Is this plan getting me anywhere? Am I following the plan I chose, or am I being sidetracked? Should I abandon this plan and try to think of another way to solve the problem?"

Look Back. Make sure that you have clearly answered the question at hand, and decide if your solution is reasonable. Also, consider what you have learned by solving the problem, not only about mathematics, but also about yourself as a doer of mathematics. At this point, (a) check your solution with the important information given in the problem, and determine if there are other possible solutions; (b) check all your computations and other work one more time; (c) think about the plan you used, and ask yourself, "Could I have solved this problem another, better way?

Would my plan work if the numbers in the problem were larger? Would my plan work in general? Could I now solve other problems similar to this one?"

Although the four phases seem to be chronological in order, it is important to remember that you might need to go back and forth between the different stages before a solution is achieved.

1.5.2 Problem-Solving Strategies

In mathematics, there are many standard strategies that can help in the process of problem solving when applied appropriately. In this section, we discuss eight of these strategies and work through a sample problem for each strategy.

Guess and Check. As the name suggests, this strategy consists of guessing the answer to a problem and then verifying whether the solution you guessed satisfies all the conditions of the problem. Once you determine the degree to which your first guess satisfied the problem, you can make a second, more informed guess and check to see if this guess satisfies the problem. This strategy works particularly well if there are only a limited number of possible answers from which to choose. At times, this strategy may not give you the solution directly. However, it will help you to eliminate some possibilities and suggest where you may look for possible solutions.

For example, you are given nine specific numbers to place without repetition in a 3-by-3 square such that the sum of the numbers in any row or column or along any diagonal is always the same. It would help you to start by guessing at some of the number positions and then work with that guess to make revised guesses. Figure 1.13 shows guesses 1, 2, and 3.

Figure 1.13

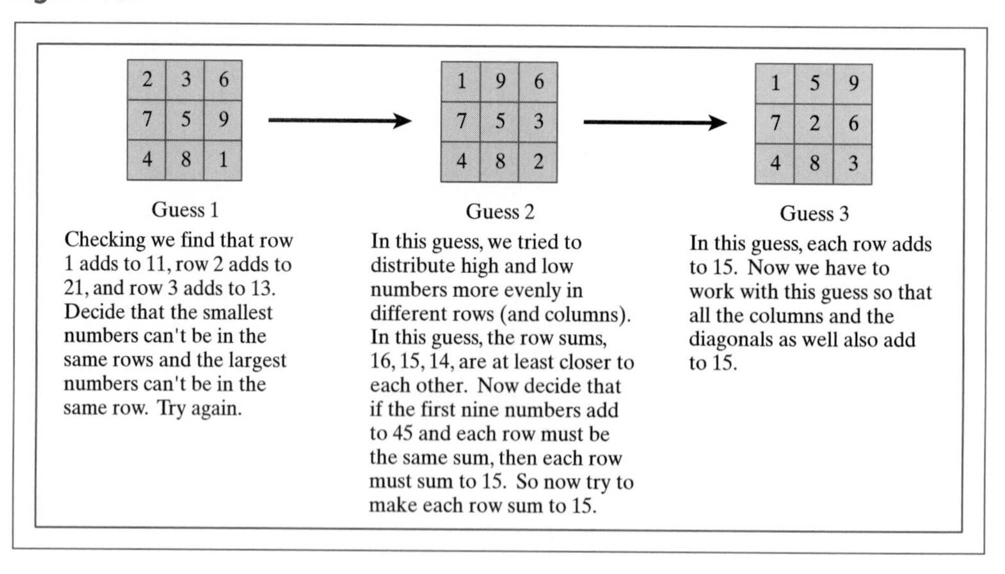

From our conclusions in Figure 1.13, we make more informed guesses, as shown in Figure 1.14.

Use a Visual Aid. There are different ways in which one can use a visual aid to look for a solution to a problem. Such visual aids include drawing a picture, using physical manipulatives, or even connecting the problem to some situation that you have experienced in your life so that you can make sense of it, using your own experience.

Drawing a picture or a diagram is particularly useful when the problem with which one is dealing involves a physical situation. A picture will help to show the

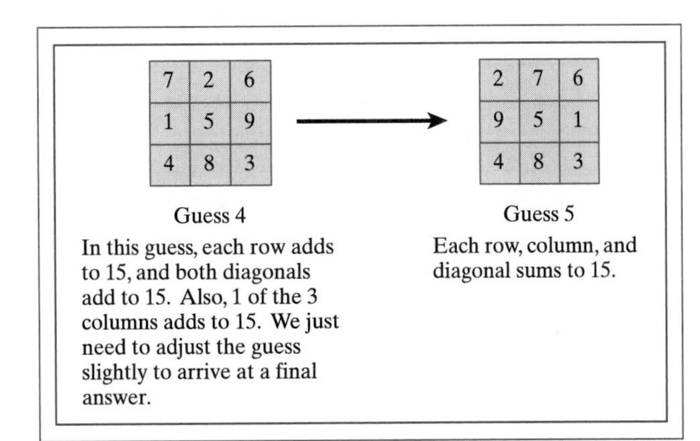

Figure 1.14

relative position of various objects in the problem. Often, it is also a useful way to organize the given information. Once you have drawn a picture, it can help you to interpret the problem and the conditions in it. It may also suggest a possible strategy to solve the problem. Although it may not be always essential, it is usually helpful for you to draw a clear and well-labeled picture. First, such a picture is easier to interpret, and second, seeing all the information in one place may make finding a solution easier.

Similarly, one may use such physical manipulatives as tiles, pencils, candy, boxes, cones, or cans to understand a problem or to look for a solution. These manipulatives may be particularly useful in case our problem involves three-dimensional objects or moving objects from one place to another when looking for a solution.

Example: John is supposed to run 4 laps around a rectangular field that measures 100 feet by 85 feet. However, he feels a little lazy and cuts across the corners each time, leaving each side 12 feet before the edge and hitting the other side 5 feet from its edge. By what distance has John cut short his run?

Solution: It is easy to see that, in this case, drawing the field and marking the distances on it would help to see the problem more clearly and to find its solution. Consider the diagram shown in Figure 1.15.

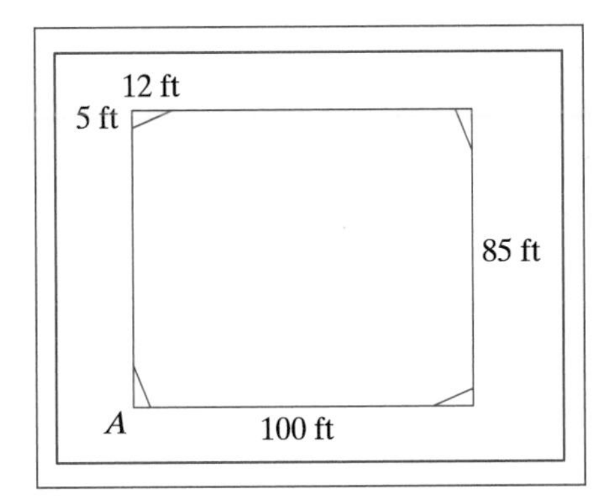

Figure 1.15

If John did not cut corners, the total distance around that he would run per lap is found by computing $2 \times (100 + 85) = 370$ feet. Each time he cuts a corner, instead of running $(12 + 5) = 17$ feet, he actually runs a distance of only 13 feet. This is determined by using the Pythagorean theorem to find the hypotenuse of the right triangle formed with side lengths of 12 and 5. The hypotenuse would be $\sqrt{12^2 + 5^2} = \sqrt{169} = 13$. Thus, finding the distance of the shortcut path we compute: $2 \times ((100 - (12 + 5) + 13 + (85 - (12 + 5) + 13)) = 354$ feet. So, in one lap, John cuts his run short by $370 - 354 = 16$ feet. Therefore, in 4 laps, he cuts his run short by $4 \times 16 = 64$ feet.

Look for a Pattern. Another useful strategy that often helps in seeking a solution to a problem is to look for patterns. This strategy is particularly useful when one is trying to generalize an idea or to look for an algebraic formula. The strategy is to look at specific cases of the solution of a problem to see if an identifiable pattern emerges. Once you have identified the patterns, it is usually a good idea to verify them with other randomly chosen instances before you feel reasonably convinced of the patterns' existence.

Example: Given n points in a plane, no three of which are collinear, how many line segments must be drawn to connect all pairs of points?

Solution: To begin the solution, it is perhaps best to draw actual points in a plane and to connect them by line segments. We should do this systematically. So, start with 2 points, then three points, then four, and so on. Then, from these trials, create a table, and try to observe a pattern to generalize a solution to the problem (see Figure 1.16).

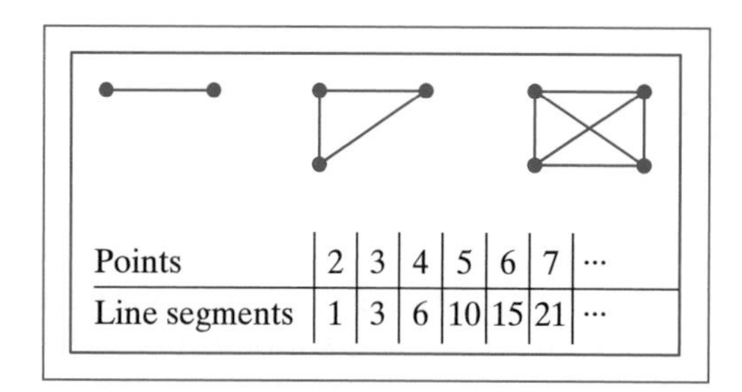

Figure 1.16

To try to establish a pattern, we can observe a pattern that can be illustrated in Table 1.8.

Table 1.8

Points	2	3	4	5	6
Line Segments	1	$1 + 2$	$1 + 2 + 3$	$1 + 2 + 3 + 4$	$1 + 2 + 3 + 4 + 5$

From this table, the pattern becomes clear. For 3 points, we have $1 + 2$ line segments; for 4 points, we have $1 + 2 + 3$ line segments, and so on. Thus, for n points, we will have $1 + 2 + 3 + \cdots + (n - 1)$ line segments. This gives us the general formula to find the number of segments that connect any number of (noncollinear if more than two) points, in pairs, to each other.

Make a Table. As we have seen in the previous problem, a table is a useful tool to use when the problem has several related components that build upon one another or when we have a list of possibilities and need to manipulate each possibility in several stages.

Example: On opening night, the director counted 56 people and noticed that 1 out of every 5 seats in the theater was empty. How many empty seats were there?

Solution: A table will facilitate finding a solution to this problem. Consider Table 1.9.

Table 1.9

Total seats	5	10	15	20	...	
Filled seats	4	8	12	16	...	56
Empty seats	1	2	3	4	...	

The table helps us organize the situation and to generate a list of possible total seats and a breakdown of how many of the total seats would be empty and how many filled. For example, if there were 5 seats total, then 1 of the 5 would be empty and 4 of the 5 would be filled. If there were 10 seats, 2 would be empty and 8 would be filled. Thus, we can use the table to project how many seats will be empty if there are 56 filled seats. We note a relationship between filled seats and empty seats in that the number of filled seats is 4 times the number of empty seats. Thus, if there are 56 filled seats, there would be $56 \div 4 = 14$ empty seats. Therefore, we can conclude that the theater must seat 70 people and that 14 of those seats were empty on opening night.

Make an Organized List. When solving problems with multiple solutions, making an organized list is a useful strategy. It is also useful when solving problems regarding various ways of combining or arranging things. Organized lists can be created within a table to facilitate the organization process.

Example: There are four infielders on the junior baseball league, and the coach is trying to determine the best batting order of the four infielders: first base, second base, third base, and shortstop. What are all the possible batting-order combinations he could try with the four players?

Solution: To make sure that we find all possible combinations, it is important that we generate a list of possibilities in an organized fashion. Therefore, let us first refer to the first-base player by using the symbol 1, second base with the symbol 2, third base with the symbol 3, and shortstop with the letter "S." Thus, one possible batting order might be (1, 2, 3, S). Another possibility would be (S, 3, 2, 1). We can continue to configure batting orders, but the best means is to organize the list so that we won't mistakenly leave out some of the possibilities. Here is a sample of the organized list:

1, 2, 3, S	2, 1, 3, S	3, 1, 2, S	S, 1, 2, 3
1, 2, S, 3	2, 1, S, 3	3, 1, S, 2	S, 1, 3, 2
1, 3, 2, S	2, 3, 1, S	3, 2, 1, S	S, 2, 1, 3
1, 3, S, 2	2, 3, S, 1	3, 2, S, 1	S, 2, 3, 1
1, S, 2, 3	2, S, 1, 3	3, S, 1, 2	S, 3, 1, 2
1, S, 3, 2	2, S, 3, 1	3, S, 2, 1	S, 3, 2, 1

By starting the list with all the possibilities where the first-base player bats first and combining the first-base player with each of the ways to arrange the remaining three batters, we begin an organized list. Continuing by considering all possibilities where the second-base player is first maintains the organization of the list. In this fashion, we find all 24 possible batting orders of the four infielders.

Work Backward. In some problems, you are given a series of related data. To work through the data to uncover a solution to your problem, you need to start with the last and most specific piece of the data and work backward through the other related pieces of data. For example, suppose you visit a friend in an unfamiliar city. Your friend gives you directions to reach her house from the highway. After spending some time with her, you are ready to return. This time, you need to reach the highway from your friend's house, but you do not have a map. Your best way to solve your problem is to try to work backward from the directions given to you earlier.

Example: April is thinking of a number. If you add 7 to the number, multiply the result by 2, and then subtract 10, you have 26. What was the original number that April thought?

Solution: We do not know what number April started with, but we do know what her result was, so we start at the end and work our way backward, unraveling the information at each stage. Starting with 26, we add back the 10 that had been subtracted to get the difference of 26 and have 36. Then, undoing the fact that this number was found by multiplying by 2, we divide the number by 2 to find 18. Finally, to undo the fact that 7 had been added to the original number, we subtract 7 from 18 to yield a solution of 11. To check that we indeed found the correct original number, start with 11, add 7 to total 18, multiply by 2 to reach 36, and then subtract 10 to end with 26. We have verified that 11 was April's original number.

Use Algebra. Using algebra is probably one strategy with which every student in high school becomes familiar to solve problems. In fact, algebra is often the only strategy that students try to use to solve a problem. Although not all problems can be solved using algebra, it certainly is a very useful problem-solving tool, particularly when it is combined with other strategies, such as drawing a picture, making a table, or looking for patterns. Just as a reminder, this strategy involves replacing an unknown quantity (or quantities) with a variable (or variables), using letters such as x or y, setting up an equation using these variables to satisfy the condition given in the problem, and then solving the equation for the variable through symbolic manipulation.

Example: Show that the sum of any five consecutive multiples of 3 is a multiple of 15.

Solution: To solve this problem, we will introduce a variable and develop an algebraic expression to show the sum of five consecutive numbers. We can write a multiple of 3 as $3m$, where m is any integer. Then, the next multiple of 3 will be $3m + 3$. Thus, five consecutive multiples of 3 can be written as $3m, 3m + 3, 3m + 6, 3m + 9, 3m + 12$. Their sum will be $3m + (3m + 3) + (3m + 6) + (3m + 9) + (3m + 12) = 15m + 30 = 15(m + 2)$, which is clearly a multiple of 15.

Solve a Simpler Problem. Sometimes, a mathematical problem may seem overwhelming simply because it has too many conditions that need to be satisfied, or because it involves big and complicated numbers, or because there are simply too many tasks that need to be done to reach the final answer. In such cases, it may be helpful to look for a simpler, related problem to solve. Once you find the solution to the easier problem, you can examine its solution to see what insight it gives you for the big, complex problem. Often, this process may make your task much easier.

Figure 1.17

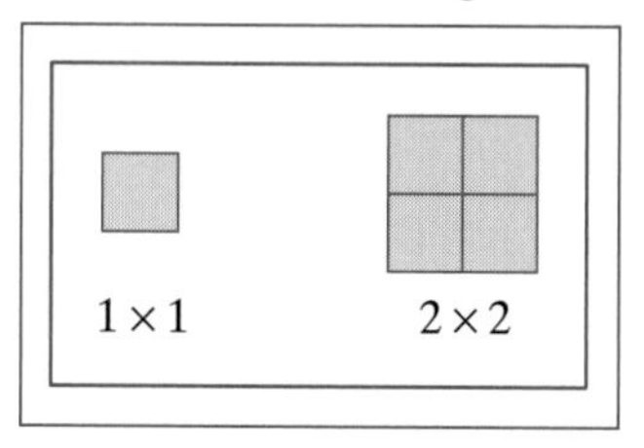

Example: How many squares are there in an 8×8 checkerboard?

Solution: To find the total number of squares on an 8×8 checkerboard may be difficult to do directly. Instead, we can try to solve a simpler problem by looking at smaller checkerboards. We can start with the 1×1 and 2×2 boards (see Figure 1.17).

Figure 1.18

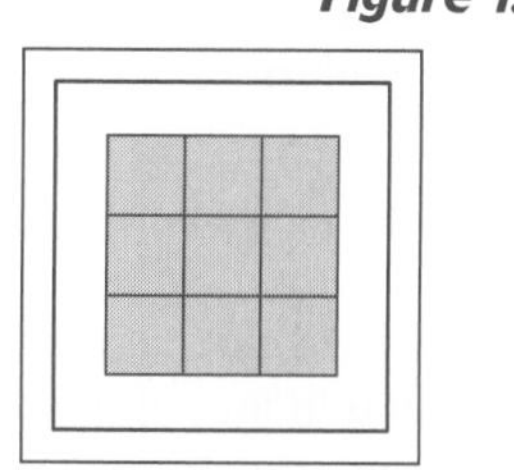

The 1×1 board clearly has just one square. The 2×2 board has 4 small (1×1) squares and 1 big (2×2) square, for a total of 5 squares. Now, we can look at a 3×3 board (see Figure 1.18). From our experience with the 2×2 board, we can conclude that counting will be much easier if we do it systematically, that is, by looking at how many 1×1 squares, how many 2×2 squares, and how many 3×3 squares there are in the 3×3 board.

In Figure 1.18, we find 9 1×1 squares, 4 2×2 squares, and 1 3×3 squares for a total of 14 squares in the 3×3 board. Now, we examine the 4×4 board in Figure 1.19. Within the 4×4 board, we find 16 1×1 squares, 9 2×2 squares, 4 3×3 squares, and 1 4×4 squares for a total of 30 squares in the 4×4 board.

Figure 1.19

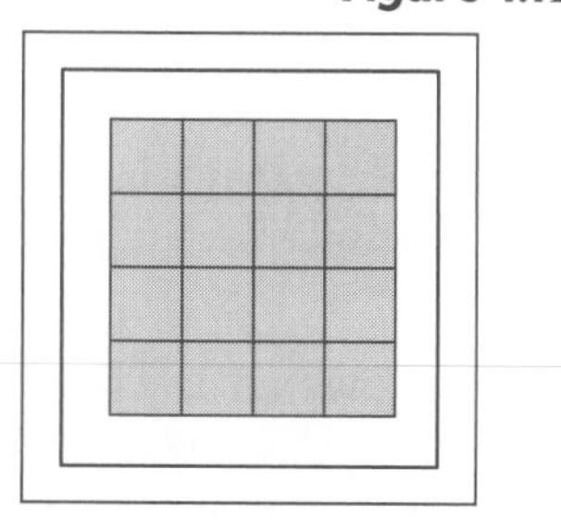

Having examined these simpler versions of the problem, we have learned a systematic means of counting the number of squares in an $n \times n$ board. Now, we can combine what we have learned from these simpler problems with the strategies of making a table and looking for a pattern (see Table 1.10) to determine the number of squares in an 8×8 board.

From the table, we see that the number of squares in the 1×1 board is 1 (the first square number). The number of squares in the 2×2 board is found by adding the first two square numbers $(1 + 4)$. The number of squares in the 3×3 board can be found by adding the first three square numbers $(1 + 4 + 9)$. The number of

Table 1.10

Size of Board	Number of Squares	Total
1×1	1	1
2×2	$1 + 4$	5
3×3	$1 + 4 + 9$	14
4×4	$1 + 4 + 9 + 16$	30
⋮	⋮	⋮
8×8	$1 + 4 + 9 + 16 + 25 + 36 + 49 + 64$	204

squares in the 4×4 boards can be found by adding the first four square numbers $(1 + 4 + 9 + 16)$. Thus, we can determine the number of squares in the 8×8 checkerboard by adding the first eight square numbers to find the sum 204.

Related Topics

Logic
Patterns

In the Classroom

In this first chapter, we emphasize the five *Process Standards* outlined in the National Council of Teachers of Mathematics *Principles and Standards for School Mathematics* (NCTM, 2000).

> *Standards are descriptions of what mathematics instruction should enable students to know and do. They specify the understanding, knowledge, and skills that students should acquire from prekindergarten through grade 12. The Content Standards—Number and Operations, Algebra, Geometry, Measurement, and Data Analysis and Probability—explicitly describe the content that students should learn. The Process Standards—Problem Solving, Reasoning and Proof, Communication, Connections, and Representation—highlight ways of acquiring and using content knowledge (p. 29).*

Some Expectations of Your Future Students:

From the *Problem Solving Standard* (NCTM, 2000, p. 52)

- build new mathematical knowledge through problem solving
- solve problems that arise in mathematics and in other contexts
- apply and adapt a variety of appropriate strategies to solve problems
- monitor and reflect on the process of mathematical problem solving

From the *Reasoning & Proof, Communication, Connections & Representation Standards* (NCTM, 2000, pp. 56, 60, 64, 67)

- recognize reasoning and proof as fundamental aspects of mathematics
- communicate their mathematical thinking coherently and clearly to peers, teachers, and others
- recognize and use connections among mathematical ideas
- create and use representations to organize, record, and communicate mathematical ideas

How Activities 1.1–1.10 Help You Develop an Adult-level Perspective on the Above Expectations:

- By stacking cereal boxes or laying patio tiles, you solve problems that may arise in everyday contexts, yet have important connections to number, geometry, and algebra. This activity also gives you the opportunity to apply different problem-solving strategies as well as to adapt a strategy that was used to solve one problem to solve another.
- By modeling situations such as the Valentine's Day Party or the Hefty Hippos puzzle, you learn to communicate your mathematical thinking coherently and clearly to each other, while gaining an opportunity to monitor formally and reflect on the process of mathematical problem solving.
- By creating and using representations to visualize a situation or to understand relationships, you learn to organize, record, and communicate mathematical ideas in activities such as Making Dice and the Tower of Hanoi.
- By exploring the mathematics in the pages of a newspaper and the activity in constructing numbers, you build new mathematical knowledge through problem solving.
- By playing games such as What's My Number? and Poison, and by exploring Family Relations, you realize that reasoning and proof are fundamental aspects of mathematics, and you develop strategies for justifying the general case by analyzing the specific.

Bibliography

Averback, W. B., & Orin, C. (1980). *Problem solving through recreational mathematics*. New York: Freeman.

Akaishi, A., & Saul, M. (1991). Exploring, learning, sharing: Vignettes from the classroom. *Arithmetic Teacher, 39*, 12–16.

Bledsoe, G. J. (1989). Hook your students on problem solving. *Arithmetic Teacher, 37*, 16–20.

Borassi, R. (1992). *Learning mathematics through inquiry*. Portsmouth. NH: Heinemann.

Brown, S. I., & Walter, M. I. (1990). *The art of problem posing* (2nd edition). Hillsdale, NJ: Erlbaum.

Brown, S. I., & Walter, M. I. (Eds.). (1993). *Problem posing: Reflections and applications*. Hillsdale, NJ: Erlbaum.

Bush, W. S., & Fiala, A. (1986). Problem stories: A new twist on problem posing. *Arithmetic Teacher, 34*, 6–9.

Campbell, P. F., & Bamberger, H. J. (1990). Implementing the *Standards*: The vision of problem solving in the *Standards*. *Arithmetic Teacher, 37*, 14–17.

Cemen, P. B. (1989). Developing a problem-solving lesson. *Arithmetic Teacher, 37*, 14–19.

Chambers, D. L. (1995). Improving instruction by listening to children. *Teaching Children Mathematics, 1*, 378–380.

Charles, R., & Lester, F. K. (1982). *Teaching problem solving: What, why and how*. Palo Alto, CA: Dale Seymour Publications.

Corwin, R. B. (1993). Doing mathematics together: Creating a mathematical culture. *Arithmetic Teacher, 40*, 336–341.

Day, R. P. (1986). A problem-solving component for junior high school mathematics. *Arithmetic Teacher, 34*, 14–17.

English, L. D., Cudmore, D., & Tilley, D. (1998). Problem posing and critiquing: How it can happen in your classroom. *Mathematics Teaching in the Middle School, 4*, 124–129.

Esty, W. W., & Teppo, A. R. (1996). Algebraic thinking, language, and word problems. In P. C. Elliott (Ed.), *Communication in mathematics: K–12 and beyond* (pp. 45–55) Reston, VA: National Council of Teachers of Mathematics.

Fortunato, I., Hecht, D., Tittle, C. K., & Alvarez, L. (1991). Metacognition and problem solving. *Arithmetic Teacher, 39*, 38–40.

Fulkerson, P. (1992). Getting the most from a problem. *Arithmetic Teacher, 40*, 178–179.

Gardiner, A. (1996). *Discovering mathematics: The art of investigation*. New York: Oxford Scientific Publications.

Garofalo, J. (1987). Metacognition and school mathematics. *Arithmetic Teacher, 34*(9), 22–23.

Garofalo, J., & Lester, F. K. (1985). Metacognition, cognitive monitoring, and mathematical performance. *Journal for Research in Mathematics Education, 16*, 163–176.

Garofalo, J., & Bryant, J. (1992). Assessing reasonableness: Some observations and suggestions. *Arithmetic Teacher, 40*, 210–212.

Goodnow, J., Hoogeboom, S., Moretti, G., Stephens, M., & Scanlin, A. (1987). *The problem solver: Activities for learning problem solving strategies*. Palo Alto, CA: Creative Publications.

Hembree, R., & Marsh, H. (1993). Problem solving in early childhood: Building foundations. In R. J. Jensen (Ed.), *Research ideas for the classroom: Early childhood mathematics* (pp. 151–170). Old Tappan, NJ: Macmillan.

Henningsen, M., & Stein, M. K. (1997). Mathematical tasks and student cognition: Classroom-based factors that support and inhibit high-level mathematical thinking and reasoning. *Journal for Research in Mathematics Education, 28*, 524–549.

Hyde, A. A., & Hyde, P. R. (1991). *Mathwise: Teaching mathematical thinking and problem solving*. Portsmouth, NH: Heinemann.

Kersh, M. E., & McDonald, J. (1991). How do I solve Thee? Let me count the ways! *Arithmetic Teacher, 39*, 38–41.

Kimball, R. L. (1991). Activities: Make your own problems—and then solve them. *Mathematics Teacher, 84*, 647–655.

Kroll, D. L., & Miller, T. (1993). Insights from research on mathematical problem solving in the middle grades. In D. T. Owens (Ed.), *Research ideas for the classroom: Middle grades mathematics* (pp. 58–77). Old Tappan, NJ: Macmillan.

Kruliks, S., & Rudnick, J. A. (1985). Developing problem-solving skills. *Mathematics Teacher, 78*, 685–692, 697–698.

Lester, F. K. (1989). Research into practice: Mathematical problem solving in and out of school. *Arithmetic Teacher, 39*, 33–35.

Lester, F. K., Jr., Masingila, J. O., Mau, S. T., Lambdin, D. V., Santos, V. M. P., & Raymond, A. M. (1994). Learning how to teach via problem solving. In D. Aichele (Ed.), *Professional development for teachers of mathematics* (pp. 152–166). Reston, VA: National Council of Teachers of Mathematics.

Lubienski, S. T. (1999). Problem-centered mathematics teaching. *Mathematics Teaching in the Middle School, 5*, 250–255.

Manouchegri, A., & Enderson, M. C. (1999). Promoting mathematical discourse: Learning from classroom examples. *Mathematics Teaching in the Middle School, 4*, 216–222.

Masingila, J. O. (1997). Let's be realistic. *Mathematics Teaching in the Middle School, 2* (3), 136–137.

Masingila, J. O. (1998). Thinking deeply about knowing mathematics. *Mathematics Teacher, 91* (7), 610–614.

Masingila, J. O., & Moellwald, F. E. (1993). Using Polya to foster a classroom environment for real-world problem solving. *School Science and Mathematics, 93* (5), 245–249.

Mason, J., Burton, L. & Stacey, K. (1985). *Thinking mathematically* (Revised edition). Menlo Park, CA: Addison-Wesley.

Matz, K. A., & Leier, C. (1992). Word Problems and the Language Connection. *Arithmetic Teacher, 39*, 14–17.

McNeal, B. (1995). Learning not to think in a textbook-baled mathematics class. *Journal of Mathematical Behavior, 14*, 205–234.

Meyer, C., & Sallee, T. (1983). *Make it simpler: A practical guide to problem solving in mathematics*. Menlo Park, CA: Addison-Wesley.

Ritchhart, R. (1999). Generative topics: Building a curriculum around big ideas. *Teaching Children Mathematics, 5*, 462–468.

Rowan, T. E., & Robles, J. (1998). Using questions to help children build mathematical power. *Teaching Children Mathematics, 4*, 504–509.

Sigurdson, S. E., Olson, A. T., & Mason, R. (1994). Problem solving and mathematics reaming. *Journal of Mathematical Behavior, 13*, 361–388.

Silver, E. A., Kilpatrick, J., & Schlesinger, B. (1990). *Thinking through mathematics: Fostering inquiry and communica-*

tion in mathematics classrooms. New York: College Entrance Examination Board.

Silverman, F. L., Winograd, K., & Strohauer, D. (1992). Student-generated story problems. *Arithmetic Teacher, 39*, 6–12.

Smith, M. S., & Stein, M. K. (1998). Selecting and creating mathematical tasks: From research to practice. *Mathematics Teaching in the Middle School, 3*, 344–350.

Stein, M. K., & Smith, M. S. (1998). Mathematical tasks as a framework for reflection: From research to practice. *Mathematics Teaching in the Middle School, 3*, 268–275.

Stimpson, V. C. (1989). Using diagrams to solve problems. *Mathematics Teacher, 82*, 194–200.

Szetela, W. (1986). The checkerboard problem extended, extended, extended. *School Science and Mathematics, 86*, 205–222.

Talton, C. F. (1988). Let's solve the problem before we find the answer. *Arithmetic Teacher, 36*(1), 40–5.

Whitin, D. J. (1987). Problem solving in action: The bulletin-board dilemma. *Arithmetic Teacher, 35*, 48–50.

CHAPTER 2 TWO

Numeration

CHAPTER OVERVIEW

Among the most important topics in elementary school mathematics are those concerned with systems of recording and naming numbers: numeration systems. In this chapter, you can learn some interesting facts about numeration and examine some concepts and procedures involving numeration ideas.

BIG MATHEMATICAL IDEAS

Generalizing, problem-solving strategies, decomposing, mathematical structure, multiple representations

NCTM PRINCIPLES & STANDARDS LINKS

Number and Operation; Problem Solving; Reasoning; Communication; Connections; Representation

TOPIC **2.1** **Characteristics of Numeration Systems**
2.2 **Base-Ten Introduction**
2.3 **Place Value and Zero**
2.4 **Decimal System and More on Base Ten**
In the Classroom
Bibliography

TOPIC 2.1 Characteristics of Numeration Systems

cornerstone **Whole Numbers**

Introduction

A numeration system is a collection of numerals with a set of characteristics that determine how to interpret those symbols and algorithms that depict methods of calculating amounts with those numerals. Characteristics associated with a variety of numeration systems include having place value, having a base, and having a zero. The Hindu-Arabic Numeration System, the prevailing system used today, has all of these characteristics and more. In this section, we list the six basic characteristics of the Hindu-Arabic Numeration System as an introduction to the sections that follow and briefly examine the characteristics of unique representation. The remaining five characteristics of the Hindu-Arabic Numeration System are examined in subsequent sections within this Topic.

Did you know that the numeration characteristic of being **additive** shows up in most of the numeration systems about which we know? Even the Primitive Numeration System, which consists of a tally-marking process that seems almost too basic to be defined by any characteristics, is an additive system. Take, for example, the amount represented in the Primitive System, ||||||. To interpret the value of these symbols, we simply add the singular values of each of the tally marks. Because each tally mark represents "1," we can interpret the number |||||| as $1 + 1 + 1 + 1 + 1 + 1 = 6$.

2.1.1 Characteristics of the Hindu-Arabic System

The Hindu-Arabic Numeration System has six key characteristics: place value, base ten, having a zero, additivity, multiplicativity, and unique representation. Although many of these characteristics are found in other numeration systems, having the combination of these six characteristics is what makes the Hindu-Arabic System singularly its own.

2.1.2 Unique Representation

One key outcome of the Hindu-Arabic System due to the combination of five of the six characteristics, those being place value, base ten, having a zero, additivity, and multiplicativity, is that we are guaranteed **unique representation.** We all agree that there is one, and only one, way to represent the amount five hundred forty-six using the Hindu-Arabic Numeration System, that is, by the symbols 546. This characteristic of unique representation, although seemingly such a simple feature, provides a clarity of number representation that is not present in all numeration systems.

Unlike the Hindu-Arabic System, the Roman Numeration System allows for multiple representation of some numbers. For example, when representing the amount 49 using the Roman Numeration System, some may use the combination of symbols XLIX, while others may simply use IL. Although multiple representations are not inherently bad, it is important in our technologically advancing world that we have a precise, unambiguous, and indisputable means of representing numbers in society.

Related Topics
Base-Ten Introduction
Place Value and Zero
Decimal System and More on Base Ten

TOPIC 2.2 *Base-Ten Introduction*

cornerstone **Whole Numbers**

Introduction

The Hindu-Arabic Numeration System that we use today is a base-ten system. This system is more sophisticated than some of the earlier systems, such as the Egyptian Numeration System, which had symbols to represent certain powers of ten. The base-ten system Hindu-Arabic System is useful in its ability to represent numbers concisely. For example, we can represent the amount four hundred fifty-nine in the Hindu-Arabic System using the symbols 459, with the 4 representing four sets of 100, the 5 representing 5 sets of ten, and the 9 representing 9 ones. Other numeration systems often require a larger set of symbols, such as the symbols ? ? ? ? ∩ ∩ ∩ ∩ ∩ ||||||||| to represent 459 in the Egyptian System. In the following sections, we explore base ten and other characteristics of our numeration system.

Numeration systems in early societies often reveal interesting aspects of the culture that prevailed and what was regarded as important in these societies. It is generally accepted that the use of base ten in many cultures evolved from the practice of using fingers of the hands for counting. In warmer climates, on the other hand, people could even use their toes for counting; this may have led to the use of the number 20 as a base in such cultures as the Maya of Central America and the Yoruba in western Africa.

For More Information Zaslavsky, C. (1996). *The Multicultural Math Classroom: Bringing in the World.* Portsmouth, NH: Heinemann.

2.2.1 Characteristics of the Egyptian Numeration System

Understanding our own numeration system can be enhanced by examining the characteristics of other systems. The Egyptian Numeration System illustrates several features that our Hindu-Arabic System shares. Note that the Egyptian Numeration System uses the following symbols: ✗ = 1000, ? = 100, ∩ = 10, | = 1.

Additive Characteristic. The Egyptian Numeration System is **additive,** meaning that the overall value of the number represented by a series of symbols is equal to the sum of the values of the individual symbols. For example, the number represented by the string of symbols ∩ ∩ ∩ ∩ ||| can be found by adding together the values of the symbols such as:

$$\cap + \cap + \cap + \cap + | + | + | = 10 + 10 + 10 + 10 + 1 + 1 + 1 = 43.$$

Our Hindu-Arabic System is also additive in that the value of the numeral 324, for example, is determined by adding together the values that each of the symbols represents as follows: $300 + 20 + 4$.

Powers of 10. All of the symbols in the Egyptian System represent various **powers of 10;** that is, the symbols are worth 1, 10, 100, 1000, 10000, and so on, which are numbers that can all be expressed as 10 or 10 times itself some number of times. For example, $10 = 10$, $100 = 10 \times 10$, $1{,}000 = 10 \times 10 \times 10$, $10{,}000 = 10 \times 10 \times 10 \times 10$, and so on. Each of these numbers could be written alternatively in exponential form such as 10^1, 10^2, 10^3, 10^4, and so on, where the exponent in this notation stands for the "power" of 10. Even the number 1 represented by the symbol "|" in the Egyptian System can be represented as a power of ten. We write the number 1 as 10^0.

Base-Ten System. Although the Egyptian System has symbols that themselves are each powers of 10, the system is considered to be as limited as a base-ten system. It is limited because there is no association of the powers of 10 with the specific placement of the digits. Consequently, the symbols alone do not provide a concise and unique means of representing large numbers even though they represent powers of 10. For example, ∩ ∩ | is the same as ∩ | ∩.

On the other hand, in the Hindu-Arabic System, **base ten** means that each digit in a numeral is associated with **a place value** that is a power of 10; that is, each of the places (or locations of a digit)—ones, tens, hundreds, thousands, and so on—is a power of 10. This association of unique powers of 10 with places allows us to write large numbers easily and uniquely, using only a few digits. See Figure 2.1.

Figure 2.1

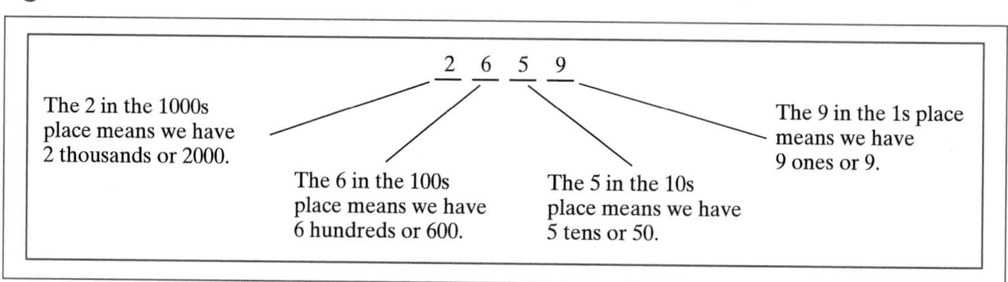

Without place value, both our Hindu-Arabic System and the Egyptian System would have awkward, nonunique representation for a number such as 2,659 (see Figure 2.2).

Figure 2.2

Without place value, 2659 may be written as:	
Egyptian	𓆼𓆼 𓍢𓍢𓍢𓍢𓍢𓍢 ∩∩∩∩∩ |||||||||
Hindu-Arabic	1000 + 1000 + 100 + 100 + 100 + 100 + 100 + 100 + 10 + 10 + 10 + 10 + 10 + 1 + 1 + 1 + 1 + 1 + 1 + 1 + 1 + 1

2.2.2 A Subtractive Characteristic Not Associated with the Hindu-Arabic System

Some numeration systems, such as the Roman Numeration System, have features that our Hindu-Arabic System does not have. For example, the Roman System has the characteristic of being **subtractive** in that sometimes, instead of adding the values of the symbols together, one must subtract the value of one digit from another to determine the value. Note that the Roman System uses the symbols: M = 1,000, D = 500, C = 100, L = 50, X = 10, V = 5, I = 1. When we see the symbols VII in the Roman System, we add the value of the symbols together: V + I + I = 5 + 1 + 1 = 7. However, when we see symbols positioned such that a smaller-valued digit precedes a larger-valued digit, as in the number IX where the symbol for 1 (I) precedes the symbol for 10 (X), we subtract the value of the smaller digit from the value of the larger digit. Thus, we interpret the number IX as X − I = 10 − 1 = 9.

2.2.3 Other Bases: The Example of Base Two

It is often useful to consider systems that operate like the Hindu-Arabic System except for having a different base. For example, consider the **base-two system** where each of the places is a power of 2 (instead of a power of 10 as in our base-ten system) and which only uses the two digits 0 and 1 (unlike the 10 digits used in our base-ten system). A base-two system is often referred to as a **binary system** (the prefix *bi* meaning "two"), particularly in the context of computer codes. In a digital computer, data and instructions are encoded as bits, which are the digits 0 and 1. The technology in a digital computer system relies on the state of an electronic circuit to represent a bit. The circuit must be capable of being in two states, one representing 1, the other 0.

An example of interpreting a base-two number in its base-ten equivalent is seen in the following. Given the base-two number 110011_{two}, we can translate the value of these symbols into terms we understand in base ten by figuring out what each symbol represents in each of the base-two places. That is,

$$\begin{aligned} 110{,}011_{two} &= (1 \times 2^5) + (1 \times 2^4) + (0 \times 2^3) + (0 \times 2^2) + (1 \times 2^1) + (1 \times 2^0) \\ &= 32 + 16 + 0 + 0 + 2 + 1 = 51_{ten} \end{aligned}$$

Similarly, we can start with a base-ten number, say 19, and determine how to represent this amount in a base-two system. To translate 19 to its base-two equivalent, we must consider the place values in base two up to the largest place that does not exceed 19. Thus, the places to consider would be $2^4 = 16$, $2^3 = 8$, $2^2 = 4$, $2^1 = 2$, $2^0 = 1$. Note that because $2^5 = 32$ exceeds 19, we can conclude that the base-two representation of 19 will be a five-digit number, and we must determine what digits (0 or 1) belong in each of the places $\underset{2^4}{\_}\ \underset{2^3}{\_}\ \underset{2^2}{\_}\ \underset{2^1}{\_}\ \underset{2^0}{\_}$. Because we can express 19 as $16 + 2 + 1$, we determine that 19 can be represented in base two by placing the digit 1 in the 2^4, 2^1, and 2^0 places and the digit 0 in the 2^3 *and* 2^2 places. Thus, 19 can be expressed as $\frac{1}{2^4}\ \frac{0}{2^3}\ \frac{0}{2^2}\ \frac{1}{2^1}\ \frac{1}{2^0}$ in base two or, more concisely written, $19 = 10011_{two}$.

Related Topics

Properties of Whole-Number Addition
Properties of Whole-Number Multiplication
Place Value and Zero
Decimal System and More on Base Ten

TOPIC 2.3 Place Value and Zero

cornerstone **Whole Numbers**
Base-Ten Introduction

Introduction

The characteristics of place value and a symbol for zero work together in the Hindu-Arabic Numeration System in that the symbol for zero serves as a placeholder when representing number amounts. However, early place-value systems did not always have a symbol for zero, which presented some limitations to those systems. In the sections that follow, we explore place value and our symbol for zero through examples.

DID YOU KNOW

The most outstanding achievement of the Babylonian Numeration System was the invention of a place-value system. The Babylonian System is an example of a numeration system in which the value of a numeral depends on its placement with respect to other numerals when writing a number. The Babylonians had two symbols, ∇, which represented 1, and $<$, which represented 10. The system was also additive, meaning that the value represented by the set of symbols was equal to the sum of each individual symbol value. For example, the number represented by the symbols $< < < \nabla\nabla$ is $10 + 10 + 10 + 1 + 1 = 32$.

The Babylonian System was also a base-60 system, so the places are all powers of 60: $60^0(60^0 = 1)$, 60^1, 60^2, and so on. Thus, to write numbers larger than 59, they used spaces to separate the powers of 60. For example, the symbols $< < \quad < \nabla\nabla$ represent $20 \times 60^1 + 12 \times 1 = 1{,}212$. The space, in this case, separates the ones place from the 60s place. In another example, the symbols $< \quad < \nabla \quad \nabla\nabla$ represent $10 \times 60^2 + 11 \times 60^1 + 2 \times 1 = 36{,}662$ (see Figure 2.3 for how these place values can be interpreted in a place-value table). Because the Babylonians implemented a place-value feature in its number system, they were able to represent fairly large numbers with only a few symbols.

$3600 = 60^2$	$\text{sixty} = 60^1$	$\text{ones} = 60^0$
$<$	$< \nabla$	$\nabla\nabla$

Figure 2.3

The limitations to the approach of leaving a space to indicate a new place include the danger of perhaps not indicating the space clearly enough. However, what is probably more limiting with this system is how to represent having a set of digits in the 60^2 place and in the ones place but no digits in the 60^1 place. Perhaps they simply left a bigger space initially to indicate that they were skipping a place. However, at some point, they decided that they needed a symbol to represent zero and introduced this new symbol, //, to their system to indicate where a place was being skipped. Thus, they, in effect, created a symbol as a placeholder, much like our present day symbol for zero.

For More Information Bunt, L. N. H., Jones, P. S., & Bedient, J. D. (1988). *The Historical Roots of Elementary Mathematics.* New York: Dover Publications.

2.3.1 Place Value

Place value refers to the value associated with a specific position a digit occupies in a number. In a place-value numeration system, the position that a digit occupies dramatically affects the value of the digit. For example, in the number 540, the digit 4 is in the tens place, so the 4 represents 4 sets of 10 or 40. On the other hand, in the number 504, the digit 4 is in the ones place and represents 4 ones. Thus, place value provides a means of representing each number amount in our number system in a unique way. See Table 2.1 for an abbreviated table of place values in the Hindu-Arabic Numeration System, showing the place values associated with the digits of the number 504.

Table 2.1

etc.	Ten Thousands Place	Thousands Place	Hundreds Place	Tens Place	Ones Place
			5	0	4

2.3.2 A Symbol for Zero

With a **symbol for zero,** 0, in the Hindu-Arabic Number System, we are able to maintain our place value, allowing for the possibility of representing numbers such as 504, meaning that there are 5 hundreds, 0 tens, and 4 ones. Without a zero, how would we represent 504? Would we leave a blank line to indicate that we are skipping the tens place, such as 5 _ 4? Alternatively, a representation of 5 4 does not make clear whether there are 5 hundreds (because a space is left) or 5 tens. Without a symbol for zero, we would need some other clear way to distinguish the place values that different digits occupy. The zero in the number 504 helps to indicate that the digit 5 really stands for 5 hundreds. The use of a symbol for zero leaves no doubt of the placement of a digit, therefore leaving no doubt as to the overall value of the number represented. Besides playing the important role of a placeholder, recall that 0 is also the cardinal number, representing the size of the empty set.

Related Topics **Decimal System and More on Base Ten**

TOPIC 2.4 Decimal System and More on Base Ten

cornerstone

Whole Numbers
Base-Ten Introduction
Place Value and Zero

Introduction

The prefix *deci* comes from the Latin word, *decem*, meaning "ten." Thus, referring to a decimal system is the same as referring to a base-ten system. Today, when we hear the word *decimal*, we might think of decimal fractions. What we will discover is that a decimal system not only allows us to write all whole numbers concisely, but it also provides the means of representing many fractional amounts as well. Consequently, our decimal system has developed over time from its beginning use as a means of representing whole-number amounts to a sophisticated system where we can represent a myriad of whole-number and fractional-number amounts.

Every numeration system has two fundamental features: a collection of symbols and a set of rules by which we combine and interpret those symbols. Our decimal system has 10 symbols: 0, 1, 2, 3, 4, 5, 6, 7, 8, and 9. These 10 symbols are also referred to as digits, which corresponds to a name used to refer to our fingers or toes. These 10 digits combined with a base of 10 provide a means of representing any whole number uniquely.

Knowledge of the decimal number system allows us to answer questions that range from the most basic to very sophisticated. For example, we might ask what digit is in the tens place of the number 23,469? The problem solver could identify the tens place and see that the digit 6 is in that place. On the other hand, a more complicated problem to solve that requires knowledge of the decimal number system would be the following. Consider what number meets the following criteria: Using some of the digits 0, 1, 2, 3, 4, 5, 6, 7, 8, 9 without using any digit more than once, determine the largest odd six-digit number with a 9 in the tens place. Through process of elimination, the problem solver can determine that the number 876,593 meets all the criteria.

2.4.1 Decimal System and Expanded Notation

The **Hindu-Arabic System** is called **a decimal system** because it is based on powers of 10. The Hindu-Arabic decimal system is best understood when numbers are written in **expanded notation.** In expanded notation, numerals are broken down, digit by digit, writing the value of each digit as a product of its face value (that is, the value of the symbol itself) times its place value (which is always a power of 10). For example, the number 6,023 can be written as $6 \times 1{,}000 + 0 \times 100 + 2 \times 10 + 3 \times 1$. Further, expressing each place value as a power of 10, the formal expanded notation of 6,023 is written as:

$$(6 \times 10^3) + (0 \times 10^2) + (2 \times 10^1) + (3 \times 10^0).$$

Note that it is commonly accepted to represent the value of the digit in the ones place as the face value times 1. Thus, expanded forms in textbooks often represent a number such as 6,023 as:

$$(6 \times 10^3) + (0 \times 10^2) + (2 \times 10^1) + (3 \times 1).$$

2.4.2 Multiplicative Characteristic

The fact that the value of digits in the Hindu-Arabic System are understood by multiplying face value times place value implies that the system is **multiplicative.** Thus, when numbers are written in expanded form, we clearly see the system as additive, multiplicative, having place value, having a base of 10, and having a zero.

Related Topics

Properties of Whole-Number Addition
Properties of Whole-Number Multiplication

In the Classroom

This chapter on numeration is the first of five chapters whose focus is on some aspect of the *Number and Operations Content Standard.* Given that "all mathematics proposed for prekindergarten through Grade 12 is strongly grounded in number" (NCTM, 2000, p. 32), we believe it is essential for you to develop a strong, connected understanding of (a) numeration systems as abstract, structured representations of number concepts (Activities 2.1–2.5), and (b) place value and base as fundamental to the structure of and operations on the base-ten numeration system we use (Activities 2.6–2.10).

Some Expectations of Your Future Students:

Pre-K–2 (NCTM, 2000, p. 78)

- count with understanding and recognize "how many" in sets of objects
- connect number words and numerals to the quantities they represent, using various physical models and representations
- use multiple models to develop initial understandings of place value and the base-ten number system
- understand the effects of adding and subtracting whole numbers

Grades 3–5 (NCTM, 2000, p. 148)

- recognize equivalent representations of the same number and generate them by decomposing and composing numbers
- understand the place-value structure of the base-ten number system
- understand the effects of multiplying and dividing whole numbers

How Activities 2.1–2.10 Help You Develop an Adult-level Perspective on the Above Expectations:

- By examining early numeration systems and comparing them and our base-ten system, you work with different representations of number.
- By exploring the historical development of number concepts and representations, you distinguish between the idea of number as concept and numeral as representation. You appreciate the need to enable children to develop this understanding through appropriate experiences and over time. For example, many ancient systems did not have the concept of, or representation for, zero.
- By understanding the symbols used in a system and how they are arranged to represent different numbers, you examine the composition of a number as sums or products of other numbers.
- By working with base blocks and in bases other than base ten, you construct a conceptual understanding of place value. You thereby develop a conceptual understanding of the procedures we employ in basic arithmetic operations.

Bibliography

Arithmetic Teacher. (February 1988). Focus issue on early childhood mathematics, *35*.

Baroody, A. J., & Wilkins, J. L. M. (1999). The development of informal counting, number, and arithmetic skills and concepts. In J. V. Copley (Ed.), *Mathematics in the early years* (pp. 48–65). Reston, VA: National Council of Teachers of Mathematics.

Bresser, R., & Holtzman, C. (1999). *Developing number sense*. Sausalito, CA: Math Solutions Publications.

Clements, D. H. (1999). Subitizing: What is it? Why teach it? *Teaching Children Mathematics, 5,* 400–405.

Flexer, R J. (1986). The power of five: The step before the power of ten. *Arithmetic Teacher, 34*(3), 5–9.

Frank, A. R. (1989). Counting skills—A foundation for early mathematics. *Arithmetic Teacher, 37,* 14–17.

Fuson, K. C., Wearne, D., Hiebert, J. C., Murray, H. G., Human, P. G., Olivier, A. I., Carpenter, T. P., & Fennema, E. (1997). Children's conceptual structures for multidigit numbers and methods of multidigit addition and subtraction. *Journal for Research in Mathematics Education, 28,* 130–162.

Gluck, D. H. (1991). Helping students understand place value. *Arithmetic Teacher, 38,* 10–13.

Harrison, M., & Harrison, B. (1986). Developing numeration concepts and skills. *Arithmetic Teacher, 33,* 18–21.

Ifrah, G. (1987). *From one to zero: A universal history of number*. New York: Penguin Books.

Krusen, K. (1991). A historical reconstruction of our number system. *Arithmetic Teacher, 38,* 46–48.

Labinowicz, E. (1985). *Learning from children: New beginnings for teaching numerical thinking*. Menlo Park, CA: Addison-Wesley.

Little, C. (1999). Counting grass. *Mathematics Teaching in the Middle Grades, 5,* 7–10.

McClain, K., & Cobb, P. (1999). Supporting students' ways of reasoning about patterns and partitions. In J. V. Copley (Ed.), *Mathematics in the early years* (pp. 112–118). Reston, VA: National Council of Teachers of Mathematics.

Menninger, K. (1992). *Number words & number symbols: A cultural history of numbers*. New York: Dover.

Nagel, N. G., & Swingen, C. C. (1998). Student's explanations of place value in addition and subtraction. *Teaching Children Mathematics, 5,* 164–170.

Payne, J. N., & Huinker, D. M. (1993). Early number and numeration. In R. J. Jensen (Ed.), *Research ideas for the classroom: Early childhood mathematics* (pp. 4–71). Old Tappan, NJ: Macmillan.

Rathmell, E. E., & Leutzinger, L. P. (1991). Implementing the *Standards*: Number representations and relationships. *Arithmetic Teacher, 38,* 20–23.

Schwartz, S. L. (1995). Enchanting, fascinating, useful number. *Teaching Children Mathematics, 1,* 486–491.

Sutton, J. T., & Urbatsch, T. D. (1991). Transition boards: A good idea made better. *Arithmetic Teacher, 38,* 4–9.

Suydam, M. N. (1986). Research report: The process of counting. *Arithmetic Teacher, 33,* 29.

Thompson, C. S. (1990). Place value and larger numbers. In J. N. Payne (Ed.), *Mathematics for the young child* (pp. 89–108). Reston, VA: National Council of Teachers of Mathematics.

Thompson, C. S. (1989). Number sense and numeration in grades K–8. *Arithmetic Teacher, 37,* 22–24.

Thompson, C. S., & Van de Walle, J. A. (1984). The power of 10. *Arithmetic Teacher, 32*(3), 6–11.

Wearne, D., & Hiebert. J. (1994). Place value and addition and subtraction. *Arithmetic Teacher, 41,* 272–274.

Weinberg, S. (1996). Going beyond ten black dots. *Teaching Children Mathematics, 2,* 432–435.

Wicket, M. S. (1997). Serving up number sense and problem solving: Dinner at the panda palace. *Teaching Children Mathematics, 3,* 476–480.

Zepp, R. A. (1992). Numbers and codes in ancient Peru: The Quipi. *Arithimetic Teacher, 39,* 42–44.

CHAPTER 3 THREE

Operations with Natural Numbers, Whole Numbers & Integers

TOPIC 3.1 Natural Numbers
3.2 Introduction to Whole Numbers
3.3 Addition with Whole Numbers
3.4 Addition Properties and Patterns (Whole Numbers)
3.5 Addition Algorithms (Whole Numbers)
3.6 Subtraction with Whole Numbers
3.7 Subtraction Algorithms (Whole Numbers)
3.8 Multiplication with Whole Numbers
3.9 Multiplication Properties and Patterns (Whole Numbers)
3.10 Multiplication Algorithms
3.11 Whole-Number Division
3.12 Division Algorithms
3.13 Introduction to Integers
3.14 Integer Addition
3.15 Integer Subtraction
3.16 Integer Multiplication
3.17 Integer Division
3.18 Order of Operations
In the Classroom
Bibliography

CHAPTER OVERVIEW

A solid understanding of addition, subtraction, multiplication, and division is crucial to being able to do mathematics, and these operations play central parts in the elementary school mathematics curriculum. In this chapter, you can learn some interesting facts about operations and will examine some concepts and procedures involving operation ideas.

BIG MATHEMATICAL IDEAS

Mathematical structure, verifying, generalizing, using algorithms

NCTM PRINCIPLES & STANDARDS LINKS

Number and Operation; Problem Solving; Reasoning; Communication; Connections; Representation

TOPIC 3.1 Natural Numbers

cornerstone **Characteristics of Numeration Systems**

Introduction

Natural numbers are the numbers we "naturally" use when we count objects. They are the most basic set of numbers to which students are introduced in elementary school. In this section, we define natural numbers.

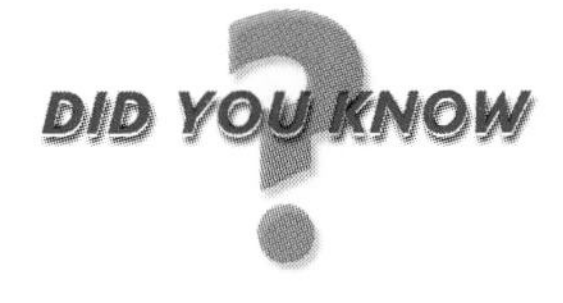

Natural numbers provide us with the most basic means of telling "how much." However, some cultures do not need many of these most basic numbers. For example, in some Australian tribes, members do not have use for numerals beyond the amount of four. In many cases, having number names for "one" and "two" are sufficient. Amounts beyond two are called "much" or "many." This reminds us that numeration systems match the needs of the cultures they serve. The more complicated numeration systems themselves began with basic elements but grew as societal demands changed.

For More Information Smith, D. E. (1958). *History of Mathematics, Volume I.* New York: Dover Publications.

The **natural numbers,** also known as the **counting numbers,** are the numbers beginning with "1" and progressing as 1, 2, 3, 4, 5, ... The ellipses (...) indicate that this set of numbers continues on in this pattern and is infinite. **The set of natural numbers,** written in set notation as $N = \{1, 2, 3, 4, 5, \dots\}$ is often simply referred to as "N." The natural numbers are used to express "how much" or "how large a set is," with the exception of sets that are empty.

Related Topics

Whole Numbers
Integers

TOPIC 3.2 Introduction to Whole Numbers

cornerstone

Natural Numbers
Sets

Introduction

The earliest roots of mathematics can probably be traced back to a need for counting and recording numbers. From as far back as historians can tell, every society has had some means of recording numerical quantities. This is true even of those societies for which no written records of numerals are available. Evidence of this fact comes from archaeological excavations from different parts of the world. Whole numbers provide the symbols we use in our Hindu-Arabic Number System for counting and recording. Within this topic, we examine the set of whole numbers and how they relate to other number representations and counting.

The Hindu-Arabic system we use today evolved slowly. It is commonly believed to have been invented by Hindu mathematicians in India in about A.D. 500. It was carried to other parts of the world by the Arabs; it took a long time before it became universally accepted.

Today we see our whole-number symbols displayed everywhere, representing the complexities of international finance and trade as well as facilitating computer programming and international internet communications. For example, the Dow Jones Industrial Average is a weighted average of stock prices for a number of prominent stocks. In spring 1999, the Dow Jones Industrial Average reached a record high of 10,000. The number 10,000, because of the large amount it represents, was significant to traders in the stock market. In computer programming, technologists are eager to have a large amount of computer memory. They often use expressions such as 500 megabytes or 10 gigabytes to describe this computer space. Without a numeration system to represent whole numbers, information could not be conveyed in this manner.

For More Information Bunt, L. N. H., Jones, P. S., & Bedient, J. D. (1988). *The Historical Roots of Elementary Mathematics*. New York: Dover Publications.

3.2.1 Number Representations

The word **number** refers to an amount or quantity. A number can be represented in several ways. One could place objects on a table to provide a **concrete representation of an amount.** For example, if you want to show the amount three, you could place three coins on a table, letting the coins physically represent the number. Traditionally, we use number **names,** such as the words *one, two, three*, and so on to describe an amount. We also use **numerals,** or symbols, such as the digits 0, 1, 2, 3, and so on as symbolic representations of number amounts.

When we refer to the number of objects in a set or a number that says "how many," this number is called a **cardinal number.** A cardinal number is usually represented by the numeral that corresponds to the amount; that is, if there are two objects, the numeral "2" can represent that amount. Also, the answer to the question, "How many objects are there?" can be answered by the cardinal number 2. On the other hand, a number that designates a comparative position or an order of some set of objects is called an **ordinal number.** For example, if I am "fourth" in line, the ordinal number "fourth" describes my position in line compared to others' positions in line. Figure 3.1 provides a summary of examples of number representations.

Figure 3.1

Number Representations

	Concrete Representation	Name	Numeral	Cardinal Number	Ordinal Number
Summary	Physical objects used to show an amount	The words we use to show an amount	The symbols we use	The amount itself	The comparative position of an object
Example 1	five blocks	Five	5 (the symbol)	5 (the amount)	the fifth block
Example 2	3 centimeters	Three	3 (the symbol)	3 (the amount)	the third centimeter

3.2.2 One-to-One Correspondence

Counting objects involves an understanding of **one-to-one correspondence.** One-to-one correspondence is the notion that you can match each item from one set with exactly one item from another set, and vice versa. For example, if six children come to a birthday party and you want to make sure each child goes home with exactly one of the six balloons you have, there are many ways you can distribute the balloons, one being according to the children's color preferences. However, in the long run, no matter how you match children with balloons, each child leaves with exactly one balloon. See Figure 3.2.

In Figure 3.2, each child has selected a favorite balloon color. However, a different pairing of child with balloon (say Ted wanted red and Abby wanted yellow and all the others stayed the same) would still result in the fact that there is exactly one balloon for each child. In that sense, we could say that there is a one-to-one correspondence between the set of children and the set of balloons.

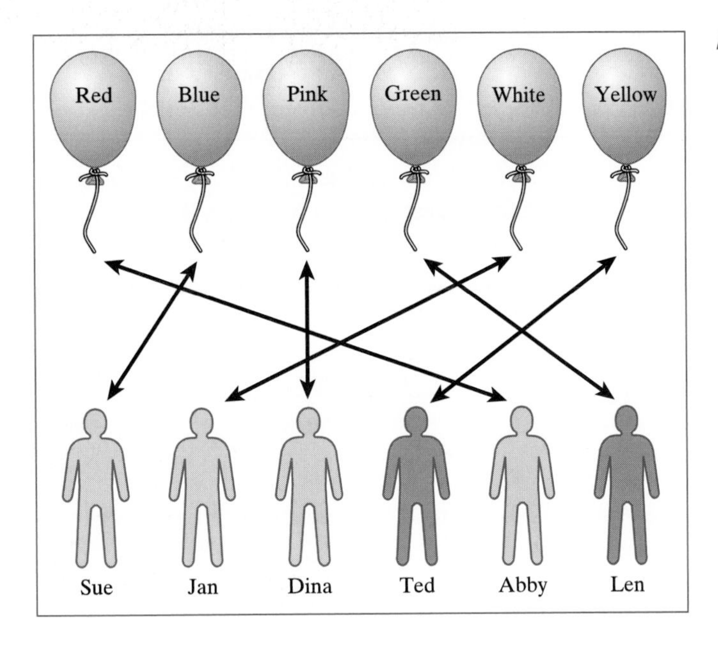

Figure 3.2

3.2.3 Whole Numbers and the Set of Whole Numbers

Whole numbers are cardinal numbers that describe the size of finite sets; that is, given a set, {a, b, c, d}, we use the whole number 4 to describe the number of elements in the set. The **set of whole numbers** is traditionally expressed as W = {0, 1, 2, 3, ... }, which indicates that the whole numbers start with zero and continue on toward infinity. Whole numbers differ from the natural numbers N = {1, 2, 3, 4, ... } in that they include the number zero, which allows for representing the cardinal number of an "empty set."

3.2.4 Whole Numbers and Counting

When we count a set of objects, we ultimately determine a whole number that states how many objects there are. Steps in counting are as follows: (a) Point to each object to be counted exactly once; (b) recite the whole numbers *in order*, beginning with 1, associating exactly one number to correspond with each object being counted; and (c) conclude the total number of objects counted by identifying the last whole number stated when pointing to the last object.

For example, if there is a group of coins on the table, we can determine how many there are by pointing to each coin exactly once and assigning consecutive numbers to each object, starting with the number one (see Figure 3.3). Each coin, then, corresponds to exactly one number. The last number stated, in this case 8, reveals the total number of coins on the table.

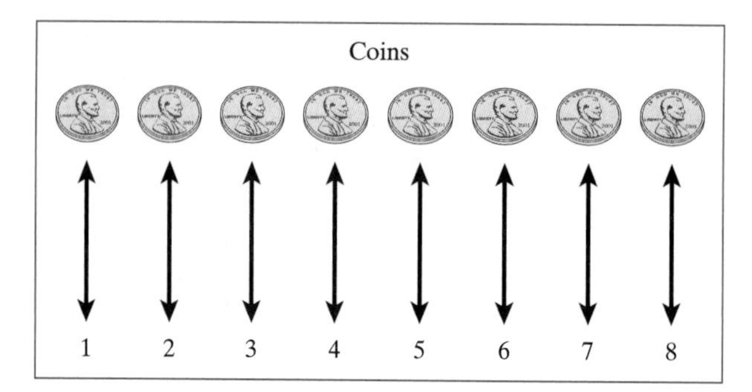

Figure 3.3

Related Topics

Properties of Whole-Number Addition
Properties of Whole-Number Multiplication
Base-Ten Introduction
Place Value and Zero
Decimal System and More on Base Ten
Integers

TOPIC 3.3 Addition with Whole Numbers

cornerstone

Whole Numbers
Place Value and Zero
Base-Ten Introduction
Decimal System and More on Base-Ten

Introduction

The roots of whole number addition stem from combining sets together to determine how many are in the new, combined set. Further, the notation for the addition process, as well as the need for an addition algorithm when adding two numbers where carrying is required, springs from the simple notion of *joining* together. In this topic, we describe the meaning of addition and provide examples of mathematical problems that call for addition.

Addition has been referred to by many names throughout history. Famous mathematician Fibonacci used the terms *composition* and *collection* as well as *addition* to refer to this operation in the 13th century. At that same time, the French used the terms *add* and *assemble* to describe the process of addition. Two centuries later, the words *join* and *summation* were frequently used. Ultimately, in the 16th century, the word *addend,* coming from the Latin phrase *numeri addendi* meaning "numbers to be added," became a popularly accepted term for the elements in an addition problem.

Some of the earliest written addition problems were written on clay tablets found in Mesopotamia. These tablets describe arithmetic calculations that depict business transactions. Addition computations were also written in sand on a sand abacus usually for the purpose of census taking or business transactions. The operation of addition itself has not changed much since Hindu-Arabic numerals began to be used.

For More Information Smith, D. E. (1958). *History of Mathematics, Volume II.* New York: Dover Publications.

3.3.1 Definition of Addition: Joining Sets

Whole-number addition is defined as the joining of two disjoint sets. **Disjoint sets** are sets that have no elements in common (other than the empty set). Suppose, then, that you have two sets, Set A and Set B, such that the sets have no common elements. When these two sets are combined, the total number of objects in the new set, more precisely expressed as $A \cup B$ and read **"A union B,"** is the sum of the number of objects in set A plus the number of objects in set B. We denote the number of elements in set A as n(A) and the number of elements in B as n(B). Therefore, the number of elements in set $A \cup B$ (or $A + B$) is $n(A \cup B) = n(A) + n(B)$. See the example

$$A = \{\text{ball, bat, glove}\} \qquad n(A) = 3$$
$$B = \{\text{doll, brush}\} \qquad n(B) = 2$$
$$A \cup B = \{\text{ball, bat, glove, doll, brush}\} \qquad n(A + B) = 5$$

$$n(A) + n(B) = n(A \cup B); \quad \text{that is, } 3 + 2 = 5$$

3.3.2 Addition Terminology and Pictorial Representation

Although addition of whole numbers is initially defined as the joining of two sets, addition can also be modeled **pictorially.** If we picture a set of objects, not necessarily expressed in set notation but as a collection of distinct, countable objects, we can

Figure 3.4

combine one set of objects with another and determine "how many altogether." For example, if I have 3 apples and pick 2 more apples, I then have 5 apples altogether. See Figure 3.4. This pictorial situation can be represented by the addition statement $3 + 2 = 5$. We refer to the numbers 3 and 2 in the addition statement as the **addends** and to the number 5 in the addition statement as the **sum.**

3.3.3 The Number-Line Model of Addition

Addition can also be represented via a **number line.** For example, on the number line below (see Figure 3.5), we see the addition of $4 + 2$ by moving along the number line, starting with zero, to the first addend, 4. Then, moving 2 more places to the right (we move right to indicate addition), we end up at 6 (the sum) on the number line.

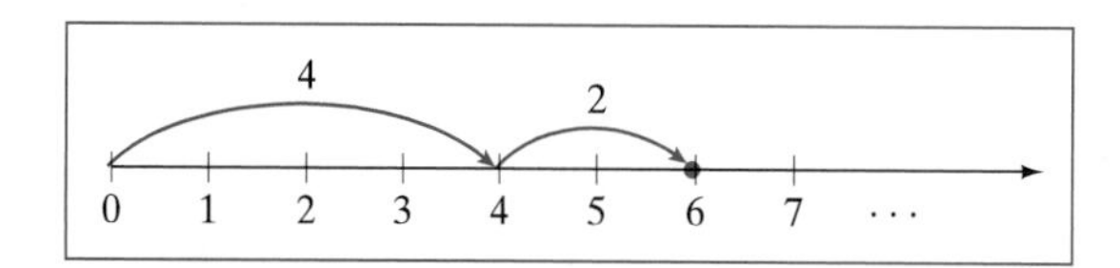

Figure 3.5

3.3.4 Addition and Mental Mathematics

Using the pictorial and number-line models, the definition of *addition* as the act of "joining together" is made clear. Ultimately, however, students need to be able to perform basic addition of whole numbers quickly in their heads, without the benefit of models. This refers to performing addition using **mental computation.** Three techniques that are useful in performing mental computations are counting up, compensation, and double plus 1 or 2.

Counting Up. After working with models, students should not only be able to picture the addition in their heads, but will have developed some mental math techniques to help them perform the operation quickly. For instance, when adding 1, 2, or 3 to any other addend, it is simple merely to say the larger addend in your head and then **count up** by 1, 2, or 3. Take, for example, adding $6 + 2$: Using mental math, the student starts by thinking (or saying aloud) the number 6 and then counts up two more numbers: 7, 8. Thus, the conclusion is that $6 + 2 = 8$.

Compensation. Other mental math techniques include the strategy of **compensation.** This strategy is useful when one of the addends is 8 or 9. Compensation essentially takes advantage of the fact that most students learn early how to add a single-digit number to 10. So, when adding an addend that is close to 10, such as adding 8 or 9, the student can ask herself or himself, "How much of the other addend would I have to add to this number (8 or 9) to get 10?" And then ask, "How much of the other addend would be 'left over' to add onto the amount of 10?"

For example, when adding the whole numbers $9 + 7$, the student may think, "I need to add 1 more to the 9 to get 10, so if I take 1 away from the 7 and add it to the 9, that leaves 6. Then I just have to add 6 to 10 to get a sum of 16." Essentially, when you "borrow" the 1 from the 7 to add to the 9, you have to "compensate" for this action by thinking of the second addend as a 6 rather than a 7. Thus the name, compensation.

Double Plus 1 or 2. Two other mental math techniques that help when adding two addends that are one or two numbers apart, such as when adding $4 + 5$ or $5 + 7$, are strategies referred to as **double plus one** or **double plus two,** respectively. Because most children easily learn the sum of an addend plus itself ($1 + 1 = 2, 2 + 2 = 4, 5 + 5 = 10$, etc.), they can use this knowledge to add other combinations of numbers. For example, when adding $4 + 5$, a student might think, "$4 + 5$ is really 1 more than $4 + 4$, and I know that $4 + 4$ is 8, so $4 + 5$ must be 9." Similarly, when the addends are two apart, as in the case of $5 + 7$, the student may mentally calculate, "Seven is 2 more than 5, and I know that $5 + 5$ is 10, so $5 + 7$ must be 2 more than 10, so it's 12."

Related Topics

Addition Properties and Patterns (Whole Numbers)
Addition Algorithms

TOPIC 3.4 Addition Properties and Patterns (Whole Numbers)

cornerstone

Whole Numbers
Addition with Whole Numbers

Introduction

There are many observable patterns associated with the addition of whole numbers. Many of these patterns can be described through properties associated with whole-number addition and can be observed in an addition table. In this topic, we demonstrate properties of whole-number addition through examples.

Properties of whole-number addition stem from properties associated with combining sets. For example, we say the union of two sets A and B, written A ∪ B, is the set of all elements in set A or in set B or in both. Thus, the union of sets is much like addition. We find when we determine the union of sets A and B that order does not matter; that is, A ∪ B yields the same result as B ∪ A. Thus, whether we start with the objects in set A and then add in the objects from set B, or whether we start with the objects in set B and then add in the objects in set A, we will still end up with the set when the two are combined. This notion is called the *Commutative Property of the Union of Sets* and is similar to the *Commutative Property of the Addition of Whole Numbers* that we will see in this section. The commutative and other properties for sets are the precursors to the properties associated with whole-number addition.

3.4.1 Whole-Number Addition Table

When students are at the point of trying to memorize their **basic addition facts,** that is, all the addition problems that involve two single-digit addends, working with an addition table is quite useful. The addition table also illustrates properties associated with the addition of elements within the set of whole numbers. See the addition table for whole numbers in Table 3.1.

3.4.2 General Patterns in the Addition Table

Students may notice a variety of patterns on the addition table. For example, diagonal rows that run from the bottom left to the top right have the same number throughout the diagonal. On the other hand, diagonal rows that run from the top

Table 3.1

+	0	1	2	3	4	5	6	7	8	9
0	0	1	2	3	4	5	6	7	8	9
1	1	2	3	4	5	6	7	8	9	10
2	2	3	4	5	6	7	8	9	10	11
3	3	4	5	6	7	8	9	10	11	12
4	4	5	6	7	8	9	10	11	12	13
5	5	6	7	8	9	10	11	12	13	14
6	6	7	8	9	10	11	12	13	14	15
7	7	8	9	10	11	12	13	14	15	16
8	8	9	10	11	12	13	14	15	16	17
9	9	10	11	12	13	14	15	16	17	18

left to the bottom right either contain only even numbers that increase as you move from left to right or contain only odd numbers that increase as you move from left to right. Students may also notice that as you move down row by row, a row begins with one larger number than the row before that, and the numbers in each row increase by one as you move across the row. Similarly, each column begins with one larger number than the column to its left, and the numbers in each column increase by one as you move down the column.

3.4.3 Additive Identity Property

When looking for more patterns in the addition table, students may immediately notice that the first row and first column match the addends that are being added to zero. Thus, at a glance, students easily see that having a zero as one of the addends yields a sum equal to the other addend. Therefore, the students have learned one of the four properties associated with whole-number addition: *zero serves as an* ***additive identity*** *for the set of whole numbers* because when zero is added to a number, it does not change the identity (or value) of that number. In brief, for any whole number, a, the additive identity of zero ensures that $a + 0 = a$ (and, similarly, that $0 + a = a$).

3.4.4 Commutative Property

Other properties of whole-number addition are evident from the addition table. For example, students might notice that the sum associated with $4 + 5$ is the same as the sum associated with $5 + 4$. Quickly, students begin to see a second property, called the **Commutative Property of Whole-number Addition** (often referred to as the **Ordering Property** in elementary textbooks). They learn that it does not matter in what order the addends appear; the sum will always be the same. So, for any two whole numbers a and b, $a + b = b + a$. The Commutative Property is very helpful when students are learning to memorize their addition facts because once they know that the sum of $6 + 7$ is 13, they automatically know that the sum of $7 + 6$ is 13. In a way, this reduces the task of memorizing basic addition facts by one-half.

3.4.5 Closure Property

The addition table also illustrates a third property, called the **Closure Property for Whole-number Addition;** that is, for any two whole numbers, a and b, $a + b = c$, where c is also a whole number. Essentially, this property says, whenever you add two whole numbers together, the sum will also be a whole number. Thus, all the sums seen within the addition table are whole numbers.

3.4.6 Associative Property

The fourth and final property of whole number addition that can be verified using the addition table is called the **Associative Property of Whole-number Addition** (often called the **Grouping Property** in elementary textbooks). This property says that when adding three (or more) whole numbers together, it does not matter which pairs of numbers you add together first. Once all three (or more) numbers are added together, the sum will always be the same.

For example, when adding $4 + 3 + 5$, we can group the 4 and 3 together and add them together. Then, take that sum, 7, and add it to 5 to reach a grand total sum of 12. This might be written as $(4 + 3) + 5 = 12$ to indicate that we are adding the 4 and 3 together first. On the other hand, we have the same sum if we choose to add the 3 and 5 together first and then add that sum, 8, to the 4, again finding a total of

12. This would be written as $4 + (3 + 5) = 12$. In general, when adding any three whole numbers a, b, and c together, the Associative Property of whole-number addition guarantees that $(a + b) + c = a + (b + c)$.

Related Topics

Addition Algorithms
Multiplication Properties and Patterns (Whole Numbers)

TOPIC 3.5 Addition Algorithms (Whole Numbers)

cornerstone

Addition with Whole Numbers
Addition Properties and Patterns

Introduction

Learning the traditional addition algorithm occurs in various stages. An **algorithm** is a set of steps or procedures to follow. The term *algorithm* grew from the word *algorism* (from the name of Arabic mathematician al-Khwārizmī), which refers to arithmetic done with Hindu-Arabic numerals that made use of a zero. An addition algorithm is a set of steps that allows us to add any two whole numbers together. Historically, addition algorithms developed as the need to add large amounts became necessary. As we show in the following sections, we begin to understand algorithms by learning to add two single-digit whole numbers and work toward being able to add two multidigited whole numbers together, performing "carrying" (or more properly "regrouping" or "renaming") when necessary.

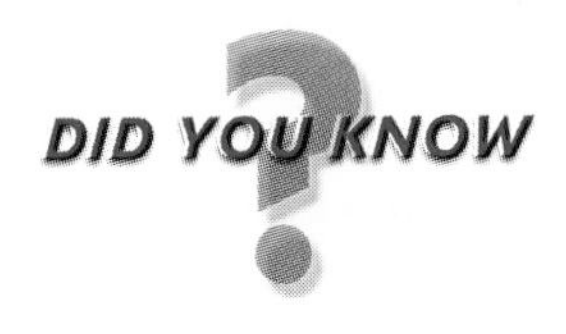

DID YOU KNOW

Hindu and Arab mathematicians performed addition problems in a variety of ways over the centuries. Their methods ultimately evolved to the methods we use today where we line addends up, one on top of the other, with like places aligned and with the sum written below a line separating the addends from the sum. However, the early work of Hindu mathematician Bhāskara, generally known as Bhāskara the Learned (c. 1114–c. 1185), shows the addition of 2, 5, 32, 193, 18, 10, and 100 as shown in Figure 3.6a in a different manner. The Arabs, led by Maximus Planudes (c. 1260–c. 1310), wrote the sum at the top of an addition problem as seen in Figure 3.6b.

Sum of the units,	2, 5, 2, 3, 8, 0, 0	20
Sum of the tens,	3, 9, 1, 1, 0	14
Sum of the hundreds,	1, 0, 0, 1	2
Sum of sums		360

Figure 3.6a

Note that the right-hand column in Figure 3.6b records the arithmetic checking method of "casting out nines" where next to each addend, after adding the digits in the addend, you record the remainder after you divide by 9. For example, with the first addend, 5687, we find $5 + 6 + 8 + 7 = 26$. Dividing 26 by 9, we find a quotient of 2 with a remainder of 8. The digit 8 is recorded to the right of 5687. Similarly, we find that 3 is the remainder when we add the digits in the addend 2343 and divide the sum by 9. The process of casting out nines is performed on the sum, 8030, as well. We check the arithmetic of adding $5687 + 2343$ by verifying that the sum of the "cast-out" amounts in the addends $(8 + 3 = 11)$, when divided by 9, has a remainder of 2, which matches the amount cast out in the sum 8030.

In whole-number addition, we often hear the expression *carrying* to indicate an exchange of ten 1s for one 10, or ten 10s for one 100, and so on. The expression *to carry* is an old one. It dates from the time when a counter was actually carried on a line abacus to the space or line above, but it was not common in English works until the 17th century. An abacus is a mechanical means of recording and combining numbers. Early versions were drawn in the sand. More-recent ver-

8030	2
5687	8
2343	3

Figure 3.6b

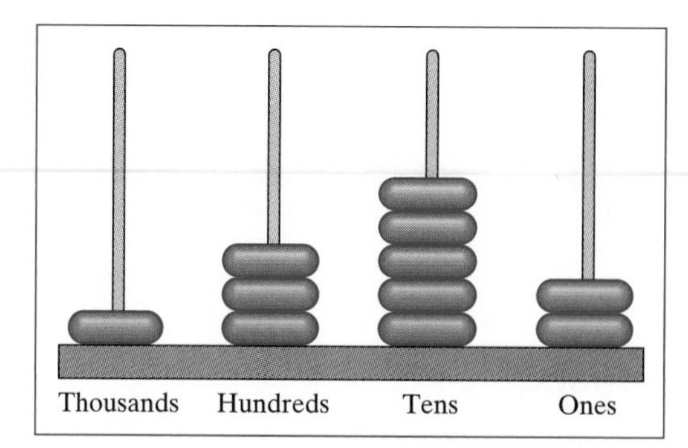

Figure 3.7

sions of the abacus keep track of place value with disks in columns that represent the different places. Figure 3.7 shows a sample place-value abacus, representing 1,352.

For More Information Smith, D. E. (1958). *History of Mathematics, Volume II.* New York: Dover Publications.

3.5.1 Addition Algorithms

Once students have mastered the basic whole-number addition facts (that is addition problems that involve adding two single-digit addends), they are ready to move forward and learn addition algorithms. **Addition algorithms** include procedures for adding a single-digit whole number to a two-digit whole number, for adding two two-digit whole numbers together, for adding a two-digit whole number to a three-digit whole number, and so on. When first introduced to such algorithms, it is useful for students to have a concrete model that helps illustrate the addition as well as reinforce the concept of place value.

3.5.2 Base-Ten Block Model

Base-ten blocks are useful manipulatives when illustrating whole-number addition. The blocks are particularly helpful because they reinforce place value and the need to add like-valued places to like-valued places, such as adding tens to tens or hundreds to hundreds.

Consider the simple addition of a two-digit addend to another two-digit addend in the problem 23 + 41. Figure 3.8 illustrates the addition problem with base-ten blocks.

Using base-ten blocks, students can model each of the addends 23 and 41 and then show, when combined, how much they have altogether. In Figure 3.8, students can see that they have 6 longs (or 6 tens) and 4 units (or 4 ones), or 64 altogether.

Once students are comfortable with the idea of modeling addition with base-ten blocks, they can then move forward to situations where some "exchanging" or "carrying" is necessary. For example, when adding 23 and 48, students soon see that they have a total of 11 units and 6 longs with their base-ten blocks (see Figure 3.9).

With this type of problem, students learn the step of trading in 10 of the units for 1 of the longs (that is trading 10 ones for 1 ten) and adding that new long to the tens column. Then, students can conclude that when they add 23 and 48, they will finish with 7 longs and 1 unit, or a sum of 71.

3.5.3 The Traditional Addition Algorithm

Once students have had many experiences with the concrete materials, they are ready to translate this knowledge to the traditional addition algorithm. For example, students should learn that when finding a sum, such as the sum of 35 and 49, they must begin by adding the numbers in the ones place together. If they find, as in this

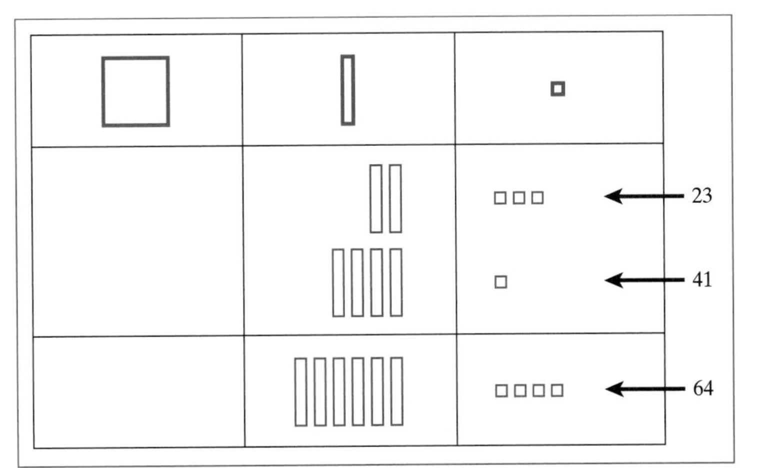

Figure 3.8

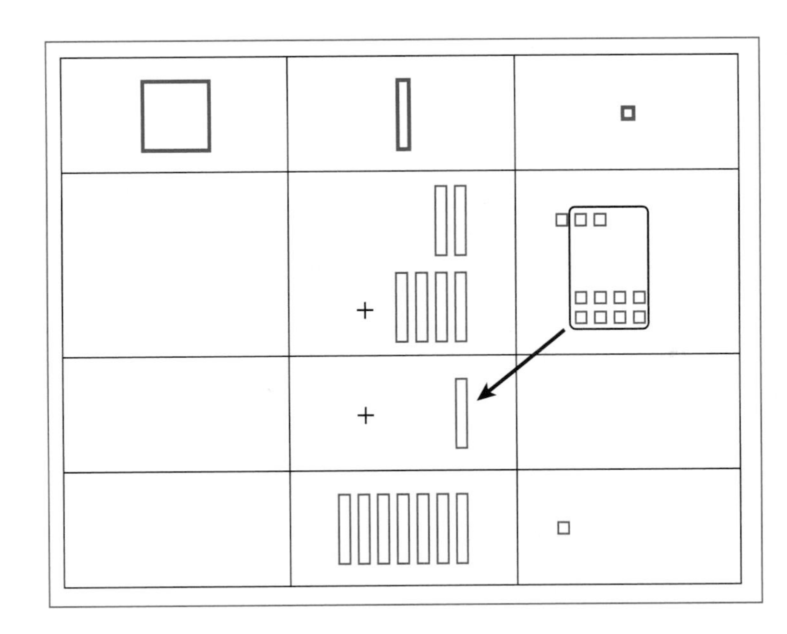

Figure 3.9

case, that they have more than 10 ones, they should "carry" a set of ten over to the tens column and determine how many ones they will have left. This number of ones that remains is written in the ones column of the sum, and then the total number of tens is determined and recorded in the tens column of the sum.

$$\begin{array}{r} {}^{1} \\ 35 \\ +49 \\ \hline 84 \end{array}$$

3.5.4 Partial Sums Algorithm

Other transitional algorithms that students may find useful when working toward learning how to add any whole number to any other whole number with multiple carrying steps is the **partial sums algorithm.** For example, when adding the whole numbers 348 and 595, students might find it useful to find the total number of units in the ones place and record that sum, find the total number of tens in the tens place and record that sum, and find the total number of hundreds in the hundreds place and record that sum. Then, once these partial sums are determined, the student can find the total sum by adding the partial sums together. This is illustrated below.

$$\begin{array}{r} 348 \\ +595 \\ \hline 13 \\ 130 \\ 800 \\ \hline 943 \end{array}$$

The partial sums algorithm is a very useful **transitional algorithm.** It reinforces the idea that in our base-ten system, we add hundreds to hundreds, tens to tens, and ones to ones. Further, students can see that when they add the ones places together and find 13, as in the example above, that this really means that they should view that 13 as being 1 ten in the tens column and 3 ones in the ones column. It also parallels nicely with a transitional multiplication algorithm we will explore in Multiplication Algorithms called partial products.

3.5.5 Scratch Algorithm

The **scratch algorithm** is a useful way to keep track of how many sets of ten you need to carry over to the next column when adding a column of addends. Adding down a column, beginning with the ones place, the student should add until a sum of 10 or greater is reached. Once the partial sum surpasses 10, the students should scratch through the last digit added in, indicating that a set of ten will have to be carried over to the next column. The student then writes, next to the number that was scratched out, how much the sum went higher than 10. The student then begins with this leftover amount and begins to add digits, continuing down the same column. Once the bottom of a column has been reached, the last amount added that was not more than 10 is written as the remaining amount that goes in the ones column. The student then counts up the number of scratches made in the ones column, carries that number to the tens column, and begins to add in that column, using the same procedure.

For example, illustrating the scratch algorithm in the problem below with the problem 548 + 287 + 344 + 539 + 488, we find that when adding down the column of numbers, the first scratch is made through the 7 in the ones column because 8 + 7 = 15. The scratch indicates that a set of ten has been reached, and the small 5 written next to the scratched-out 7 indicates that a remainder of 5 is left when you take away that set of ten.

$$\begin{array}{r} {}^{2\;3\;3} \\ 5\;4\;8 \\ \not{2}_0\;\not{8}_5\;\not{7}_5 \\ 3\;4\;4 \\ 5\;\not{3}_2\;\not{9}_8 \\ +\;\not{4}\;8\;\not{8} \\ \hline 2\;2\;0\;6 \end{array}$$

Continuing, the 5 is added to the 4 for a sum of 9. This sum is added to the 9 for a sum of 18. At this point, a second scratch must be made and a small 8 written next to the scratch to indicate that a set of ten has been reached, with a remainder of 8. This 8 is added to the final 8 in that column for a total of 16. The 6 is recorded in the ones place because we have reached the end of the column, and a scratch is made through the 8. Then the number of scratches is found to be 3. Then a 3 is carried to the next column, and the process is continued.

3.5.6 The Scratch Algorithm in Other Bases

The scratch algorithm can be useful when adding numbers in other bases. For example, when adding $1023_{four} + 2332_{four} + 2121_{four} + 1233_{four}$, the scratches in each column represent sets of four and the small number written next to each scratch represents the amount that the partial sum has gone higher than 4.

$$\begin{array}{r} {\scriptstyle 2\;2\;3\;\;2} \\ 1\;0\;\not{2}_0\;3 \\ \not{2}_1\;\not{3}_2\;3\;\not{2}_1 \\ 2\;1\;\not{2}_1\;1 \\ +\;\not{1}\;\not{2}\;\not{3}\;\not{3}_{four} \\ \hline 2\;0\;1\;0\;1_{four} \end{array}$$

Related Topics

Subtraction Algorithms
Multiplication Algorithms
Division Algorithms

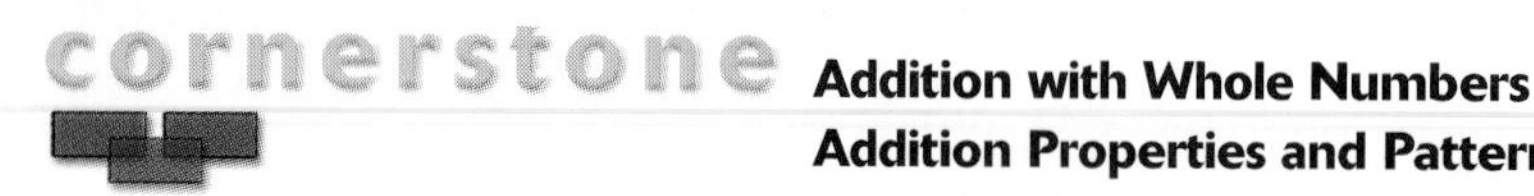

TOPIC 3.6 Subtraction with Whole Numbers

cornerstone **Addition with Whole Numbers**
Addition Properties and Patterns (Whole Numbers)

Introduction

Subtraction is considered the inverse operation to addition. It is often associated with the notion of having a set of objects, taking something away from the set, and determining how much of the set remains. We will find, however, that there are other representations of the concept of subtraction. In the following sections, we examine three interpretations of subtraction through sample story problems.

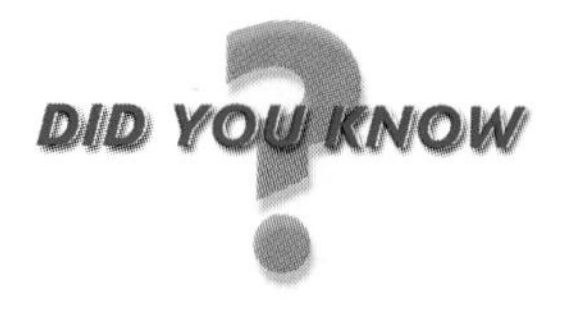

DID YOU KNOW

The names and process of the numbers used in subtraction have varied greatly over the centuries. Even today, the formal terms of subtraction are seldom used outside classroom instruction. For example, we might say *deduct what I owe and pay me the rest,* using the word *deduct* for *subtract* and the *rest* for the *difference.* Mathematician Fibonacci, in the 13th century, used the expression *extract* instead of *subtract,* and spoke of *extraction* instead of *subtraction.* In the 16th century, the words *detract* and *subduct* were used for *subtract.*

Today's terms *minuend* and *subtrahend* are actually abbreviations of the Latin expressions *numerus minuendus* ("number to be diminished") and *numerus subtrahendus* ("number to be subtracted"). The words *difference* and *remainder* have never been popular in spite of the fact that they are commonly found in textbooks today.

For More Information Smith, D. E. (1958). *History of Mathematics, Volume II.* New York: Dover Publications.

3.6.1 Definition of Terms

Whole-number subtraction is the inverse operation of whole-number addition. In any subtraction statement, such as $a - b = c$, we say that a is the **minuend,** b is the **subtrahend,** and c is the **difference.** Unlike addition, which is viewed only as the act of joining, subtraction has three interpretations. One can conceive of subtraction as the act of *taking away*, the act of *finding a missing addend*, or the act of *comparing two sets*.

3.6.2 Take-Away Subtraction

In a **take-away** view of subtraction, subtraction is called for when there is a starting amount, something is removed from that amount, and you want to determine how much remains. For example, suppose Ben has 5 letters to mail and drops 1 on the way to the mailbox. To determine how many letters he has left, the scenario suggests that we start with 5 letters, remove 1 of those 5 from the set, and find that there are 4 letters remaining. Consequently, the answer can be found by performing the subtraction $5 - 1 = 4$.

3.6.3 Missing-Addend Subtraction

Missing-addend subtraction is the appropriate interpretation when we know what amount we start with, know what we want to end up with, but do not know the difference between the starting and the ending amount. For example, if Kate has 9 dollars but needs 12 dollars to buy the new doll she wants, how many more dollars does she need? We can think of this problem as an addition situation, saying, "I have to add 3 dollars to the 9 dollars I already have to then have a total of 12 dollars." Thus, there is the sense of finding the missing addend, or missing amount, $[9 + \_ = 12]$ that has to be added to reach a certain total. The required operation is subtraction because to find the missing amount, we subtract $12 - 9 = 3$.

3.6.4 Comparison Subtraction

Finally, there is the **comparison subtraction** interpretation. In this situation, two amounts are compared, and we ask how much larger one amount is than the other. For example, suppose Chris has 8 concert tickets and Allison has 3 concert tickets. How many more tickets does Chris have than Allison? In this case, we are determining *how much more* 8 is than 3, and so we subtract $8 - 3 = 5$.

3.6.5 A Pictorial Model

Like addition, subtraction can be modeled pictorially. We can illustrate the subtraction problem $7 - 4 = 3$ with the pictorial model Figure 3.10 below. The picture indicates that we started with 7 letters and removed 4 from the set. This leaves 3 letters as the difference in this take-away, pictorial model.

Figure 3.10

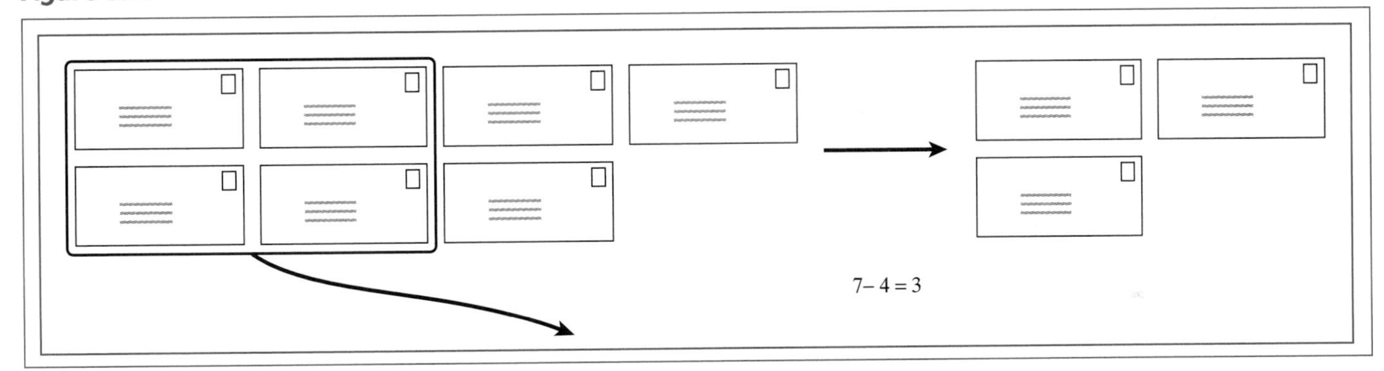

3.6.6 Number-Line Model

Subtraction can also be modeled with a number line. In Figure 3.11, we show the problem $8 - 2$, beginning at 8 on the number line, moving two notches to the left (left to indicate the action of taking away), and ending at the point 6 on the number line.

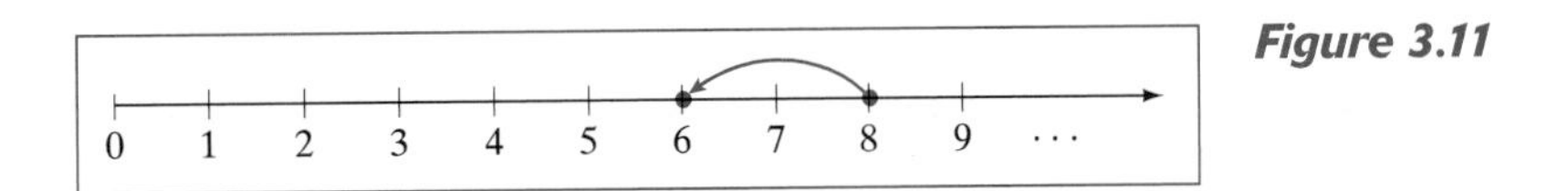

Figure 3.11

3.6.7 Subtraction and Mental Mathematics

Parallel to the learning of whole-number addition facts, students use mental-computation techniques to be able to perform subtraction facts quickly in their heads. Two useful techniques used when performing mental subtraction are *counting back* and *compensation*.

Counting Back. A mental math strategy similar to counting up is a strategy called **counting back.** This strategy is useful in subtraction situations where the subtrahend is 1, 2, or 3. For example, to subtract $6 - 2$ mentally, it is appropriate to think of the minuend, 6, and then count backward two numbers (because the subtrahend is 2)—5, 4—to determine that $6 - 2$ is 4.

Compensation. Another subtraction strategy is **compensation,** which was discussed in the addition of whole numbers. When the subtrahend is a number close to 10, we can subtract 10, but we must make sure to compensate for the change. For example, in the problem 15 − 8, we know that 15 − 10 = 5, but because we subtracted more than necessary by 2, we must add 2 to 5 to determine the difference of 7.

Related Topics

Subtraction Algorithms (Whole Numbers)

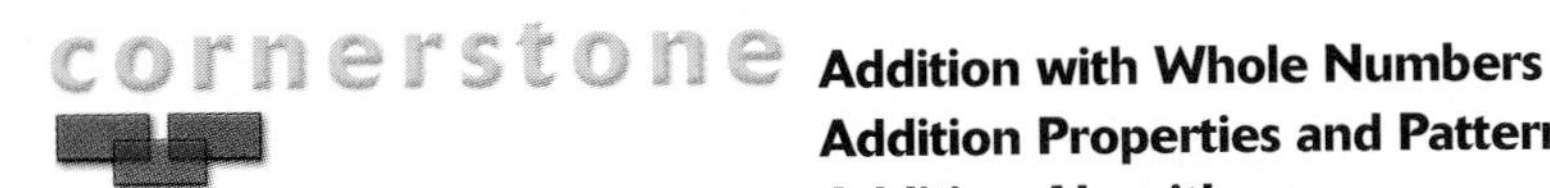

TOPIC 3.7 Subtraction Algorithms (Whole Numbers)

cornerstone

Addition with Whole Numbers
Addition Properties and Patterns (Whole Numbers)
Addition Algorithms
Subtraction with Whole Numbers

Introduction

Subtraction algorithms are related to addition algorithms in that both can involve exchanging. In addition, we traditionally exchange 10 of one place (such as 10 ones) for 1 of a larger place (such as one set of ten). This process is usually referred to as *carrying*. In subtraction, there is usually the need to exchange one item in a larger-valued place (such as 1 one-hundred) for 10 of a smaller-valued place (such as 10 sets of ten) to be able to subtract. This process is often referred to as "borrowing" (or more properly "renaming" or "regrouping").

Calculations involving subtraction were first used in business transactions. There are four or five methods of subtraction in common use today. The borrowing method traditionally used in our schools is only one popular method. Another method used is the **equal additions method.** Essentially, in this method, you add equal amounts to the minuend and the subtrahend of the subtraction problem in a manner that makes the subtraction easier by not requiring regrouping. Instead, the equal additions method draws heavily on understanding place-value relationships between adjacent columns. An example is:

$$\begin{array}{r} 842 + 10 \text{ ones} \\ -\underline{356} + 1 \text{ ten} \end{array} \rightarrow \begin{array}{r} 84_{12} + 10 \text{ tens} \\ -\underline{366} + 1 \text{ hundred} \end{array} \rightarrow \begin{array}{r} 8_{14}\,{}_{12} \\ -\underline{466} \\ 486 \end{array}$$

In this problem, we see that the same amount was added to each number at each stage. In the first stage, 10 was added to each. In the top number, the 10 was added to the ones place so that there is 12 in the ones place. In the bottom number, the amount 10 was added to the tens place making the tens place 6 instead of 5. In the second adding stage, 100 was added to both numbers. The 100 was recorded as 10 tens and added to the tens column making it 14 sets of ten. The amount 100 was added to the bottom number in the hundreds place making it 400. Then, the subtraction was performed, place by place, without any regrouping required.

3.7.1 Connections to Addition

Subtraction algorithms for whole numbers expand ideas developed in basic subtraction facts. The development of subtraction algorithms generally progresses from subtracting a one-digit whole number from a two-digit whole number, requiring neither borrowing (or exchanging) to subtract any whole number from any other larger whole number nor multiple exchanges.

3.7.2 Subtraction Using Base-Ten Blocks

Like addition algorithms, subtraction algorithms are best introduced using concrete models such as the base-ten blocks. Illustrated in Figure 3.12 is the subtraction 68 − 54. Borrowing is not necessary.

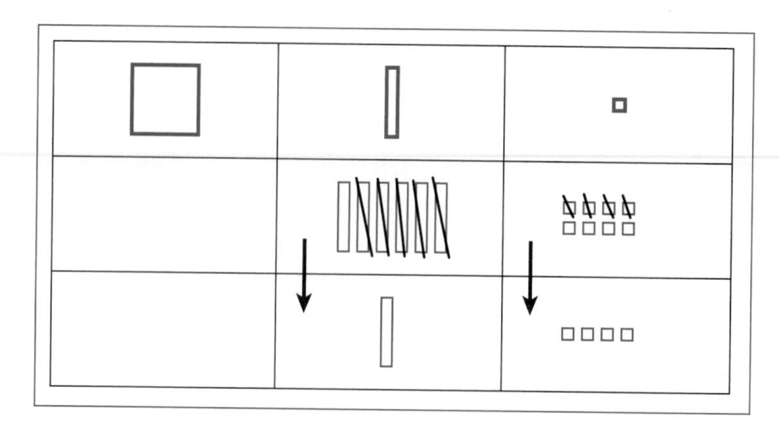

Figure 3.12

Once we observe that subtraction involves subtracting tens from tens, ones from ones, and so on, we then *exchange* one block for 10 of the next smaller block to perform the necessary subtraction. For example, in the problem 81 − 54, there are not enough units in the ones column to take away 4 units. See Figure 3.13.

In Figure 3.13, we see the step of *exchanging* 1 of the longs in the tens column for 10 of the units so that 7 longs are in the longs column and 11 units are in the ones column. Then the subtraction of 54, that is, 5 longs and 4 units, is illustrated, leaving a difference of 2 longs and 7 units, or 27.

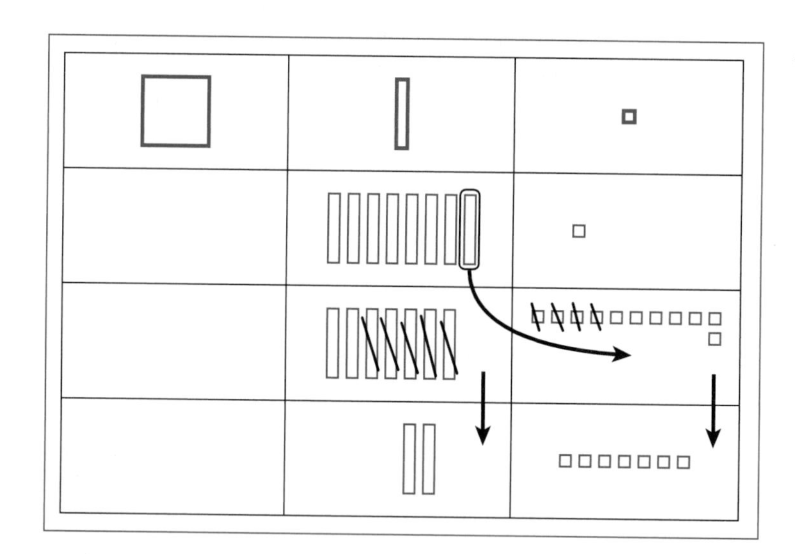

Figure 3.13

3.7.3 Notation for the Subtraction Algorithm

The subtraction algorithm can be performed without using concrete models. However, we need to be aware of the exchanges of larger blocks for the next smaller blocks that must take place to complete the subtraction. Notation associated with the formal subtraction algorithm provides a means of recording when such an exchange has taken place. For example, the exchange required in the subtraction problem 81 − 54 can be illustrated using a scratch mark to note that 1 of the tens in 81 is being exchanged for 10 ones, leaving 7 tens and making a total of 11 units in the ones column.

$$\begin{array}{r} {}^{7}\,{}^{11} \\ 8\not{1} \\ -54 \\ \hline 27 \end{array}$$

The algorithm used to subtract a two-digit whole number from a two-digit whole number involving one act of borrowing also works with more-complicated problems involving multiple trades. For example, to subtract 3,456 from 5,289, we follow the traditional algorithm, recording each exchange.

$$\begin{array}{r} ^{1\,18} \\ 5\not{2}\not{8}9 \\ -3496 \\ \hline 1793 \end{array} \qquad \begin{array}{r} ^{11} \\ ^{4\;\not{1}\;18} \\ \not{5}\not{2}\not{8}9 \\ -3496 \\ \hline 1793 \end{array}$$

Exchange 1 ⟶ Exchange 2

In this problem, as in all whole-number subtraction problems, we begin with the ones place. Because 6 can easily be subtracted from 9, the subtraction is done, and we move into the tens place. In the tens place, we need to subtract 9 from 8. Because this poses a problem, we exchange 1 set of one hundred for 10 sets of ten. We record this exchange by scratching through the 2 in the hundreds place and writing a 1 above the 2, and then by scratching through the 8 in the tens place and writing 18 above the 8. Thus, after Exchange 1, there is 1 set of one hundred and 18 sets of 10. Now the subtraction of 9 tens from 18 tens can be done and the difference recorded. Next, when faced with subtracting 4 one hundreds from 1 one hundred, we must borrow again. The picture illustrates that 1 set of one thousand is exchanged for 10 sets of one hundred, leaving 4 one thousands and 11 one hundreds. The subtractions in the hundreds and thousands columns are completed to then yield an overall difference of 1,793.

3.7.4 Subtraction in Base Six

The subtraction algorithm can be performed using numbers in other bases, for example, in base six. Note that in base six, each column represents a power of six. Therefore, an exchange of a block from one column (or place) is equal to 6 blocks in the next smaller column (or place).

For example, consider the steps in the subtraction of $421_{six} - 145_{six}$:

$$\begin{array}{r} ^{1\;\;11_{six}} \\ 4\,\not{2}\,\not{1}_{\,six} \\ -\,1\,4\,5_{\,six} \\ \hline 2\,3\,2_{\,six} \end{array} \qquad \begin{array}{r} ^{11_{six}} \\ ^{3\;\;\not{1}\;\;11_{six}} \\ \not{4}\,\not{2}\,\not{1}_{\,six} \\ -\,1\,4\,5_{\,six} \\ \hline 2\,3\,2_{\,six} \end{array}$$

Exchange 1 ⟶ Exchange 2

Starting with the ones place, we need to subtract 5 from 1. Because this is not possible in whole-number subtraction, we borrow one set of 6 from the next higher place (the sixes place). Thus, in the first exchange, we see that one set of 6 is borrowed, leaving 1 in the sixes place. That set of six is brought to the ones place, added to the 1 that is already there, and is expressed 11_{six} (note that 11_{six} is worth 7). Then the subtraction of $11_{six} - 5_{six}$ yields 2_{six} in the ones column.

Then when we move to the sixes place to subtract 4 from 1, we find that we must perform another exchange. In the third column, 1 set of 36 (or 6 squared) is borrowed from the 4 that are there, leaving 3 sets. One of these is the exchanged for 6 sets of six. Adding these 6 sets of six to the 1 that is already there yields 7 or 11_{six} in the sixes place. Subtracting, we end up with a difference of 3 in the middle column. Finally, the 1 set of 36 is subtracted from the 3 sets of 36 in the third column, leaving 2 as the difference in the third column.

Related Topics

Multiplication Algorithms
Division Algorithms

TOPIC 3.8 Multiplication with Whole Numbers

cornerstone **Addition with Whole Numbers**

Introduction

Multiplication of whole numbers is initially seen as a *shortcut* to adding multiple amounts of the same number together. This is most often introduced as repeated addition. However, multiplication can be seen as having more than this one interpretation. In this topic, we describe three interpretations of multiplication and illustrate multiplication through various models.

The term *multiplication* comes from the Latin phrase *numerus multiplicandus* meaning "the number to be multiplied." A few Latin writers suggested the term *multiplicate* as the process of multiplication. In the 14th century, the word *multiplier* was used to refer to the terms being multiplied together, as were *multiplicans* and *mutlipliant*.

The word *product* stems from the Latin word *producere* meaning "to lead forth" or "result." In the 15th century, the word *factor*, from the Latin root *factus*, became the popular term, as did the word *product*.

The process of multiplication was referred to in Egyptian times as *duplation*, which was the repetition of a same number over and over a given number of times. It is speculated that the Greeks multiplied on wax tables, but most evidence indicates that multiplication was done most extensively throughout history on some sort of abacus.

For More Information Smith, D. E. (1958). *History of Mathematics, Volume II*. New York: Dover Publications.

3.8.1 Definition and Terminology

When two whole numbers a and b are multiplied together and the result is a whole number, c, this is written $a \times b = c$. The whole numbers a and b are called **factors,** and c is called the **product.** Multiplication of whole numbers builds, somewhat, on the addition of whole numbers in that the one key way to view multiplication is as **repeated addition.** Multiplication, though, can also be viewed as an **array** and as a **Cartesian product.**

3.8.2 Repeated-Addition Multiplication

Multiplication is often viewed as repeated addition. We can use repeated addition to solve such problems as the following: "If Frank wants to give three pieces of candy to each of his five party guests, how many pieces of candy will he need?" We can answer this question by adding 3 together five times: $3 + 3 + 3 + 3 + 3 = 15$. See Figure 3.14. We have repeatedly added 3. Ultimately, we find that adding 3 five times can be translated to 5 times 3, or $5 \times 3 = 15$.

Figure 3.14

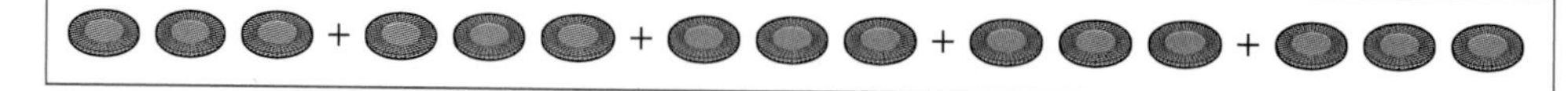

3.8.3 Array Multiplication

Multiplication can often be viewed as determining the number of elements in an array. An **array** is a rectangular collection of objects, placed in equal-size rows and columns. For example, if the classroom has 6 rows of desks with 4 desks in each row, the total

Figure 3.15

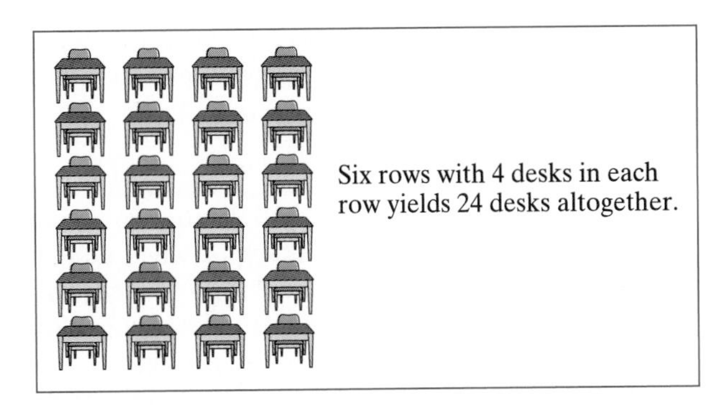

number of desks can be found by multiplying $4 \times 6 = 24$. Note that the picture associated with this problem is a rectangular collection of desks comprised of rows and columns. See Figure 3.15. It is valuable to be able to see multiplication in an array situation because array multiplication, defined as the number of rows times number of columns, is connected to finding the area of a rectangle by multiplying length $\times$ width.

3.8.4 Cartesian Product

Finally, multiplication can be seen as a Cartesian-product situation. A **Cartesian product** (or "cross-product" as it is sometimes called) is the set of different combinations that can be created when pairing each item from one set with each item from another set; that is, if set $A = \{a, b, c\}$ and set $B = \{x, y\}$, then the Cartesian product of sets A and B, written $A \times B = \{(a, x), (a, y), (b, x), (b, y), (c, x), (c, y)\}$. Notice that the size of set A, written $n(A)$, times the size of set B, written $n(B)$, is equal to the size of $A \times B$, written $n(A \times B)$.

Consider this more concrete situation. Suppose the local diner offers a soup-and-sandwich lunch special with your choice of one of two soups, {onion and tomato}, combined with your choice of one of three sandwiches, {cheese, turkey, and veggie}, how many different lunch specials can be ordered? To solve this problem, we can pair each of the soup choices with each of the sandwich choices as follows: {(onion soup, cheese sandwich), (onion soup, turkey sandwich), (onion soup, veggie sandwich), (tomato soup, cheese sandwich), (tomato soup, turkey sandwich), (tomato soup, veggie sandwich)}. Ultimately, we find that there are six possible lunch combinations, which can be seen as the product of the number of soup choices times the number of sandwich choices, or $2 \times 3 = 6$.

3.8.5 Tree Diagram

There are a number of ways to model multiplication. First, when illustrating the Cartesian-product situation described above, a **tree-diagram model** is quite useful. In Figure 3.16, we see the modeling of the combinations of soup-and-sandwich

Figure 3.16

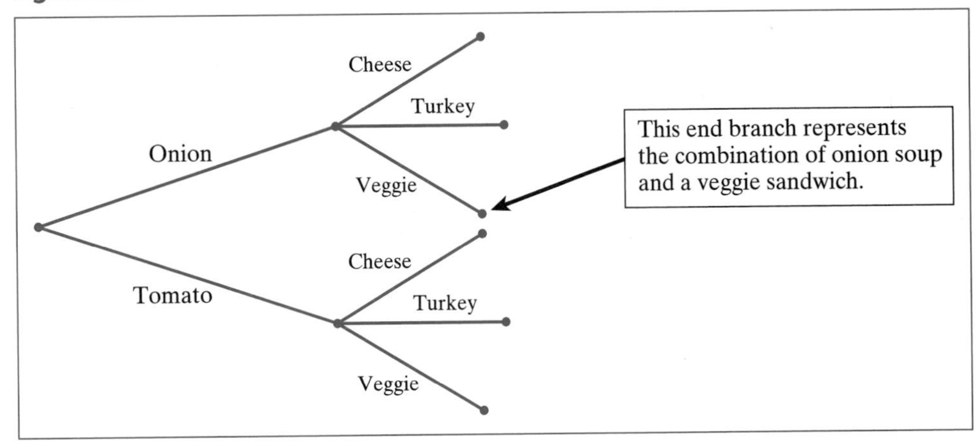

specials, using a tree diagram. Note that the number of ending branches of the tree indicate the total number of possible combinations.

3.8.6 Array Model and Area Model

As previously alluded to, multiplication is often modeled through an **array** drawing of rows and columns. For example, when determining how many flower seeds you were able to plant if you had 9 rows of seeds with 8 seeds in each row, the model in Figure 3.17a clearly illustrates this situation. However, if the rectangular region in a multiplication situation is not an arrangement of discrete objects and rather is described by some linear dimensions of measure, then a better model to represent the situation would be a rectangular **area model.** For example, if a floor tile measures 8 cm by 9 cm, the total square footage of the floor tile $8 \times 9 = 72$ sq cm. See Figure 3.17b.

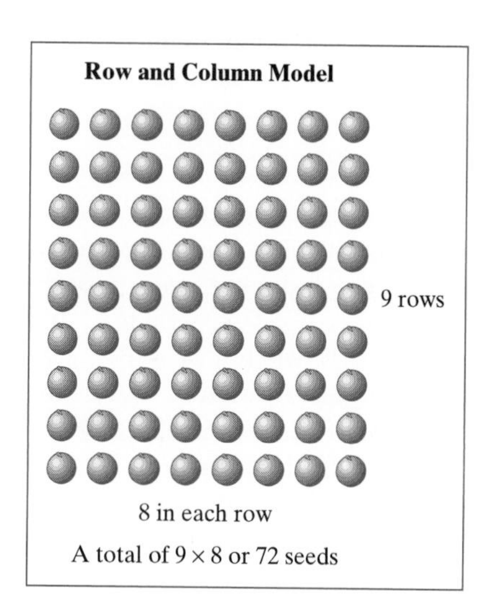

Figure 3.17a

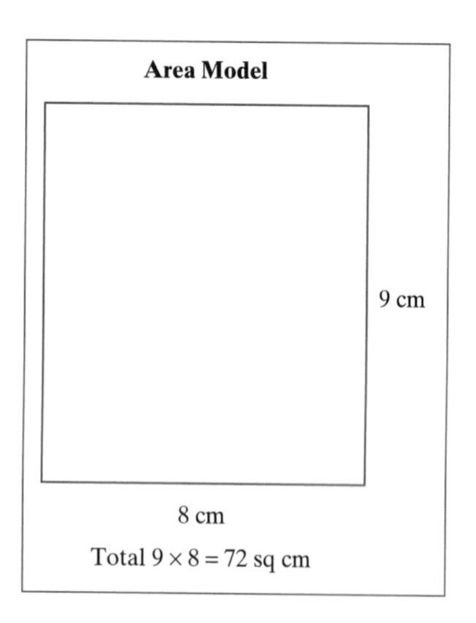

Figure 3.17b

3.8.7 Number-Line Model

The repeated addition view of multiplication can be illustrated using a set model as was seen in Figure 3.14. However, it is also possible to illustrate repeated addition using a **number-line model.** For example, if we place six 3-inch strips of trim around a hexagonal picture frame and need to know how many inches of trim we need, we would view this as repeated addition and model this multiplication on the number line that shows an arrow that started at zero and jumped three notches six times to end up on the amount 18. See Figure 3.18.

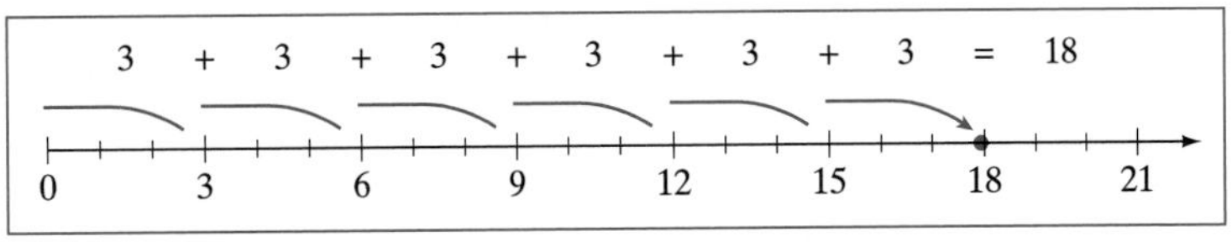

$3 + 3 + 3 + 3 + 3 + 3 = 18$

Figure 3.18

Related Topics

Multiplication Properties and Patterns (Whole Numbers)
Multiplication Algorithms

TOPIC 3.9 Multiplication Properties and Patterns (Whole Numbers)

cornerstone

Addition Properties and Patterns
Multiplication and Whole Numbers

Introduction

Like properties of whole-number addition, properties of whole-number multiplication stem from properties associated with operations on sets. Properties of multiplication influenced the development of the multiplication table. Today's multiplication table provides a convenient summary of the basic multiplication facts and illustrates many patterns of whole-number multiplication. In this topic, we describe and provide examples of the properties of whole-number multiplication.

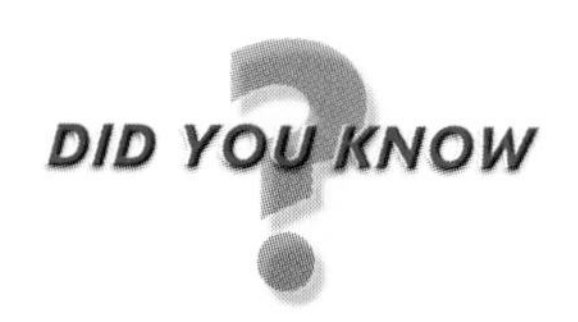

The square form of the multiplication table was known as the Pythagorean Table. Many mistakenly thought Pythagoras himself created this multiplication table, but it was confirmed that later Pythagoreans were the inventors. It was found in the 10th-century work of Anicius Manlius Severinus Boethius and a work attributed to Beda Venerabilis (c. 710).

Table 3.2

1	2	3	4	5	6	7	8	9
2	4	6	8	10	12	14	16	18
3	6	9	12	15	18	21	24	27
4	8	12	16	20	24	28	32	36
5	10	15	20	25	30	35	40	45
6	12	18	24	30	36	42	48	54
7	14	21	28	35	42	49	56	63
8	16	24	32	40	48	56	64	72
9	18	27	36	45	54	63	72	81

Figure 3.19

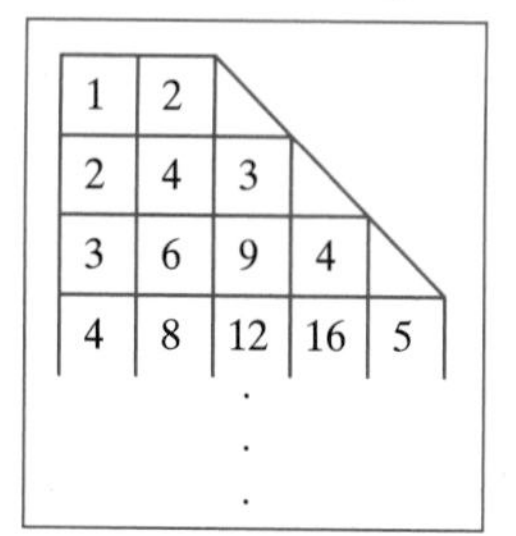

A second standard form of the multiplication table was the triangular array (see Figure 3.19). It appeared first in a Prague manuscript in the 14th century, but several variations have been noted throughout history.

For More Information Smith, D. E. (1958). *History of Mathematics, Volume II*. New York: Dover Publications.

3.9.1 Multiplication Table

When learning basic multiplication facts, that is, the set of multiplication problems involving two one-digit factors, working with a multiplication table is quite useful. The multiplication table illustrates patterns and properties associated with the multiplication of elements within the set of whole numbers.

Table 3.3

x	0	1	2	3	4	5	6	7	8	9
0	0	0	0	0	0	0	0	0	0	0
1	0	1	2	3	4	5	6	7	8	9
2	0	2	4	6	8	10	12	14	16	18
3	0	3	6	9	12	15	18	21	24	27
4	0	4	8	12	16	20	24	28	32	36
5	0	5	10	15	20	25	30	35	40	45
6	0	6	12	18	24	30	36	42	48	54
7	0	7	14	21	28	35	42	49	56	63
8	0	8	16	24	32	40	48	56	64	72
9	0	9	18	27	36	45	54	63	72	81

3.9.2 Zero Property of Multiplication

Many patterns can be observed in the whole-number multiplication table. For example, observe that the first row and the first column of the multiplication table are all zeros. This illustrates the **Zero Property of Whole-number Multiplication** that says any whole number times zero will yield a product of zero. Formally, for any whole number, a, $a \times 0 = 0$ and $0 \times a = 0$.

3.9.3 Multiplicative Identity

Notice that the second row and the second column of the multiplication table above match the set of factors that head the rows and columns of the table. This illustrates the **Identity Property of Multiplication;** that is, from the table one can see that the whole number, 1, is the **multiplicative identity,** which means that for any whole number, a, $a \times 1 = a$ and $1 \times a = a$.

3.9.4 Commutative Property

Other properties of whole-number multiplication are evident from the multiplication table. For example, notice that when you look at the product associated with 4×3, it is the same product associated with 3×4. Thus, like with the addition of whole numbers, multiplication of whole numbers has the property of being commutative. Essentially, the **Commutative Property of Whole-number Multiplication** states that it does not matter in what order the factors appear; the product will always be the same. So, for any two whole numbers, a and b, $a \times b = b \times a$.

3.9.5 Closure Property

The multiplication table illustrates a fourth property of whole-number multiplication called the **Closure Property;** that is, for any two whole numbers, a and b, $a \times b = c$, where c is also a whole number. Essentially, this property says, whenever you multiply two whole numbers together, the product will also be a whole number. All of the products seen within the multiplication table are whole numbers.

3.9.6 Associative Property

Another property of whole-number multiplication is the **Associative Property.** This property says that when multiplying three (or more) whole numbers together, it does not matter which pairs of numbers you multiply together first; once all three (or more) numbers are multiplied together, the product will always be the same. For example, when multiplying $4 \times 2 \times 6$, we can group the 4 and 2 together and multiply them first, taking that product, 8, and multiply it by 6 for a total product of 48. This might be written as $(4 \times 2) \times 6 = 48$ to indicate that the 4 and 2 are multiplied together first. On the other hand, we get the same product if we multiply the 2 and 6 together first and then multiply 4 times that product, again arriving at a total of 48. This would be written as $4 \times (2 \times 6) = 48$. In general, when multiplying any three whole numbers, a, b, and c, together, the Associative Property of whole-number multiplication guarantees that $(a \times b) \times c = a \times (b \times c)$.

3.9.7 Distributive Property of Multiplication over Addition

The final property that is associated with the multiplication of whole numbers is the **Distributive Property of Multiplication over Addition.** The property states that for any whole numbers, a, b, and c, $a \times (b + c) = (a \times b) + (a \times c)$. The Distributive Property can be put to use to find the product of 5×14. It might be easier to view the factor 14 as $10 + 4$ because instead of performing the multiplication of 5×14, you can compute $5 \times (10 + 4) = (5 \times 10) + (5 \times 4)$. The products 5×10 and 5×4 can easily be found using mental math. The sum of the products $50 + 20 = 70$ can also be found quickly using mental math.

Related Topics
Multiplication Algorithms
Division Algorithms

TOPIC 3.10 *Multiplication Algorithms*

cornerstone

Addition Algorithms
Multiplication with Whole Numbers
Multiplication Properties and Patterns

Introduction

Learning the traditional multiplication algorithm proceeds through various stages. We initially learn to multiply two single-digit whole numbers together to find a product and work toward being able to multiply two multidigited whole numbers together, performing "carrying" when necessary. In the following sections, a variety of multiplication algorithms are displayed with examples to illustrate the algorithms in use.

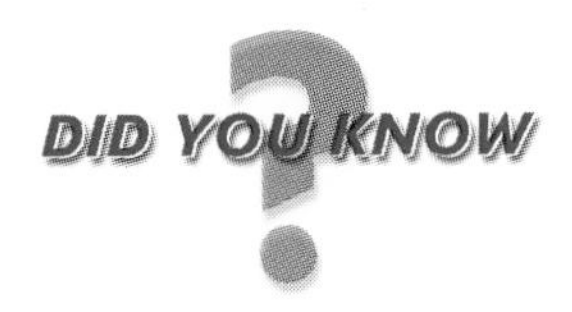

DID YOU KNOW?

Multiplication algorithms developed as the need to add similar amounts to each other repeatedly became necessary. The first real methods of multiplication started with Italian mathematician Luca Pacioli (A.D. 1494) who came up with eight plans for multiplication algorithms. Our common method of multiplying stemmed from a plan that took the form shown in Figure 3.20.

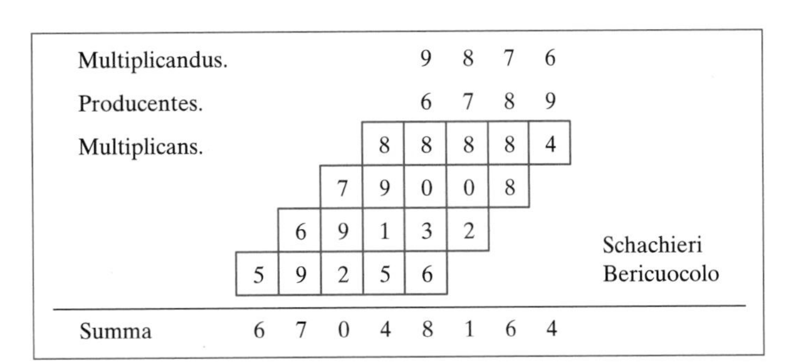

Multiplicandus.					9	8	7	6
Producentes.					6	7	8	9
Multiplicans.				8	8	8	8	4
			7	9	0	0	8	
		6	9	1	3	2		
	5	9	2	5	6			
Summa	6	7	0	4	8	1	6	4

Schachieri Bericuocolo

Figure 3.20

Like the algorithm in Figure 3.20, our current multiplication algorithm provides a procedure for finding the product of the numbers that are written in columnar form, with partial products written underneath the two factors and the final sum written at the very bottom.

For More Information Smith, D. E. (1958). *History of Mathematics, Volume II.* New York: Dover Publications.

3.10.1 Multiplication Algorithm

Developing the multiplication algorithm begins with working from known, basic whole-number multiplication facts and leads to multiplying a one-digit whole number times a two-digit whole number and beyond. As with addition and subtraction, concrete models, such as base-ten blocks, can assist the learning of such algorithms.

3.10.2 Base-Ten Blocks

The multiplication of 16×13 can be visualized using base-ten blocks. The base-ten blocks provide a means of illustrating the array model of multiplication, with 16 rows having 13 unit blocks in each row. See Figure 3.21.

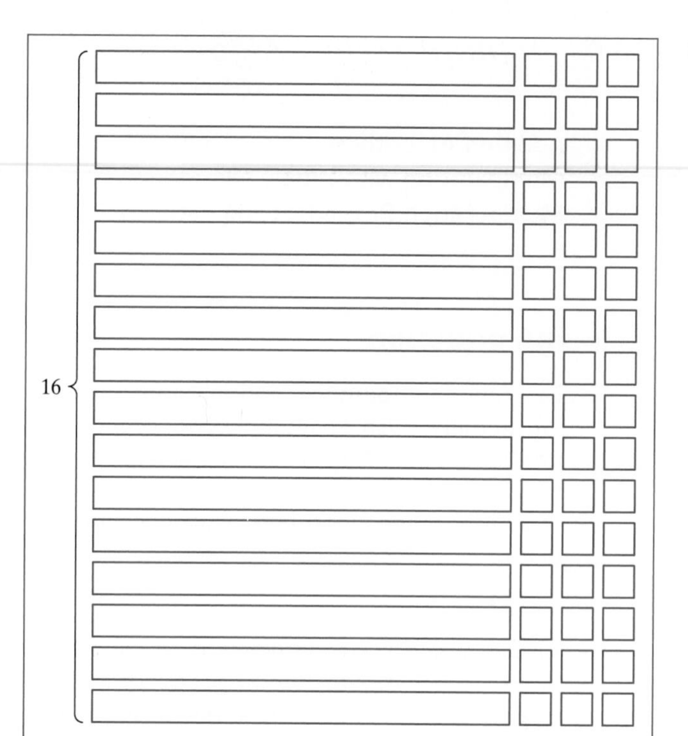

Figure 3.21

3.10.3 Partial-Products Multiplication Algorithm

To further develop the partial-products multiplication algorithm, regroup the base-ten blocks from Figure 3.21 as seen in Figure 3.22; that is, where possible, begin with the upper left-hand corner exchange 100 unit blocks in for a flat or hundreds block. Then, starting from the top to the right of the flat, in each column, trade in sets of 10 units for

Figure 3.22

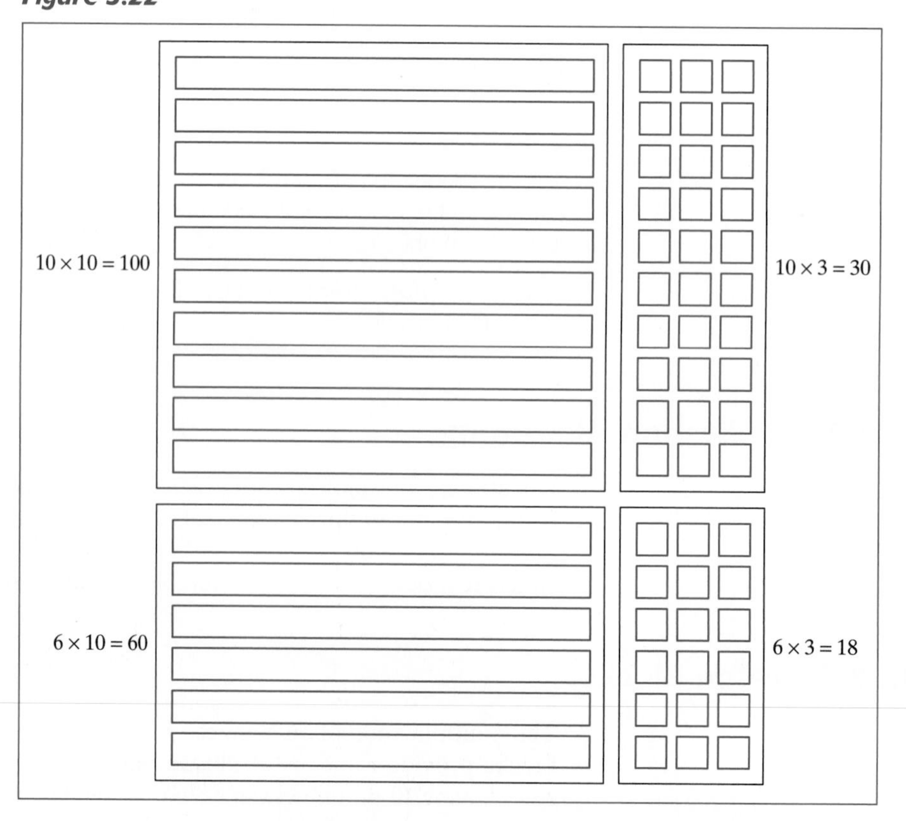

as many longs as possible. Do the same for the rows starting with the row beneath the flat. Once all possible exchanges are complete, we can quickly count up, the total number in the product of 16×13.

This concrete model of array multiplication lays a visual foundation for the **partial-products multiplication algorithm.** Seen within the array above are the partial products found in the traditionally written version of partial products, illustrated below.

$$\begin{array}{rl} 13 & \\ \times 16 & \\ \hline 18 & (6 \times 3) \\ 60 & (6 \times 10) \\ 30 & (10 \times 3) \\ 100 & (10 \times 10) \\ \hline 208 & \end{array}$$

Note that the partial products are those found by multiplying the ones-place digit of one factor times the ones-place digit of the other factor (3×6), multiplying the ones-place digit of one factor times the value of the tens-place digit of the other factor and vice versa $(3 \times 10$ and $10 \times 6)$ and finally, multiplying the value of the tens-place digit by the value of the tens-place digit in each of the factors (10×10). Using this transitional multiplication algorithm, we see that when multiplying, for example, 3×6, the product 18 must be recorded as 8 ones in the ones column and 1 in the tens column. Recognizing that such products have to be written in terms of tens and ones leads nicely to the carrying involved in the standard algorithm.

3.10.4 Standard Multiplication Algorithm

Using the transitional partial-products algorithm, we can better understand performing multiplication using the **standard algorithm of multiplication.** Ultimately, using this algorithm, we can easily perform complex multiplications such as 23×345, doing the necessary carrying:

$$\begin{array}{rl} 345 & \\ \times\ 23 & \\ \hline 15 & (3 \times 5) \\ 120 & (3 \times 40) \\ 900 & (3 \times 300) \\ 100 & (20 \times 5) \\ 800 & (20 \times 40) \\ 6000 & (20 \times 300) \\ \hline 7935 & \end{array}$$

Partial Products

$$\begin{array}{r} {}^{1}3\,{}^{11}4\,5 \\ \times\ 23 \\ \hline 1035 \\ 6900 \\ \hline 7935 \end{array}$$

Standard Algorithm

In the problem 23×345, the standard algorithm requires the multiplication of 3 times the number 345, carrying when necessary, and the multiplication of 20 times 345, again carrying when necessary. Then these two products are added together to yield the total product. This standard algorithm makes good use of the Distributive Property of multiplication over addition in that we can find the product of 23×345 as $(20 + 3) \times 345 = (20 \times 345) + (3 \times 345)$.

The details of the standard algorithm followed in the box above goes as follows. When the multiplication 3×345 is carried out, the product $3 \times 5 = 15$ is recorded with the 5 written in the ones column below the multiplication line and the set of ten carried to the tens column and recorded as 1 set of ten. Then the $3 \times 4 = 12$ (really representing 3×40) is found, and the 1 that was carried is added to that to make a total of 13. The 3 is recorded in the tens column, and the 1 is carried to the hundreds column. Finishing, the product 3 × 3 (really 3 × 300) is found to be 9, and the 1 that was carried is added to that for a total of 10. The 10 is recorded with the 0 in the hundreds place and the 1 in the thousands place. A similar procedure is followed for 20 × 345, with the added step of putting a zero in the ones place as a placeholder.

Related Topics

Division Algorithms

TOPIC 3.11 Whole-Number Division

cornerstone **Multiplication with Whole Numbers**

Introduction

Division can be thought of as the inverse operation to multiplication. Much of division is connected to multiplication. For example, in long division we estimate the answer to a division problem by *guessing* what we would have to multiply the divisor by to get as close to the starting dividend as possible. Although the two operations are strongly related, the properties associated with multiplication of whole numbers do not hold for the division of whole numbers. Within this topic, we will see that this division process can be interpreted in two different ways using a variety of models to illustrate the division process.

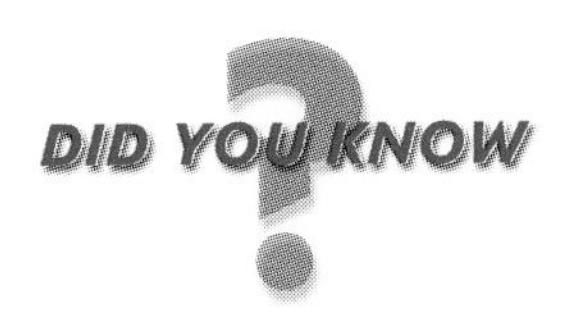

Probably the oldest and most widely accepted definition of **division** is *the operation of seeking a number which, multiplied by the divisor, is equal to the dividend.* Historically, division has been referred to as partitioning. However, in the Middle Ages, division performed on the abacus was often based on the operation of subtraction.

Early mathematicians gave names only to two of the numbers used in division, the *numerus dividendus* ("the number to be divided") and the *numerus divisor* ("the number doing the dividing"). Names such as *answer* or *result* were commonly used for quotient. The name for the remainder of a division problem has varied quite a bit over time. Medieval Latin writers used *numerus residuus, residuus,* and *residua* to refer to the remainder.

Division was considered the most difficult operation. In the late 15th century, Italian mathematician Pacioli remarked, "If a man can divide well, everything else is easy, for all the rest is involved therein."

For More Information Smith, D. E. (1958). *History of Mathematics, Volume II.* New York: Dover Publications.

3.11.1 Definition and Terminology

In the division of the whole number a by the whole number b, where $a \div b = c$, a is called the **dividend,** b is called the **divisor,** and c is called the **quotient.** Division is thought of as the inverse operation of multiplication. Consequently, division can be viewed as finding the **missing factor** in a multiplication problem. Generally, there are two situations in which division is found: **repeated subtraction** (sometimes called measurement division) and **sharing** (also known as partitioning).

3.11.2 Repeated Subtraction

Repeated subtraction is the view of division presented in the following problem. *If the general store has 36 balloons and plans to give four balloons to as many customers as possible when they enter the store, how many lucky customers will receive balloons?* The action in solving this problem is that of starting with 36 balloons and subtracting 4 at a time until none are left. Arithmetically this can be seen as:

$$36 - 4 = 32, 32 - 4 = 28, 28 - 4 = 24, 24 - 4 = 20, 20 - 4 = 16,$$
$$16 - 4 = 12, 12 - 4 = 8, 8 - 4 = 4, 4 - 4 = 0$$

From this, we see that we can subtract 4 a total of 9 times. The conclusion, then, is that $36 \div 4 = 9$.

3.11.3 Sharing

Division may be needed in a different type of situation. Consider this example: *If Eva has 28 marbles that she wants to put in 4 different bags, having the same number in each bag, how many can she put in each bag?* This scenario presents the idea of sharing the 28 marbles equally among four different sites. This is much like distributing the marbles, saying, *one in the first bag, one in the second bag, one in the third bag, one in the fourth bag, another in the first bag, another in the second bag, another in the third bag*, and so on. The image is that of doling things out one at a time to a fixed number of sets until they are all gone (see Figure 3.23).

In Step 1, one marble is placed in each bag. In Step 2, a second marble is placed in each bag. Continuing in this manner of partitioning the marbles, all marbles have been distributed at the end of Step 7. Altogether, each bag received 7 marbles. Thus, $28 \div 4 = 7$.

Figure 3.23

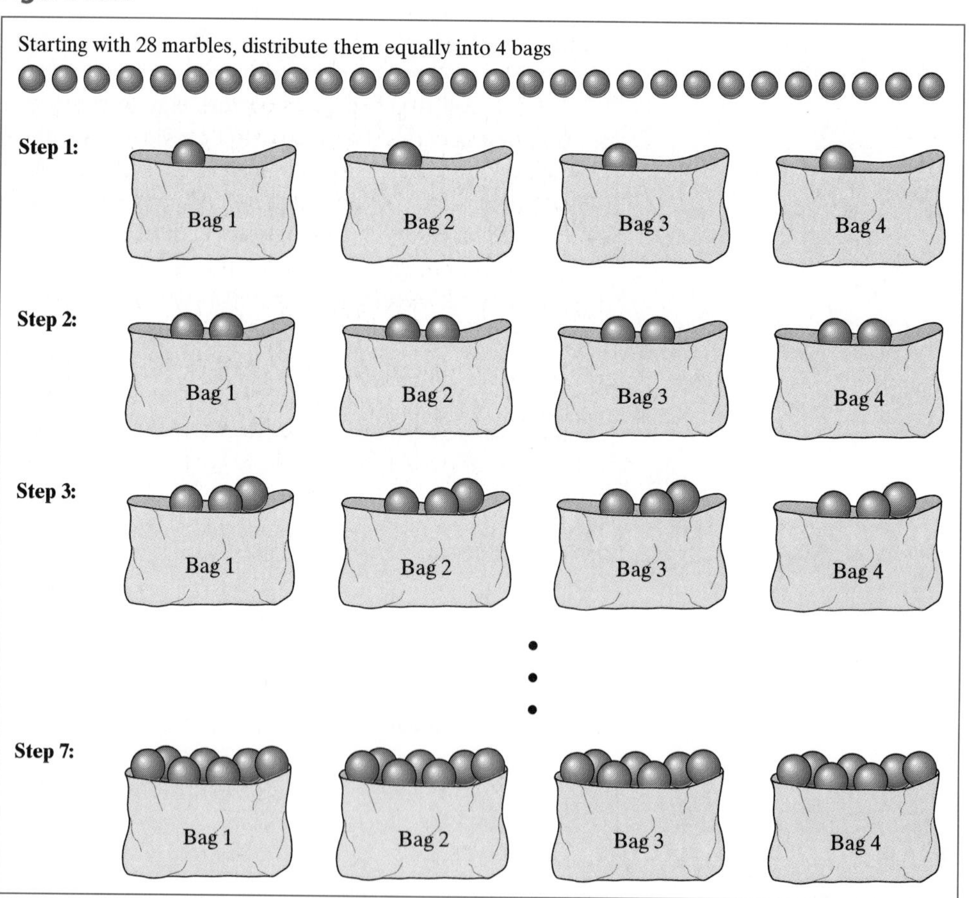

3.11.4 Repeated Subtraction Versus Partitioning

The distinction between repeated-subtraction division and sharing division can be explained in this way. When we divide, we start with an initial amount. With repeated-subtraction division, we know how many items we want to have in each set; we simply do not know how many sets we will have. On the other hand, with partitioning division, we know among how many sets we want to divide our items. We simply do not know how many items will be in each set.

Initial Amount ÷ the number in each set = the number of sets (repeated subtraction)
Initial Amount ÷ the number of sets = the number in each set (partitioning)

3.11.5 Discrete Models of Division

We can model whole-number division by using any kind of counters or set of objects (such as counting chips) and by physically dividing a total number of them into a certain number of piles, determining how many in each pile. This type of division was modeled with marbles in Figure 3.23. Likewise, we can take away a fixed number of items from a larger pile and determine how many of these fixed-size sets we can create from the original set. In Figure 3.24, the division problem $15 \div 5 = 3$ is modeled, using chips to represent the problem.

Figure 3.24

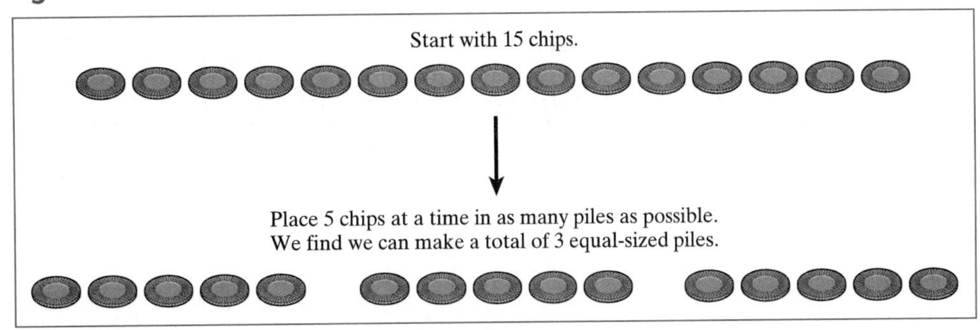

3.11.6 The Array Model

The division problem $15 \div 5 = 3$ can also be modeled in an array. In Figure 3.25, given the task of placing 15 chips in equal-sized rows, we find that if there are 5 chips in each row, then there must be 3 rows. This model is very useful when solving a division problem, viewing it from the missing-factor approach.

Figure 3.25

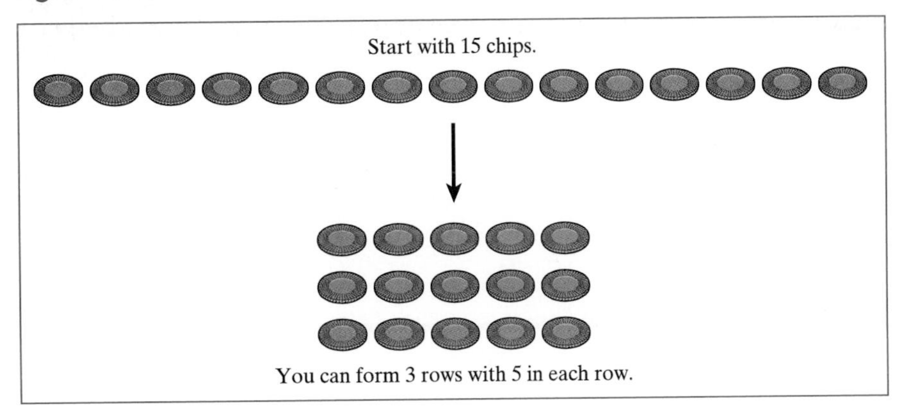

3.11.7 The Number-Line Model

We can also illustrate division using a number-line model. For example, we can show the problem $24 \div 4 = 6$ on a number line. The picture in Figure 3.26 shows that if you take a section of a number line that is 24 units in length and divide it into 4 equal lengths, each of these smaller lengths is 6 units long.

Figure 3.26

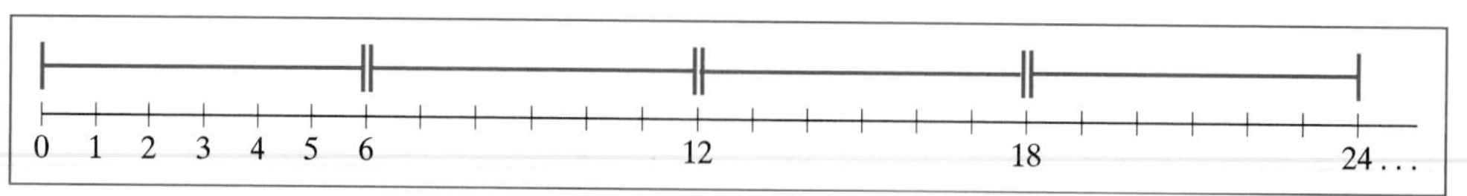

3.11.8 Division by Zero

One key issue to note with division is that division by zero is undefined; that is, you cannot solve a division problem where the divisor is zero. We can understand why this is so by thinking about division as finding a missing factor. For example, when we solve the problem $14 \div 7 = ?$, we can ask, "What can I multiply 7 times to have a product of 14?" The answer would be 2 because $2 \times 7 = 14$. Thus, $14 \div 7 = 2$.

If we have a situation where the divisor is 0, such as in the problem $10 \div 0 = ?$, restating the problem as a missing-factor situation makes no sense; that is, we ask the question, "What number times zero yields a product of 10?" There is no such number because according to the Zero Property of whole-number multiplication, zero times any whole number is zero. Therefore, there is no logical solution to this problem. Hence, division by zero is not defined.

Related Topics **Division Algorithms**

TOPIC 3.12 Division Algorithms

cornerstone **Whole-Number Division**

Introduction

The traditional division algorithm builds on the other operations. We first learn to divide with a single-digit divisor and work toward being able to divide a multidigited whole number by a multidigited divisor. In the following sections, we explore the division algorithm by looking at concrete models and various transitional division algorithms.

Division algorithms developed as the need to break amounts into smaller, same-sized amounts became necessary. The oldest form of division is the one used by the Egyptians, which was a process based upon what they termed *duplation* and *mediation*. See the table below, which is used for division by 8. For example, to divide 19 by 8, we can arrange the division as follows: Take $2 \times 8 = 16$, $\frac{1}{2}$ of $8 = 4$, and so on, and select the numbers in the right-hand column of the table that have 19 for their sum. For this example, $16 + 2 + 1 = 19$. The quotient, then, is $2 + \frac{1}{4} + \frac{1}{8} = 2\frac{3}{8}$.

1	8
2	16
$\frac{1}{2}$	4
$\frac{1}{4}$	2
$\frac{1}{8}$	1

Note that the left column consists of numbers that represent the result when the numbers in the right column are divided by 8. Thus, this duplation and mediation table is designed specifically for division by 8.

In the late Middle Ages, a method of division that consisted of using the factors of the divisor was developed and known as *per repiego*. For example, the problem $216 \div 24$ reduces to $216 \div 8 \div 3$. The goal was to reduce the problem to one-digit divisors that could be found more easily.

In the middle of the 16th century, an algorithm called the *English short division* was developed. It was based on recognizing products in the columns of the multiplication table and working backward to find the related factors. This method is also known as *division by the column*. The division-by-the-column method is not unlike our traditional algorithm that relies on estimation of products to find the quotient.

For More Information Smith, D. E. (1958). *History of Mathematics, Volume II*. New York: Dover Publications.

3.12.1 Concrete Models of the Algorithm

The division algorithm help us divide a two-digit dividend by single-digit divisor, as well as to perform more-complicated division calculations such as dividing a four-digit dividend by a three-digit divisor. Initially, it is best to introduce the division algorithms using concrete models such as the base-ten blocks. For example, 369 ÷ 3 can be modeled using base-ten blocks, as seen in Figure 3.27.

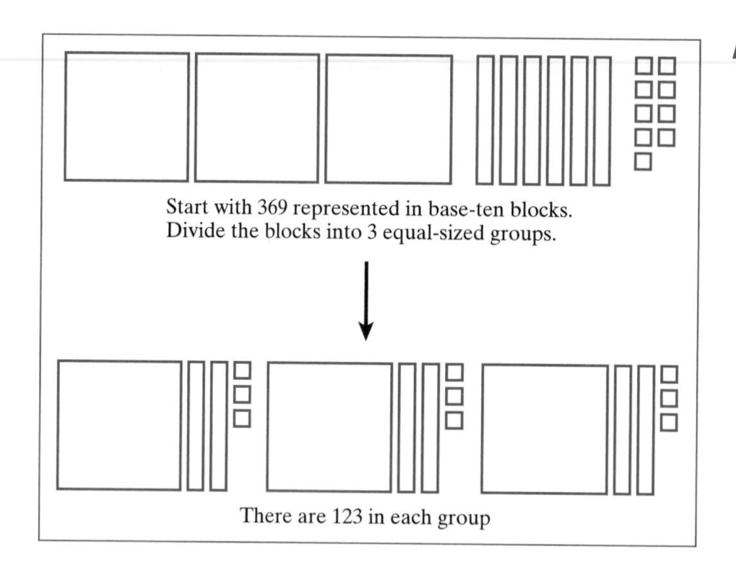

Figure 3.27

We can readily connect the modeled division to the traditional, written representation of the long-division algorithm. The problem 369 ÷ 3 = 123 can be recorded in this manner:

$$\begin{array}{r} \left.\begin{array}{r} 3 \\ 20 \\ 100 \end{array}\right\} 123 \\ 3\overline{)369} \\ -300 \\ \hline 69 \\ -60 \\ \hline 9 \\ -\ \ 9 \\ \hline 0 \end{array}$$

Starting with the 369 in the form of base-ten blocks, we first ask how many flats can we put in each of the three groups? The answer is 1, and it is recorded above the hundreds column of the number 369. Then the question is asked, "When we put one hundred in each of the three groups, how much did we put out altogether?" The answer is 300. The 300 is written under the 369 and is subtracted to leave a difference of 69.

Next, we ask the question, "How many longs can we put in each of the three groups?" The answer is 2, and it is recorded above the tens column of the number 369. Then the question is asked, "When we put 2 tens in each of the three groups, how much did we put out altogether?" The answer is 60. The 60 is written under the 69 and subtracted to leave a difference of 9.

We then ask, "How many units can we put in each of the three groups?" The answer is 3, and it is recorded above the ones column of the number 369. Then the question is asked, "When we put 3 units in each of the three groups, how much did we put out altogether?" The answer is 9. The 9 is written under the 9 and is subtracted to leave a difference of 0.

3.12.2 Scaffolding

There are transitional division algorithms that help students build toward the traditional algorithm. **Scaffolding** is a means of writing the solution to a long-division problem in stages, almost like finding "partial quotients." We then add those partial quotients together. For example, when computing $1284 \div 4 = 321$, we can illustrate the division using scaffolding as follows:

$$\begin{array}{r} 321 \\ \hline 1 \\ 20 \\ 300 \\ 4\overline{)1284} \\ -\ 1200 \\ \hline 84 \\ -\ 80 \\ \hline 4 \\ -\ 4 \\ \hline 0 \end{array}$$

3.12.3 Another Transitional Algorithm

We may also use variations of another transitional division algorithm. For example, we may develop the ability to "guess" the correct number of times a divisor goes into a dividend. See the example below of another transitional algorithm, illustrating the problem $681 \div 21$:

$$\begin{array}{rcl} & \text{Guess} & \\ 21\overline{)\ 681} & \downarrow & \\ -420 & (20 \times 21) & \\ \hline 261 & & \\ -210 & (10 \times 21) & \Big\} \ 20 + 10 + 2 \\ \hline 51 & & \\ -42 & (2 \times 21) & \\ \hline 9 & & \end{array}$$

So,

$$\begin{array}{r} 32\text{ R }9 \\ 21\overline{)\ 681} \end{array}$$

3.12.4 Links to Estimating Products

One thing to note about performing the division algorithm is the strong connection to multiplication estimation. In the long run, to perform long division, we need to have developed a keen ability to estimate how many times a divisor goes into a dividend to find the quotient.

We can use any number of estimation techniques to determine the quotient. One useful technique is front-end estimation where we consider only the left-most digit (or the two left-most digits) of the dividend and divisors, making all other digits zero. For example, when performing the long division $4568 \div 13$, we can use front-end estimation and think of the problem $4000 \div 10$ and then make a guess at the first number in the divisor by asking, "How many times does 10 go into 4?" Once we determine that it will not go into 4 a whole-number amount of times, we continue, asking next, "How many times does 10 go into 40?" We see that 4 might be a good guess for the left-most digit of the quotient. We check our guess and adjust if necessary.

Related Topics

Rational Numbers

TOPIC 3.13 Introduction to Integers

cornerstone
Whole Numbers
Whole Numbers and the Set of Whole Numbers

Introduction

The set of integers is the set of numbers that includes the natural numbers, the negatives of the natural members and zero. They allow us to represent arithmetic situations in which a negative amount occurs, such as a negative bank-account balance or temperatures below zero. In this section, we examine the relationship between integers and other sets of numbers.

Julia Bowman Robinson, born in 1919, made considerable contributions to the field of mathematical research called number theory. In particular, she dedicated much of her research life to solving a problem about a particular Diophantine equation, a type of equation with only integer solutions. Julia Bowman Robinson became the first female mathematician to be elected to the National Academy of Sciences and the first woman nominated for the presidency of the American Mathematical Society.

For More Information Cooney, M. P. (Ed.) (1996). *Celebrating Women in Mathematics and Science*. Reston, VA: National Council of Teachers of Mathematics.

3.13.1 Natural Numbers and Whole Numbers

Natural numbers are the numbers 1, 2, 3, 4, 5, ... that we use to count objects. Thus, these numbers are also called **counting numbers.** When we add the number zero (0) to this group of natural numbers, we have the group of numbers 0, 1, 2, 3, 4, 5, ... that we call the **whole numbers.**

3.13.2 Positive and Negative Numbers

The natural numbers are all **positive numbers;** that is, they can be represented on a number line as greater than, or to the right of, zero. **Negative numbers,** on the other hand, are numbers that are less than, or to the left of, zero on a number line. See Figure 3.28.

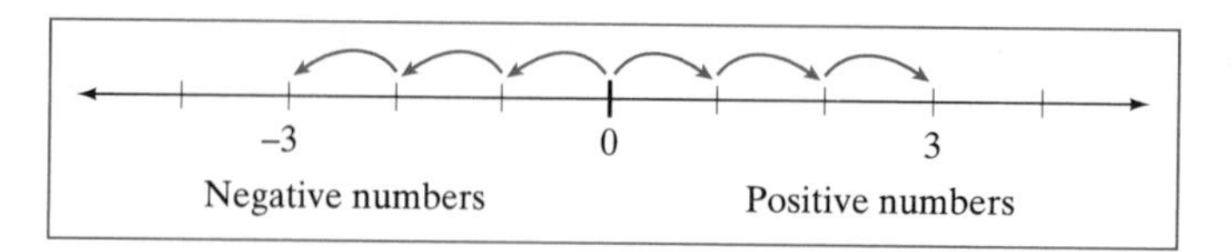

Figure 3.28

Notice that if we move three units to the right of zero, we can represent the distance away from zero with the positive number 3. If we move three units to the left of zero, the distance away from zero is still a distance of three, however we indicate that we are to the left of zero by using the negative number -3.

3.13.3 Notation for Positive and Negative Numbers

Positive numbers are represented simply by using the number that represents how far to the right of zero that number is; that is, the positive number six is represented with the symbol 6. Negative numbers, on the other hand, are represented by using a negative sign ($-$) preceding the number to indicate that the amount is negative or represents an amount to the left (or less than) zero. For example, negative 8 is represented as -8. In some situations, you might see a positive number represented with a positive sign (+) in front of the number, such as +4 to represent positive 4. However, in most cases, numbers are assumed to be positive when no preceding sign is used.

3.13.4 Opposites

We say that two numbers are **opposites** if they have different signs associated with them. For example, 4 and −4 are opposites because there is a negative sign associated with −4 and an implied positive sign associated with 4. In general, the numbers n and $-n$ are opposites of each other. The negative sign itself is sometimes expressed as *the opposite*; that is, $-n$ might be read as *the opposite of* n. Thus, $-(-n)$ could be read as *the opposite of negative* n. Therefore, $-(-n) = n$.

3.13.4 Absolute Value

The **absolute value** of a number is the number of units the number is from zero, no matter whether it is to the right or to the left of zero on a number line. For example, the absolute value of 5 is 5 because it is five units away from zero. On the other hand, the absolute value of −5 is also 5 because −5 is five units away from zero. The notation for the absolute value of a number, n, is $|n|$.

In general, if n is a positive number, then $|n| = n$ and $|-n|$ is n. A different way of saying this is, for any number, n, $|n| = n$ when n is positive and $|n| = -n$ when n is negative. For example, suppose $n = -3$. Then $|-3| = -(-3)$, that is, the negative or opposite of negative three, which is 3. Yet another way to think of the absolute value of a number, n, is $|n| =$ maximum of n and $-n$. So, $|-3| =$ maximum of -3 and $-(-3)$ or 3.

3.13.6 Integers and the Set of Integers

Integers are numbers that include all natural numbers, all negatives of natural numbers (the opposites of the natural numbers), and the number zero; that is, **the set of integers** $= \{\ldots, -4, -3, -2, -1, 0, 1, 2, 3, 4, \ldots\}$. The set of integers is denoted, I. As indicated by the ellipses (...), there is an infinite number of integers. Often, we refer to special subsets of the integers, such as the **set of positive integers** $= \{1, 2, 3, 4, 5, \ldots\}$, (denoted I^+), and the **set of negative integers** $= \{\ldots, -4, -3, -2, -1\}$, (denoted I^-).

Related Topics	
	Rational Numbers
	The Set of Rational Numbers

TOPIC 3.14 Integer Addition

Whole-Number Addition Properties and Patterns
Whole-Number Addition
Integers

Introduction

Integer addition involves adding positive integers to positive integers, negative integers to negative integers, and combining positive and negative integers. Performing addition with integers presents interesting and varied mathematical algorithms and models. In this section, we will learn traditional integer addition formulas and will illustrate why these procedures make sense.

When mathematician Carl Friedrich Gauss was just 10 years old, his teacher wanted to keep the unruly boy busy. He asked Gauss to find the sum of the integers from 1 to 100. Gauss completed the task in a few minutes. Astounded, the teacher asked how he found the sum so quickly. Gauss explained that when he added 1 + 100, the sum was 101. He found the same sum when adding 2 + 99, 3 + 98, 4 + 97, and so on. Thus, with the 100 numbers, there would be 50 pairs that would sum to 101, so the total sum would be 50 times 101 or 5,050. Gauss's observation led to a formula for finding the sum of the first n positive integers: $\text{Sum} = \frac{n(n+1)}{2}$.

For More Information Reimer, L., & Reimer, W. (1990). *Mathematicians Are People Too: Stories from the Lives of Great Mathematicians, Volume 1*. Palo Alto, CA: Dale Seymour Publications.

3.14.1 Properties of Integer Addition

The addition of integers has a number of properties. Most of these properties correspond directly to the properties of whole-number addition; that is, like in the addition of whole numbers, the addition of integers has (1) an additive identity of zero, (2) the Commutative Property that says that the order of the addends does not affect the sum, (3) the Associative Property that says that when adding three or more integers, the way you choose to group numbers to add two at a time does not affect the sum, and (4) the Closure Property that says that whenever you add two integers, the sum is also an integer. (See Addition Properties and Patterns (Whole Numbers) for illustrative examples of these properties.)

3.14.2 The Additive Inverse Property

The set of integers has an additional property that the set of whole numbers does not have. The set of integers has the **Additive Inverse Property** that says, for any integer, a, and its **inverse** (or opposite), $-a$, $a + (-a) = 0$ and $-a + a = 0$. For example, given the integer, 5, and its inverse, -5, it is true that $5 + (-5) = 0$ and $-5 + 5 = 0$.

3.14.3 Addition Rules for Integers

When adding integers, there are several addition rules that can be used. These rules apply when adding two positive integers, when adding a positive integer to a negative integer, and when adding two negative integers.

Rule for Adding Two Positive Integers. Because the set of positive integers is equivalent to the set of natural numbers, adding two positive integers is the same as adding two natural numbers; that is, for two positive integers a and b, the sum is found by computing $a + b$. For example, $3 + 6 = 9$.

Rule for Adding a Positive Integer to a Negative Integer. To add a positive integer to a negative integer, let's first suppose that a and b are positive integers. Then we can let $-b$ represent a negative integer. To find the sum of $a + (-b)$, there are two cases to consider.

In the first case, $|a| > |-b|$. Then, the rule for integer addition says that $a + (-b) = a - b$. For example, $5 + (-2) = 5 - 2 = 3$.

In the second case, $|a| < |-b|$. In this situation, the rule for integer addition says that $\mathrm{a} + (-\mathrm{b}) = -(|-\mathrm{b}|-\mathrm{a})$. For example, $4 + (-7) = -(7 - 4) = -3$.

Rule for Adding Two Negative Integers. Given that a and b are positive integers, we can let $-a$ and $-b$ represent negative integers. The rule for adding two negative integers says that $-a + (-b) = -(|a|+|b|)$. For example, $-3 + (-6) = -(3 + 6) - 9$.

3.14.4 Models of Integer Addition

Because the rules for integer addition can seem somewhat complicated, it is helpful to see how integer addition can be modeled. Two models that are useful are the set model and the number-line model.

Set Model of Integer Addition. Suppose we want to illustrate integer addition using the set model. We can use two-color counters—chips that are white on one side and black on the other—to illustrate examples of integer addition. In all examples, let the white chips represent positive numbers and the black chips represent negative numbers. See Figure 3.29 for an example of integer representations using two-color counters.

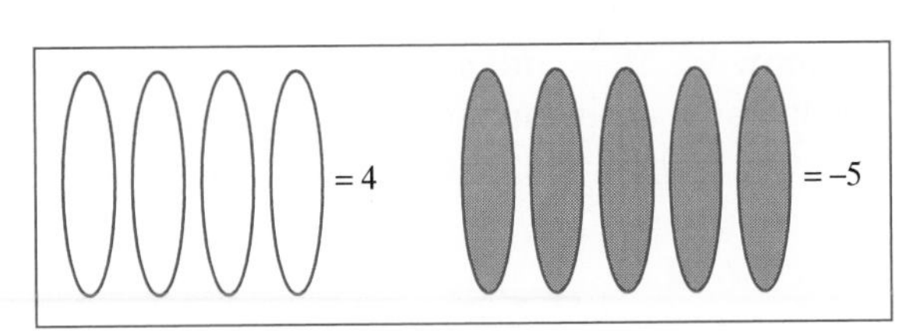

Figure 3.29

Example 1: To illustrate the addition of 4 + (−5,) we can combine the representations of the integers shown in Figure 3.29 as done in Figure 3.30.

When we combine 4 white counters with 5 black counters, we next observe that we can pair up black counters with white counters (see Figure 3.31); that is, using the

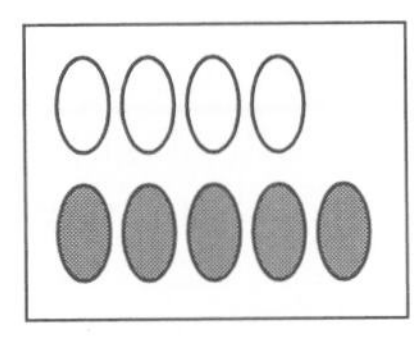

Figure 3.30

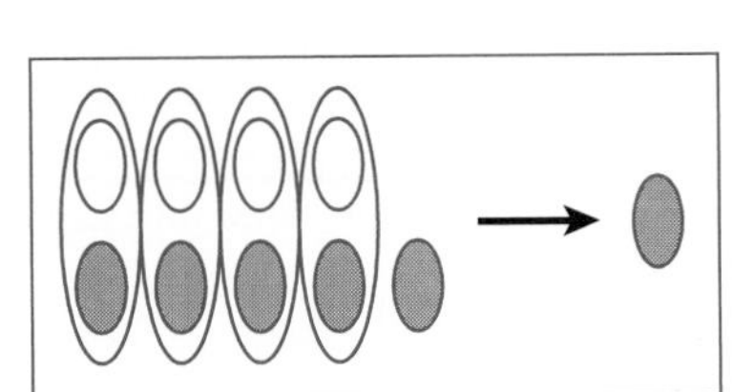

Figure 3.31

Additive Inverse Property of integer addition, we note that because $1 + (-1) = 0$, then each pairing of a black counter with a white counter represents a sum of zero. Thus, removing these sets of zero from the collection of counters does not affect the total value represented by the counters. Therefore, once all pairs of counters that represent zero are removed, we find that one black counter remains. So, the sum $4 + (-5) = -1$.

Example 2: When adding two negative integers, the two-color counters are also useful for modeling the addition. Adding $-2 + (-3)$, we demonstrate the addition combining 2 black counters with 3 more black counters for a total of 5 black counters (see Figure 3.32). Thus, the sum is -5.

Figure 3.32

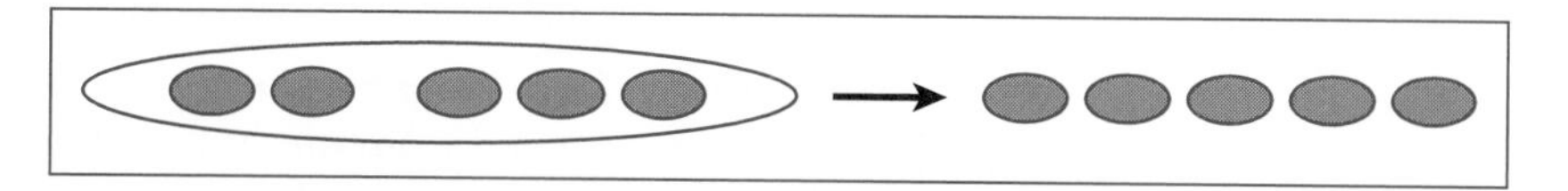

The Number-Line Model of Integer Addition. Integer addition can be clearly modeled on a number line. The rules involved in using the number line to illustrate integer addition are: (1) Always start with your arrow on zero; (2) your arrow always points toward the positive direction (to the right); (3) positive-integer addends tell you to move your arrow forward; and (4) negative integers tell you to move backward.

Example 1: Illustrating the addition problem $6 + (-4)$ on the number line, we begin with our arrow on zero and facing right as shown in Figure 3.33.

Figure 3.33

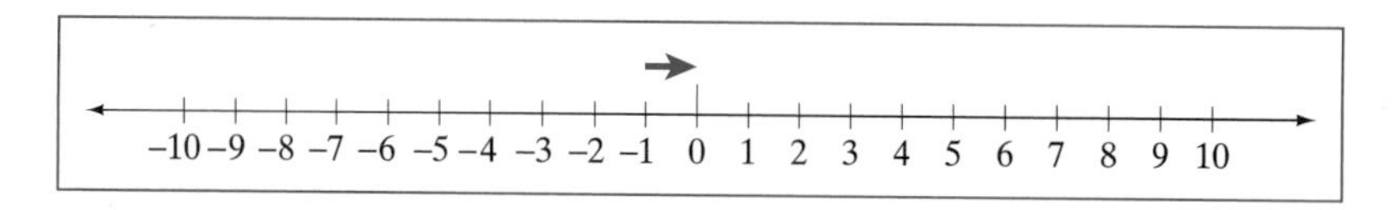

Show the first addend by advancing the arrow forward 6 units. Then show the second addend by moving the arrow backward 4 units. The arrow stops at 2. (See Figure 3.34.) Therefore, $6 + (-4) = 2$.

Figure 3.34

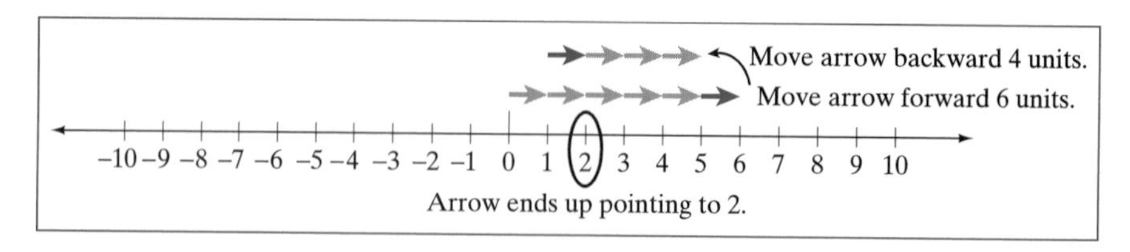

Example 2: Next, we illustrate the problem $-8 + 3$. Beginning with our arrow on zero and facing toward the right, we show the first addend by moving the arrow backward 8 units. Then, show the second addend by moving the arrow forward 3 units. The arrow stops at -5. (See Figure 3.35.) Therefore, $-8 + 3 = -5$.

Figure 3.35

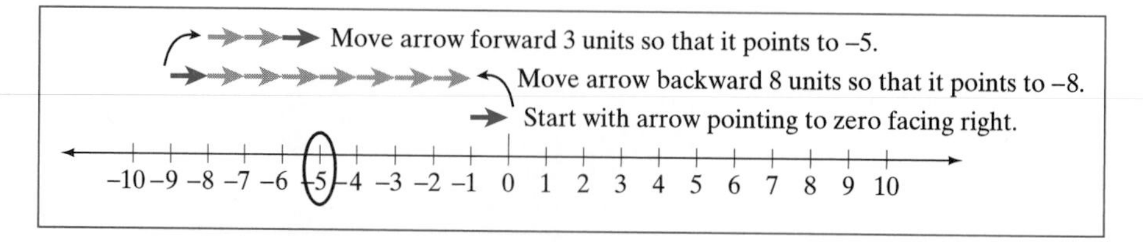

Example 3: We can illustrate the addition of two negative integers on the number line. Showing $-3 + (-5)$, we begin with our arrow pointing to zero and facing toward the right. We represent the first addend by moving our arrow backward 3 units. We represent our second addend by moving the arrow backward another 5 units. The arrow lands on -8, showing that $-3 + (-5) = -8$. See Figure 3.36.

Figure 3.36

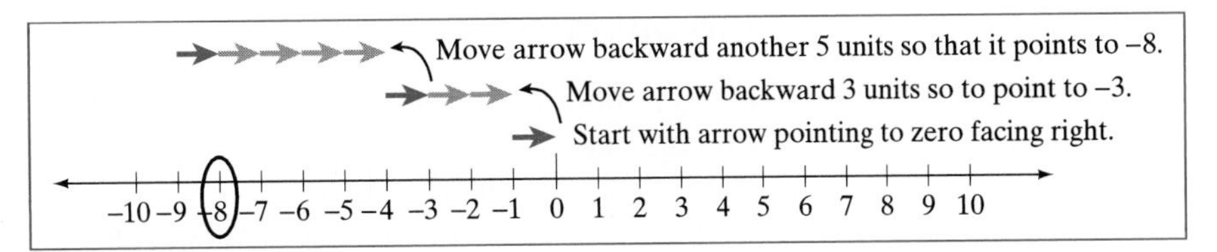

Related Topics

- **Order of Operations**
- **Integer Subtraction**
- **Integer Multiplication**
- **Integer Division**

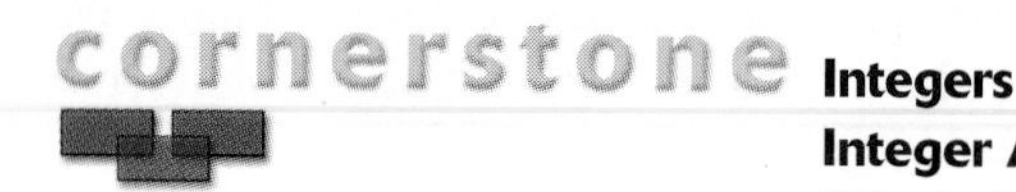

TOPIC 3.15 Integer Subtraction

Integers
Integer Addition
Whole-Number Subtraction

Introduction

Integer subtraction involves finding the difference between combinations of positive and negative integers. To some, the idea of subtracting a negative amount makes little sense, particularly when we are told that subtracting a negative number is like adding a positive number. In this section, models such as the number-line model and the set model help to explain this abstract notion.

DID YOU KNOW

The use of negative numbers has not been found in the writings of ancient Egyptians, Babylonians, Hindus, or Greeks. The Chinese, however, in as early as 200 B.C., made the first distinction and use of negative numbers as subtrahends in subtraction problems. Centuries later, the Chinese began to show more distinction between positive and negative numbers by writing positive numbers in red and negative numbers in black.

For More Information Smith, D. E. (1958). *History of Mathematics, Volume II*. New York: Dover Publications.

3.15.1 Closure Property for Integer Subtraction

Most of the properties that hold for integer addition do not hold for integer subtraction; that is, integer subtraction is not commutative, is not associative, does not have an identity, and does not have an inverse feature. (See Integer Addition and Properties of Whole-Number Addition for more details about these properties.) However, integer subtraction does have the property of being closed. The **Closure Property of integer subtraction** states that the difference, $a - b$, between any two integers, a and b, is also an integer. For example, the difference between the integers 7 and -9, $7 - (-9) = 16$, where 16 is also an integer.

3.15.2 Rules for Integer Subtraction

When subtracting an integer from another integer, there are rules to follow, depending on whether the minuend (starting amount) and the subtrahend (amount being subtracted) are positive or negative integers.

Rules for Subtraction Involving Two Positive Integers. When both the minuend and the subtrahend (see Subtraction With Whole Numbers for more details) are positive, then there are two cases to consider. First, when the minuend is greater than the subtrahend, then positive-integer subtraction works the same as the subtraction of whole numbers. For example, $7 > 5$; thus, $7 - 5 = 2$.

When the subtrahend is greater than the minuend, the difference is found by subtracting the minuend from the subtrahend and applying a negative sign to that answer; that is, for two positive integers a and b where $a < b$, $a - b = -(b - a)$. For example, given the integers 3 and 8, the difference $3 - 8 = -(8 - 3) = -5$.

Rules for Subtraction Involving Two Negative Integers. When finding the difference between two negative integers, there is a specific rule we can use to rewrite the subtraction problem as an integer addition problem. The rule states that given any two negative integers expressed as $-a$ and $-b$, $-a - (-b) = -a + b$.

Essentially, the rule says that subtracting a negative number is equivalent to adding the (positive) opposite of that number. For example, $-5 - (-6) = -5 + 6$. Likewise, $-7 - (-4) = -7 + 4$. To compute the answers, then, to these problems, we can use the rules for integer addition that apply when one addend is positive and one addend is negative (see Integer Addition for details). Thus, $-5 - (-6) = -5 + 6 = 1$, and $-7 - (-4) = -7 + 4 = -3$.

Rules for Subtracting a Negative Integer from a Positive Integer. When subtracting a negative integer from a positive integer, we can use essentially the same method described above for rewriting the subtraction problem as an integer addition problem; that is, given a positive integer, a, and a negative integer, $-b$, $a - (-b) = a + b$. Thus, when subtracting -7 from 4, we can rewrite the subtraction problem as $4 - (-7) = 4 + 7$. To complete the computation, we can simply add the two positive integers to find the answer $4 - (-7) = 4 + 7 = 11$.

Rules for Subtracting a Positive Integer from a Negative Integer. When subtracting a positive integer from a negative integer, we can rewrite the subtraction problem as an integer addition problem and then follow the rules for integer addition. The rule for rewriting such a subtraction statement says, given a positive integer, a, and a negative integer, $-b$, $-b - a = -b + (-a;)$ that is, subtracting a positive integer is equivalent to adding the (negative) opposite of that integer. For example, given the integers -7 and 9, $-7 - 9 = -7 + (-9)$. To determine the final computational answer, we can use the rule for adding two negative integers (see Integer Addition for details and examples). Thus, $-7 - 9 = -7 + (-9) = -16$.

3.15.3 Models of Integer Subtraction

The rules for integer subtraction are closely related to rules for integer addition. When we model integer subtraction using the set model and the number-line model, we can begin to see the connections and to understand the integer subtraction processes better.

The Set Model of Integer Subtraction. Suppose we want to illustrate integer subtraction using the set model. We can use two-color counters, chips that are white on one side and black on the other, to illustrate examples of integer subtraction. In all examples, let the white chips represent positive numbers and the black chips represent negative numbers. See Figure 3.37 for examples of integer representations using two-color counters.

Figure 3.37

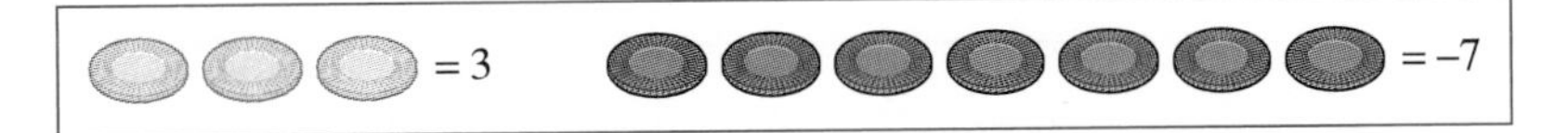

Example 1: Suppose we want to first model the subtraction problem $-6 - (-4)$. We can begin by placing 6 black chips out to represent the beginning amount, or minuend. We can then take away 4 black chips to model the subtraction of the -4. See Figure 3.38.

Figure 3.38

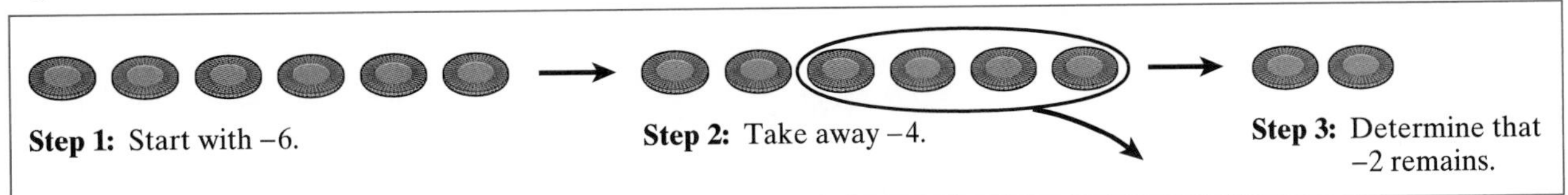

Example 2: A more-complicated example arises when we subtract $-7 - 3$. To model this problem, we begin with 7 black chips to represent -7. However, there are no white chips that we can remove to model the subtraction of positive 3. To work with this model, we must make use of the Additive Inverse Property (see Integer Addition); that is, if we simultaneously add 1 white chip and 1 black chip to the set of chips, we are essentially adding zero because an integer plus its opposite equal zero. Therefore, we can model the subtraction problem $-7 - 3$ as seen in Figure 3.39.

Figure 3.39

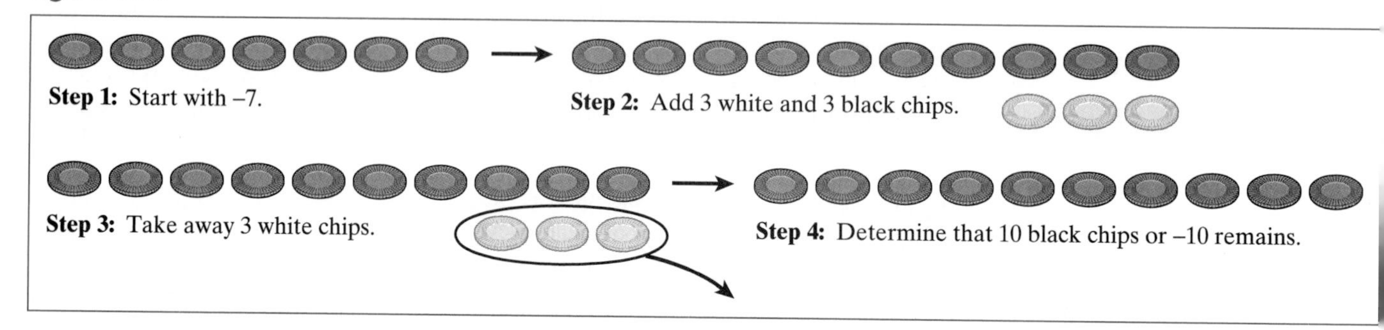

The model of integer subtraction in Figure 3.37 suggests a relationship between integer subtraction and integer addition. The rule for subtracting $-7 - 3$ tells us to rewrite the subtraction problem as the addition problem $-7 + (-3)$ and then add the negative integers. The set model allows us to see the conversion of the subtraction problem to an integer addition problem; that is, we ultimately add the 3 new black chips to the original 7 black chips to find a total of 10 black chips, or an answer of -10.

Example 3: A similar result happens when we subtract a negative integer from a positive one. In the problem $2 - (-4)$, we can model the problem with chips (see Figure 3.40) and find that in the end we have 6 white chips, or 6 as an answer.

The Number-Line Model of Integer Subtraction. Integer subtraction can be clearly modeled on a number line. The rules involved in using the number line to illustrate integer subtraction are: (1) Always start with your arrow on zero; (2) your arrow points to the negative (left) direction when subtracting; (3) a positive integer tells you to move your arrow forward; and (4) a negative integer tells you to move backward.

Example 1: To illustrate the subtraction $5 - 7$, we begin with our arrow on zero and facing forward. (See Figure 3.41.) We move forward to positive 5. We then turn the arrow to face the negative direction because we are about to subtract. We finally move the arrow forward 7 because 7 is positive. The result shows that the arrow ends at -2.

Example 2: Illustrating the subtraction of a negative number can be done by following the same rules. Subtracting $3 - (-2)$, we begin at zero and facing right and move forward 3 units. We then change directions to prepare to subtract. Finally we

Figure 3.40

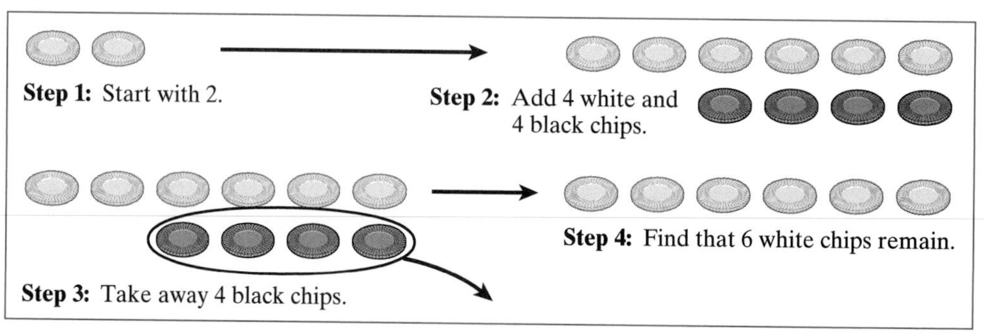

Figure 3.41

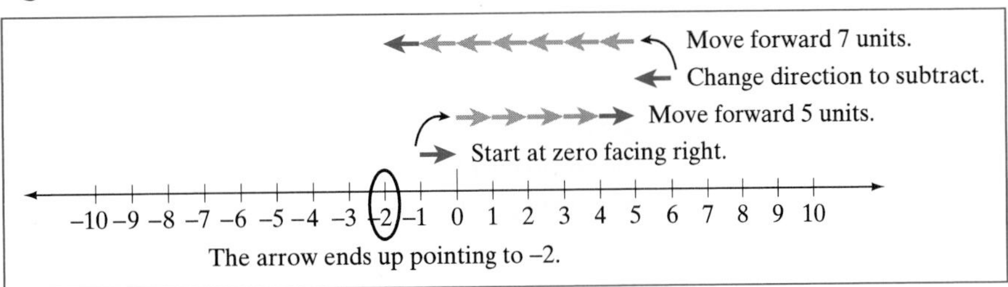

move backward 2 units because the integer subtracted is negative. (See Figure 3.42.) The arrow ends pointing to positive 5, and this is our answer.

Figure 3.42

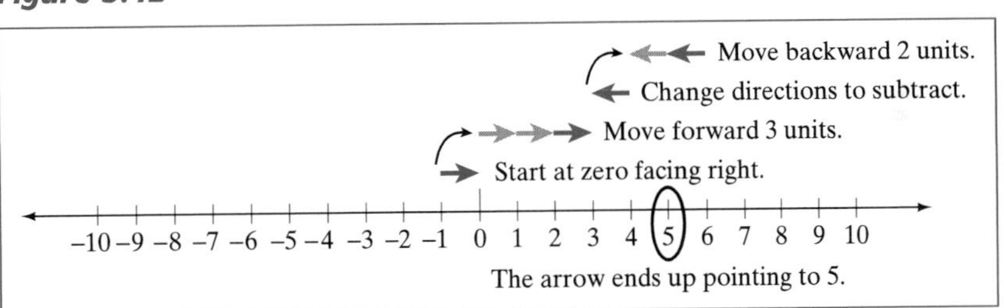

Example 3: We can illustrate $-8 - (-3)$ on the number line. Figure 3.43 shows the arrow beginning on zero and facing right and then moving backward 8 units to show -8. Then the arrow changes directions for subtraction and moves backward 3 units for the -3. The end result is a difference of -5.

Figure 3.43

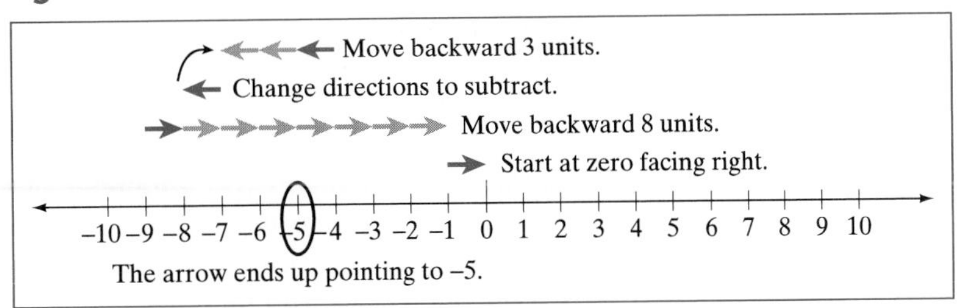

Related Topics

Order of Operations
Integer Multiplication
Integer Division

TOPIC 3.16 Integer Multiplication

cornerstone

Integers
Integer Addition
Whole-Number Multiplication
Properties of Whole-Number Multiplication

Introduction

Multiplying integers together combines the basic fundamental steps of multiplying whole numbers with the interesting addition of having negative factors. People know the basic rule that a negative number times a negative number is a positive number. In this section, models are used to bring understanding to this and other rules for the multiplication of integers.

The rules for multiplying negative numbers by negative numbers makes sense when you analyze the pattern of product found by multiplying a given integer by a series of positive and negative integers. Observe:

$$4 \times -3 = -12$$
$$3 \times -3 = -9$$
$$2 \times -3 = -6$$
$$1 \times -3 = -3$$
$$0 \times -3 = 0$$
$$-1 \times -3 = 3$$
$$-2 \times -3 = 6$$
$$-3 \times -3 = 9$$
etc....

As the first factor decreases by 1, the product of the first factor and -3 increases by 3. Thus, the pattern suggests that a negative times a negative should yield a positive number to continue the pattern.

3.16.1 Properties of Integer Multiplication

The multiplication of integers has a number of properties. These properties correspond directly to the properties of whole-number multiplication; that is, as in the multiplication of whole numbers, the multiplication of integers has (1) a multiplicative identity of one, (2) the Commutative Property that says that the order of the factors does not affect the product, (3) the Associative Property that says that when multiplying three or more integers, the way you choose to group numbers to multiply two at a time does not affect the product, (4) the Closure Property that says that whenever you multiply two integers, the product is also an integer, (5) the Zero Property of multiplication that says that zero times any integer yields a product of zero, and (6) the Distributive Property of multiplication over addition that shows us two ways to view the multiplication of an integer by the sum of two integers. (See Multiplication Properties and Patterns [Whole Numbers] for illustrative examples of these properties.)

3.16.2 Multiplication Rules for Integers

When multiplying integers, there are several multiplication rules that can be used. These rules apply when multiplying two positive integers, when multiplying a positive integer by a negative integer, and when multiplying two negative integers.

Rule for Multiplying Two Positive Integers. Because the set of positive integers is the set of whole numbers (less zero), the rule for multiplying two positive integers is to perform the multiplication like whole-number multiplication. Thus, choosing the positive integers, 6 and 8, we find $6 \times 8 = 48$.

Rule for Multiplying a Positive Integer by a Negative Integer. When multiplying a positive integer by a negative integer, the product is always the negative (opposite) of the product of the absolute values of the two factors; that is, given a positive integer, a, and a negative integer, $-b$, the product $a \times -b = -(|a| \times |-b|) = -(a \times b)$. For example, given -5 and 4, we find $5 \times (-4) = -(5 \times 4) = -20$.

Rule for Multiplying Two Negative Integers. When multiplying two negative integers, the product will always be positive. The rule for multiplying negative integers states that the product of two negative integers is the product of the absolute values of each integer; that is, given two negative integers expressed as $-a$ and $-b$, the product $-a \times (-b) = |-a| \times |-b| = a \times b$. For example, the product of -6 and -12 is found by $-6 \times (-12) = |-6| \times |-12| = 6 \times 12 = 72$.

3.16.3 Models of Integer Multiplication

Multiplication of integers can be modeled using a set model and a number-line model. These models help provide a visual means of understanding the processes used in integer multiplication.

The Set Model of Integer Multiplication. Suppose we want to illustrate integer multiplication using the set model. We can use two-color counters, chips that are white on one side and black on the other, to illustrate examples of integer multiplication. In all examples, let the white chips represent positive numbers and the black chips represent negative numbers. See Figure 3.44 for examples of integer representations using two-color counters.

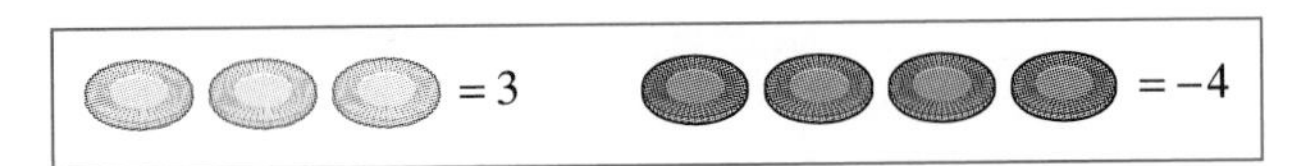

Figure 3.44

Rules for Working with the Set Model for Multiplication. The rules for working with the two-colored counters are as follows: (1) The absolute value of the first factor tells you how many chips to put out at a time (in one set); (2) the absolute value of the second factor tells you how many sets of counters to put out at a time; (3) the sign of the first factor tells you which side of the counter to have face up (white for positive, black for negative); and (4) the sign of the second factor tells you whether to maintain the face-up color, determined by the first factor (if a positive sign), or to change the face-up color to its opposite (if a negative sign).

Example 1: To illustrate the multiplication $3 \times (-4)$, we know that we will put down 3 counters (chips) at a time and we will put down 4 sets of three chips. Because 3 is positive, those chips will be placed white-side up. (See Figure 3.45.) Because the

Figure 3.45

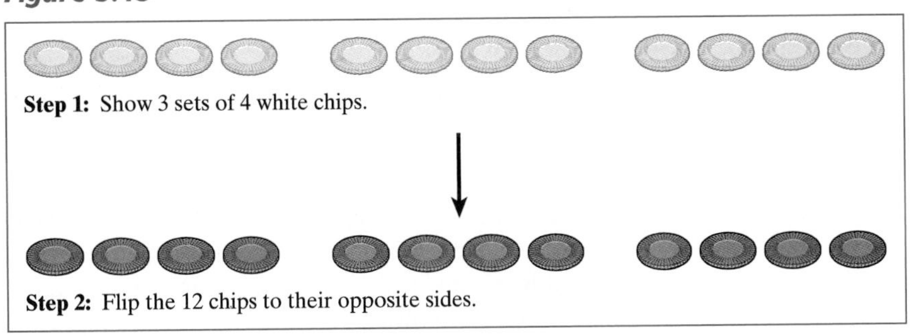

second factor is negative, we then know to flip the chips to their opposite side. The end result is 12 black chips that represent the product -12.

Example 2: To show the product $-2 \times (-5)$, we would illustrate two sets of 5, starting out black-side up because of the negativity of the first factor. (See Figure 3.46.) Then, we flip the counters to their opposite side because of the negativity of the second factor. The result is 10 white chips, or positive 10.

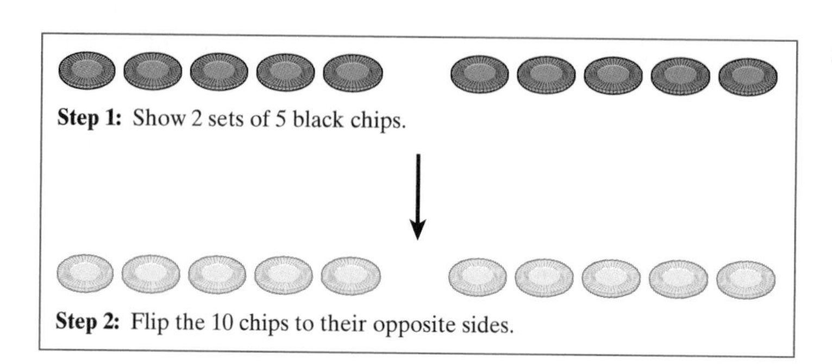

Figure 3.46

This model provides a frame of reference for seeing how a negative times a negative yields a positive. In essence, it shows that multiplying by a second factor that is negative causes you to "flip" or change the sign, regardless of whether you start with a positive or a negative first factor.

Example 3: If your first factor is negative and your second factor is positive, then you would start with counters black-side up, and you would not flip them because the second factor would not change the sign. See Figure 3.47 for a depiction of the product $-3 \times 5 = -15$.

Figure 3.47

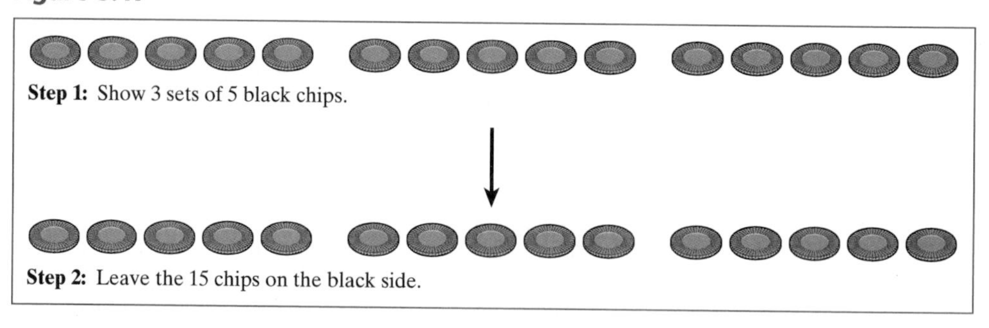

The Number-Line Model of Integer Multiplication. Integer multiplication can be clearly modeled on a number line. The rules involved in using the number line to illustrate integer multiplication are: (1) Always start with your arrow on zero; (2) your arrow points in the positive direction (to the right) if the first factor is positive or in the negative direction (to the left) if the first factor is negative; (3) the arrow moves forward if the second factor is positive and moves backward if the second factor is negative; (4) the absolute value of the first factor tells you how many units to move your arrow each time; and (5) the absolute value of the second factor tells you how many times to move your arrow.

Example 1: To illustrate -3×6 on the number line, we begin with our arrow pointing left because the first factor, -3, is negative. We will move our arrow forward because the second factor, 6, is positive. We will move our arrow 3 units six times. See Figure 3.48.

Example 2: To show the product $4 \times (-3)$ on the number line, begin with the arrow facing right because 4 is positive, and then move the arrow backward because -3 is negative. We will move the arrow 4 units at a time, moving the arrow three times. This shows a product of -12. See Figure 3.49.

Figure 3.48

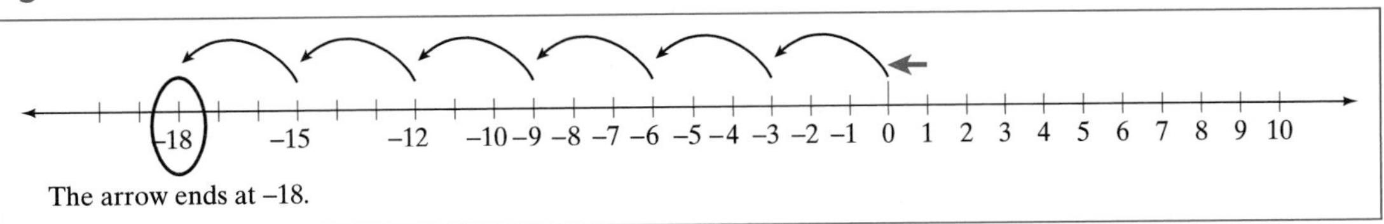

Figure 3.49

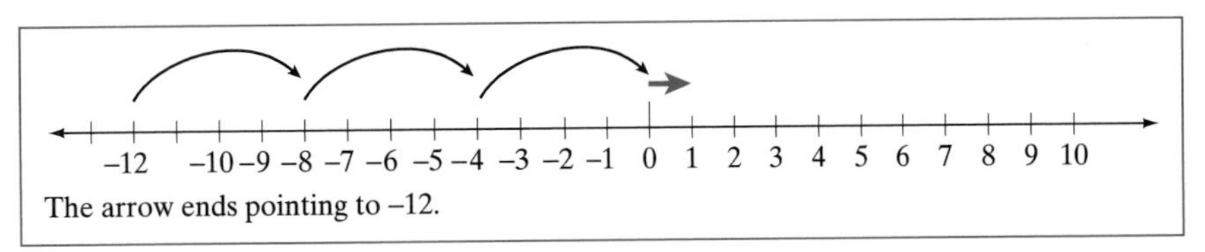

Example 3: The product $-2 \times (-3) = 6$ can also be shown on the number line. Figure 3.50 illustrates that we begin with the arrow at zero and facing left because -2 is negative. We will move the arrow backward because -3 is negative. We will move the arrow 2 units at a time, moving it three times.

Figure 3.50

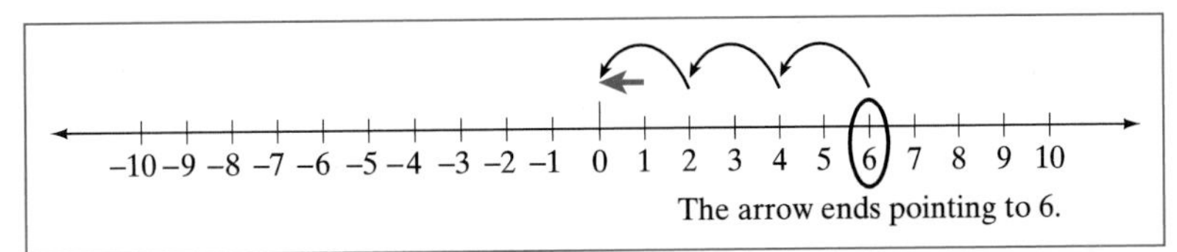

Related Topics **Integer Division**

TOPIC 3.17 Integer Division

cornerstone

Integers
Integer Addition
Integer Subtraction
Integer Multiplication
Whole-Number Division

Introduction

The division of integers builds from basic rubrics of division and the absolute values of integers. In this section, we explore rules for dividing positive and negative dividends by positive and negative divisors. These rules are further explored through the number line and set models.

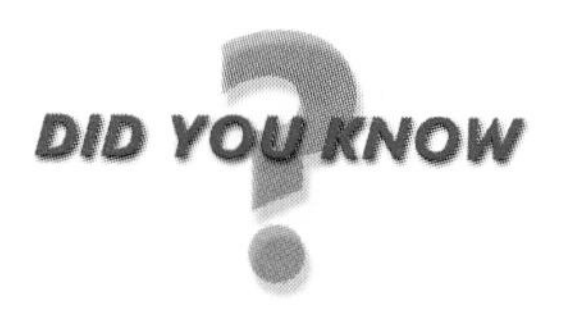

The concept of absolute value was given its name by a mathematician named Karl Weierstrass who did most of his mathematical work in the mid- to late 19th century. The term originated from the context of imaginary numbers and complex numbers when referring to what mathematicians call the "norm."

For More Information Smith, D. E. (1958). *History of Mathematics, Volume II*. New York: Dover Publications.

3.17.1 Rules for Dividing Integers

When we divide an integer by an integer, the answer, or quotient, is sometimes an integer. Other times, the answer comes out with a remainder or a fractional part. To understand basic integer division and the rules for integer division, we will look solely at examples where the quotients are also integers. See Whole-Number Division for a review of the terms quotient, divisor, and dividend.

There are three different rules that we can use when dividing integers. They are closely related to the rules for multiplication of integers. The rules pertain to cases where both the dividend and the divisor are positive, where both the dividend and the divisor are negative, and where the dividend and the divisor have different signs.

Rule for Dividing Integers When Both Dividend and Divisor are Positive. When the dividend and divisor in a division problem are both positive, the division is computed as if the dividend and divisors were whole numbers; that is, for integers, 40 and 5, the quotient of $40 \div 5 = 8$.

Rule for Dividing Integers When Both Dividend and Divisor are Negative. If both the dividend and the divisor are negative, then the quotient will be positive; that is, a negative number divided by a negative number is equal to the quotient of the absolute values of the two numbers. Essentially, the rule says that $-a \div (-b) = |-a| \div |-b| = a \div b$. For example, $-30 \div -5 = 30 \div 5 = 6$.

Rule for Dividing Integers When the Dividend and Divisor Have Opposite Signs. When the dividend and the divisor have opposite signs, the quotient will always be negative. The quotient is the negative of the quotient that is found by dividing the absolute value of the quotient by the absolute value of the divisor. Thus, for any integers, a and b, where exactly one of a or b is negative, $a \div b = -(|a| \div |b|)$. For example, $54 \div (-9) = -(54 \div 9) = -6$. Similarly, $-72 \div 8 = -(72 \div 8) = -9$.

3.17.2 The Set Model for Illustrating Integer Division

Probably the best model for illustrating integer division is the set model. By following rules similar to those in the set model for integer multiplication, we can illustrate how to divide integers using two-color counters.

Rules for Working with the Set Model for Integer Division. The rules for working with the two-colored counters to illustrate division of integers are as follows: (1) The absolute value of the dividend tells you how many chips to put out at the beginning; (2) the absolute value of the divisor tells you into what-sized sets to break the starting amount; (3) the sign of the dividend tells you which side of the counter to have face up at the beginning (white for positive, black for negative); and (4) the sign of the divisor tells you whether to maintain the face-up color determined by the dividend (if a positive sign) or change the face-up color to its opposite (if a negative sign).

Example 1: To illustrate $-12 \div 4$ with the two-color counters, we start with 12 counters out facing black-side up because the dividend is a negative 12. We then break these 12 counters into groups of 4, determining how many groups we can form. Because the divisor is positive, we leave the counters black-side up. Thus, the quotient is shown to be negative 3 because we have 3 sets of black counters. See Figure 3.51 below.

Figure 3.51

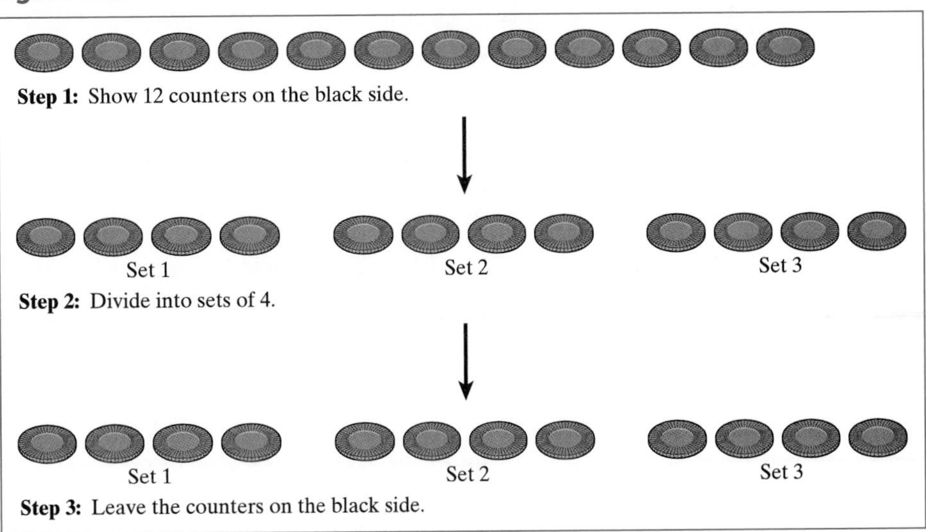

Example 2: Illustrating how to determine the quotient of $24 \div (-6)$, we begin with 24 counters, facing white-side up as seen in Figure 3.52. We divide these counters into sets of 6, finding there are 4 sets. We then flip the counters to their opposite sides because the divisor is negative. Thus, the result is 4 sets of black counters, representing a quotient of negative 4.

Example 3: To find the quotient of $-21 \div (-3)$, we begin with 21 counters facing black-side up as shown in Figure 3.53. We then break the 21 counters into sets of 3. We find that there are 7 such sets. Then, because the divisor is negative, we flip the counters over to their opposite sides. Thus, we have 7 sets of white counters, so the quotient is positive 7.

Figure 3.52

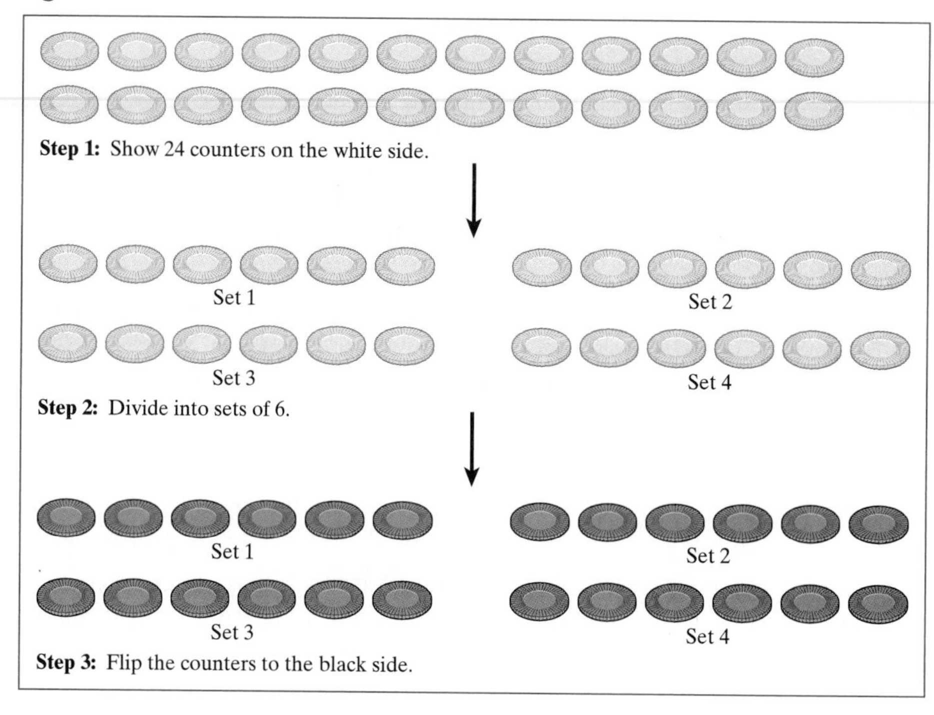

Figure 3.53

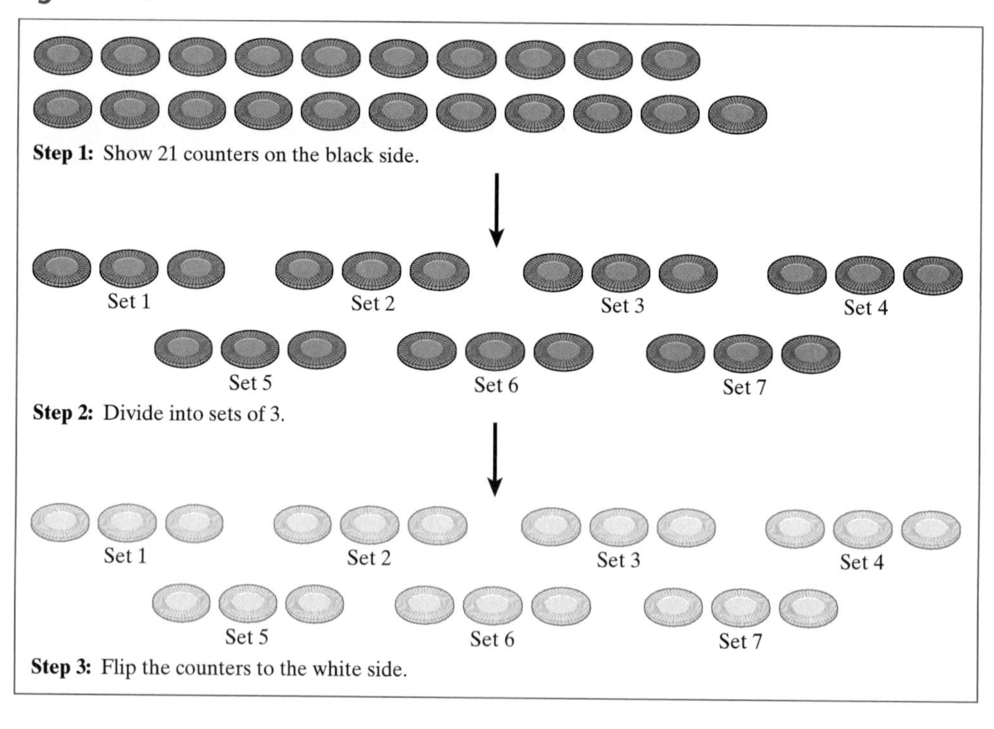

Related Topics

Order of Operations

Order of Operations

cornerstone

Integers
Integer Addition
Integer Subtraction
Integer Multiplication
Integer Division

Introduction

Order of operations is a topic that explains which operations are computed before others. In this section, we examine how agreement to this order of performing operations affords the ability to compute arithmetic problems and to determine a universally accepted answer.

Many people associate the phrase Please Excuse My Dear Aunt Sally" with remembering that the order of operations is Parentheses, Exponents, Multiplication & Division, and Addition & Subtraction.

3.18.1 Definition of the Term Order of Operations

The phrase **order of operations** refers to the sequence we follow when performing a series of operations in a particular computation. For example, if you were to compute the problem, $30 - 20 \div 2 \times 5 + 3$, you would find different answers, depending on the order in which you performed the operations. You might think that you simply work from left to right, performing each operation as you come to it in the problem. So in the example $30 - 20 \div 2 \times 5 + 3$, you might think, $30 - 20$ is 10, then $10 \div 2$ is 5, then 5×5 is 25, and then $25 + 3$ is 28. However, this is not the case; there are specific rules we must follow.

3.18.2 The Rules of Order of Operations

There is a hierarchy of which operations should take place before others. Figure 3.54 shows a schematic design of levels of operations.

As Figure 3.54 indicates, in a series of operations, you perform any operation enclosed in parentheses first. For example, in the problem $2 + (5 \times 4)$, you find the product $5 \times 4 = 20$ first, then add 2 to 20 for an answer of 22. If the computation were $9 - 2^2 + (5 \times 4)$, you would still perform the parenthetical operation first. Next, you would find the value of $2^2 = 4$. Then, once these first two levels of computation had been performed, the problem of solving $9 - 2^2 + (5 \times 4)$ becomes $9 - 4 + 20$. Because addition and subtraction are operations on the same level, you perform the operations as you encounter them from left to right; that is, you find $9 - 4 = 5$ and then add 5 to 20 for a final answer of 25.

Some problems do not have any parentheses, such as the problem $30 - 20 \div 2 \times 5 + 3$ that was mentioned above. When faced with this, you perform higher-level operations first; that is, in this problem, you perform all multiplications and divisions first (in the order in which they appear) before moving on to additions and subtractions. Thus, the solution to $30 - 20 \div 2 \times 5 + 3$ can be thought of in this way: Perform the division $20 \div 2 = 10$ first; then take that quotient of 10 and multiply by 5 for a product of 50. The problem is then reduced to $30 - 50 + 3$, which would finally be completed by performing the $30 - 50 = -20$ first, followed by $-20 + 3 = -17$.

Figure 3.54

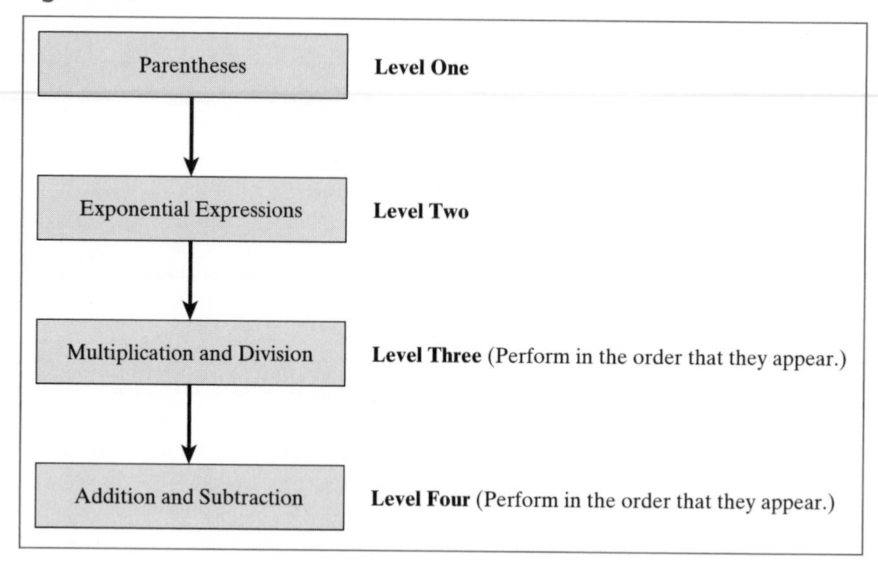

Notice that if when solving $30 - 20 \div 2 \times 5 + 3$ you did not follow the order of operation sequence, you could generate a variety of answers depending on the order you chose to compute each step. For example, if you simply perform the multiplication before the division—that is, you do not go in order from left to right when doing level-three operations—you would find the solution to be $30 - 20 \div 10 + 3 = 30 - 2 + 3 = 28 + 3 = 31$. Other variations from the order-of-operations rules would yield other incorrect answers to the computational problem. Thus, following the universally accepted order of operations is important so that we can all agree on the correct answer in computational situations.

3.18.3 The Calculator and Order of Operations

Most calculators are automatically programmed to follow the correct order of operations; that is, if you punched into your calculator the problem $30 - 20 \div 2 \times 5 + 3$ that we have examined above, the calculator would perform the operations of division and multiplication in that order first and then perform the subsequent subtraction and addition pieces, finding an answer of -17. Try this problem in your calculator.

Related Topics

Exponents

In the Classroom

This chapter continues to focus on the *Number and Operations Content Standard.* In its focus on the four arithmetic operations as they act on familiar sets of numbers, you will continue to construct a better understanding of number as concept. Given that "as they progress from prekindergarten through grade 12, students should attain a rich understanding of ... how numbers are embedded in systems that have structures and properties" (NCTM, 2000, p. 32), we believe you must develop an understanding of (a) properties associated with a set of objects that are acted on by an operation, within which sets of numbers acted on by familiar operations are specific examples (Activities 3.1–3.2); (b) the meanings we associate with familiar arithmetic operations and corresponding models of each (Activities 3.3–3.5); and (c) the relationship between the numerical representation of a number and the procedures used to compute familiar operations (Activities 3.6–3.12).

Some Expectations of Your Future Students:

Pre-K–2 (NCTM, 2000, p. 78)

- understand various meanings of addition and subtraction of whole numbers and the relationship between the two operations
- understand situations that entail multiplication and division, such as equal grouping of objects and sharing equally
- develop and use strategies for whole-number computations, with a focus on addition and subtraction

Grades 3–5 (NCTM, 2000, p. 148)

- explore numbers less than zero by extending the number line and through familiar applications
- understand various meanings of multiplication and division
- understand and use properties of operations, such as the distributivity of multiplication over addition
- develop fluency in adding, subtracting, multiplying, and dividing whole numbers

How Activities 3.1–3.12 Help You Develop an Adult-level Perspective on the Above Expectations:

- By identifying various actions associated with each arithmetic operation, you construct meaningful models of the latter. For example, comparing, taking away, and adding onto are different ways a child might think about a problem that adults solve by subtraction.
- By exploring sets of numbers and the result of familiar arithmetic operations on them, you continue to understand number as concept. For example, the set $\{0, 1, 2, 3, \dots\}$ of whole numbers is not closed under subtraction $2 - 3 = -1$, which is not in the set. The "take-away" model of subtraction may generate a need for the concept of negative numbers.
- By explaining why different algorithms for the same arithmetic operation work, you analyze strategies for computation and deepen your understanding of the numerical representation of the numbers and its composition in terms of place values. For example, you understand why multiplication is commutative and distributes over addition.
- By solving as well as creating nonroutine, multistep, multioperation problems, you stretch your thinking about familiar operations as well as the numbers they operate on.

Bibliography

Abel, J., Allinger, G. D., & Andersen, L. (1987). Popsicle sticks, computers, and calculators: Important considerations. *Arithmetic Teacher 34*, 8–12.

Allinger, G D., & Payne, J. N. (1986). Estimation and mental arithmetic with percent. In H. L. Schoen (Ed.), *Estimation and mental computation* (pp. 141–155). Reston, VA: National Council of Teachers of Mathematics.

Baek, J. (1998). Children's invented algorithms for multidigit multiplication problems. In L. J. Morrow (Ed.), *The teaching and learning of algorithms in school mathematics* (pp. 151–160). Reston, VA: National Council of Teachers of Mathematics.

Baker, A., & Baker, J. (1991). *Math's in the mind: A process approach to mental strategies*. Portsmouth, NH: Heinemann.

Baroody, A. J. (1984). Children's difficulties in subtraction: Some causes and cures. *Arithmetic Teacher, 32*(3), 14–19.

Baroody, A. J. (1985). Mastery of the basic number combinations: Internalization of relationships or facts? *Journal for Research in Mathematics Education, 16*, 83–98.

Baroody, A. J, & Standifer, D. J. (1993). Addition and subtraction in the primary grades. In R. J. Jensen (Ed.), *Research ideas for the classroom: Early childhood mathematics* (pp. 72–102). Old Tappan, NJ: Macmillan.

Battista, Michael T. (1983). A complete model for operations on integers. *Arithmetic Teacher, 30*, 26–31.

Beattie, Ian D. (1986). Modeling operations and algorithms. *Arithmetic Teacher, 33*, 23–28.

Bobis, J. (1991). Using a calculator to develop number sense. *Arithmetic Teacher, 38*, 42–45.

Burns, M. (1989). Teaching for understanding: A focus on multiplication. In P. R. Trafton (Ed.), *New directions for elementary school mathematics* (pp. 123–133). Reston, VA: National Council of Teachers of Mathematics.

Burns, M. (1991). Introducing division through problem-solving experiences. *Arithmetic Teacher, 38* (8), 14–18.

Burns, M. (1992). *About teaching mathematics: A K–8 resource*. Sausalito, CA: Math Solutions Publications.

Burns, M. (1995). *Math by all means: Multiplication, grade 3*. Sausalito, CA: Math Solutions Publications.

Carpenter, T. P., Carey, D. A., & Kouba, V. L. (1990). A problem-solving approach to the operations. In J. N. Payne (Ed.), *Mathematics for the young child* (pp. 111–131). Reston, VA: National Council of Teachers of Mathematics.

Carroll, W. M., & Porter, D. (1997). Invented strategies can develop meaningful mathematical procedures. *Teaching Children Mathematics, 3*, 370–374.

Cemen, P. B. (1993). Teacher to teacher: Adding and subtracting integers on the number line. *Arithmetic Teacher, 40*, 388–389.

Chang, L. (1985). Multiple methods of teaching the addition and subtraction of integers. *Arithmetic Teacher, 33*, 14–19.

Clark, F. B., & Kamii, C. (1996). Identification of multiplicative thinking in children in grades 1–5. *Journal for Research in Mathematics Education, 27*, 41–51.

Corwin, R. B. (1989). Multiplication as original sin. *Journal of Mathematical Behavior, 8*, 223–225.

Crowley, M. L., & Dunn, M. L. (1985). On multiplying negative numbers. *Mathematics Teacher, 78*, 252–256.

Curcio, F. R., Sicklick, F., & Turkel, S. B. (1987). Divide and conquer: Unit strips to the rescue. *Arithmetic Teacher, 35*, 6–12.

Graeber, A. O., & Baker, K. M. (1992). Little into big is the way it always is. *Arithmetic Teacher, 39*, 18–21.

Graeber, A. O., & Tanenhaus, E. (1993). Multiplication and division: From whole-numbers to rational numbers. In D. T. Owens (Ed.), *Research ideas for the classroom: Middle grades mathematics* (pp. 99–117) Old Tappan, NJ: Macmillan.

Gutstein, E., & Romberg, T. A. (1995). Teaching children to add and subtract. *Journal of Mathematical Behavior, 14*, 283–324.

Hall, W. D. (1983). Division with base-ten blocks. *Arithmetic Teacher, 31*, 21–23.

Hankes, J. E. (1996). An alternative to basic-skills remediation. *Teaching Children Mathematics, 2*, 452–457.

Hendrickson, A. D. (1986). Verbal multiplication and division problems: Some difficulties and some solutions. *Arithmetic Teacher, 33*, 26–33.

Hopkins, M. H. (1987). Number facts—or fantasy. *Arithmetic Teacher, 34*, 38–42.

Huinker, D. M. (1989). Multiplication and division word problems: Improving students' understanding. *Arithmetic Teacher, 37*(2), 8–12.

Kamii, C. (1994). *Young children continue to reinvent arithmetic: 3rd grade*. New York: Teachers College Press.

Kamii, C., & Joseph, L. (1988). Teaching place value and double column addition. *Arithmetic Teacher, 35*(6), 48–52.

Kamii, C., & Lewis, B. A. (1993). The harmful effects of algorithms in primary arithmetic. *Teaching K–8, 23*(5), 36–38.

Kamii, C., Lewis, B. A., & Booker, B. M. (1998). Instead of teaching missing addends. *Teaching Children Mathematics, 4*, 458–461.

Kamii, C., Lewis, B. A., & Livingston, S. (1993). Primary arithmetic: Children inventing their own procedures. *Arithmetic Teacher, 41*, 200–203.

Kouba, V. L., & Franklin, K. (1993). Multiplication and division: Sense making and meaning. In R. J. Jensen (Ed.), *Research ideas for the classroom: Early childhood mathematics* (pp. 103–126). Old Tappan, NJ: Macmillan.

Leutzinger, L. P. (1999). Developing thinking strategies for addition facts. *Teaching Children Mathematics, 6*, 14–18.

Leutzinger, L. P. (1999). *Facts that last: A balanced approach to memorization* [one volume for each operation]. Chicago, IL: Creative Publications.

Lindquist, M. M. (1987). Estimation and mental computation: Measurement. *Arithmetic Teacher, 34*, 16–18.

Lobato, J. E. (1993). Making connections with estimation. *Arithmetic Teacher, 40*, 347–351.

Mahlios, J. (1988). Word problems: Do I add or subtract? *Arithmetic Teacher, 36*, 48–52.

Meyer, R. A., & Riley, J. E. (1986). Multiplication Games. *Arithmetic Teacher, 33*, 22–25.

Ohanian, S., & Burns, M. (1995). *Math by all means: Division, grades 3–4*. Sausalito, CA: Math Solutions Publications.

Pearson, E. S. (1986). Summing it all up: Pre-1900 algorithms. *Arithmetic Teacher, 33*, 38–41.

Quintero, A. H. (1986). Children's conceptual understanding of situations involving multiplication. *Arithmetic Teacher, 33* (3), 34–37.

Rathmell, E. C. (1978). Teaching thinking strategies to teach the basic facts. In M. N. Suydam (Ed.), *Developing computational skills* (pp. 13–38). Reston. VA: National Council of Teachers of Mathematics.

Rathmell, E. C., & Trafton, P. R. (1990). Whole number computation. In J. N. Payne (Ed.), *Mathematics for the young child* (pp. 153–172). Reston, VA: National Council of Teachers of Mathematics.

Remington, J. (1989). Introducing multiplication. *Arithmetic Teacher, 37*, 12–14, 60.

Reys, B. J. (1991). *Developing number sense in the middle grades: Addenda series, grades 5–8*. Reston, VA: National Council of Teachers of Mathematics.

Reys, R. E., & Nohda, N. (Eds.). (1994). *Computational alternatives for the twenty-first century: Cross-cultural perspectives from Japan and the United States*. Reston. VA: National Council of Teachers of Mathematics.

Reys, B. J., & Reys, R. E. (1990). Implementing the *Standards*: Estimation—Direction from the *Standards. Arithmetic Teacher, 37*, 22–25.

Reys, B. J., & Reys, R. E. (1999). Computation in the elementary curriculum: Shifting the emphasis. *Teaching Children Mathematics, 5*, 236–241.

Rightsel, P. S., & Thornton, C. A. (1985). 72 addition facts can be mastered by mid-grade 1. *Arithmetic Teacher, 33*, 8–10.

Schwartz, S. L., & Curcio, F. R. (1995). Learning mathematics in meaningful contexts: An action-based approach in the primary grades. In P. A. House (Ed.), *Connecting mathematics across the curriculum* (pp. 116–123). Reston, VA: National Council of Teachers of Mathematics.

Sowder, J. T. (1990). Mental computation and number sense. *Arithmetic Teacher, 37*, 18–20.

Stanic, G. M. A., & McKillup, W. D. (1989). Developmental algorithms have a place in elementary school mathematics instruction. *Arithmetic Teacher, 36*(5), 14–16.

Sundar, V. K. (1990). Thou shalt not divide by zero. *Arithmetic Teacher, 37*, 50–51.

Thompson, C. S., & Hendrickson, A. D. (1986). Verbal addition and subtraction problems: Some difficulties and some solutions. *Arithmetic Teacher, 33*(7), 21–25.

Thompson, C. S., & Van de Walle, J. A. (1980). Transition boards: Moving from materials to symbols in addition. *Arithmetic Teacher, 28*(4), 4–8.

Thompson, C. S., & Van de Walle, J. A. (1981). Transition boards: Moving from materials to symbols in subtraction. *Arithmetic Teacher, 28*(5), 4–9.

Thompson, F. (1991). Two-digit addition and subtraction: What works? *Arithmetic Teacher, 38*, 10–13.

Thornton, C. A. (1990). Strategies for the basic facts. In J. N. Payne (Ed.), *Mathematics for the young child* (pp. 133–151). Reston, VA: National Council of Teachers of Mathematics.

Thornton. C. A., & Smith, P. (1988). Action research: Strategies for learning subtraction facts. *Arithmetic Teacher, 35*(8), 9–12.

Thornton, C. A., & Toohey, M. A. (1984). *A matter of facts: Addition, subtraction, multiplication, division*. Palo Alto, CA: Creative Publications.

Trafton, P. R., & Zawojewski, J. S. (1990). Implementing the *Standards*: Meaning of operations. *Arithmetic Teacher, 38*, 18–22.

Tucker, B. F. (1989). Seeing addition: A diagnosis-remediation case study. *Arithmetic Teacher 36*(5), 10–11.

Van de Walle, J. A. (1991). Redefining computation. *Arithmetic Teacher, 38*(5), 46–51.

Van de Walle, J. A., & Thompson, C. S. (1985). Partitioning sets for number concepts, place value, and long division. *Arithmetic Teacher, 32*(5), 6–11.

Van Lehn, L. (1986). Arithmetic procedures are induced from examples. In J. Hiebert (Ed.), *Conceptual and procedural knowledge: The case of mathematics* (pp. 133–179). Hillsdale, NJ: Erlbaum.

Weiland, L. (1985). Matching instruction to children's thinking about division. *Arithmetic Teacher, 33*(4), 34–45.

Whitin, D. J. (1994). Exploring estimation through children's literature. *Arithmetic Teacher, 41*, 436–441.

Whitman, N. C. (1992). Activities: Multiplying integers. *Mathematics Teacher, 85*, 34–38, 45–51.

CHAPTER 4 FOUR

Number Theory

TOPIC 4.1 Prime and Composite Numbers
4.2 Prime Factorization and Tree Diagrams
4.3 Fundamental Theorem of Arithmetic
4.4 Divisibility
4.5 Greatest Common Divisor
4.6 Least Common Multiple
4.7 Clock Arithmetic & Modular Arithmetic
4.8 Patterns
In the Classroom
Bibliography

CHAPTER OVERVIEW

Number theory is a branch of mathematics that involves the study of numbers and, in particular, the natural numbers. In this chapter, you can learn interesting facts related to number theory and will examine some concepts and procedures involving number theory ideas.

BIG MATHEMATICAL IDEAS

Conjecturing, decomposing, verifying, problem-solving strategies, multiple representations

NCTM PRINCIPLES & STANDARDS LINKS

Number and Operation; Problem Solving; Reasoning; Communication; Connections; Representation

TOPIC 4.1 *Prime and Composite Numbers*

cornerstone

Integer Division
Divisibility

Introduction

There are many interesting theorems associated with prime numbers and composite numbers. Learning to distinguish between prime and composite integers requires determining divisors associated with an integer. In this section, we explain the definitions of prime and composite and examine a method for finding prime numbers.

There are a number of interesting conjectures involving prime numbers. Once such conjecture is **Goldbach's conjecture,** which states that any even number greater than 4 can be expressed as the sum of two odd prime numbers. For example, $6 = 3 + 3$, $8 = 3 + 5$, and $10 = 5 + 5$. See if you can find how to write the even numbers from 12 to 20 as the sum of two odd prime numbers! This conjecture has never been proven, so it cannot be called a theorem.

For More Information Smith, D. E. (1958). *History of Mathematics, Volume II.* New York: Dover Publications.

4.1.1 Definitions of Prime and Composite

Every positive integer greater than 1 is designated as either prime or composite. A **prime number** is a positive integer that has exactly two unique factors—the number 1 and the number itself. For example, the number 7 is prime because its only factors are 1 and 7. A **composite number** is a positive integer that has more than 2 unique factors. For example, the number 12 is composite because it has as factors the numbers: 1, 2, 3, 4, 6, and 12. Note that the number 1 is considered neither prime nor composite. It has exactly 1 unique factor—itself.

4.1.2 Sieve of Eratosthenes

An interesting method of generating the set of prime numbers up through the first 100 integers is by using the **Sieve of Eratosthenes.** The idea behind the Sieve of Eratosthenes is that multiples of a given prime number will necessarily be composite because that given prime number will be a factor of that resulting multiple. Therefore, the multiple will have more than the two factors of 1 and itself. Thus, if we systematically start with the first prime number, 2, cross out all multiples of 2, then do the same for the next prime number, 3, cross out all multiples of three, and continue in this fashion, we will ultimately cross out all composite numbers and leave only the prime numbers. In this sense, we "sift out" the prime numbers. Figure 4.1 shows all the multiples of 2 crossed out.

	2	3	~~4~~	5	~~6~~	7	~~8~~	9	~~10~~
11	~~12~~	13	~~14~~	15	~~16~~	17	~~18~~	19	~~20~~
21	~~22~~	23	~~24~~	25	~~26~~	27	~~28~~	29	~~30~~
31	~~32~~	33	~~34~~	35	~~36~~	37	~~38~~	39	~~40~~
41	~~42~~	43	~~44~~	45	~~46~~	47	~~48~~	49	~~50~~
51	~~52~~	53	~~54~~	55	~~56~~	57	~~58~~	59	~~60~~
61	~~62~~	63	~~64~~	65	~~66~~	67	~~68~~	69	~~70~~
71	~~72~~	73	~~74~~	75	~~76~~	77	~~78~~	79	~~80~~
81	~~82~~	83	~~84~~	85	~~86~~	87	~~88~~	89	~~90~~
91	~~92~~	93	~~94~~	95	~~96~~	97	~~98~~	99	~~100~~

Figure 4.1

Continuing the Sieve of Eratosthenes, Figure 4.2 shows all multiples of 2 and 3 crossed out.

Finally, Figure 4.3 shows the Sieve of Eratosthenes completed up through the number 100, revealing that the prime numbers less than 100 are 2, 3, 5, 7, 11, 13, 17, 19, 23, 29, 31, 37, 41, 43, 47, 53, 59, 61, 67, 71, 73, 79, 83, 89, and 97.

	2	3	~~4~~	5	~~6~~	7	~~8~~	~~9~~	~~10~~
11	~~12~~	13	~~14~~	~~15~~	~~16~~	17	~~18~~	19	~~20~~
~~21~~	~~22~~	23	~~24~~	25	~~26~~	~~27~~	~~28~~	29	~~30~~
31	~~32~~	~~33~~	~~34~~	35	~~36~~	37	~~38~~	~~39~~	~~40~~
41	~~42~~	43	~~44~~	~~45~~	~~46~~	47	~~48~~	49	~~50~~
~~51~~	~~52~~	53	~~54~~	55	~~56~~	~~57~~	~~58~~	59	~~60~~
61	~~62~~	~~63~~	~~64~~	65	~~66~~	67	~~68~~	~~69~~	~~70~~
71	~~72~~	73	~~74~~	~~75~~	~~76~~	77	~~78~~	79	~~80~~
~~81~~	~~82~~	83	~~84~~	85	~~86~~	~~87~~	~~88~~	89	~~90~~
91	~~92~~	~~93~~	~~94~~	95	~~96~~	97	~~98~~	~~99~~	~~100~~

Figure 4.2

	2	3	~~4~~	5	~~6~~	7	~~8~~	~~9~~	~~10~~
11	~~12~~	13	~~14~~	~~15~~	~~16~~	17	~~18~~	19	~~20~~
~~21~~	~~22~~	23	~~24~~	~~25~~	~~26~~	~~27~~	~~28~~	29	~~30~~
31	~~32~~	~~33~~	~~34~~	~~35~~	~~36~~	37	~~38~~	~~39~~	~~40~~
41	~~42~~	43	~~44~~	~~45~~	~~46~~	47	~~48~~	~~49~~	~~50~~
~~51~~	~~52~~	53	~~54~~	~~55~~	~~56~~	~~57~~	~~58~~	59	~~60~~
61	~~62~~	~~63~~	~~64~~	~~65~~	~~66~~	67	~~68~~	~~69~~	~~70~~
71	~~72~~	73	~~74~~	~~75~~	~~76~~	~~77~~	~~78~~	79	~~80~~
~~81~~	~~82~~	83	~~84~~	~~85~~	~~86~~	~~87~~	~~88~~	89	~~90~~
~~91~~	~~92~~	~~93~~	~~94~~	~~95~~	~~96~~	97	~~98~~	~~99~~	~~100~~

Figure 4.3

Related Topics

Prime Factorization and Tree Diagrams

TOPIC 4.2 *Prime Factorization and Tree Diagrams*

cornerstone

Divisibility
Prime and Composite Numbers

Introduction

Breaking composite integers into their prime factorizations is akin to finding the fundamental and unique composition of an integer. In this section, we learn the definition of prime factorization and examine a method of finding an integer's prime factorization using a tree diagram.

We can determine the number of divisors that an integer has by examining the prime factorization of the number. One interesting fact about the number of divisors is this: Only square numbers have an odd number of divisors. To uncover why this is the case, examine the prime factorization of these square numbers: 4, 9, 16, 25, 36, 49, 64, 81, and 100. Try to uncover a pattern and explain why square numbers are the only numbers with an odd number of divisors.

4.2.1 Definition of Prime Factorization

There are many ways to factor integers. *Factoring* means that we express a given integer as a product of two or more numbers. For example, 24 can be factored many ways, such as $24 = 2 \times 12, 24 = 3 \times 8, 24 = 2 \times 2 \times 6$. An integer can also be expressed solely as a product of prime numbers. When we express a number as a product of prime numbers, we call this expression the **prime factorization** of the number. For example, the prime factorization of 24 would be $24 = 2 \times 2 \times 2 \times 3$. As another example, the prime factorization of 30 would be $2 \times 3 \times 5$. Note that because prime numbers only have 1 and itself as factors, it makes little sense to talk about the prime factorization of prime numbers; that is, to express 7 as a product of prime numbers is impossible because the only way to factor 7 is $7 = 1 \times 7$ and because the number 1 is not prime.

4.2.2 Tree Diagrams

A useful means of determining the prime factorization of a number is to make a tree diagram. A tree diagram essentially starts with a number and then branches out to factor that number as the product of two other numbers. If either of these other numbers are not prime, then the tree can branch further to express those numbers as the product of two numbers. The tree continues branching in this manner until the end of all the branches are prime numbers. See how Figure 4.4 illustrate a tree diagram for finding the prime factorization of 36.

Figure 4.4

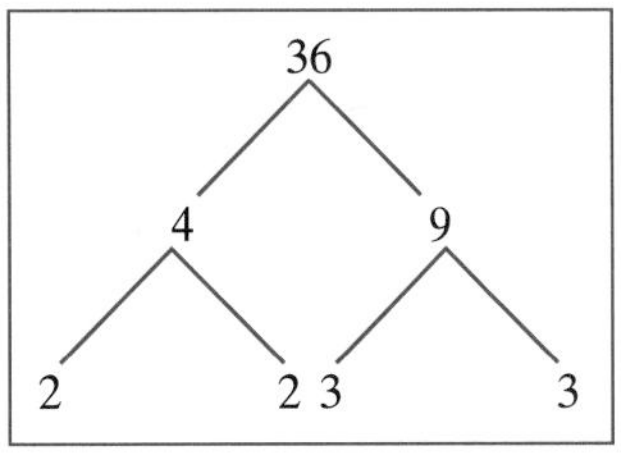

36
4 9
2 2 3 3

Figure 4.5

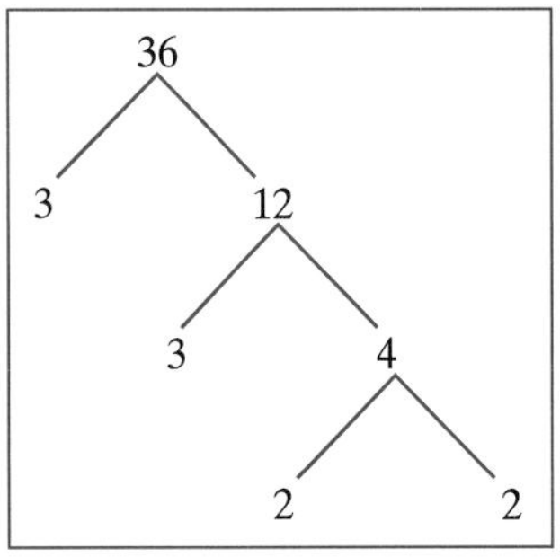

36
3 12
3 4
2 2

From the tree diagram, we take the prime numbers that appear at the ends of the branches and multiply them together to form the prime factorization. Thus, the prime factorization of 36 is $2 \times 2 \times 3 \times 3$.

If you begin your tree diagram with a different first factoring, your prime factorization will always end the same. For example, Figure 4.5 shows a different tree diagram for factoring the number 36, but notice that the prime numbers that result at the end of each final branch are still the same, confirming that the prime factorization of 36 is $2 \times 2 \times 3 \times 3$.

4.2.3 Notation of Prime Factorization

Note that order of factors does not alter the prime factorization itself; that is, expressing the prime factorization of 36 as $3 \times 2 \times 3 \times 2$ is equivalent to the prime factorization expressed by $2 \times 2 \times 3 \times 3$. However, it is good form to express prime factorization in ascending order of prime factors. Thus, the form $2 \times 2 \times 3 \times 3$ is preferable for the prime factorization of 36. Prime factorizations are also frequently expressed using exponential notation; that is, because 2×2 can be written as 2^2 and because 3×3 can be expressed as 3^2, the prime factorization of 36 can be expressed as $2^2 \times 3^2$. This is a very useful and popularly used notation for prime factorization.

Related Topics

Fundamental Theorem of Arithmetic

TOPIC 4.3 Fundamental Theorem of Arithmetic

cornerstone

Prime and Composite Numbers
Prime Factorization and Tree Diagrams

Introduction

In mathematics, two very basic premises that underlie key theorems are existence and uniqueness. The Fundamental Theorem of Arithmetic makes a claim about the existence and uniqueness of the prime factorization of integers. This very brief section describes this very important theorem in number theory.

DID YOU KNOW

The fact that there are an infinite number of prime numbers is an intriguing notion. A proof of this claim can be simply stated by making use of the **Fundamental Theorem of Arithmetic.** The proof is what is called an indirect proof, where we assume that there is a finite number of prime numbers and show that this cannot be true. Thus, the logical conclusion will be that there must be an infinite number of primes.

Proof: Assume that there is a largest prime and call it p_n. Then, using all primes less than p_n, $p_1, p_2, p_3, \ldots p_{n-1}$, we can look at the composite integer, n, that is one more than the product of all the primes: $n = (p_1 \times p_2 \times p_3 \times \cdots \times p_{n-1} \times p_n) + 1$.

We know that $p_1 | (p_1 \times p_2 \times p_3 \times \cdots \times p_{n-1} \times p_n)$. Therefore, p_1 cannot divide $(p_1 \times p_2 \times p_3 \times \cdots \times p_{n-1} \times p_n) + 1$. Similar arguments can be made for why none of the other primes $p_2, p_3, \ldots, p_{n-1}, p_n$ can be divisors of n. Therefore, none of the prime numbers are divisors of n. However, the Fundamental Theorem of Arithmetic assures us that every composite integer has a unique prime factorization. If none of the primes $p_1, p_2, p_3, \ldots p_{n-1}, p_n$ are factors of n, then this conflicts with the Fundamental Theorem of Arithmetic. Therefore, there must be an infinite number of prime numbers.

For More Information Bunt, L. N. H., Jones, P. S., & Bedient, J. D. (1988). *The Historical Roots of Elementary Mathematics.* New York: Dover Publications.

The Fundamental Theorem of Arithmetic states that every composite integer has a unique prime factorization, without considering order of factors or notational style as changing the prime factorization itself. Thus, for example, the prime factorization of 20 is $2 \times 2 \times 5$. Although you can rearrange the order of factors, such as $2 \times 5 \times 2$, or write the factorization in exponential form, such as $2^2 \times 5$, the primes involved and the number of each prime involved remain the same. Thus, the prime factorization is unique in this sense.

This theorem is important because the prime factorization of a number can help generate a complete list of factors for a number, determine the number of factors for a number, and help determine the least common multiple of two numbers and the greatest common divisor of two numbers. Take, for example, the integer 90. Its unique prime factorization is $2 \times 3^2 \times 5$. We can use the prime factorization to help uncover all the factors of 90. The factors of 90 would include 1, each of the primes in the prime factorization, and all product combinations of 2 or more of the prime factors in the prime factorization. So, for 90 the factors would be: 1, 2, 3, 5, 2×3, 3^2, 2×5, 3×5, 2×3^2, $2 \times 3 \times 5$, $3^2 \times 5$, and $2 \times 3^2 \times 5$; that is, the factors of 90 are 1, 2, 3, 5, 6, 9, 10, 15, 18, 30, 45, and 90. Because the prime factorization of 90 is unique, we can be certain that we are generating all of the factors of 90 because no other combinations of the prime factors is possible.

Related Topics

Greatest Common Divisor
Least Common Multiple

TOPIC 4.4 Divisibility

cornerstone **Integer Division**

Introduction

The concept of divisibility centers around determining the integers that divide into other integers evenly with no remainder. In this section, we are introduced to the formal definition of divisibility, to concepts related to divisibility, and to several theorems of divisibility. We also explore a variety of divisibility tests that allow us to determine quickly and accurately which integers are divisors of other integers.

Euclid of Alexandria (330–275 B.C.) did most of his mathematical work in the area of number theory. One of his most interesting findings was a means of classifying numbers by the number of proper divisors they had. He said that an integer is considered **perfect** if it is equal to the sum of its proper divisors. Note that a **proper divisor** of a given integer is any divisor except for the integer itself. Thus, the number 6 is perfect because the sum of its proper divisors $1 + 2 + 3$ is 6. Integers that are not perfect are classified as either deficient or abundant. An integer is considered **deficient** if the sum of its proper divisors is less than the integer, and an integer is **abundant** if the sum of its proper divisors is greater than the integer itself, so 8 $(8 < 1 + 2 + 4)$ is deficient, but 12 $(12 > 1 + 2 + 3 + 4 + 6)$ is abundant.

For More Information Reimer, L., & Reimer W. (1995). *Mathematicians Are People Too: Stories from the Lives of Great Mathematicians, Volume 2*. Palo Alto, CA: Dale Seymour Publications.

4.4.1 Definition of Divisibility

When we say that an integer a is **divisible** by an integer b, we mean that there is some other integer, c, such that $a \div b = c$ with no remainder. In other words, a is divisible by b if there is some other integer, c, such that $c \times b = a$. For example, we can say that 24 is divisible by 8 because $24 \div 8 = 3$ or $3 \times 8 = 24$. On the other hand, we say 10 is not divisible by 3 because there is no integer, c, such that $c \times 3 = 10$.

4.4.2 Divisibility Notation

When we talk about divisibility, instead of saying that a is divisible by b, we often use the expression, b divides a. The notation for divisibility mirrors this expression; that is, we use the notation $b|a$, which is read "b divides a" to suggest that a is divisible by b. Thus, it would be appropriate to write $3|27$, which reads "3 divides 27," meaning that 27 is divisible by 3. If a number is not divisible by another number, we use the symbol $\nmid$ to read "does not divide." For example, we can say that $3 \nmid 14$.

4.4.3 Divisibility Theorems

There are several theorems that stem from the definition of divisibility. Given integers a, b, c, and k:

Theorem 1: If $a|b$, then $a|k \times b$. Essentially, this theorem says that if an integer divides another integer, then it will divide any multiple of that integer, too. For example, because $3|6$, then $3|2 \times 6$, $3|4 \times 6$, $3|10 \times 6$, and so on.

Theorem 2: If $a|b$ and $a|c$, then $a|(b + c)$. This theorem states that if an integer divides two separate integers, then it will also divide the sum of those two integers. For example, because $4|12$ and $4|16$, then $4|(12 + 16)$.

Theorem 3: If $a|b$ and $a|c$, then $a|(b - c)$. Like Theorem 2, this theorem states that if an integer divides two separate integers, then it will also divide the difference between those two integers. For example, because $5|35$ and $5|20$, then $5|(35 - 20)$.

4.4.4 False, But Often Believed Divisibility Theorems

There are also false theorems that people often mistakenly believe are true; that is, people tend to believe that the reverse argument of the theorems above are also true even though this is not always the case. Given integers *a, b, c,* and *k*:

False Theorem 1: If $a|k \times b$, then $a|k$ and $a|b$. This false theorem is the inverse of Theorem 1 above. Although an integer divides the product of two integers, it may not divide each integer individually. For example, $6|(4 \times 9)$, but $6 \nmid 4$ and $6 \nmid 9$.

False Theorem 2: If $a|(b + c)$, then $a|b$ and $a|c$. Many people assume this is true because it is the inverse of Theorem 2. However, consider this example. We know that $3|(4 + 8)$, but $3 \nmid 4$ and $3 \nmid 8$.

False Theorem 3: If $a|(b - c)$, then $a|b$ and $a|c$. This is the inverse of Theorem 3 and is not necessarily true. For example, $8|(70 - 6)$, but $8 \nmid 70$ and $8 \nmid 6$.

4.4.5 Related Terms

We can use the notion of divisibility to describe sets of numbers. For example, we can describe the set of **even numbers** this way. An integer, *a*, is even if *a* is divisible by 2 (that is, $2|a$). Similarly, we can describe **odd numbers** by saying that an integer, *a*, is odd if it is not divisible by 2 (that is, $2 \nmid a$). For example, 8 is even because 8 is divisible by 2. On the other hand, 7 is odd because 7 is not divisible by 2.

The concept of divisibility also suggests a number of related mathematical vocabulary. For example, if $b|a$, then we say that *b* is a divisor of *a*. This makes sense because one way of defining divisibility is to say that $b|a$ if there is some integer, *c*, such that $a \div b = c$. We clearly see *b* acting as a divisor, particularly where *b* divides evenly with no remainder. If $b|a$, we can also refer to *b* as a **factor** of *a*. When we consider the alternative way of describing divisibility, that is, *a* is divisible by *b* if there is some other integer, *c*, such that $c \times b = a$, *b* acts as a factor. Thus, the words *divisor* and *factor* can be used interchangeably when talking about divisibility.

The concept of divisibility is also related to the concept of **multiple**. We say *a* is a multiple of *b* if there is some whole number, *c*, such that $c \times b = a$. For example, 30 is a multiple of 6 because we can identify the whole number 5 such that $5 \times 6 = 30$. So, when we know that $b|a$, that is, *b* divides *a*, we can also claim that *a* is a multiple of *b*.

4.4.6 Divisibility Tests

There are some divisibility tests that are designed to be an efficient means of determining whether or not a given integer is divisible by a particular number without having to perform long division to check. What follows are the descriptions of divisibility tests for divisibility by 2, 3, 4, 5, 6, 7, 8, 9, 10, 11, and 12.

Divisibility Test for 2. The test for divisibility by 2 states that a number is divisible by 2 if the digit in the ones place is 0, 2, 4, 6, or 8. For example, the number 6,598 is divisible by 2 because the ones place digit is an 8. Therefore, it passes the test for divisibility by 2. On the other hand, the number 309 is not divisible by 2 because the ones place digit is a 9. Thus, it fails the divisibility test for 2.

Divisibility Test for 5. The test for divisibility by 5 states that a number is divisible by 5 if the digit in the ones place is 0 or 5. For example, at a glance we can tell that the number 324,880 is divisible by 5 because the ones place digit is a 0. Similarly, we say that the number 7,899 fails the divisibility test for 5 because the ones place is neither 5 or a 0. Thus, 7889 is not divisible by 5.

Divisibility Test for 10. The test for divisibility by 10 states that a number is divisible by 10 if the digit in the ones place is a zero. For example, we can easily tell that 956 is not divisible by 10 but that 74,330 is divisible by 10 by checking to see whether or not the digit in the ones place is a zero.

There is another way we can think about devising a test for divisibility by 10. Because $10 = 2 \times 5$, both 2 and 5 are divisors of 10. We could say that if a number passes both the tests for divisibility by 2 and divisibility by 5, then it should be divisible by 10. Recall that to be divisible by 2, the ones place must be 0, 2, 4, 6, or 8. To be divisible by 5, the ones place must or 0 or 5. To pass both tests simultaneously (see Figure 4.6), we can see that the overlap between the two tests would be having a zero in the ones place, which is the originally stated divisibility test for 10. This alternative way of thinking about the divisibility test for 10 as being a combination of passing the tests for 2 and 5 will be crucial in developing divisibility tests for other numbers such as for 6 and 12 (see these explained further in this section).

Figure 4.6

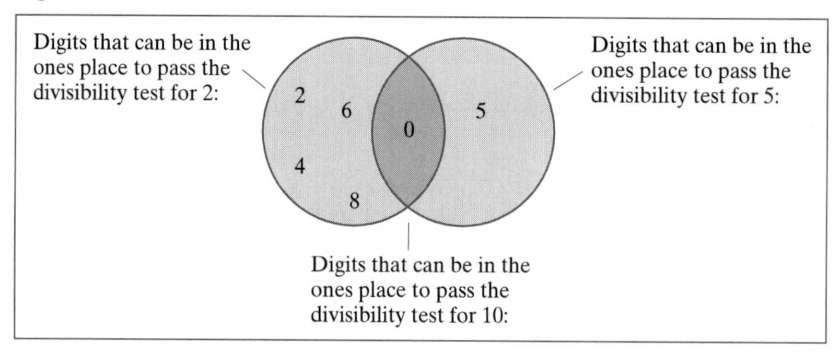

Divisibility Test for 3. The test for divisibility by 3 states that a number is divisible by 3 if the sum of the digits in the number is divisible by 3. For example, the number 786 is divisible by 3 because $7 + 8 + 6 = 21$, and 21 is a number that is divisible by 3. Using the test, we can confirm that 1255 is not divisible by 3 because $1 + 2 + 5 + 5 = 13$, and 13 is not divisible by 3.

Divisibility Test for 9. The test for divisibility by 9, which is very similar to the test for divisibility by 3, states that a number is divisible by 9 if the sum of the digits in the number is divisible by 9. For example, the number 22,356 is divisible by 9 because $2 + 2 + 3 + 5 + 6 = 18$, and 18 is a number that is divisible by 9. On the other hand, 8991 is not divisible by 9 because $8 + 9 + 9 + 2 = 28$, and 28 is not divisible by 9.

Divisibility Test for 4. The test for divisibility by 4 states that a number is divisible by 4 if the number formed by the two right-most digits is divisible by 4; then the entire number is divisible by 4. For example, in the number 78,924, the number formed by the two right-most digits, 24, is divisible by 4. Therefore, the number 78,924 is also divisible by 4. However, the number 44,433 is not divisible by 4 because the number 33 is not divisible by 4.

Divisibility Test for 8. The test for divisibility by 8 is similar to the test for divisibility by 4. The test for divisibility by 8 states that a number is divisible by 8 if the number formed by the three right-most digits is divisible by 8. For example, because the number 568 is divisible by 8, the number 123,568 is divisible by 8. Further, the number 6,790 is not divisible by 8 because 790 is not divisible by 8.

Divisibility Test for 6. The test for divisibility by 6 states that a number is divisible by 6 if it passes the divisibility tests for 3 and 2. In other words, a number is divisible by 6 if the sum of the digits is divisible by 3 AND if the digit in the ones place is a 0, 2, 4, 6, or 8. For example, the number 777 is not divisible by 6 because even though the sum of the digits, $7 + 7 + 7 = 21$, is divisible by 3 (that is, it passes the test for divisibility by 3), the digit in the ones place is a 7. Therefore, it is not divisible by 2. Consequently, 777 is not divisible by 6 because it did not pass both tests. On the other hand, the number 516 is divisible by 6 because the sum of the digits $5 + 1 + 6 = 12$ is divisible by 3 and the digit in the ones place is a 6.

Divisibility Test for 12. The divisibility test for 12 states that a number is divisible by 12 if it passes the divisibility tests for 3 and 4; that is, a number is divisible by 12 if the sum of the digits is divisible by 3 and the number formed by the two right-most digits is divisible by 4. To verify, we find that for the number 912, the sum of the digits, $9 + 1 + 2 = 12$, is divisible by 3 and that the number formed by the two right-most digits, 12, is divisible by 4. Therefore, because 912 passes both the test for divisibility by 3 and the test for divisibility by 4, then 912 is divisible by 12.

You might be tempted to say that another test for divisibility by 12 might be passing the divisibility tests for 2 and for 6. This does not work, however, because the numbers 2 and 6 are not relatively prime. For two integers, a and b, we say that a and b are **relatively prime** if the only factor they have in common is 1. When we develop new tests for divisibility, we can create new tests by combining other tests as long as (a) the product of the numbers associated with each divisibility test equals the number for which we are designing a new test, and (b) the numbers associated with each of the combined divisibility tests are relatively prime. Thus, because $4 \times 3 = 12$ and 4 and 3 are relatively prime, a possible divisibility test for divisibility by 12 would be passing the tests for divisibility by 3 and by 4. On the other hand, even though $2 \times 6 = 12$, because 2 and 6 are not relatively prime—that is, they not only share 1 as a common factor, but also 2 as a common factor—passing the tests for 2 and 6 would not be sufficient to say that a number passes the divisibility test for 12. Note that the number 6 is divisible by 2 and is also divisible by 6, but it is not divisible by 12. Therefore, this counterexample shows that the only viable test for divisibility by 12 is the one involving the tests for divisibility by 3 and 4.

Divisibility Test for 7. The test for divisibility by 7 states that a number is divisible by 7 if the number found by taking the difference between the number formed by all the digits except the digit in the ones place, and twice the digit in the ones place is divisible by 7. For example, given the number 791, we first find the number formed by all the digits except the digit in the ones place: 79. Then we subtract twice the digit in the ones place from this number: $79 - (2 \times 1) = 79 - 2 = 77$. Because 77 is divisible by 7, the original number, 791, is divisible by 7. Looking to another example, the number 1,235 is not divisible by 7 because when we take 123 and subtract (2×5), we get $123 - 10 = 113$. Because 113 is not divisible by 7, 1,235 is not divisible by 7.

Divisibility Test for 11. The test for divisibility by 11 states that a number is divisible by 11 if the difference found by subtracting the sum of all the digits in the even-powered places from the sum of the digits in all the odd-powered places is divisible by 11. Recall that because the Hindu-Arabic Number System is a base-ten system, then each place can be expressed as a power of ten (the hundreds place is 10^2, the thousands place is 10^3, etc.). For all the places where the exponent (power) is odd, as in the thousands place, we add the digits to find the sum of those digits. We do the same for digits in even-powered places. So, for example, given the number 128,457, we add the digits in the odd-powered places: $5 + 8 + 1 = 14$, and then add the digits in the even-numbered places: $7 + 4 + 2 = 13$. We then find the difference between those sums: $14 - 13 = 1$. Because the number 1 is not divisible by 11, the number 128,457 is not divisible by 11. On the other hand, checking the number 837,298, because $(9 + 7 + 8) - (3 + 2 + 8) = 11$, and 11 is divisible by 11, the number 837,298 is also divisible by 11.

Related Topics

Prime and Composite Numbers
Greatest Common Divisor
Least Common Multiple

TOPIC 4.5 Greatest Common Divisor

cornerstone

Prime and Composite Numbers
Prime Factorization and Tree Diagrams
Fundamental Theorem of Arithmetic

Introduction

The greatest common divisor is a key concept that lays the foundation for meaningful work with fractions. In particular, the greatest common divisor, is useful in reducing fractions to their simplest form. In this section, we define and illustrate notation for the greatest common divisor. We also illustrate two methods for determining the greatest common divisor of two or more integers.

Mathematician Euclid of Alexandria's work in number theory included the development of an algorithm for finding the greatest common divisor. In fact, the algorithm is referred to as the Euclidean algorithm in his honor.

For More Information Reimer, L., & Reimer, W. (1995). *Mathematicians Are People Too: Stories from the Lives of Great Mathematicians, Volume 2.* Palo Alto, CA: Dale Seymour Publications.

4.5.1 Definition of Greatest Common Divisor

The **greatest common divisor** (GCD) of two or more integers is the largest divisor that is shared by each of the numbers. Because the words *divisor* and *factor* can be used interchangeably, the greatest common divisor is often referred to as the **greatest common factor** (GCF). So, given the numbers 24 and 30, we can see that they have several factors in common: 1, 2, 3, and 6. Because the largest divisor that they both have is 6, we say the greatest common divisor of 24 and 30 is 6.

4.5.2 Relatively Prime

For two integers, a and b, we say a and b are **relatively prime** if the only factor they have in common is 1; that is, a and b are relatively prime if the GCD $(a, b) = 1$. For example, GCD $(4, 9) = 1$ because they have no divisors in common other than 1.

4.5.3 Notation for GCD

The notation GCD (a, b) is read "the greatest common divisor of a and b." Thus, using the example above, we can say the GCD $(24, 30) = 6$. We can find the greatest common divisor that more than two numbers share. To find GCD (12, 30, 45), we look for the largest factor that is shared by all three integers. Looking at factors of each number and finding the ones they have in common, we can determine that GCD $(12, 30, 45) = 3$.

4.5.4 Methods for Finding the GCD

There are a number of methods for finding the GCD of two or more integers. These methods include finding the intersection of their sets of divisors, the Euclidean algorithm, and using prime factorization.

Method 1: Finding the Intersection of Sets of Divisors. One way to find the greatest common divisor of two or more numbers is, first, to list the divisors of each number and then, once you locate all the divisors they have in common, you can determine the largest from this set. For example, to find the GCD (48, 56), we can first list the divisors of 48 and the divisors of 56.

$$\text{The set of divisors of } 48 = \{1, 2, 3, 4, 6, 8, 12, 16, 24, 48\}.$$
$$\text{The set of divisors of } 56 = \{1, 2, 4, 7, 8, 14, 28, 56\}.$$

Now we can determine the set of divisors that they have in common $= \{1, 2, 4, 8\}$. From this set, we determine that 8 is the largest divisor they share. Therefore, GCD $(48, 56) = 8$.

Method 2: The Euclidean Algorithm. A second means of determining the greatest common divisor is the **Euclidean algorithm**. The algorithm is based on the division algorithm that says that the GCD of two positive integers, a and b, will be the same as both the GCD of b and the remainder found when a is divided by b; that is, given any two positive integers, a and b, a can be expressed as $a = bq + r$, where $0 \leq r < b$. Then it follows that the GCD $(a, b) = \text{GCD}(b, r)$. The Euclidean algorithm builds on this division algorithm and suggests a set of repetitive steps that lead to the identification of the GCD (a, b):

Let a and b be any two positive integers. Then using the division algorithm, we can find a series of natural numbers $q_1, q_2, q_3, \ldots q_n$ and $r_1, r_2, r_3, \ldots r_{n-1}$ such that $r_{i+1} < r_i$ and:

$$\begin{aligned} a &= bq_1 + r_1 \\ b &= r_1q_2 + r_2 \\ r_1 &= r_2q_3 + r_3 \\ &\vdots \\ r_{n-3} &= r_{n-2}q_{n-1} + r_{n-1} \\ r_{n-2} &= r_{n-1}q_n \end{aligned}$$

Then, GCD $(a, b) = r_{n-1}$.

For example, to find the GCD (672, 456) using the Euclidean algorithm:

$$\begin{aligned} 672 &= 456 \times 1 + 216 \\ 456 &= 216 \times 2 + 24 \\ 216 &= 24 \times 9 \end{aligned}$$

Now that we have reached a point where there is no remainder, we can determine that the GCD (672, 456) is the remainder in the second-to-last step; that is, GCD $(672, 456) = 24$. Note that this algorithm allows us to determine the GCD of only two integers at a time.

For a second example, find the GCD (20,753, 2136)

$$\begin{aligned} 20{,}753 &= 2136 \times 9 + 1529 \\ 2136 &= 1529 \times 1 + 607 \\ 1529 &= 607 \times 2 + 315 \\ 607 &= 315 \times 1 + 292 \\ 315 &= 292 \times 1 + 23 \\ 292 &= 23 \times 12 + 16 \\ 23 &= 16 \times 1 + 7 \\ 16 &= 7 \times 2 + 2 \\ 7 &= 2 \times 3 + 1 \\ 3 &= 1 \times 3 \end{aligned}$$

Therefore, the GCD $(20, 751, 2136) = 1$; that is, we can conclude that 21,751 and 2136 are relatively prime.

Method 3: Using Prime Factorizations. Another way to find the greatest common divisor of two or more numbers is, first, to find the prime factorization of each number and then to identify which prime factors and how many of each prime factor they both have. Then, the GCD will be the product of all the prime factors they share. For example, to find the GCD (72, 60), first find the prime factorizations of 72 and 60.

$$\text{The prime factorization of } 72 = 2 \times 2 \times 2 \times 3 \times 3.$$
$$\text{The prime factorization of } 60 = 2 \times 2 \times 3 \times 5.$$

Now, we determine the primes and the number of each prime that their factorizations have in common: 2, 2, 3. Multiplying these primes together, we find $2 \times 2 \times 3 = 12$. Therefore, GCD $(72, 60) = 12$.

Related Topics

Simplifying Fractions

TOPIC 4.6 Least Common Multiple

cornerstone

Prime and Composite Numbers
Prime Factorization and Tree Diagrams

Introduction

The least common multiple of two integers is a key concept in number theory. The least common multiple is often a good choice when finding a common denominator for two fractions when adding or subtracting fractions. In this section, we introduce the definition and notation for least common multiple and demonstrate two methods for determining the least common multiple between two or more numbers. We also explore a key relationship between the least common multiple and the greatest common divisor of two or more integers.

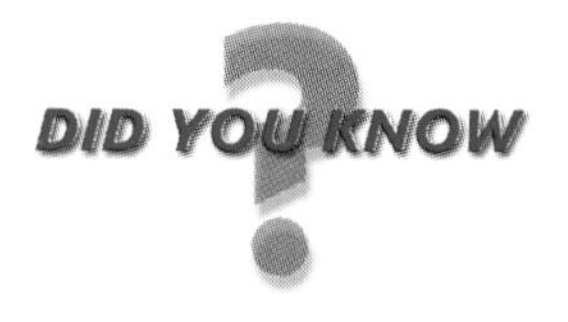

Female mathematician Sophie Germain (1776–1831) began her mathematics career disguising herself as a man because women were not wholly accepted as mathematicians at that time. She initially became recognized for her work in number theory, particularly on the strides she made toward solving Fermat's last theorem. This theorem has fascinated mathematicians for centuries and was not completely proven until 1993 by British mathematician Andrew Wiles, a professor at Princeton University.

For More Information Reimer, L., & Reimer, W. (1995). *Mathematicians Are People Too: Stories from the Lives of Great Mathematicians, Volume 2.* Palo Alto, CA: Dale Seymour Publications.

4.6.1 Definition of Least Common Multiple

The **least common multiple** (LCM) of two or more positive integers is the smallest nonzero multiple that is common to each number. Every positive integer has an infinite number of multiples. The set of nonzero **multiples** of an integer, a, is the set of numbers found by multiplying a times the set of positive integers; that is, the set of multiples of $a = \{a, 2a, 3a, 4a, 5a, 6a, \ldots\}$. Even though there is an infinite number of multiples for each number and an infinite number of multiples that two or more numbers share, there is an identifiable smallest multiple that they share. For the numbers 6 and 10, we can find that even though they share multiples $30, 60, 90, 120, \ldots$, the smallest multiple they have in common is 30.

4.6.2 Notation for LCM

The notation LCM (a, b) is read "the least common multiple of a and b." For the example given above, we can say that the LCM $(6, 10) = 30$. We can find the least common multiple that more than two numbers share. To find LCM $(6, 8, 10)$, we are looking for the smallest multiple shared by all three integers. Looking at multiples of each number and finding the ones they have in common, we can determine that LCM $(6, 8, 10) = 120$.

4.6.3 Methods for Finding the LCM

There are a number of methods for finding the LCM of two or more integers. These methods include finding the intersection of their sets of multiples and using prime factorization.

Method 1: Finding the Intersection of Sets of Multiples. One way to find the least common multiple of two or more numbers is, first, to list the multiples of each number and then, once you identify what these sets look like, you can determine what they have in common. Finally, you can determine the smallest from this set of multiples that they share. For example, to find the LCM (14, 10), we can first illustrate the set of multiples of 14 and the set of multiples of 10.

$$\text{The set of multiples of } 14 = \{14, 28, 42, 56, 70, 84, 98, 112, 126, 140, 154, \dots\}.$$
$$\text{The set of multiples of } 10 = \{10, 20, 30, 40, 50, 60, 70, 80, 90, 100, 110, 120, 130, 140, 150, 160, \dots\}.$$

Now we can observe a pattern of multiples that they share and determine the set of common multiples to be $= \{70, 140, 210, 280, \dots\}$. From this set of common multiples, we determine that 70 is the least common multiple. Therefore, LCM $(14, 10) = 70$.

Method 2: Using Prime Factorizations. Another way to find the least common multiple of two or more numbers is, first, to find the prime factorization of each number and then to identify both the prime factors that appear in either number and the largest number of times that each prime factor appears in either number. The LCM will be the product of all the prime factors identified through this process. For example, to find the LCM (60, 72), first find the prime factorizations of 60 and 72.

$$\text{The prime factorization of } 60 = 2 \times 2 \times 3 \times 5.$$
$$\text{The prime factorization of } 72 = 2 \times 2 \times 2 \times 3 \times 3.$$

Now, determine all the primes that appear in 2, 3, or 5. Then, determine the largest number of times each prime appears in either number (2 appears three times in 72, 3 appears twice in 72, and 5 appears once in 60). So, we take three 2s, two 3s, and one 5. Multiplying these primes together, we find $2 \times 2 \times 2 \times 3 \times 3 \times 5 = 360$. Therefore, LCM $(72, 60) = 360$.

In a second example, we find the LCM (10, 45, 21):

$$\text{The prime factorization of } 10 = 2 \times 5.$$
$$\text{The prime factorization of } 45 = 3 \times 3 \times 5.$$
$$\text{The prime factorization of } 21 = 3 \times 7.$$

We find that the primes 2, 3, 5, and 7 appear in one or more of the factorizations. Looking for the highest number of times these primes appear in any one factorization, we find that 2 appears once in 10, 3 appears twice in 45, 5 appears once in both 10 and 45, and 7 appears once in 21. Multiplying these primes together the appropriate number of times, we find $2 \times 3 \times 3 \times 5 \times 7 = 630$. Therefore, LCM $(10, 45, 21) = 630$.

4.6.4 Relationships Between the LCM and GCD

There is a strong relationship between the LCM and the GCD. Given two integers a and b, $\text{GCD}(a, b) \times \text{LCM}(a, b) = a \times b$. This relationship is most easily seen if we examine the prime factorizations of two numbers, writing the prime factorizations in exponential form. Consider the numbers 24 and 40. Let us first find the prime factorization of each number.

$$24 = 2^3 \times 3$$
$$40 = 2^3 \times 5$$

Note that the GCD $(24, 40) = 2^3$ and the LCM $(24, 40) = 2^3 \times 3 \times 5$. When we multiply the LCM $\times$ GCD, we have $2^3 \times 2^3 \times 3 \times 5$, which equals $24 \times 40 = 960$.

Related Topics

Adding and Subtracting Fractions

Clock Arithmetic & Modular Arithmetic

cornerstone **Integer Division**

Introduction

Most of us learned to tell time on a standard clock. Doing so, we had to, albeit most often subconsciously, perform clock arithmetic. Unlike the infinite set of integers, a clock offers a finite set of integer numbers on which to perform operations. Like clock arithmetic, modular arithmetic deals with a finite set of numbers. In this section, we explore the fundamental concepts of clock and modular arithmetic and learn to perform the four basic operations in these settings.

One of the greatest names in the history of mathematics comes from the area of number theory. German mathematician Carl Friedrich Gauss (1777–1855) is a noted number theorist and is often called the Prince of Mathematicians. Much of his work in number theory was published in *Disquisitiones Arithmeticae* when he was only 24 years old. In this publication, his work with the congruence of numbers and modular arithmetic is highlighted.

Modular arithmetic has been used by mathematicians working for national security groups such as the National Security Agency (NSA) in developing and interpreting secret codes. As a result, many secret messages could be sent and received safely because of codes that involved various congruence sequences for equating and unraveling information presented in disguised form.

For More Information Reimer, L., & Reimer, W. (1990). *Mathematicians Are People Too: Stories from the Lives of Great Mathematicians, Volume 1.* Palo Alto, CA: Dale Seymour Publications.

4.7.1 Telling Time

When we tell time, we usually refer to the hours of 1 o'clock, 2 o'clock, 3 o'clock, ..., 12 o'clock, and then we start over again at 1 o'clock, differentiating morning and afternoon times using A.M. and P.M. If it is 8 o'clock in the morning and someone tells you to meet them in six hours, we know that the time to meet is 2:00 P.M. How do we figure out the time to meet? Some people might simply count up six hours from 8 A.M., counting 9 o'clock, 10 o'clock, 11 o'clock, 12 o'clock, 1 o'clock, 2 o'clock, knowing that once you reach 12 o'clock, you loop back, starting again at 1 o'clock.

4.7.2 Clock Arithmetic

Rather than counting up to find a time, we can perform **clock arithmetic**. The key behind clock arithmetic is that time is **cyclic** in nature; that is, the clock itself is fixed, and there is a finite set of times on the clock that loop around in cycles. Because the clock is cyclic, another way of finding out what time it is 6 hours after 8 o'clock is simply to add the 6 and 8 for a sum of 14. Knowing that 12 is the highest hour on the clock, we can determine the "real time" associated with 14. To do this, answer the question, "How much more than 12 did I go?" In essence, in doing so, we are finding, after we divide out a set of 12 hours, the amount that remains. Thus, to find what 10 o'clock plus 10 more hours is, we find the sum, $10 + 10 = 20$. Dividing 20 by 12, we find a remainder of 8. Therefore, $10 + 10$ in clock arithmetic is 8.

4.7.3 Modular Arithmetic

Modular arithmetic works much like the way clock arithmetic works. In modular arithmetic, we say that integers that have the same remainders when divided by an integer, m, are congruent in that modulo. Formally, we say for integers a, b, and m that *a and b are congruent modulo m*, written $a \equiv b \bmod m$, if a and b have the same remainders when divided by m. Another way of stating this definition is as follows: a is congruent to b (mod m) if m divides $(a - b)$. Let us check both versions of this definition. For example, in mod 5, the numbers 6 and 11 are congruent because $6 \div 5 = 1$ *with a remainder of 1* and $11 \div 5 = 2$ *with a remainder of 1*. Usually, given an integer, we like to identify to which of the remainders in the modulo that integer is congruent. Thus, we can say $6 \equiv 1 \pmod 5$ and $11 \equiv 1 \pmod 5$ because 1 is the common remainder. We can also check that 11 and 6 are congruent (mod 5) by checking that 5 does divide the difference $(11 - 6)$ or that $5|5$.

Modulo. When performing modular arithmetic, the **modulus** you are working within, say *modulo m*, defines a fixed set of numbers that comprise the set of possible remainders when you divide by m. For example, in modulo 5, written mod 5, the set of possible remainders when you divide any integer by 5 are the integers: 0, 1, 2, 3, and 4.

Addition in Modular Arithmetic. To add numbers in a particular modulo, you simply find the sum and then translate that sum to the appropriate remainder in that modulo. For example, in mod 4, adding $9 + 8$, we get 17. However, dividing by 4 and finding the remainder, we determine that $17 \equiv 1 \pmod 4$. Note that because $9 \equiv 1 \pmod 4$ and $8 \equiv 0 \pmod 4$, it is clear that adding $9 + 8$ in mod 4 is equivalent to adding $1 + 0$ in mod 4, which equals 1.

Subtraction in Modular Arithmetic. To find $2 - 4$ in mod 5, we might use the missing-addend approach for subtraction; that is, we can ask, "$4 + x = 2 \pmod 5$?" Because we know that 7 has a remainder of 2 when divided by 5, we can try to form a sum of 7 and conclude that $x = 3$ because $4 + 3 = 7$ and $7 \equiv 2 \pmod 5$.

An Addition Table for Mod 5. We can create an addition table for mod 5 using the remainders 0, 1, 2, 3, and 4 as the potential addends (see Table 4.1). All sums are written as 0, 1, 2, 3, and 4, having reduced each sum to its "simplest" congruent form.

Note that the addition table for mod 5 can assist in finding a missing sum when solving a subtraction problem in mod 5. For example, to find $1 - 2$ in mod 5, look for a sum of 1 within the table that is associated with an addend of 2. Trace along the table to find the missing addend of 4, showing that $4 + 2 = 1 \pmod 5$.

Table 4.1

+	0	1	2	3	4
0	0	1	2	3	4
1	1	2	3	4	0
2	2	3	4	0	1
3	3	4	0	1	2
4	4	0	1	2	3

Multiplication in Modular Arithmetic. We can multiply numbers in modulo. For example, 4×8 in mod 6 is 2 because $32 \div 6 = 5$ with a remainder of 2. Similarly, 7×9 in mod 6 is 3 because $63 \div 6 = 10$ with a remainder of 3.

Division in Modular Arithmetic. Division in modular arithmetic can be viewed as a missing-factor problem. For example, to solve $3 \div 4$ in mod 5, we can seek the missing factor, x, such that $4 \times x = 3$ in mod 5. Because $8 \equiv 3 \pmod 5$, we can say that $x = 2$ because $4 \times 2 = 8$.

A Multiplication Table for Mod 5. We can create a multiplication table for mod 5 using the remainders 0, 1, 2, 3, and 4 as the potential factors (see Table 4.2). All products are written as 0, 1, 2, 3, and 4, having reduced each product to its "simplest" congruent form.

Note that solving division problems using the missing-factor approach is made easier using the multiplication table. For example, to solve $1 \div 4$ in mod 5, look for

Table 4.2

x	0	1	2	3	4
0	0	0	0	0	0
1	0	1	2	3	4
2	0	2	4	1	3
3	0	3	1	4	2
4	0	4	3	2	1

the product 1 within the table; then find where a factor of 4 is associated with that product. Then, tracing along the table, locate the other missing factor of 4. Thus, $1 \div 4 = 4 \pmod 5$.

Related Topics

Divisibility

TOPIC 4.8 *Patterns*

Introduction

Patterns exist all around us. From the musical rhythms we hear, the days of the week that we follow, to the designs in sunflowers and pinecones, we observe and feel comfort in patterns on a daily basis. Some mathematicians define *mathematics* as the science of patterns. Mathematical patterns describe much of the work around us, providing a means of predicting and generalizing from a fixed set of examples. In this section, we explore types of patterns and the process of generalizing a pattern.

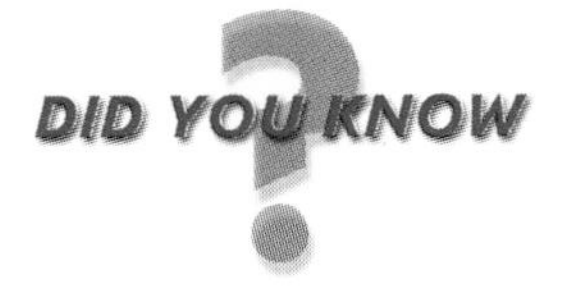

Leonardo of Pisa, better known as Fibonacci (c. 1170–after 1240) introduced a mathematical pattern that describes many phenomena in nature. This type of pattern has become known as a Fibonacci sequence. Many have been fascinated by Fibonacci sequences, and over time, many applications of the sequences have been uncovered. For example, botanists have found that the patterns of leaf buds on many stems, the spirals of seeds in the heads of sunflowers, the petals on artichokes, and the scales on pineapples all follow these sequences. (Fibonacci sequences are discussed later in this section.)

For More Information Reimer, L., & Reimer, W. (1995). *Mathematicians Are People Too: Stories from the Lives of Great Mathematicians, Volume 2*. Palo Alto, CA: Dale Seymour Publications.

4.8.1 Definition of a Pattern

A **pattern** is a design or a relationship that is predictable in some manner or a schematic that can be followed. In mathematics, we can observe a wide range of number patterns and geometric patterns. For example, a number pattern might be $1, 2, 3, 1, 2, 3, 1, 2, 3, \ldots$ which is a sequence of numbers that follow the same scheme: They follow an ordering that repeats so that we can predict subsequent numbers by observing previous numbers.

An example of a geometric pattern is:

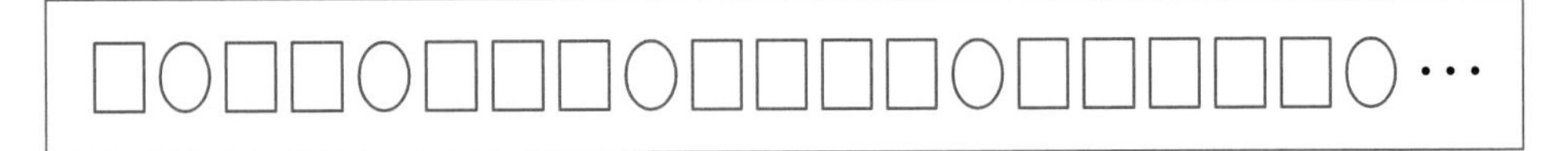

Because there is an identifiable relationship among the shapes in succession, we can describe how the pattern of shapes would continue; that is, we predict that following the last circle would be six squares, which would be followed by one circle, followed by seven squares, and so on.

4.8.2 Arithmetic Sequences

A *sequence* is a listing of numbers, separated by commas. An **arithmetic sequence** is a pattern of numbers where the difference between any two terms of the sequence is the same amount; that is, we can determine some amount, a, such that the $(n+1)st$ term is found by finding the sum of the nth term and a. Examples of arithmetic sequences include:

$$1, 3, 5, 7, 9, \ldots$$
$$3, 6, 9, 12, 15, \ldots$$
$$40, 30, 20, 10, 0, -10, \ldots$$

Notice that in each of the examples, the difference between any two successive terms is constant. In the first example, the difference is always 2; in the second example, it is always 3; and in the last example, the difference is always 10.

4.8.3 Geometric Sequences

A **geometric sequence** is a pattern of numbers where each term is a constant multiple of the preceding term; that is, we can determine some amount, a, such that the $(n{+}1)st$ term is found by multiplying the nth term by a. Examples of geometric sequences include:

$$1, 2, 4, 8, 16, 32, \ldots$$
$$3, 9, 27, 81, \ldots$$
$$1, .1, .01, .001, .0001, \ldots$$

In every example, each term differs from the next by the same factor. In the first example, each successive term is found by multiplying the previous term by 2; in the second example, each successive term is found by multiplying the previous term by 3; and in the last example, each successive term is found by multiplying the previous term by 0.1.

4.8.4 Growing Patterns

Many patterns are **growing patterns,** where each successive term changes from the previous term, but the change differs with each term. For example, consider the pattern: $1, 3, 6, 10, 15, 21, \ldots$. Notice that the difference between the first two terms is 2, the difference between the next two terms is 3, the difference between the next two terms is 4, and so on. Thus, there is an identifiable pattern among the series of numbers, and that pattern grows from term to term.

Consider another growing pattern $1, 3, 12, 60, \ldots$, where each successive term is found by multiplying the previous term by an amount that changes with each term. The second term is found by multiplying the first term by 3; the third term is found by multiplying the second term by 4; the fourth term is found by multiplying the third term by 5; and so on.

4.8.5 Fibonacci Sequences

A **Fibonacci sequence** is a sequence where each successive term is found by summing the two previous terms. An example of a Fibonacci Sequence is 1, 1, 2, 3, 5, 8, 13, 21, 34, ... Notice that adding $1 + 1 = 2$, adding $1 + 2 = 3$, adding $2 + 3 = 5$, $3 + 5 = 8$, and so on.

4.8.6 Finding the nth Term

When we generate a sequence of numbers in a mathematical problem, we often strive to explain the pattern by describing the **nth term,** that is, the **general term.** In describing the nth term, we provide a generalization for the pattern. Finding the nth term is strongly related to inductive reasoning (refer to the topic on Logic).

Consider this mathematical problem and how finding a generalization of a pattern of numbers can help solve the problem:

One day I called two friends and invited them to a party. The next day, each of my two friends called and invited two more friends to the party. The next day, each of the people who had been invited thus far called and invited two more friends to the party. If this went on for 15 days, how many people were actually called and invited to the party on the 15th day?

Table 4.3

Day	Number Invited
1	2
2	4
3	8
4	16
5	32
.	.
.	.
n	2^n

To solve this problem, we can try to establish a pattern of how many people were invited on the nth day. We can use a table to help facilitate this pattern (see Table 4.3).

From the table, we observe the pattern that each day, the number of people invited on that day increased by a multiple of 2. Thus, we note that on the nth day, the number of people invited can be expressed as 2^n. Therefore, we can find that on the 15th day, 2^{15}, or 32,768, people were called and invited. What a party this is going to be!

4.8.7 Figurate Numbers

Figurate numbers are numbers that are associated with polygons such as triangles, squares, and pentagons. Figurate numbers generate a pattern related to the "size" of the shape. **Triangular numbers** are found by examining the triangular shapes shown in Figure 4.7.

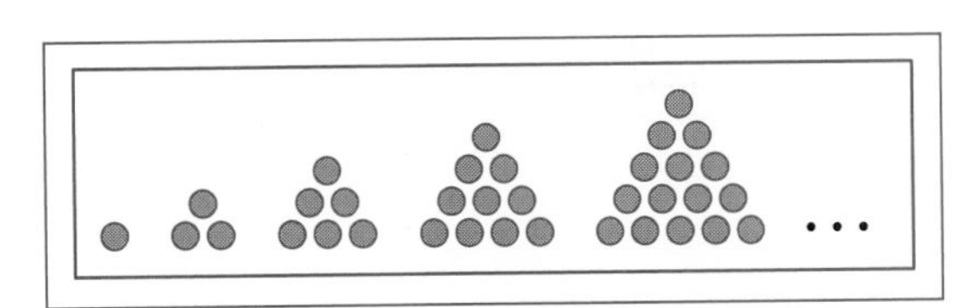

Figure 4.7

Notice that the first "triangle" is formed with 1 dot, the second triangle is formed with 3 dots, the third triangle is formed with 6 dots, the fourth with 10 dots, and so on. Thus, we say that the sequence of triangular numbers is 1, 3, 6, 10, 15, ... As we examine the number of dots in each row of a given triangle, we observe that the number in each row increases by 1 as we move from top to bottom. The third triangular number is found by adding $1 + 2 + 3 = 6$. The fourth triangular is found by adding $1 + 2 + 3 = 4$. Thus, the nth triangular number will be found by adding $1 + 2 + 3 + \cdots + n = \dfrac{n(n + 1)}{2}$.

Square numbers are found by examining the "size" of square shapes as shown in Figure 4.8.

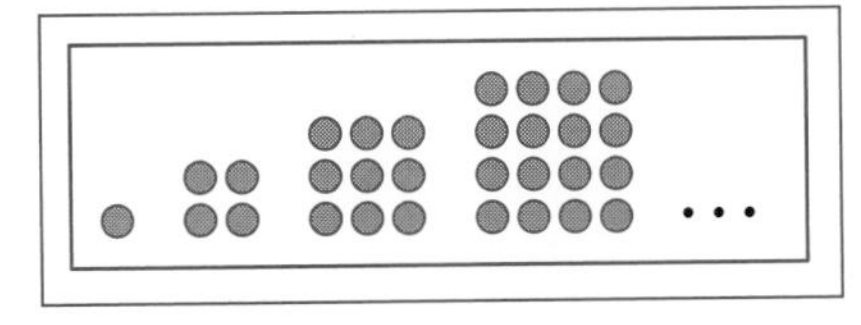

Figure 4.8

Looking at the square shapes and at the total number of dots required to make each square, we generate the list of square numbers: 1, 4, 9, 16, ... The first square number is 1^2 or 1, the second square number is 2^2 or 4, the third square number is 3^2 or 9, and so on. Thus, the nth square number is n^2.

Pentagonal numbers are found by examining the pentagonal shapes found in Figure 4.9.

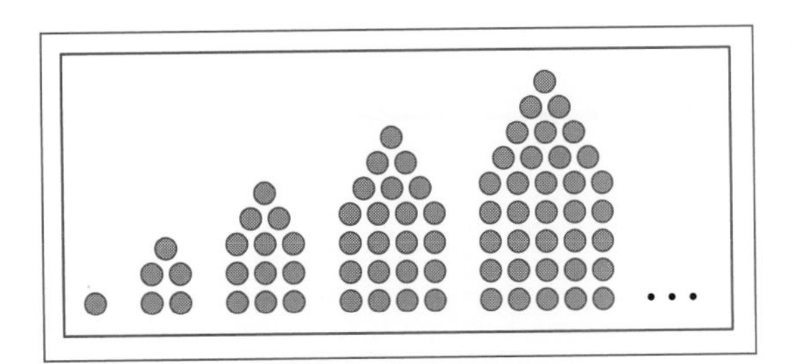

Figure 4.9

Pentagonal numbers include the numbers 1, 5, 12, 22, 35, ... We can examine the fourth pentagon and notice that it can be described as the combination of the fourth square figure and the third triangular figure (see Figure 4.10).

Thus, we can generalize and describe the nth pentagonal figure as the combination of the nth square figure and the (n − 1)st triangular figure. Therefore, in general, the nth pentagonal figure can be found by $n^2 + \frac{(n-1)n}{2}$.

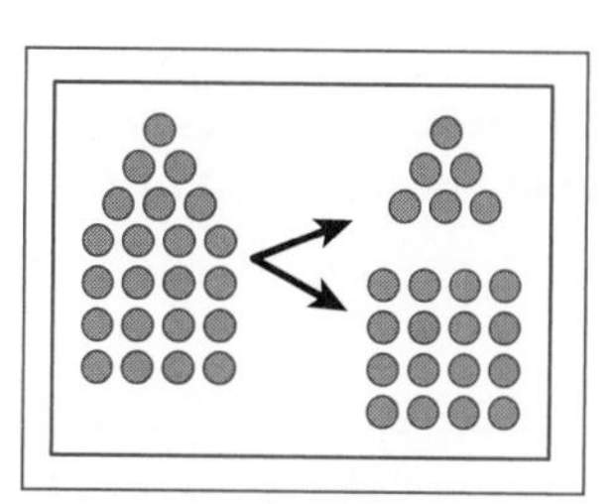

Figure 4.10

Related Topics

Logic
Problem Solving

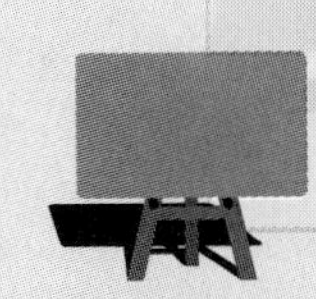

In the Classroom

While the focus on the *Number and Operations Content Standard* continues, this chapter also begins to emphasize the *Reasoning and Proof Process Standard.* "By developing ideas, ... justifying results, and using mathematical conjectures ... with different expectations of sophistication—at all grade levels, students should see and expect that mathematics makes sense" (NCTM, 2000, p. 56). We believe that you should make such mathematical sense of (a) natural numbers by analyzing their factors and developing and justifying conjectures (Activities 4.1–4.9) and (b) patterns that build on the theories that you have developed and whose formalization requires your creative use of operations with natural numbers (Activities 4.10–4.16).

Some Expectations of Your Future Students:

Pre-K–2 (NCTM, 2000, p. 78)

- develop understanding of the relative position and magnitude of whole numbers and of ordinal and cardinal numbers and their connections
- develop a sense of whole numbers and represent and use them in flexible ways, including relating, composing, and decomposing numbers
- develop fluency with basic number combinations for addition and subtraction

Grades 3–5 (NCTM, 2000, p. 148)

- describe classes of numbers according to characteristics such as the nature of their factors
- identify and use relationships between operations, such as division as the inverse of multiplication, to solve problems
- develop fluency with basic number combinations for multiplication and division

How Activities 4.1–4.16 Help You Develop an Adult-level Perspective on the Above Expectations:

- By developing and organizing your knowledge of the factors of natural numbers, you learn to classify numbers according to their possible factorizations.
- By seeing common patterns in examples of composite numbers that have structurally similar prime factorization, you begin to generalize the idea of prime numbers to other more abstract ways of composing and decomposing numbers.
- By working with the concept of the uniqueness of the prime factorization of a composite number, you construct ways to determine both the greatest common factor and the least common multiple of two numbers.
- By creating tests for determining divisibility in different bases and by exploring modular arithmetic, you continue to develop your sense of numeration systems and the additive and multiplicative structures embedded in them. You begin to use the inverse operations of addition (or multiplication) to solve problems involving subtraction (or division) in modular arithmetic, thus becoming more flexible in your thinking of each operation.
- By analyzing different number patterns, you continue to develop ways of composing and decomposing numbers using the four operations. You learn to identify the common pattern and construct the general form of the nth number.

Bibliography

Bezuszka, S. J. (1985). A test for divisibility by primes. *Arithmetic Teacher, 33*, 36–38.

Brown, G. W. (1984). Searching for patterns of divisors. *Arithmetic Teacher, 32*, 32–34.

Dearing, S. A., & Holtan, B. (1987). Factors and primes with a T square. *Arithmetic Teacher, 34*, 34.

Dockweiler, C. J. (1985). Palindromes and the "Law of 11." *Arithmetic Teacher, 33*, 46–47.

Eisen, A. P. (1999). Exploring factor sets with a graphing calculator. *Mathematics Teaching in the Middle School, 5*, 78–82.

Graviss, T., & Greaver, J. (1992). Extending the number line to make connections with number theory. *Mathematics Teacher, 85*, 418–420.

Hudson, F. M. (1990). Are the primes really infinite? *Mathematics Teacher 83*, 660.

Lamb, C. E., & Hutcherson, L. R. (1984). Greatest common factor and least common multiple. *Arithmetic Teacher, 31*, 43–44.

Lefton, P. (1991). Number theory and public-key cryptography. *Mathematics Teacher, 84*, 54–62.

Martinez, J. G. R. (1988). Helping students understand factors and terms. *Mathematics Teacher, 81*, 747–751.

Olson, M. (1991). Activities: A geometric look at greatest common divisor. *Mathematics Teacher, 84*, 202–208.

CHAPTER 5 FIVE

Data & Chance

CHAPTER OVERVIEW

A great many events in the world around us involve uncertainty and chance. It is easy to find examples from business, education, law, medicine, and everyday experience. Two examples come readily to mind: (1) The weather forecaster on TV says, "There is a 70% chance of rain tomorrow"; (2) sometimes with only a small portion of votes counted, newscasters are able to project winners of political elections and final percentages of votes with considerable accuracy. How was the 70% figure obtained? How can newscasters attain such accuracy with so little information? The branches of mathematics called probability and statistics were developed to help us deal with situations involving uncertainty and chance in a precise and objective manner. In this chapter, you can learn interesting facts about probability and statistics and will examine some concepts and procedures involving probability and statistics ideas.

BIG MATHEMATICAL IDEAS

Data and chance, independence/dependence, multiple representations, mathematical modeling

NCTM PRINCIPLES & STANDARDS LINKS

Data Analysis, Statistics, and Probability; Problem Solving; Reasoning; Communication; Connections; Representation

TOPIC 5.1 Probability Notions
5.2 Equally Likely Outcomes
5.3 Mutually Exclusive and Complementary Events
5.4 Multistep Experiments
5.5 Counting Principles
5.6 Statistics Notions
5.7 Measures of Central Tendency
5.8 Data Dispersion
5.9 Representing Data
In the Classroom
Bibliography

TOPIC 5.1 Probability Notions

cornerstone **Fractions**

Introduction

Probability situations are all around us. Whether we are listening to the weather reporter explain the chances of it raining that day, or whether we play the lottery hoping to win, even though we know we are not likely to win, probabilities influence the decisions we make. In this section, we examine probability terms and introduce a formula for computing simple probabilities. We also see several examples of how a simulation of a given experiment helps to model a situation and facilitate finding the probability of certain outcomes from the experiment.

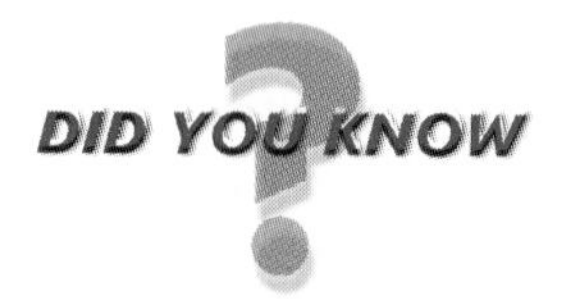

Mathematician Pierre de Fermat is credited for fully developing the study of probability. He corresponded often with mathematician Blaise Pascal in the mid-1600s to test his theories and to compare results of probability experiments.

For More Information Reimer, L., & Reimer, W. (1995). *Mathematicians Are People Too: Stories from the Lives of Great Mathematicians, Volume 2*. Palo Alto, CA: Dale Seymour Publications.

5.1.1 Probability Terminology

Probability is about degrees of uncertainty and chance. An **experiment** is any activity in which different results can occur. These results are called **outcomes.** There is a numerical chance (probability) associated with each outcome. The list of all possible outcomes is considered the **sample space** of an experiment. Any subset of the sample space is considered an **event.**

For example, suppose we have a hat containing 5 yellow marbles, 6 blue marbles, 3 white marbles, and 4 red marbles. An experiment associated with the hat and marbles is drawing a marble out of the hat and noting the color of the marble. The possible outcomes associated with the experiment are: drawing a yellow marble, drawing a blue marble, drawing a white marble, and drawing a red marble. Thus, the sample space is yellow, blue, white, and red. We can consider possible events as being: drawing either a red or a blue marble, or drawing anything but a white marble.

5.1.2 Computing a Simple Probability

A **probability** is the computed likeliness that a given event will happen. Probabilities can be expressed as fractions between 0 and 1, as decimals between 0.00 and 1.00, or as percentages between 0% and 100%. To compute the probability of a particular event occurring, we usually create a fraction where the numerator depicts the number of ways the event can occur and the denominator represents the total number of ways any outcome in the sample space can occur; that is, given an event, A, the probability of event A occurring, denoted P(A), is found by:

$$P(A) = \frac{\text{number of ways A can occur}}{\text{total number of outcomes}}$$

As you can see, the numerator is always ≥ 0, the denominator is always > 0, and the numerator is always $\leq$ denominator.

For example, in the marble experiment where there are 5 yellow marbles, 6 blue marbles, 3 white marbles, and 4 red marbles in a hat and a marble is drawn from the hat, we can compute the probability of drawing a blue marble by:

$$\text{P(drawing a blue marble)} = \frac{\text{number of blue marbles}}{\text{total number of marbles}} = \frac{6}{18} = \frac{1}{3}$$

In a given experiment, we can determine a **probability distribution,** which is a list of all possible outcomes to the experiment along with the probabilities associated with each outcome. For example, in the marble experiment previously described, we can form the probability distribution illustrated in Table 5.1.

Note that each of the probabilities listed in the probability distribution in Table 5.1 are fractions between 0 and 1 and that the sum of all the probabilities listed in the probability distribution is 1. In general, the probability of an event is between 0 and 1 (inclusive), and the sum of all the probabilities in the probability distribution is 1.

Table 5.1

Outcomes	Probability
yellow	$\frac{5}{18}$
blue	$\frac{6}{18}$
white	$\frac{3}{18}$
red	$\frac{4}{18}$
	1

Table 5.2

Outcomes	Probability
yellow	$\frac{5}{18}$
blue	$\frac{6}{18}$
white	$\frac{3}{18}$
red	$\frac{4}{18}$
	1

5.1.3 Simulation with a Spinner

Figure 5.1

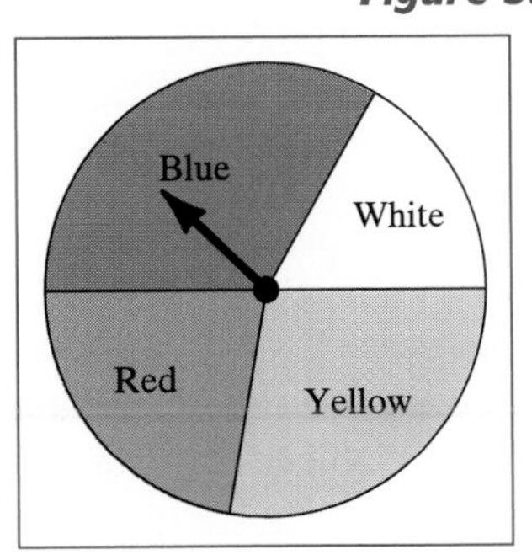

Because probabilities describe what could happen in an experiment, it is often useful to act out or model the experiment with what we call a **simulation** of the experiment. Let us imagine a simulation of the probabilities of outcomes associated with spinning a spinner as shown in Figure 5.1. By spinning the spinner in Figure 5.1 many times (e.g., 100 times, or 1,000 times, or even 10,000 times), we can approximate (that is, simulate) an experiment involving drawing marbles from a hat. Note that the probability distribution of the spinner in Figure 5.1 is the same as that shown in Table 5.1. (See Table 5.2 for the probability distribution associated with the spinner.)

The probabilities for each outcome on the spinner are determined by figuring what part of the whole each outcome represents. For example, because one-sixth of the circle is white, the probability of spinning the arrow and having it land on a white section is one-sixth. Likewise, because the blue section of the spinner is twice as large as the white section, the blue section is two-sixths (or one-third) of the spinner. Therefore, the probability of spinning blue is one-third. Finally, because the red and yellow sections on the spinner are four-eighteenths and five-eighteenths of the whole circle, respectively, the probabilities of spinning red and yellow are four-eighteenths and five eighteenths, respectively.

Related Topics

Equally Likely Outcomes
Mutually Exclusive and Complementary Events
Multistep Experiments

TOPIC 5.2 Equally Likely Outcomes

cornerstone

Fractions
Decimals
Decimals and Fractions
Probability Notions

Introduction

Sometimes, the chance of some outcome occurring is very high and other times very low. However, in some settings, the chances of different outcomes occurring are equal. For example, whenever we flip a coin before a game to decide who goes first, we find this a fair method to decide because the coin is just as likely to land on heads as it is on tails. Thus, one person does not have a better chance than the other to win. In this section, we explore situations where different outcomes are equally likely and how to determine the probabilities for equally likely outcomes. We also compare theoretical versus empirical probabilities.

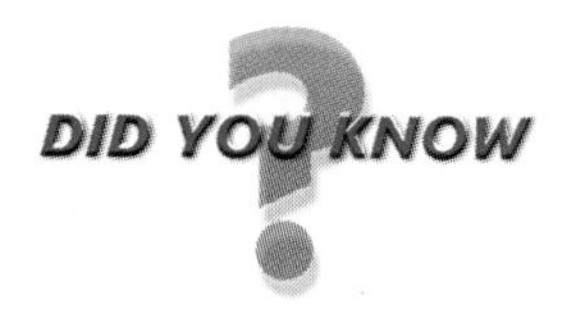

Jakob Bernoulli was a Swiss mathematician from a family of mathematicians. The Bernoulli family made many contributions to the theory of probability in the mid- to late 1600s. Jakob Bernoulli is most noted for being the first person to develop the fundamental concept behind empirical probability. This concept grew into a theory known as the theory of large numbers, which states that when you conduct an experiment a large number of times, the data gathered from outcomes of the experiment begin to stabilize as the number of times the experiment is conducted grows larger and larger.

For More Information Reimer, L., & Reimer, W. (1995). *Mathematicians Are People Too: Stories from the Lives of Great Mathematicians, Volume 2.* Palo Alto, CA: Dale Seymour Publications.

5.2.1 Equally Likely

In probability experiments, the probability assigned to each outcome provides a numerical expression of how **likely** the outcome is to occur. For example, if, in an experiment, there are two outcomes, Yes and No, and the probability of the outcome being Yes is $\frac{4}{5}$ and the probability of the outcome being No is $\frac{1}{5}$, then we say that the outcome Yes is more likely to occur than No because the probability is higher.

In some experiments, all outcomes are **equally likely;** that is, they have the same probability of occurring. For example, when rolling a single die (where the numbers rolled represent the number of dots on the top face), the sample space associated with the experiment is 1, 2, 3, 4, 5, and 6. In a fair die, each outcome is equally likely because you are just as likely to roll a 1 as a 2, as a 3, as a 4, as a 5, or as a 6. Because there are six possible outcomes and each outcome is equally likely, then the probability associated with each outcome is $\frac{1}{6}$.

5.2.2 Certain and Impossible Events

A **certain event,** E, is an event that must occur; that is, every element of the sample space satisfies E. A certain event has probability equal to 1. For example, suppose you flip a regular coin. What is the probability of the coin landing on heads or tails? Clearly, both of the outcomes are in the sample space, and one of the outcomes must occur. Thus, the probability of a coin landing on heads or tails is 1.

On the other hand, in a given experiment, we can describe events that are considered **impossible events.** Impossible events are events that are not part of the sample space, and, therefore, the probability of an impossible event is not defined. For example, if you have 5 red marbles and 5 blue marbles in a hat, we might ask the question, "What is the probability of drawing a green marble?" You may be tempted to claim that the probability of drawing a green marble in this situation is zero. However, because there are no green marbles, green will not be a part of the sample space, and thus the probability of this event is not defined.

5.2.3 Theoretical Probability

When we consider an experiment and determine the probabilities associated with each outcome based on the likeliness of each one occurring, we are determining a theoretical probability. For example, when rolling a single die, each of the outcomes 1, 2, 3, 4, 5, and 6 has a $\frac{1}{6}$ chance of occurring because there are six possible outcomes and they are all equally likely. The probability $\frac{1}{6}$ is a **theoretical probability** because it is the probability associated with what should happen given the nature of the experiment of rolling a die.

5.2.4 Empirical Probability

An **empirical probability** is a probability that is determined from data collected by conducting multiple rounds of an experiment; that is, given an experiment, if the experiment is conducted and data from the experiment are collected, we can use that data to determine the probability of the different outcomes occurring.

For example, suppose we actually roll a single die 100 times and reach these results: The number one occurs 11 times, the number two occurs 24 times, the number three occurs 14 times, the number four occurs 13 times, the number five occurs 26 times, and the number six occurs 12 times. From this experiment, we determine the empirical probabilities associated with each roll of that single die as shown in Figure 5.2.

In the experiment, the probabilities of each roll have been determined by actually rolling the die. Notice that these **experimental probabilities** vary from the theoretical probabilities. Comparing empirical and theoretical probabilities, we can confirm the fundamental notion behind probability: Probability is all about chance and what **should** happen, not necessarily about what **does** actually happen.

5.2.5 Relating Empirical and Theoretical Probabilities

Note that the empirical probabilities determined in the experiment of rolling a die were based on only 100 rolls of a die. If we were to actually roll a die 1,000 times, 100,000 times, or even 1,000,000 times, we would likely produce data that determined empirical probabilities that were more closely matched to the theoretical probability associated with this experiment; that is, as the number of times, n, that

Figure 5.2

Outcomes	Probability
1	$\frac{11}{100}$
2	$\frac{24}{100}$
3	$\frac{14}{100}$
4	$\frac{13}{100}$
5	$\frac{26}{100}$
6	$\frac{12}{100}$

we conduct an experiment becomes large, the closer empirical probabilities mirror theoretical probabilities. In fact, as *n* becomes so large as to approach infinity, empirical probabilities will equal theoretical probabilities.

5.2.6 Expected Value

Once we have a probability distribution of numerical outcomes and related probabilities, whether theoretical or empirical probabilities, we can compute the **expected value** of an experiment. The expected value of an experiment is essentially what the average outcome of an experiment would be over time where the outcomes are number values.

To compute the expected value of an experiment with outcomes $O_1, O_2, O_3, O_4, \ldots O_n$ and respective probabilities $p_1, p_2, p_3, p_4, \ldots p_n$, we determine the expected value, E,

$$E = (O_1 \times p_1) + (O_2 \times p_2) + (O_3 \times p_3) + (O_4 \times p_4) + \cdots + (O_n \times p_n)$$

For example, consider the probability distribution associated with spinning a spinner where the possible outcomes are 2, 4, 6, 8, and 10 and the probabilities associated with each outcome are those listed in Table 5.3.

Table 5.3

Outcomes	Probability
2	0.125
4	0.25
6	0.30
8	0.20
10	0.125

Given this probability distribution, we can compute the expected value:

$$\begin{aligned} E &= (2 \times 0.125) + (4 \times 0.25) + (6 \times 0.30) + (8 \times 0.20) + (10 \times 0.125) \\ &= 0.25 + 1 + 1.8 + 1.6 + 1.25 \\ &= 5.93 \end{aligned}$$

What this expected value tells us is that if we spin the spinner over and over again for a large number of times, the overall average of the outcomes we generate should be close to 5.93.

Related Topics
Mutually Exclusive and Complementary Events
Multistep Experiments

TOPIC 5.3 Mutually Exclusive and Complementary Events

Probability Notions
Equally Likely Outcomes
Adding Rational Numbers
Sets
Set Notation

Introduction

If two friends each enter a contest, they may wonder what the chances are that one of them will win. Thus, they are not just interested in the likeliness of one of them winning, but they also want to know the chances of one or the other winning. Intuitively, we understand that the chance of one or the other winning is greater than the chance of exactly or only one of them winning. In this section, we compute the probability of one event or another occurring. We explain how to distinguish between mutually exclusive events and non-mutually exclusive events. Further, given an event, we describe what the complement of that event is.

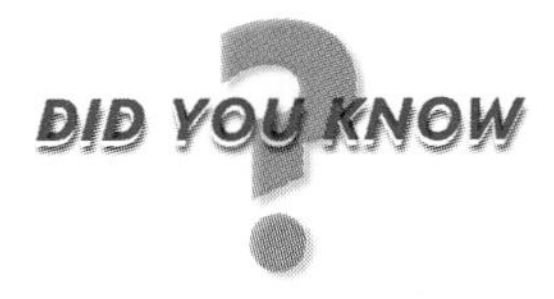

Many people like to play the lottery. They hold onto the hope that one day might be their lucky day and they will win. In reality, in a lottery where you are to choose 6 numbers from a set of 45 numbers, the chance that your set of 6 numbers will be the winning set is approximately 1 in 8 million!

5.3.1 Mutually Exclusive Events

Two events, A and B, are said to be **mutually exclusive events** if they cannot occur at the same time. For example, when rolling a die, let us consider the events A and B where A = rolling a 4 and B = rolling a 3. Events A and B are mutually exclusive because it is impossible to roll a 4 and a 3 at the same time in rolling a single ordinary die.

5.3.2 Non-Mutually Exclusive Events

Two events, A and B are said to be **non-mutually exclusive events** if they can occur at the same time. For example, when drawing a card from an ordinary deck of 52 cards, consider two events A and B where event A = drawing an ace and event B = drawing a diamond. Events A and B are non-mutually exclusive because it is possible, when drawing a card, to draw both an ace and a diamond at the same time; that is, if we draw the ace of diamonds, both events A and B have occurred simultaneously.

5.3.3 Understanding "Or" Probability Situations

In some cases, we need to determine the probability of either one or another outcome occurring. Given two events, A and B, the probability of event A or event B occurring, denoted P(A or B) or $P(A \cup B)$, is:

P(A or B) = P(A) + P(B), when A and B are mutually exclusive events.
P(A or B) = P(A) + P(B) − P(both A and B), when A and B are not mutually exclusive events.

Example 1: When rolling a single die, we might want to find the probability of rolling either a 5 or a 2. Because the events of rolling a 5 and rolling a 2 are mutually exclusive, we compute the probability of rolling a 5 or a 2 by:

$$\text{P(rolling 5 or 2)} = \text{P(rolling 5)} + \text{P(rolling 2)} = \frac{1}{6} + \frac{1}{6} = \frac{2}{6} = \frac{1}{3}.$$

Example 2: Suppose we roll a single die and we want to find the probability of rolling an even number or rolling a number less than 3. The events A and B, where A = rolling an even number and B = rolling a number less than 3, are non-mutually exclusive because it is possible to roll a number that is both even and less 3 at the same time if you roll the number 2. Thus, when computing the probability of A or B occurring, we use the formula for non-mutually exclusive events:

P(rolling an even number or a number less than 3) = P(rolling even) + P(rolling < 3) − P(rolling an even number less than 3).

Because there are six numbers on the die and three of them are even, the probability of rolling an even number is $\frac{3}{6}$. Further, because there are two numbers on the die that are less than 3, the probability of rolling a number less than 3 is $\frac{2}{6}$. Finally, because there is only 1 number on the die that is both even and less than 3, the probability of rolling an even number less than 3 is $\frac{1}{6}$. Therefore, P(rolling an even number or a number less than 3) $= \frac{3}{6} + \frac{2}{6} - \frac{1}{6} = \frac{4}{6} = \frac{2}{3}$.

Example 3: When drawing a card from an ordinary deck of cards, what is the probability of drawing either a diamond or a club? Because the events drawing a diamond and drawing a club are mutually exclusive, computing, we find:

$$\begin{aligned} \text{P(drawing a diamond or a club)} &= \text{P(drawing a diamond)} + \text{P(drawing a club)} \\ &= \frac{13}{52} + \frac{13}{52} \\ &= \frac{26}{52} = \frac{1}{2}. \end{aligned}$$

Example 4: When drawing a card from an ordinary deck of cards, what is the probability of drawing a heart or a queen? Because the events drawing a heart and drawing a queen are non-mutually exclusive, we compute:

$$\begin{aligned} \text{P(drawing a heart or queen)} &= \text{P(drawing a heart)} + \text{P(drawing a queen)} \\ &\quad - \text{P(drawing the queen of hearts)} \\ &= \frac{13}{52} + \frac{4}{52} - \frac{1}{52} \\ &= \frac{16}{52} = \frac{4}{13}. \end{aligned}$$

5.3.4 Using a Venn Diagram to Look at Mutually Exclusive Events

A Venn-diagram representation of two events in a given experiment can help determine whether or not the two events are mutually exclusive or non-mutually exclusive. For example, consider an experiment involving spinning the following spinner (see Figure 5.3):

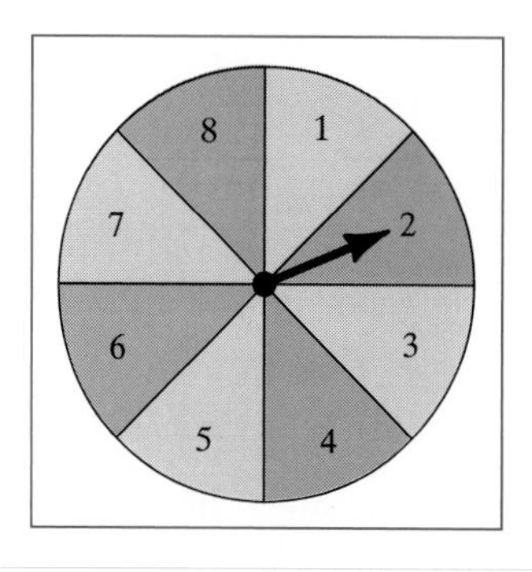

Figure 5.3

If we want to determine the probability of spinning an odd number or a number that is a multiple of 3, we can determine whether these events are mutually exclusive or not by forming the Venn diagram shown in Figure 5.4.

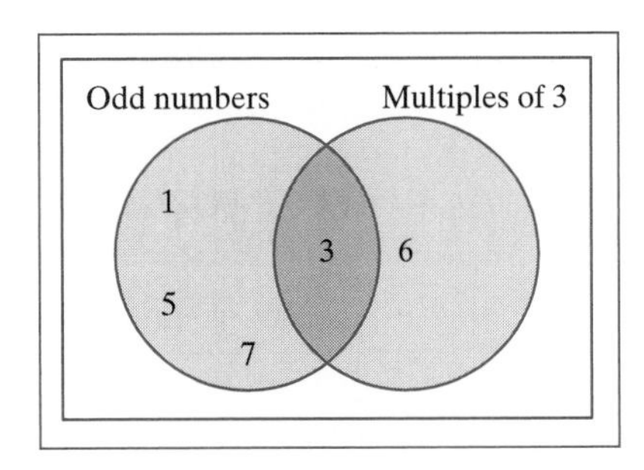

Figure 5.4

The overlap in the Venn diagram shows that the two events are not mutually exclusive. Therefore, we can determine the probability of spinning an odd number or a multiple of 3 by:

$$\begin{aligned} P(\text{odd or multiple of 3}) &= P(\text{odd}) + P(\text{multiple of 3}) - P(\text{odd and multiples of 3}) \\ &= \frac{4}{8} + \frac{2}{8} - \frac{1}{8} = \frac{5}{8}. \end{aligned}$$

5.3.5 Interpreting an "Or" Situation

Figure 5.5

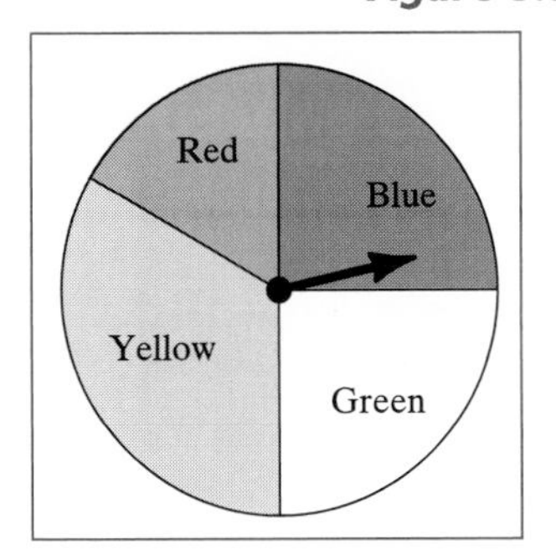

In some probability questions, it is not explicitly clear that we are dealing with an "or" situation, even when it is, in fact, an "or" situation. For example, consider the spinner shown in Figure 5.5.

If we ask the question, "What is the probability of spinning a primary color?" we can interpret the question as an "or" situation, asking instead, "What is the probability of spinning a red or a blue or a yellow?" Because these events are mutually exclusive, we can solve the problem:

$$\begin{aligned} P(\text{spinning a primary color}) &= P(\text{spinning red or blue or yellow}) \\ &= P(\text{red}) + P(\text{blue}) + P(\text{yellow}) \\ &= \frac{1}{6} + \frac{1}{4} + \frac{1}{3} = \frac{2}{12} + \frac{3}{12} + \frac{4}{12} = \frac{9}{12} = \frac{3}{4}. \end{aligned}$$

As another example, suppose a number, x, is selected at random from a set of numbers $\{1, 2, 3, \ldots 8\}$. What is the probability that $x < 5$?

This question can be rephrased as an "or" situation asking, "What is the probability of selecting a 1 or 2 or 3 or 4?" Thus, because these are mutually exclusive events, we solve by finding $P(1) + P(2) + P(3) + (P4) = \frac{1}{8} + \frac{1}{8} + \frac{1}{8} + \frac{1}{8} = \frac{4}{8} = \frac{1}{2}$.

5.3.6 Table of Data Simulation

Sometimes, we are given empirical data from which to determine the answer to an "or" probability situation. Suppose we had 1,000 people volunteer to participate in a study. From a questionnaire, we learn that some of the male and female participants in the study are smokers and some are nonsmokers. Consider the table of data of male and female smokers and nonsmokers provided in Table 5.4.

If we randomly choose one person from the pool of 1,000 volunteers to be the first person to be interviewed, what is the probability that the person selected is

Table 5.4

	Smokers	Nonsmokers	Total
Female	198	245	443
Male	310	247	557
Total	508	492	1,000

either a female or a nonsmoker? Using data from the table, and noting that the events "being a female" and "being a nonsmoker" are not mutually exclusive, we compute:

$$\begin{aligned}\text{P(female or nonsmoker)} &= \text{P(female)} + \text{P(nonsmoker)} - \text{P(nonsmoking female)}\\ &= \frac{443}{1000} + \frac{492}{1000} - \frac{245}{1000}\\ &= \frac{690}{1000}\\ &= \frac{69}{100}.\end{aligned}$$

A Venn-diagram (see Figure 5.6) representation of the male and female smokers and nonsmokers illustrates that the events "being female" and "being a nonsmoker" are not mutually exclusive. The striped section indicates nonsmokers and the polka-dotted section indicates females. The overlap is seen as the 245 nonsmoking females (refer to Figure 5.6).

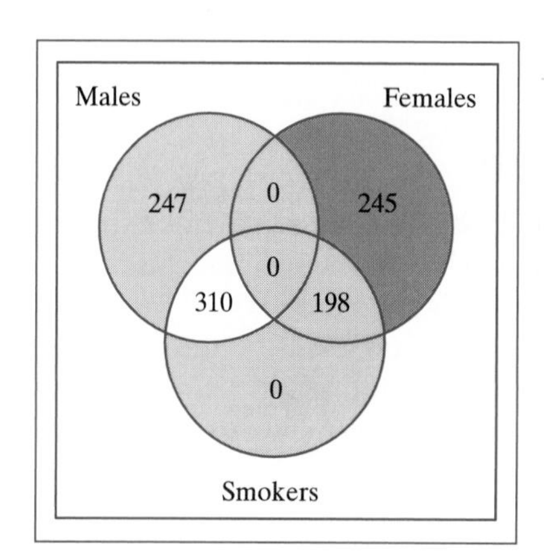

Figure 5.6

5.3.7 Complementary Events

Given an event A, we say the **complement** of event A, denoted $\overline{A}$, is the sample space of an experiment less the event A. Stated another way, the complement of event A is the set of all events in the sample space *except* event A. For example, when rolling a die, if we know event A = rolling a 3, then the complement of A is $\overline{A}$ = rolling anything but 3. Note that sometimes the complement of A, $\overline{A}$, is referred to as "not A."

The probability of a complementary event, $\overline{A}$, denoted $P(\overline{A})$, is closely related to the probability of event A, P(A): $P(\overline{A}) + P(A) = 1$. Thus, given the probability of an event A, we can determine the probability of its complement, or vice versa, using this relationship as follows:

$$P(\overline{A}) = 1 - P(A)$$
$$P(A) = 1 - P(\overline{A})$$

For example, given a hat with 4 green marbles, 3 blue marbles, and 2 black marbles, and the experiment of drawing a marble out of the hat, suppose A is the event drawing a black marble. Then $\overline{A}$ = not drawing a black marble, or $\overline{A}$ = drawing a green or blue marble. Consequently, because $P(A) = \frac{2}{9}$, then $P(\overline{A}) = 1 - \frac{2}{9} = \frac{7}{9}$.

5.3.8 Odds In Favor and Odds Against

To determine the **odds in favor** of an event, E, occurring, we compute the ratio, $\frac{P(E)}{P(\overline{E})}$ [or $P(E): P(\overline{E})$]. To compute the **odds against** an event, E, occurring, we compute the ratio, $\frac{P(\overline{E})}{P(E)}$ [or $P(\overline{E}) : P(E)$].

Example 1: Leroy has a chance for 5 different summer jobs, 3 of which are at resort areas. If he selects a job at random, find the odds in favor of the job being at a resort and the odds against its being at a resort.

The event at hand is E = selecting a job at a resort. Thus, $P(E) = \frac{3}{5}$ and $P(\overline{E}) = \frac{2}{5}$. To determine the odds in favor, we find $\frac{P(E)}{P(\overline{E})} = \frac{3/5}{2/5} = \frac{3}{2}$ or 3 : 2. To determine the odds against, we compute $\frac{P(\overline{E})}{P(E)} = \frac{2/5}{3/5} = \frac{2}{3}$ or 2 : 3.

Example 2: You have a "loaded" coin where there is a two-thirds chance of flipping heads and a one-third chance of flipping tails. What are the odds in favor of flipping heads? What are the odds against flipping heads?

$$\text{Odds in Favor of flipping heads} = \frac{\frac{2}{3}}{\frac{1}{3}} = \frac{2}{1} \text{ or } 2:1.$$

$$\text{Odds Against flipping heads} = \frac{\frac{1}{3}}{\frac{2}{3}} = \frac{1}{2} \text{ or } 1:2.$$

Related Topics

Multistep Experiments

Multistep Experiments

cornerstone

Probability Notions
Equally Likely Outcomes
Mutually Exclusive and Complementary Events
Multiplying Rational Numbers

Introduction

I flip a coin five times, and I wonder how likely it is that it will land on heads each time. Problems like this center around multistep experiments. The probability of a coin landing on heads on one flip is much different than the probability of this happening five times in a row. In this section, we examine probabilities associated with multiple events, determining when these events are independent from each other and when they are dependent in a way that affects the probability of both events occurring.

Did you know that in a group of 15 people, there's a 25% chance that at least two people will have the same birthday? Further, if you are in a group of 25 people, there's almost a 60% chance that at least two people will have the same birthday. In fact, in a group of 35 people, there's more than an 80% chance that at least two people will have the same birthday. Next time you are in a group of 35 people, test the probability!

5.4.1 Independent Versus Dependent Events

When an experiment involves multiple events, to compute the probability of both events occurring, we must first determine whether those events are independent or dependent. Two events, one to follow the other, are **independent** if the occurrence of the first event does not affect the probability of the occurrence of the second event. Two events, one to follow the other, are **dependent** if the occurrence of the first event affects the probability of the occurrence of the second event.

For example, suppose I am going to roll a die twice. What is the probability that we will roll a 3 on the first roll and an even number on the second roll? Because the die is "fixed," that is, does not change after the first roll, the events of rolling a 3 and then rolling an event number are independent.

On the other hand, suppose we have a hat with 6 red marbles and 4 white marbles. We are to draw two marbles out of the hat, one right after the other, without replacing the first marble. What is the probability that we draw a red marble on the first draw and a white marble on the second draw? Because, after the first draw, the marble is not replaced, we have reduced the number of marbles in the hat. This will affect the probability of drawing a white marble on the second draw. Therefore, these two events, drawing a red marble followed by drawing a white marble, are dependent.

5.4.2 Conditional Probability

When the probability of event B is affected by the occurrence of event A, we express the probability of B, given that A has occurred, read **B given A,** using the notation P(B|A). We call a situation involving P(B|A) a **conditional probability.**

For example, if we have a hat with 6 red marbles and 4 white marbles, as previously described, and we draw 2 marbles out, one at a time, without replacing the first marble, we can determine the probability of drawing the white marble on the second draw using conditional probability. If we have drawn a red marble on the first draw and have not replaced the marble, then we are left with just 9 marbles in the hat, 5 being red and 4 being white. Thus, the probability of drawing a white marble on the second draw is affected by the fact that one red marble has been removed from the hat. Thus, the P(drawing a white | a red marble has been drawn) $= \frac{4}{9}$.

5.4.3 Understanding "And" Probability Situations

In probability situations where we determine the probability of two events, A and B, both occurring, denoted P(A and B) or P(A ∩ B), we compute the probability of these "and" probability settings by:

$$P(\text{A and B}) = P(\text{A}) \times P(\text{B}) \text{ when A and B are independent events.}$$
$$P(\text{A and B}) = P(\text{A}) \times P(\text{B}|\text{A}) \text{ when A and B are dependent events.}$$

Figure 5.7

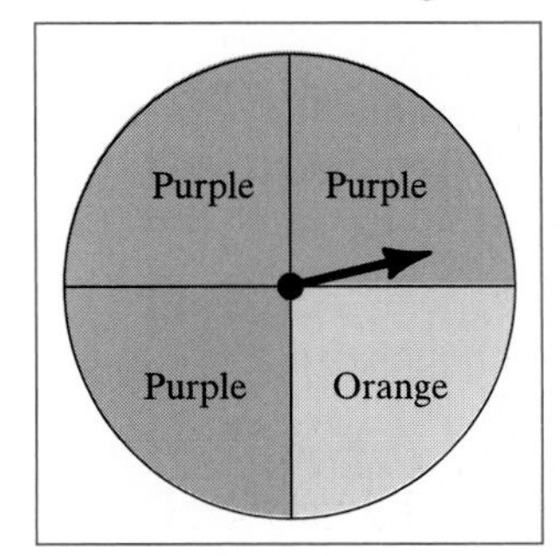

Example 1: Suppose we are experimenting with the spinner shown in Figure 5.7 and we spin the spinner twice. We may ask the question, "What is the probability of spinning red twice in a row?" This is equivalent to determining the probability of spinning a red and a red. Because what happens on the first spin does not affect the probabilities associated with the second spin, the two events are independent. Therefore, the probability of spinning a red and then another red is found by:

$$P(\text{spinning red and red}) = P(\text{red}) \times P(\text{red}) = \frac{1}{4} \times \frac{1}{4} = \frac{1}{16}.$$

Example 2: Suppose we revisit the situation where we have a hat with 6 red marbles and 4 white marbles and we draw two marbles, without replacing the first marble drawn. What is the probability of drawing a red marble followed by a white marble? The probability of drawing a red marble on the first draw is $\frac{6}{10}$. If we draw a red marble on the first draw and do not replace the marble, then we are left with just 9 marbles in the hat, 5 being red and 4 being white. Thus, the probability of drawing a white marble on the second draw is affected by the fact that one red marble has been removed from the hat. Thus, the P(drawing a white | a red marble has been drawn) $= \frac{4}{9}$. Thus, to compute the probability of drawing the red marble and the white marble as described, we compute:

$$\begin{aligned} P(\text{red first and white second}) &= P(\text{red}) \times P(\text{white}|\text{red}) \\ &= \frac{6}{10} \times \frac{4}{9} \\ &= \frac{24}{90} = \frac{4}{15}. \end{aligned}$$

Example 3: Suppose you are going to draw four cards from a deck of cards, one right after the other, replacing cards into the deck after each draw. What is the probability of drawing four queens? We can think of this as an "And" the situation of finding the probability of drawing a queen and a queen and a queen and a queen. Because we are replacing the cards after each draw, each event is independent of the others. Therefore we compute:

$$\begin{aligned} &P(\text{queen and queen and queen and queen}) \\ &\quad = P(\text{queen}) \times P(\text{queen}) \times P(\text{queen}) \times P(\text{queen}) \\ &\quad = \frac{4}{52} \times \frac{4}{52} \times \frac{4}{52} \times \frac{4}{52} \\ &\quad = \frac{256}{7{,}311{,}616} \\ &\quad = \frac{1}{28{,}651}. \end{aligned}$$

Example 4: We are going to draw 5 cards from a deck of cards, one right after the other, without replacing cards after they have been drawn. What is the probability of drawing five hearts? As an "And" situation, we must find the probability of drawing a heart and a heart and a heart and a heart and a heart. In this case, because cards are not being replaced after each draw, the events in this multistep experiment are dependent. Thus, we determine the probability by:

P(heart and heart and heart and heart and heart) =
P(heart) × P(heart | 1 heart has been drawn) × P(heart | 2 hearts have been drawn) ×
P(heart | 3 hearts have been drawn) ×
P(heart | 4 hearts have been drawn)

$$= \frac{13}{52} \times \frac{12}{51} \times \frac{11}{50} \times \frac{10}{49} \times \frac{9}{48}$$

$$= \frac{33}{66{,}640}.$$

Notice that the conditional probability associated with the second draw indicates that after the first card has been drawn, there are only 51 cards left. Also, because we're assuming that the first card drawn was a heart, then only 12 hearts remain after the first draw. Thus, the probability for drawing a heart, given that one heart has already been drawn, is $\frac{12}{51}$. The determination of the conditional probabilities for the third, fourth, and fifth draws follow a similar process.

5.4.4 Tree Diagrams

Tree diagrams can illustrate probabilities associated with multistep experiments by showing the possible combination of outcomes at each step of multistep experiments. For example, suppose we are working with the spinner in Figure 5.8.

Figure 5.8

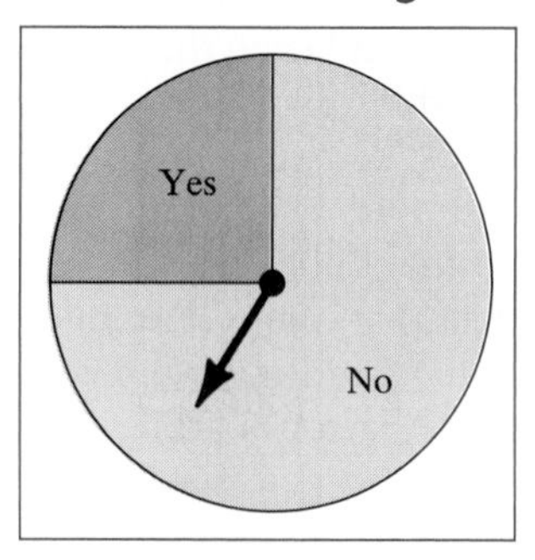

As seen in the spinner, P(Yes) = $\frac{1}{4}$, and P(No) = $\frac{3}{4}$. If we spin the spinner three times, we can illustrate the various probabilities associated with the three spins and the different combinations of outcomes by using a **tree diagram.** Note that the events associated with each spin of this multistep experiment are independent because the spinner is not altered from spin to spin.

Figure 5.9 illustrates the tree diagram for the probabilities of three spins of this spinner. Highlighted in the diagram is the path that we could follow to determine the probability of spinning a "Yes" followed by a "No" followed by a "Yes." We would compute the probability:

$$P(\text{Yes and No and Yes}) = P(\text{Yes}) \times P(\text{No}) \times P(\text{Yes})$$

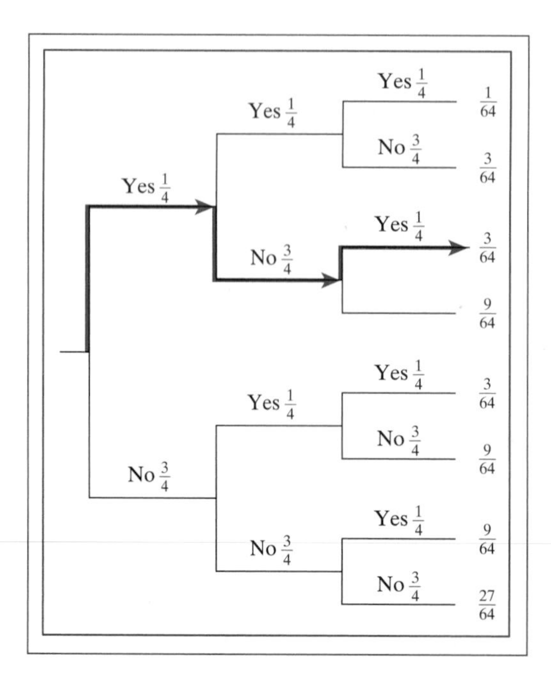

Figure 5.9

$$= \frac{1}{4} \times \frac{3}{4} \times \frac{1}{4}$$
$$= \frac{3}{64}$$

Related Topics

Counting Principles
Whole-Number Multiplication

TOPIC 5.5 Counting Principles

cornerstone

Multiplying Whole Numbers
Rational-Number Multiplication

Introduction

When it comes to determining how many ways to arrange books on a shelf or how many ways to put people into groups, we can use counting techniques that make the counting process simpler and easier. In this section, we examine ways to compute combinations, permutations, and basic counting principles.

Mathematician Blaise Pascal is known for developing Pascal's triangle, a triangular arrangement of numbers that has many interesting patterns. One notable pattern in the triangle is the row-by-row pattern of numbers that show the number of combinations you can make from n items, selecting r at a time. For example, looking at row 4 of Pascal's triangle shown in Figure 5.10, the numbers across represent, in order, the number of combinations you can make (1) choosing 0 from a set of 3 objects, (2) choosing 1 from a set of 3 objects, (3) choosing 2 from a set of 3 objects, and (4) choosing 3 from a set of 3 objects. Those possible combinations, in order, are 1, 3, 3, 1.

Row 3 of Pascal's triangle shows the combinations you can make with 2 objects; that is, the first number, 1, is the number of combinations you can create with 2 objects, selecting 0 at a time. The second number, 2, is the number of combinations you can create with 2 objects, selecting 1 at a time. Finally, the last number, 1, is the number of combinations you can create with 2 objects, selecting 2 at a time. In general, the nth row in Pascal's triangle represents the number of combinations you can make with $n - 1$ objects, selecting 0 at a time, 1 at a time, 2 at a time, ... $3n - 1$ at a time. For more information on determining combinations, see the section entitled Computing Combinations in this topic.

For More Information Reimer, L., & Reimer, W. (1990). *Mathematicians Are People Too: Stories from the Lives of Great Mathematicians, Volume 1*. Palo Alto, CA: Dale Seymour Publications.

5.5.1 Factorial

Given a whole number, n, we compute the factorial of n, read ***n* factorial** and denoted ***n!***, by multiplying n times all the nonzero whole numbers less than n; that is:

$$n! = n \times (n - 1) \times (n - 2) \times (n - 3) \times \cdots \times 3 \times 2 \times 1.$$

For example, $4! = 4 \times 3 \times 2 \times 1 = 24$. Likewise, $10! = 10 \times 9 \times 8 \times 7 \times 6 \times 5 \times 4 \times 3 \times 2 \times 1 = 3{,}628{,}800$. Note that $0!$ is defined to be equal to 1. Note that most calculators have a factorial button, often labeled $n!$, that compute the factorial value for the value of n that is input into the calculator.

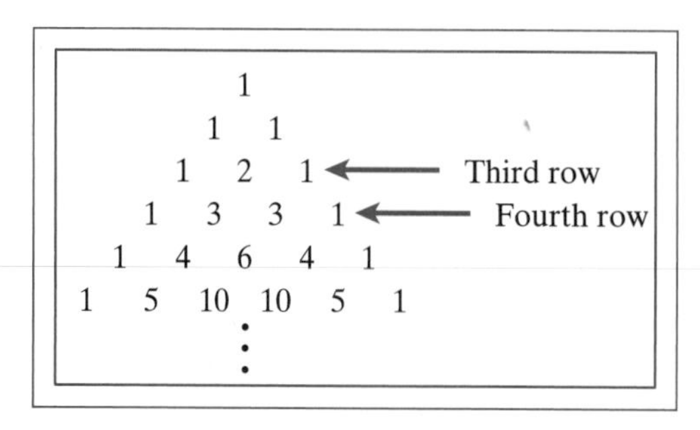

Figure 5.10

5.5.2 Definition of Permutations

Given a set of objects, we can compute the number of unique ways to arrange those objects in different orders. The arrangements of a set of objects are called the **permutations** of the objects. For example, given the set of objects = {a, b, c}, we can determine six permutations of these objects: abc, acb, bac, bca, cab, cba.

Suppose we have the set of objects = {a, b, c, d} and, selecting two at a time, we want to determine all the permutations of two items from the set. There are 12 such permutations: ab, ba, ac, ca, ad, da, bc, cb, bd, db, cd, dc.

5.5.3 Computing Permutations

Given whole numbers n and r, where $n \geq r$, we can determine the number of permutations of n thing taken r at a time, denoted $P(n, r)$, by the formula:

$$P(n, r) = \frac{n!}{(n - r)!}.$$

For example, how many permutations of 6 objects, taken 2 at a time, are there? We compute: $P(6, 2) = \frac{6!}{(6 - 2)!} = \frac{6!}{4!} = \frac{6 \times 5 \times 4 \times 3 \times 2 \times 1}{4 \times 3 \times 2 \times 1} = \frac{30}{1} = 30.$

Note that $P(n, n) = n!$ For example, the number of permutation of 5 objects selected 5 at a time is found by $P(5, 5) = 5! = 720$.

5.5.4 Permutation Notation and the Calculator

Many calculators have the capability of computing permutations, given your input values for n and r. Some calculators will have a button labeled $P(n, r)$ or ${}_nP_r$. Although each calculator works in its own way, most calculators require that you input the value for n and then punch the ${}_nP_r$ button and then will wait for you to input the r value. Once the r value is input, you then punch the "equal to" button, and the calculator will display the result.

5.5.5 Definition of Combinations

Given a set of objects, we can compute the number of unique ways to group those objects into certain-sized sets. These groupings of a set of objects are called the **combinations** of the objects. For example, given the set of objects = {x, y, z}, we can determine three ways to combine two of the objects at a time: xy, xz, yz. When we are finding the combinations of objects, we are not concerned with the different orderings or permutations of the groups of objects, but simply the different groupings.

Suppose we have the set of objects = {w, x, y, z} and we want to determine all the different combinations we can form taking three objects at a time. There are four such combinations: wxy, wxz, wyz, xyz.

5.5.6 Computing Combinations

Given whole numbers n and r, where $n \geq r$, we can determine the number of combinations of n things taken r at a time, denoted $C(n, r)$, by the formula:

$$C(n, r) = \frac{n!}{r!\,(n - r)!}.$$

For example, how many combinations of 8 objects, taken 3 at a time, are there? We compute: $C(8, 3) = \frac{8!}{3!\,(8 - 3)!} = \frac{8!}{3!\,5!} = \frac{8 \times 7 \times 6 \times 5 \times 4 \times 3 \times 2 \times 1}{3 \times 2 \times 1 \times 5 \times 4 \times 3 \times 2 \times 1} = \frac{56}{1} = 56.$

Note that $C(n, n) = 1$. For example, the number of combinations of 5 objects selected 5 at a time is 1.

5.5.7 Combination Notation and the Calculator

Many calculators have the capability of computing combinations given your input values for n and r. Some calculators will have a button labeled $C(n, r)$ or ${}_nC_r$. Although each calculator works in its own way, most calculators require that you input the value for n and then punch the ${}_nC_r$ button and then will wait for you to input the r value. Once the r value is input, you then punch the "equal to" button and the calculator will display the result.

5.5.8 Fundamental Counting Principle

In some cases, we need to count the number of ways that a certain situation can occur, and the method required does not involve permutations or combinations. For these types of problems, we can create a solution using the Fundamental Counting Principle.

The **Fundamental Counting Principle** says that given events A and B, and letting n(A) represent the number of ways for A to occur and n(B|A) represent *the number of ways for B to occur, given that A has occurred*, then the number of ways for A and B to both occur, denoted n(A and B) is found by:

$$\text{n(A and B)} = \text{n(A)} \times \text{n(B|A)}$$

Example 1: Suppose we are creating license-plate codes that are comprised of three letters of the alphabet, with no letter being used more than once in a given code. To count the number of different license-plate codes we can develop, we can use the Fundamental Counting Principle. We can compute the number of codes by:

$$\begin{aligned}\text{n(codes)} &= \text{n(choices for first letter)}\\ &\times \text{n(choices for second letter|one letter has already been selected)}\\ &\times \text{n(choices for third letter|two letters have already been selected).}\\ &= 26 \times 25 \times 24 = 15{,}600.\end{aligned}$$

Note that there are 26 letters in the alphabet, so there are 26 possible choices for the first letter in the code. Once one letter has been used, there are only 25 letters left to choose from for the second letter in the code. Finally, once the first two letters have been chosen, only 24 letters remain to choose for the third letter in the code. Thus, we find the total number of possible codes by multiplying these numbers together.

Example 2: Within a certain area code, we want to figure out all of the 7-digit phone numbers the phone company can distribute in a certain town where phone numbers all begin with the prefix 358. Thus, phone numbers in that area would all look like: 358 - ____. We essentially need to determine the number of ways to fill in the last four digits of the phone number.

Thinking in terms of the Fundamental Counting Principle, to solve this problem we need to find the number of ways to create the first missing digit *and* the number of ways to create the second missing digit *and* the number of ways to create the third missing digit *and* the number of ways to create the fourth missing digit. Note that in this scenario, it is okay to use the same digit more than once in the last four digits in a given phone number. Because we have 10 digits (0, 1, 2, 3, 4, 5, 6, 7, 8, 9) from which to choose for each of the four digits, we can conclude that we have $10 \times 10 \times 10 \times 10$ or 10,000 different phone numbers to create.

Related Topics
Multistep Experiments

TOPIC 5.6 Statistics Notions

cornerstone

Percents
Fractions
Decimals

Introduction

Of all the branches of mathematics, statistics is probably the most visible one in the everyday world. If you pick up any newspaper or magazine today, you are likely to come across various surveys done on different categories of population or data records such as those in sports, medicine, or the corporate world. In this section, we define statistics and terms associated with statistical studies, including hypotheses, populations, and random sampling.

The beginning of the modern theory of statistics and statistical analysis can be traced back to the time of the Industrial Revolution. Statistical analysis began at this time for practical reasons. It was at this time that mass production through large factories began. This created urban populations with all the attendant problems of overpopulation, disease, unemployment, and large population changes. Social scientists of that time were searching for tools with which to study these changes and were unable to come to grips with the fast-changing social scenario. Then two 17th-century Englishmen, John Graunt and William Petty, studying the death records of English cities, came across surprising patterns of these data. Based on these data, the two made some startling observations. These observations laid the foundations of the study of statistics in the modern sense.

For More Information Reimer, L., & Reimer, W. (1995). *Mathematicians Are People, Too: Stories from the Lives of Great Mathematicians, Volume 2*. Palo Alto, CA: Dale Seymour Publications.

5.6.1 Statistics and Statistical Data

Statistics is a collection of methods for planning experiments, gathering data, and then organizing, analyzing, interpreting, and representing data, with the intent of drawing conclusions and making decisions based on the data.

A piece of **data** is a fact or figure that provides some information, usually gathered either to support a position or for some sort of decision-making purpose. **Statistical data** is usually numerical information that can be presented in a variety of forms, including whole numbers, fractions, decimal numbers, or percents. Examples of data in use are seen in the following statements:

More than 50% of all marriages end in divorce.

The Dow Jones Average went up 15 points today.

Fourth-fifths of college students own their own cars.

The government spends an average of $1.5 million per year on miscellaneous expenditures.

5.6.2 Hypotheses

A **hypothesis** is a statement that is used as a premise for investigating and for proving whether something is true or false. Investigating a hypothesis involves gathering and analyzing data. Examples of hypotheses are the statements, "Most American college students who own their own cars own domestic cars," "Smoking causes cancer," and "The average family has 1.5 pets." Each of these statements is stated as if it is true.

However, to prove their truth or falseness, we must gather information. For example, to determine whether or not most American college students own domestic cars, the investigator would have to survey a sample of college students and use this data to determine whether or not it is appropriate to conclude that most American college students own domestic cars.

5.6.3 Populations, Samples, and Types of Sampling

In a statistical study, the **population** is the complete collection of items (people, test scores, etc.) to be studied. For example, in studying the hypothesis, "Most American college students who own their own cars own domestic cars," the population for the study would be all American college students who own their own cars.

Because it would be impossible to contact each and every American college student to ask him (or her) if he owns a car and whether or not the car is domestic, we would probably draw on a subgroup of college students to survey who would be representative of the entire population. This group of items drawn from a population is called a **sample.**

Samples can be determined in a variety of ways. In **random sampling,** a subset of the population is selected in a way that each member of the population has an equal chance of being selected. For example, when you are called on the telephone to be polled about your choice of presidential candidates, your phone number was selected randomly by a computer-generated program. Thus, anyone with a phone number on the computer's list had a chance of being called.

In some instances, when sampling from a population, we need to subdivide that population into at least two different subgroups that share the same characteristics. In this situation, we are engaged in **stratified sampling.** For example, suppose we are studying the number of hours that college students prepare for exams. We may want to make sure we include or differentiate between freshmen, sophomores, juniors, and seniors. Thus, the population of college students is stratified into four categories, and samples are drawn, proportionally, from each of the categories. Within each category, random sampling can be applied so that each freshmen, each sophomore, each junior, and each senior has an equal chance of being selected. Note that in this particular stratified sampling, we are guaranteeing that the sample we select will include students from each class.

Systematic sampling is a method of sampling where we identify a starting point and then select every kth element in the population. For example, in a graduating class, you might select every fourth graduate exiting the graduation ceremony to survey about a current events issue.

In **cluster sampling,** we divide a population area into sections and then randomly select a few of those sections. Then we choose all the members from the selected sections. For example, if we want to survey high school students about school lunches, we can randomly select certain tables in the cafeteria and survey all the students sitting at those tables.

Finally, there is **convenience sampling,** where we simply use the sample that is available to us. For example, if a veterinarian wants to study the effectiveness of a new type of flea collar, she might use her own patients to test the flea collar because these pets are readily available.

Measures of Central Tendency
Data Dispersion
Representing Data

TOPIC 5.7 *Measures of Central Tendency*

cornerstone **Statistics Concepts**

Introduction

In many statistical situations, we want to know what happens in general or on average. For example, we might want to know the average income of women in the United States as compared to the average income of men, or we might want to know the most popularly selling car during the last five years. Both of these situations are asking to find some sort of measure of central tendency or some sense of what is happening on average. In the following sections, we explore a variety of ways of determining a measure of central tendency, including finding the mean, the median, and the mode of a set of data.

Two important early instances of the use of statistics are provided by Florence Nightingale and Gregor Mendel. Florence Nightingale (1820–1910), the famous English nurse who worked with death records in hospitals during the Crimean War, used statistics to show that change was desperately needed in the conditions of the hospitals. The Austrian monk, Gregor Mendel (1822–1911), maintained records of his botany experiments and, from the results, laid the foundations of the science of genetics.

For More Information Smith, D. E. (1958). *History of Mathematics, Volume II*. New York: Dover Publications.

5.7.1 Mean

In statistical studies, we often want to determine the mean of the data. The **mean** is one way of determining what the data imply on average. To compute the mean, denoted, $\bar{x}$, we divide the sum of individual numerical data items, $x_1, x_2, x_3, \ldots, x_n$ by the total number of data items, n; that is:

$$\text{Mean} = \bar{x} = \frac{x_1 + x_2 + x_3 + \cdots + x_n}{n}$$

For example, we might want to determine the mean test score on a final exam in a history course. Thirty students took the final exam, and the scores earned were the following:

44, 51, 60, 61, 62, 66, 66, 68, 70, 71,
71, 71, 74, 75, 78, 79, 79, 80, 81, 82,
83, 83, 84, 84, 85, 85, 87, 88, 90, 93

To determine the mean, we find the sum of the test scores and divide by 30:

$$\bar{x} = \frac{44 + 51 + 60 + \cdots + 90 + 93}{30} = \frac{2183}{30} = 72.8.$$

Notice that even though the raw scores were given as whole numbers, the mean shown is a decimal number.

In some statistical situations, you may not have the raw data itself but may have the data presented in a table such as a frequency table (refer to the section on Frequency Tables in the Representing Data topic for more details on developing frequency tables). We can compute the mean of the data presented in a frequency table.

Suppose, instead of raw data form, the data on history final exam scores are presented in a frequency table as shown in Table 5.5. The class column indicates how

Table 5.5

Class	Frequency
40–49	1
50–59	1
60–69	6
70–79	9
80–89	11
90–99	2
	30

scores were classed or categorized as scores between 40 and 49, between 50 and 59, and so on, and the frequency column depicts how many test scores fell within that categorization with the total frequency being 30, the total number of test scores.

Because we do not know, for example, the exact scores in the 60–69 category, we can determine a representative score for that category by computing the mean of the high and the low score represented by the category; that is, finding the mean of 60 and 69, $\frac{60+69}{2} = \frac{129}{2} = 64.5$, we can let the score 64.5 represent the average score earned within that category. We call this representative score for the class the **class mark.** Using the class marks and the frequency of each class, we can approximate the sum of test scores in class and the sum of the all test scores as shown in Table 5.6.

Table 5.6

Class	Frequency	Class Mark	Class Mark × Frequency
40–49	1	44.5	44.5
50–59	1	54.5	54.5
60–69	6	64.5	387.0
70–79	9	74.5	670.5
80–89	11	84.5	929.5
90–99	2	94.5	189.0
	30		Sum of All Scores 2,275.0

We then compute the mean from the frequency table by dividing the sum of the scores determined in the last column by the total frequency. Computing, we find: $\frac{2275}{30} = 75.8$. You may note that the mean that was determined by using raw data scores, 72.8, differs from the mean that was determined by using the frequency table, 75.8. Of course, the mean from the frequency table is an approximation. However, if raw data are not available but a frequency table is, the mean from the frequency table usually provides a reasonable representation of the true mean.

5.7.2 Mode

Another way to examine data and determine what occurs on average is to find the mode of the data. The **mode** of a set of data is the score, or item, that occurs most often. For example, consider the 30 test scores from the history final discussed:

44, 51, 60, 61, 62, 66, 66, 68, 70, 71,
71, 71, 74, 75, 78, 79, 79, 80, 81, 82,
83, 83, 84, 84, 85, 85, 87, 88, 90, 93

We look at the scores and notice that 71 occurred most often. Therefore, 71 is mode of this data set. In some data sets, there may be more than one mode. If there are two scores that occur the same number of times and more than any others, we say the data set is **bimodal.** If there are more than two modes, we say the data is **multimodal.** Finally, if no score occurs more than any others, we say there is **no mode.**

If raw data are not available but we have a frequency table of data, we can approximate the mode. Recall the frequency table for the history final exam scores (see Table 5.7).

Table 5.7

Class	Frequency
40–49	1
50–59	1
60–69	6
70–79	9
80–89	11
90–99	2
	30

We look to the frequency table and note the class that has the highest frequency. In this example, the class 80–89 has the highest frequency. We then find the class mark, 84.5, of that class and let that class mark represent the mode of the data in this frequency table. As in the case where the mean of raw scores and the mean found in a frequency table differ, so do the modes found from raw data and from frequency tables.

5.7.3 Median

The **median** of a set of data is the middle score or the data item when data items are written in increasing order. Determining the median differs, depending on whether there is an even or an odd number of data items. Suppose, for example, we have 15 quiz scores:

39, 42, 59, 64, 66, 70, 77, **80,** 81, 82, 83, 83, 86, 90, 99

Because the scores are already written in increasing order, the middle score can be identified within the data set as the eighth score in the list, or 80.

Suppose instead of 15 scores, we have the following 16 scores:

39, 42, 59, 64, 66, 70, 77, **80, 81,** 82, 83, 83, 86, 90, 99, 100

Table 5.8

Class	Frequency
40–49	1
50–59	1
60–69	6
70–79	9
80–89	11
90–99	2
	30

There is no single middle score among this data set because there is an even number of scores. Thus, to determine the median, we find the middle two scores and determine the average of those two scores. In this example, the middle two scores are 80 and 81. So, the median is $\frac{80 + 81}{2} = \frac{161}{2} = 80.5$.

When data are provided in a frequency table instead of in raw data form, we can determine the median. For example, consider the data in the frequency table of history test scores as revisited in Table 5.8.

To determine the median, because there is an even number of scores, we need to average the middle two scores—that is, the class marks associated with the 15th and 16th scores. Using the frequency column, beginning at the top, we can add the frequencies until we get a sum of 15 and a sum of 16 to find into which classes the 15th and 16th scores fall. Adding we find that both the 15th and 16th scores fall within the class 70–79. Therefore, the average of the class marks for the 15th and 16th scores would be 74.5 ($\frac{70 + 79}{2} = 74.5$). Thus, the median for the data in this frequency table is 74.5.

5.7.4 Range and Midrange

Given a set of raw data, we can determine the range of scores. The **range** is computed by subtracting the high score from the low score; that is, range = high − low. Thus, given the following set of data scores:

44, 51, 60, 61, 62, 66, 66, 68, 70, 71,
71, 71, 74, 75, 78, 79, 79, 80, 81, 82,
83, 83, 84, 84, 85, 85, 87, 88, 90, 93

Table 5.9

Class	Frequency
40–49	1
50–59	1
60–69	6
70–79	9
80–89	11
90–99	2
	30

we compute the range: Range = 93 − 44 = 49. Thus, the data scores in this example have a range of 49.

The **midrange** of a data set can be found by finding the average of the high and low scores; that is, Midrange $= \frac{\text{high} + \text{low}}{2}$. Using the same data set used to compute the range, we find that the midrange $= \frac{93 + 44}{2} = \frac{137}{2} = 68.5$.

Both the range and the midrange can be approximated from a frequency table. Given a frequency table, such as the one shown in Table 5.9, we can assume the low score to be the lowest number in the range of the first class and the high score to be the highest number in the range of the last class. In this example, the low score would be 40 and the high score would be 99. Computing the range and the midrange with these scores, we find, range = 99 −40 = 59, and the Midrange = $\frac{40 + 99}{2} = \frac{139}{2} = 69.5$.

5.7.5 Weighted Average

With some data, we can compute an average where different data items are given different weights or values. Given data items $x_1, x_2, x_3, \ldots x_n$, with their corresponding weights $w_1, w_2, w_3, \ldots w_n$, we can compute the weighted average of these n data items by:

$$\text{Weighted Average} = \frac{(w_1 \times x_1) + (w_2 \times x_2) + (w_3 \times x_3) \ldots (w_n \times x_n)}{(w_1 + w_2 + w_3 + \ldots + w_n)}$$

For example, suppose a student's grade is based on quizzes, exam 1, exam 2, a project, a portfolio, a reflective writing journal, and a final exam. Joan's score on each item and the weight that each item will have on the overall grade determination are shown in Table 5.10.

Table 5.10

Assessment Item	Score	Weight on Grade
Quiz average	77	14
Exam 1	76	15
Exam 2	65	15
Project	86	13
Portfolio	89	13
Reflective Writing	95	10
Final Exam	79	20

To determine Joan's grade, or weighted average, we compute:

$$\begin{aligned} \text{Weighted Average} &= [(77 \times 14) + (76 \times 15) + (65 \times 15) + \\ &\quad (86 \times 13) + (89 \times 13) + (95 \times 10) + (79 \times 20)]/(14 + \\ &\quad 15 + 15 + 13 + 13 + 10 + 20) + \\ &= (1078 + 1140 + 975 + 1118 + 1157 + 950 + 1580)/100 \\ &= 7998/100 \\ &= 79.98 \end{aligned}$$

Related Topics

Data Dispersion
Representing Data

TOPIC 5.8 *Data Dispersion*

cornerstone **Statistics Concepts**
Measures of Central Tendency

Introduction

Statistics provide a means not only of determining what happens on average, but also of determining how an individual score compares to the average score or to the rest of the data scores. For example, when students receive their SAT scores, they are told their individual score as well as the percentile that their score represents. This percentile score explains how high their score was compared to the other students who took the SATs. In the following sections, we explore how to compute percentile scores as well as other ways of showing how data scores are dispersed, including finding the variance and standard deviation.

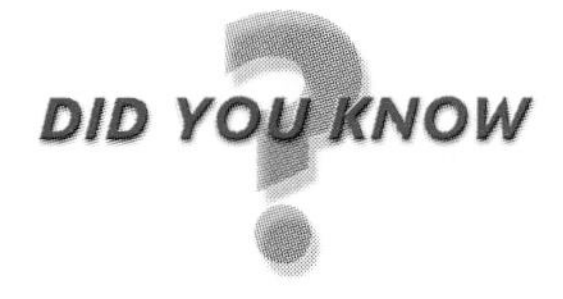

DID YOU KNOW

One of the most common concepts in statistics is the normal distribution curve. This curve, often called the normal curve or Gaussian curve (after Carl Friedrich Gauss, who found a large number of uses for it) occurs in a large number of real-life situations. In 1835, a Belgian astronomer and statistician, Lambert Adolph Quételet, made a startling discovery. In a study of the mental and physical traits of human beings, a large amount of data that had been gathered by the artists of the Renaissance period, Quételet found that almost all these traits were distributed along a normal curve. Height, the size of the brain or the head, intelligence, the sizes of different limbs were all distributed normally in any one national type. In fact, the same is true for many plant and animal groups as well. What is disturbing for modern social scientists are the data that are not normally distributed such as the records for income, higher education, and access to such facilities as proper medical care. Instead of being normally distributed, these situations will follow skewed curves, reflecting the fact that in most societies, the benefits of prosperity are often limited to a select few.

For More Information Smith, D. E. (1958). *History of Mathematics, Volume II*. New York: Dover Publications.

5.8.1 Variance and Standard Deviation

Beyond needing to know the mean of data, we often want to know how the complete set of data varies about the mean. Two measures that provide such information about how data are dispersed around the mean are variance and standard deviation. In short, the **variance** is the mean of the squared deviation from the mean. Because the variance is determined by the square of the deviation, it is measured in units that are the square of the original units. To make the measure of deviation from the mean expressed in the same units as the original data, we often use the **standard deviation,** which is the square root of the variance. The standard deviation measures the average amount by which data items vary from the mean.

Given a set of data items, $x_1, x_2, x_3, \ldots x_n$, with a mean of $\bar{x}$, we compute the variance of the data, denoted s^2, by:

$$s^2 = \frac{(x_1 - \bar{x})^2 + (x_2 - \bar{x})^2 + (x_3 - \bar{x})^2 + \cdots + (x_n - \bar{x})^2}{n}$$

The standard deviation of the data, denoted s, is the square root of the variance; that is:

$$s = \sqrt{\frac{(x_1 - \bar{x})^2 + (x_2 - \bar{x})^2 + (x_3 - \bar{x})^2 + \cdots + (x_n - \bar{x})^2}{n}}$$

To illustrate, suppose we have the following set of high daily temperatures for 5 different cities: 20°, 35°, 70°, 77°, 80°. To find the variance and standard deviation, we first must determine the mean:

$$\bar{x} = \frac{20° + 35° + 70° + 77° + 80°}{5} = \frac{282}{5} = 56.4°.$$

Then, using the mean, we compute the variance:

$$\begin{aligned} s^2 &= \frac{(20 - 56.4)^2 + (35 - 56.4)^2 + (70 - 56.4)^2 + (77 - 56.4)^2 + (80 - 56.4)^2}{5} \\ &= \frac{(-36.4)^2 + (-21.4)^2 + (13.6)^2 + (20.6)^2 + (23.6)^2}{5} \\ &= \frac{1324.96 + 457.96 + 184.96 + 424.36 + 556.96}{5} \\ &= \frac{2949.2}{5} \\ &= 589.8 \text{ sq degrees.} \end{aligned}$$

From the variance we compute the standard deviation:

$$s = \sqrt{589.8} = 24.3 \text{ degrees.}$$

5.8.2 Z-Scores

For any individual data item in a data set, we can determine how much that item varies from the mean by computing its z-score. A **z-score** is the number of standard deviations that a given data value, x, is above or below the mean. We compute a z-score, denoted z, by:

$$z = \frac{x - \bar{x}}{s}$$

where $\bar{x}$ is the mean and s is the standard deviation. Thus, using the data scores 20°, 35°, 70°, 77°, 80° with mean $\bar{x} = 56.4°$, standard deviation $s = 24.3°$, we can find the z-scores associated with each of the data values:

$$\text{The z-score for } 20° = \frac{20 - 56.4}{24.3} = \frac{-36.4}{24.3} = -1.50°.$$

$$\text{The z-score for } 35° = \frac{35 - 56.4}{24.3} = \frac{-21.4}{24.3} = -0.88°.$$

$$\text{The z-score for } 70° = \frac{70 - 56.4}{24.3} = \frac{13.6}{24.3} = 0.56°.$$

$$\text{The z-score for } 77° = \frac{77 - 56.4}{24.3} = \frac{20.6}{24.3} = 0.85°.$$

$$\text{The z-score for } 80° = \frac{80 - 56.4}{24.3} = \frac{23.6}{24.3} = 0.97°.$$

The negative z-score indicates that the given score is so many standard deviations below the mean, and positive z-scores indicate that a given score is so many standard deviations above the mean. For example, the score 20 is one-and-one-half standard deviations below the mean, and the score 77 is 0.85 standard deviations above the mean.

5.8.3 Normal Distribution of Data

If we have a set of data and form a graphical representation of the data using two axes, where the horizontal axis represents the data scores and the vertical axis represents the frequency of those scores, we form what is called a **distribution curve** (see Figure 5.11). When the distribution curve is a smooth, bell-shaped curve, we say the data has a **normal distribution.**

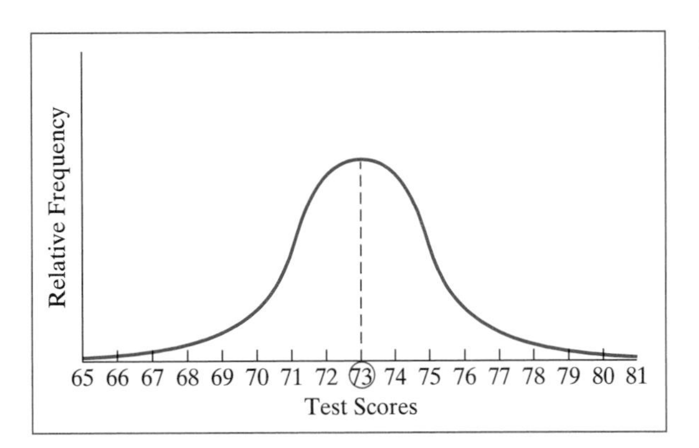

Figure 5.11

When a data set has a normal distribution, the mean, the median, and the mode are all the same. In a normal distribution, about 68% of the data scores will fall within one standard deviation of the mean, about 95% of the data scores will fall within two standard deviations of the mean, and about 99.7% of the data scores will fall within three standard deviations of the mean. See Figure 5.12.

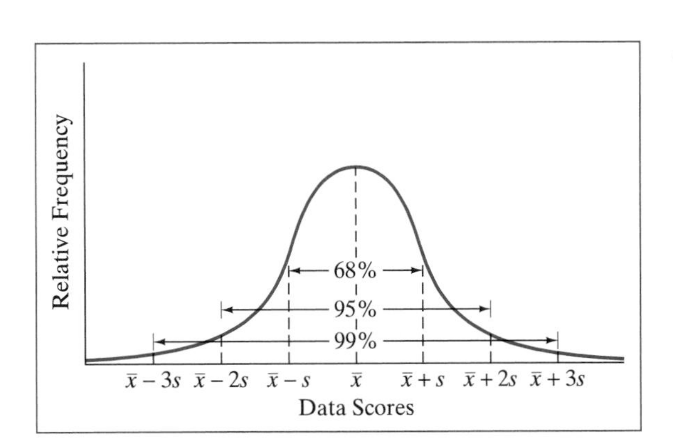

Figure 5.12

5.8.4 Skewed and Bimodal Data Distributions

A distribution of data is **skewed** if it is not symmetric (like the normal distribution) and extends to one side more than the other. Data that are skewed to the left are said to be **negatively skewed,** and data that is skewed to the right are said to be **positively skewed.** See Figure 5.13 for a comparison of normal and skewed data.

When a set of data is negatively skewed, the mean and the median are to the left of the mode, thus the phrase *skewed to the left*. When a set of data is positively skewed, the mean and the median are to the right of the mode, matching the phrase *skewed to the right*.

In some situations, the data are bimodal. In such cases, the data distribution curve would appear somewhat as in Figure 5.14, showing two points where the curve peaks to its highest position.

5.8.5 Quartiles

One way of showing how data are dispersed is to break the data into four equal parts called **quartiles.** The first quartile, denoted Q_1, represents the point in a collection of data scores, listed in increasing order, that separates the bottom one-fourth of the scores from the top three-fourths of the scores. The second quartile, denoted

Figure 5.13

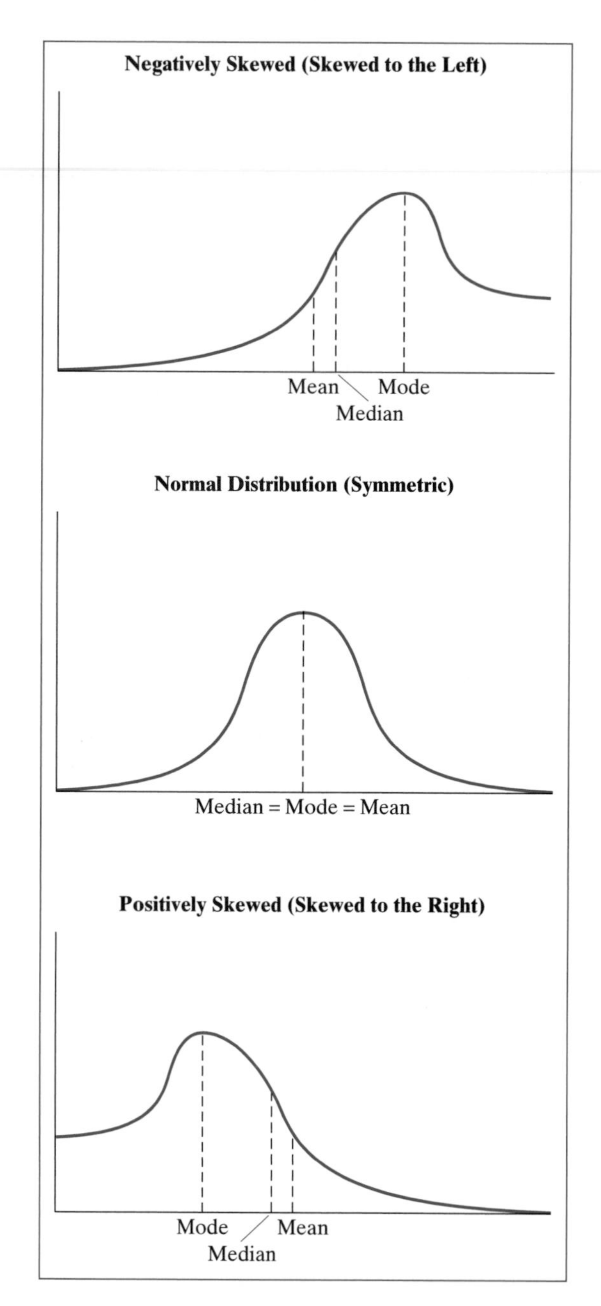

Figure 5.14

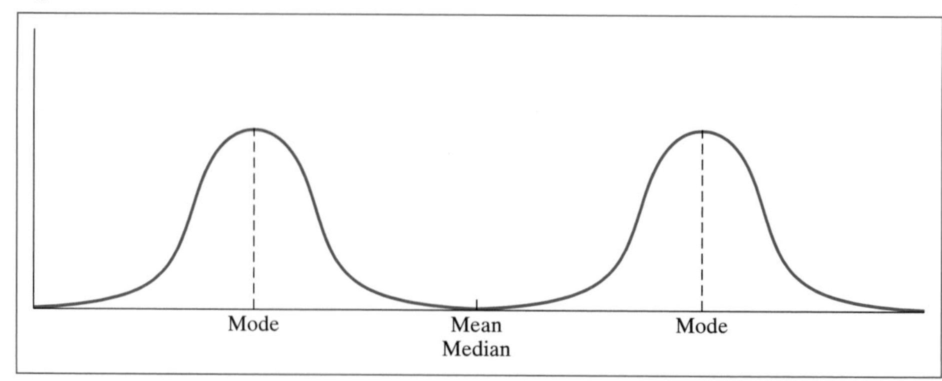

Q_2, is the point that separates the top half of the scores from the bottom half. Note that the median and the second quartile are the same number. Finally, the third quartile, denoted Q_3, is the point that separates the bottom three-fourths of the scores from the top one-fourth.

Consider the following 25 data scores presented in increasing order:

20, 29, 30, 46, 48, 55, 60, 67, 68, 70, 77, 77, 78, 79, 80, 81, 84, 84, 85, 86, 87, 88, 90, 96, 97

To identify the quartiles, we begin by finding the median, or second quartile. Because there is an odd number of scores, Q_2 is the 13th score in the order, or the value 78. Now, considering just the 12 scores below the second quartile, we identify the first quartile by finding the median of the bottom half of the scores: 20, 29, 30, 46, 48, 55, 60, 67, 68, 70, 77, 77. Because there is an even number of data items in the bottom half, we find the middle two scores and average them to determine the first quartile. Thus, $Q_1 = \frac{55 + 60}{2} = \frac{115}{2} = 57.5$.

To find Q_3, we focus on the top half of the scores, 79, 80, 81, 84, 84, 85, 86, 87, 88, 90, 96, 97 and find the median of these. Thus, $Q_3 = \frac{85 + 86}{2} = \frac{171}{2} = 85.5$.

5.8.6 Percentiles

Related to quartiles is the concept of percentiles; that is, given the whole set of data, or 100% of the data, we can look to each data point and determine the percentile score of that data item. The **percentile** of a data score depicts what percent of the data scores lies below that data score. In general, the kth percentile value, denoted P_k, is the point where k% of the data scores are less than that value. For example, the 10th percentile, P_{10}, would be the value where 10% of the data scores fall below the value.

Given a particular data value, x, **we can determine the percentile of that score x**, that is, what percent of the data values lies below x, by:

$$\text{percentile of score } x = \frac{\text{number of scores less than } x}{\text{total number of scores}} \times 100.$$

Consider the data set of 25 scores:

20, 29, 30, 46, 48, 55, 60, 67, 68, 70, 77, 77, 78, 79, 80, 81, 84, 84, 85, 86, 87, 88, 90, 96, 97

Selecting the data score 85, let us compute the percentile of the score 85. To compute, we count the number of scores less than 84, there are 18 such scores, and use the formula:

$$\text{percentile of } 85 = \frac{18}{25} \times 100 = 72.$$

Thus, the score 85 is the 72 percentile, which means that 72% of the data scores in this set are below 85.

It is interesting to note that the 25th percentile is the same as the first quartile; that is, $P_{25} = Q_1$. Similarly, $P_{75} = Q_3$. Finally, $P_{50} = Q_2 =$ median.

To find the value of the kth percentile, given n data scores, we compute:

$$V = \frac{k}{100} \times n.$$

Now if V is computed to be a whole number, then the value of the kth percentile is halfway between the Vth data element and $(V + 1)$st element in the sequence of the original data set. If the V is not a whole number when computed, then we round V up to the next larger whole number to find the position of the kth percentile.

For example, using the 25 data scores:

20, 29, 30, 46, 48, 55, 60, 67, 68, 70, 77, 77, 78, 79, 80, 81, 84, 84, 85, 86, 87, 88, 90, 96, 97

we might want to determine the 80th percentile of the data. To do so, we compute: $V = \frac{80}{100} \times 25 = 20$. Because 20 is a whole number, we find the 20th and 21st data scores in the list of data and find the average of those scores. Finding the scores 86 and 87, we conclude that 86.5 is the 80th percentile.

In a second example, we find the 10th percentile. Computing $V = \frac{10}{100} \times 25 = 2.5$, we round the value 2.5 up to the next whole number, 3. Then, locating the 3rd data score, 30, we conclude that $P_{10} = 30$.

5.8.7 Box-and-Whisker Plots

A **box-and-whisker plot** is a graphical means of illustrating the dispersion of data, using high and low scores and quartiles. Working with the following 25 data scores:

20, 29, 30, 46, 48, 55, 60, 67, 68, 70, 77, 77, 78, 79, 80, 81, 84, 84, 85, 86, 87, 88, 90, 96, 97

we have found that $Q_1 = 57.5$, $Q_2 = 78$, and $Q_3 = 85.5$, and we determine the high score to be 97 and the low score to be 20. We form a box-and-whisker plot as shown in Figure 5.15.

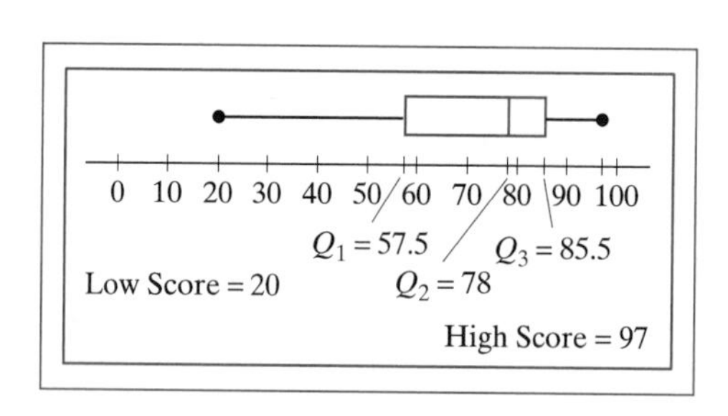

Figure 5.15

First, we develop a number line that is labeled, spanning the scores in the data set. We form a rectangular box stretching between Q_1 and Q_3. We then divide that rectangular box into two pieces, using the point where Q_2 is located. With the box completed, we create whiskers that extend from the box to the points that indicate the high score and low score. This visual display of the high and low scores combined with the quartiles provides an immediate picture of where the top one-fourth of the scores lie, where the bottom one-fourth of the scores lie, and where the middle point of the scores lies.

As a second example, suppose for a given data set the following statistics are determined: low score = 10, $Q_1 = 20$, $Q_2 = 40$, and $Q_3 = 70$, and high score = 90. The box-and-whisker plot for this data set is shown in Figure 5.16.

From the illustration in Figure 5.16, we can see that one-fourth of the scores fall between 40 and 70. We can also conclude that one-fourth of the scores fall between 70 and 90. We can also see that half of the scores fall below 40. Thus, even without the original data, we can draw many conclusions about the dispersion of data from a box-and-whisker plot alone.

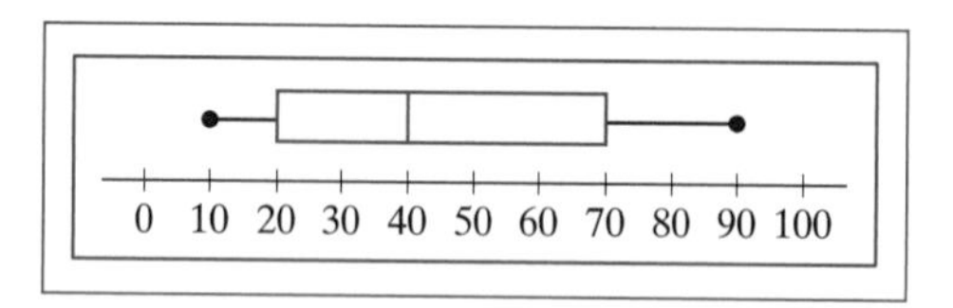

Figure 5.16

Related Topics Representing Data

TOPIC 5.9 Representing Data

cornerstone **Statistical Concepts**
Measures of Central Tendency
Data Dispersion

Introduction

Data can be represented in a variety of ways to present clear information discovered in a statistical study. However, statistics and the way they are represented can sometimes be used to mislead. In the 19th century, English prime minister Benjamin Disraeli stated that "there are three kinds of lies: lies, damned lies, and statistics." He was alluding to the fact that sometimes statistics can be used to deceive the general population. In this section, we examine a number of means of representing data, including graphs such as pictographs, bar graphs, histograms, line graphs, scatterplots, stem-and-leaf plots, and circle graphs.

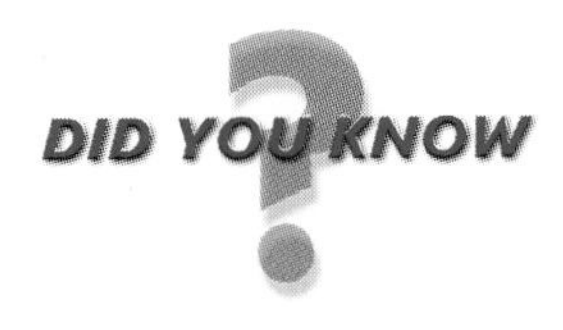

Sometimes, without sufficient statistics, we can draw incorrect conclusions. Take, for example, the well-known headline, "Dewey Defeats Truman," printed in the *Chicago Daily Tribune* in 1948. This announcement about the winner of the presidential election involving Harry Truman and Thomas Dewey went to press before the final votes were counted. The article mistakenly naming Dewey as the winner was based on statistics gathered from voters as they exited the voting booths that election day. Even though the exit polls indicated that Dewey would win by a reasonable margin, this sample of data proved insufficient.

5.9.1 Pictographs

A **pictograph** is a graph in which pictures are used to represent data. There is usually a key that accompanies a pictograph that describes what a single picture represents. For example, consider the pictograph shown in Figure 5.17.

We see from the key that each picture of a person represents 1,000 students. From the pictograph, we can infer that 2,000 students participated in a summer library reading program in Indiana; 4,000 students participated in a program in

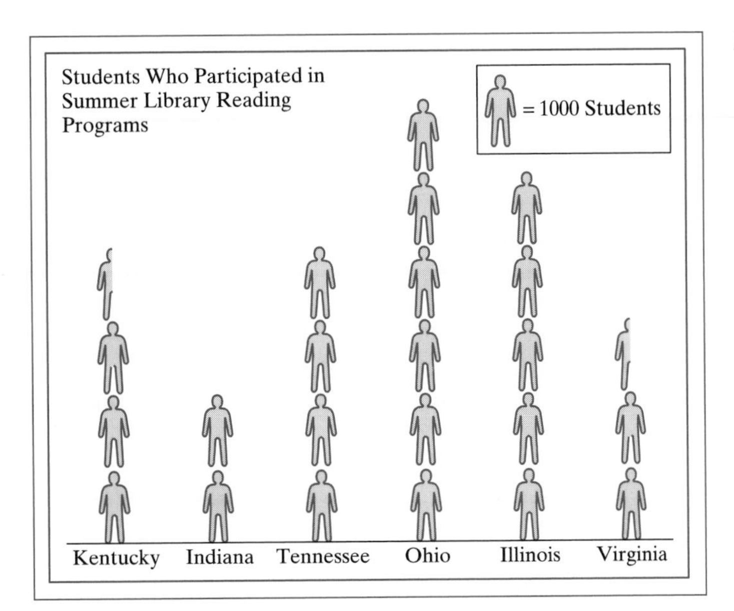

Figure 5.17

Tennessee. Notice that for Kentucky, there are three full pictures of a person plus a picture of half of a person. We can interpret this part of the pictograph as indicating that 3,500 students participated in the program in Kentucky.

5.9.2 Bar Graphs

A **bar graph** provides a means of representing data that are categorized in some way. Typically, the horizontal scale represents the categories, numerical or nonnumerical, and the vertical scale represents the frequency of each category. For example, examine the bar graph depicted in Figure 5.18 that represents the favorite colors of elementary students.

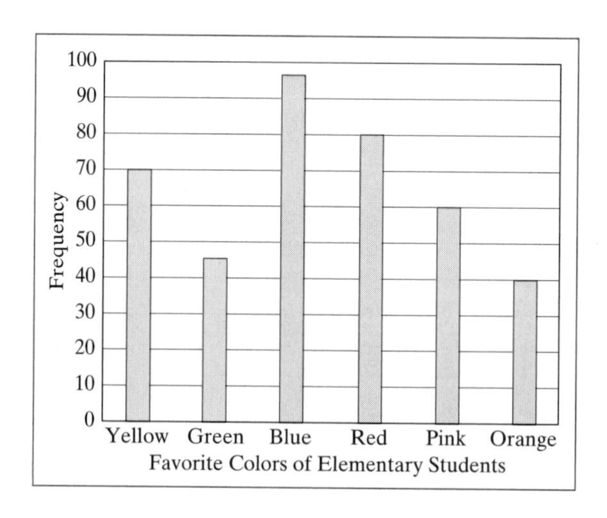

Figure 5.18

From the bar graph, we can see that blue is the most popular color among a sample of elementary students. From the vertical scale, we can see that of the students surveyed, 95 selected blue as their favorite color.

Suppose we choose to label the vertical scale in Figure 5.18 differently; that is, instead of the scale increasing in increments of 10, the scale increases in increments of 40 (see Figure 5.19).

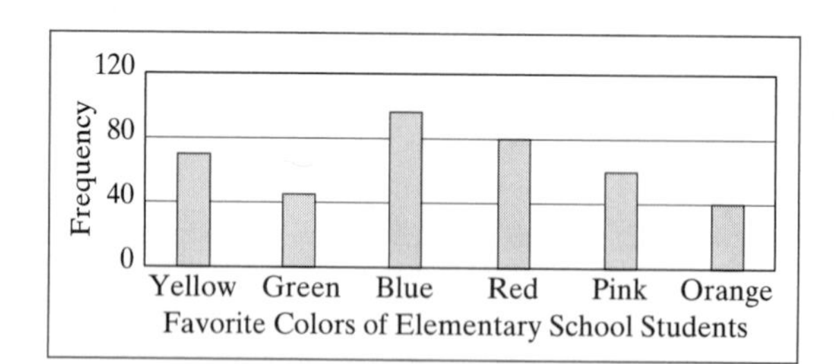

Figure 5.19

Because the scale has changed, the differences in favorite color choices are not as striking. Thus, this example shows that the choice of scale can dramatically affect the appearance of a bar graph and the information you wish to display in the graph.

Bar graphs can also be presented so that the frequency is depicted along the horizontal axis. This type of graph would be referred to as a **horizontal bar graph,** as opposed to the more typical **vertical bar graphs** where the frequency is shown along the vertical axis as in Figures 5.18 and 5.19. An example of a horizontal bar graph is seen in Figure 5.20, where we illustrate the same data as shown in Figure 5.18 but from the horizontal perspective.

A final possibility to consider with bar graphs are the concepts of **double bar graphs** or **triple bar graphs.** In these types of bar graphs, we can illustrate comparative data at several levels. For example, suppose we want to compare the number of alien sightings in various regions of the country across several decades. We can create a triple bar graph, like the one shown in Figure 5.21, to depict all of this information concisely.

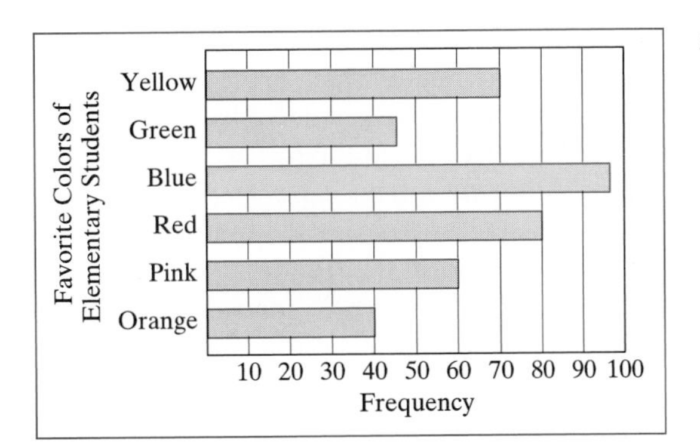

Figure 5.20

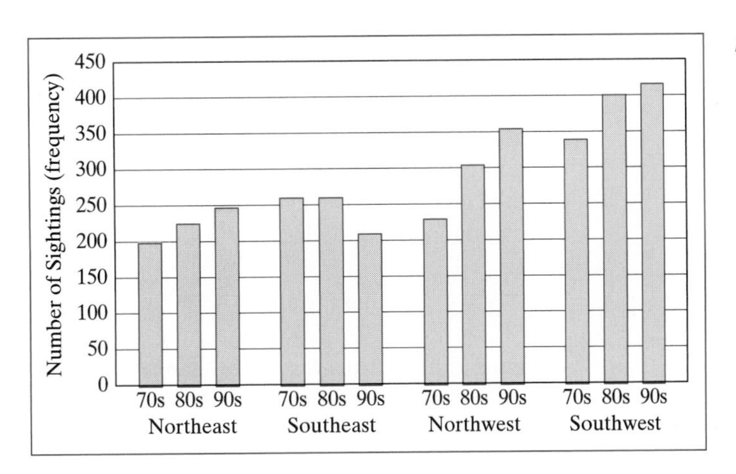

Figure 5.21

5.9.3 Frequency Tables

A way of categorizing raw data so that it is easier to interpret is to develop a **frequency table.** A basic frequency table has two columns, one denoting categories, or **classes,** of the data and one denoting the **frequency.** The classes are most often based on the raw data at hand and by how many categories into which you wish to break the data. For example, suppose the following set of data was gathered from 30 people who were surveyed about how many months they were engaged before they were married.

2	3	10	12	8	9	19	6	8	6
5	1	9	18	10	7	4	6	12	5
7	9	10	12	8	6	8	6	10	3

The data, as presented, are not very informative. However, if we break the data into categories, such as 0–2 months, 3–5 months, 6–8 months, and so on and then count how many responses fall into each category, we can see which category had the most responses and be better able to draw conclusions from the data. Using the categories suggested, we can form the frequency table shown in Table 5.11.

Table 5.11

Months Engaged	Frequency
0–2	2
3–5	5
6–8	11
9–11	7
12–14	3
15–17	0
18–20	2
	30

Looking at the classes, we refer to the first number in each class as the lower class limit and the last number as the upper class limit. Thus, in the class 9–11, 9 is the lower class limit, and 11 is the upper class limit. Notice that the size of each class is the same. This is important when developing a frequency table. As a result, the difference between the lower class limit of one class and the lower class limit of the class that follows is always the same. The same can be said of the difference between any two consecutive upper class limits. In this case, the difference between any two consecutive class limits is 3, indicating that each class is of size three.

Suppose you want to break the data into a particular number of, say n, classes. You can determine the size of each class by first determining the **range** of the data to be represented and then dividing the range by n.

Table 5.12

Months Engaged	Frequency
1–4	5
5–8	13
9–12	10
13–16	0
17–21	2
	30

For example, if we decide we want to break the previous set of data into 5 classes, we first find the range, which is the high score minus the low score, in this case to be $19 - 1 = 18$. Then, we divide 18 by 5 and find 3.6. Rounding the quotient 3.6 up to the next higher whole number, 4, we then know to form classes with 4 in each class to guarantee that we will end up with 5 classes. Starting with the lowest data score, 1, we begin our first class, forming the first class 1–4 so that the class size is 4. Continuing in this manner, we form the frequency table shown in Table 5.12.

Note that no matter how many classes into which you break the data, the total frequency, in this case 30, always reflects the total number of data items included in the study.

5.9.4 Histograms

Histograms provide a means of representing data from a frequency table in a way that shows a consistent flow from one class to another. Unlike a bar graph where categories can be nonnumerical (like colors) and are shown as separate from each other along the horizontal axis, a histogram illustrates only numerical data and has a horizontal axis, whose scale is not labeled by the separate categories but by what we call the class mark. The class mark is the average of a particular class and serves as a representative of that class. To find the class mark, we compute:

$$\text{Class Mark} = \frac{\text{lower class limit} + \text{upper class limit}}{2}$$

Figure 5.13 illustrates a frequency table with a class-mark column included. Notice, for example, that the first class mark was found by: $\frac{1+4}{2} = 2.5$. Also note that the difference between consecutive class marks is the same as the difference between consecutive upper class limits and lower class limits.

We will use these class marks to label the horizontal scale for our histogram. We let the vertical scale represent the frequency from the frequency table. From the frequency Table 5.13, we note that our frequency scale needs to span frequencies from 0 to 13. Thus, we can form a histogram that represents data in our frequency table as in Figure 5.22.

Table 5.13

Months Engaged	Frequency	Class Mark
1–4	5	2.5
5–8	13	6.5
9–12	10	10.5
13–16	0	14.5
17–21	2	18.5

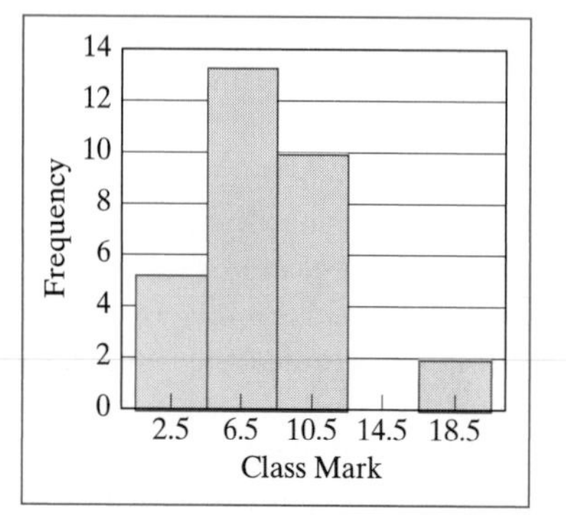

Figure 5.22

Notice that in a histogram, unlike the bar graph, the bars showing the frequency for each class are adjoining; that is, there are no gaps between the bars. This illustrates the fact that in a frequency table, one category blends naturally into the next, leaving no gaps. In a frequency table, a data item fits into one and only one category, and there is no possibility that a data item will not belong in one of the categories because we have left no gap between classes.

5.9.5 Line Graphs and Frequency Polygons

Line graphs, also known as **frequency polygons,** provide a different means of representing the same type of data that you can illustrate with a histogram. Line graphs (frequency polygons) illustrate frequency across a set of continuous numerical data. Because the scale along the horizontal axis of a histogram is designed so as to show the continuity of numerical classes without gaps or breaks, we can form a line graph from any histogram.

From the histogram in Figure 5.22, we have formed a corresponding line graph in Figure 5.23. Notice that to form the line graph, we identify the midpoint of the top side of each of the rectangles in the histogram with a point and then connect the points to form the line graph.

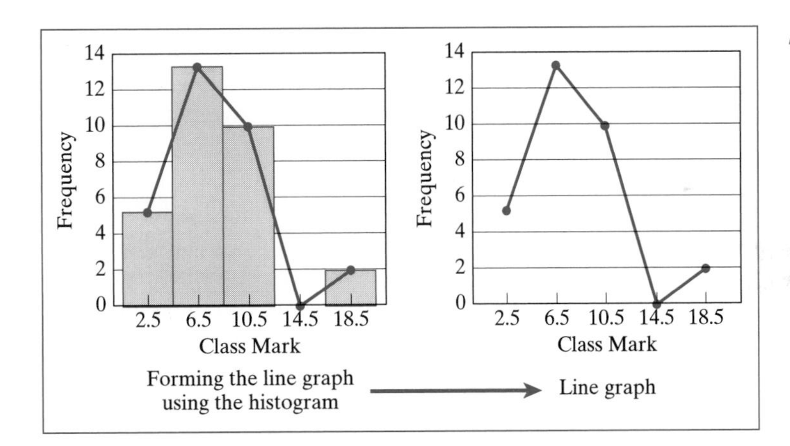

Figure 5.23

5.9.6 Line Graphs and Scatter Plots

A line graph does not have to be formed from a histogram. Whenever the horizontal axis represents continuous numerical data, such as time, temperature, and distances, we can use lines to connect key data points. For example, when you have gathered **paired data,** that is, two pieces of corresponding data items from each source, we can compare those data items using a line graph. This is much like graphing coordinate points in the plane (see topics on Coordinate Graphing).

For example, suppose you have gathered data on the size of the freshmen class at Jones University for the years 1990–1999 to attempt to note any trends across the decade. For each year, there is a corresponding data piece that indicates the size of the freshmen class that year. This paired data may be presented as ordered pairs (x,y) or in a table as shown in Table 5.14.

Table 5.14

Year	Number of Freshmen
1990	2500
1991	2480
1992	2400
1993	2200
1994	2350
1995	2600
1996	2550
1997	2700
1998	2780
1999	2900

Using these paired data, we can let the horizontal axis represent the year and the vertical axis represent the sizes of the freshman classes. Doing this, we can place points that correspond to each pair of data and then connect the points, forming the line graph as seen in Figure 5.24.

From a line graph, we can observe changes and trends quite readily. For example, we can observe that the size of the freshman class dropped more dramatically from 1992 to 1993 than it dropped from 1990 to 1991. We can also observe that the most dramatic increase occurred between 1994 and 1995. We make this observation based

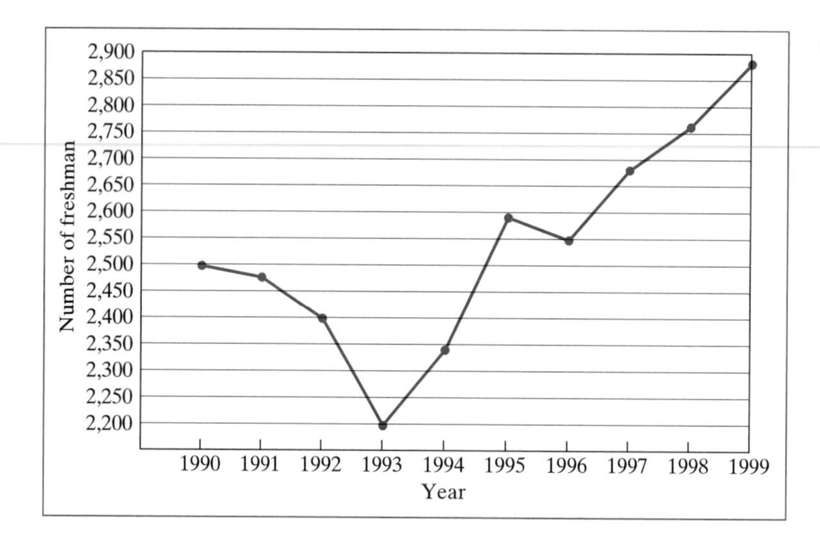

Figure 5.24

on the steepness of the slope of the line connecting the points between those two years.

Scatterplots are also formed from paired data, but they are different from line graphs. A line graph is formed when you have, for each numerical category that appears in the horizontal scales, exactly one corresponding numerical value along the vertical scale. Thus, when showing the points that correspond to each set of paired data, we would never have two or more points in the same row or column in the graph. For example, in the year 1995 in the data set in Table 5.14, there is only one corresponding piece of data, the 2600 students in the freshman class.

A **scatterplot** represents paired data that are collected from a number of sources through which we hope to determine whether there is some connection between the two types of data collected. For example, suppose we want to determine whether there is any relationship between the number of hours a student studies for an exam and the resulting exam score. We suspect that the more students study, the higher the exam scores will be, but we need to gather data from students.

Consider the paired data shown in Table 5.15 that were gathered from 25 students. Each student provided two pieces of data, an exam score and the amount of time spent studying (in hours).

If we plot the pairs of points, labeling the horizontal axis with the range of exam scores and labeling the vertical axis with the range of study times. Plotting each point of paired data, we form the scatterplot shown in Figure 5.25.

Once we have plotted the points of our scatterplot, we try to determine whether there is any relationship between time spent studying and a score on the exam. Through the set of points, we draw a line that best shows the trend of the points. We call this line the **line of best fit.** A line of best fit is determined by imposing a line through the plotted points so that most points lie on the line, just above the line, or just below the line. Figure 5.26 demonstrates a possible line of best fit for the scatterplot.

Notice that although most of the points cluster around the line of best fit, either above or below the line, there are some data points that lie far away from the line. For example, notice the point that represents the student who earned a grade of 85

Table 5.15

Exam Score	55	60	99	87	66	88	45	77	69	81	76	75	90	92	30	78	85	96	70	25	59	92	80	84	74
Study Time	2	3	6	4	3	4	1	3.5	2.5	3	2.5	2.5	7	6.5	1	3	3.5	6	2.5	2	2	6	4	1	4

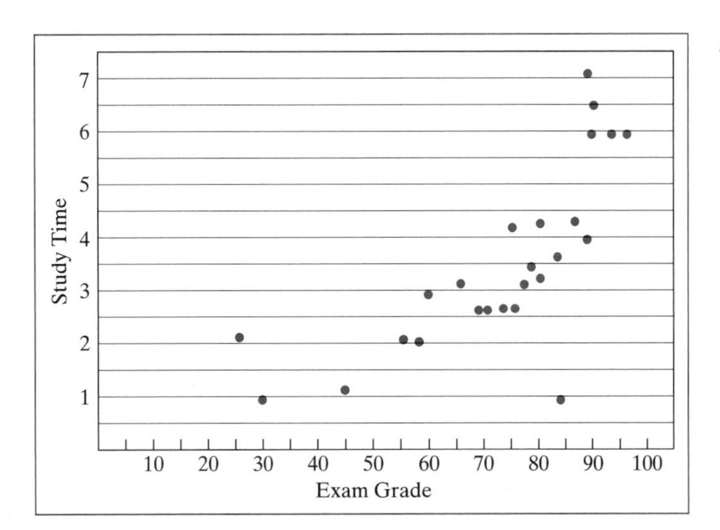

Figure 5.25

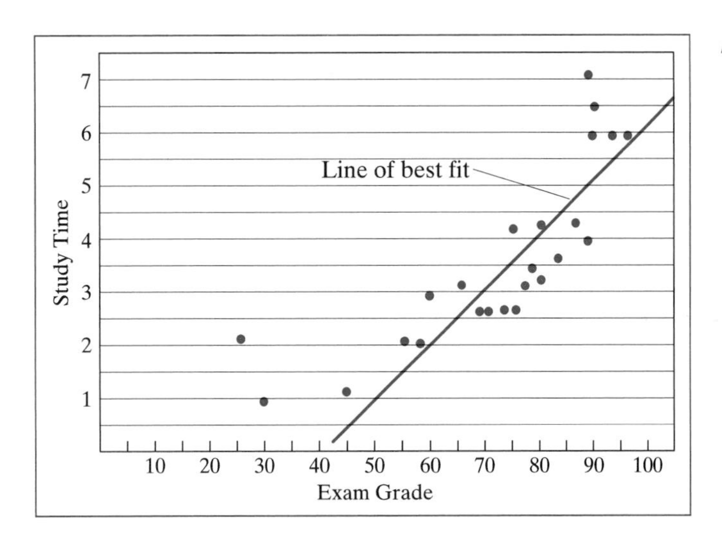

Figure 5.26

on the exam, but only studied 1 hour. Because this point lies far from the line of best fit, we call this point an **outlier.** There are usually some explanations as to why some points are outliers. For example, the student who only studied 1 hour but scored an 85 on the exam may have taken the course before and thus may not require as much study time before the exam.

With scatterplots, we usually try to see if there is some sort of **correlation** between the two types of data items that are being compared. We say that there is a **positive correlation** if the line of best fit has a positive slope, that is, if the line tends upward as we trace it from left to right.

The line of best fit in Figure 5.26 has a positive slope, therefore we can conclude from these data that there is a positive correlation between the amount of time spent studying and exam scores.

If the line of best fit has a negative slope—that is, it tends downward as we trace the line from left to right—then we say that there is a **negative correlation.** If we cannot identify any line of best fit among the points, then we say there is no correlation (see Figure 5.27).

5.9.7 Circle Graphs/Pie Charts

When we want to represent data so that we can immediately see how one category of data compares to the other categories of data, we can use a **circle graph,** also known as a **pie chart.** To form a circle graph (pie chart), we need to determine what

Figure 5.27

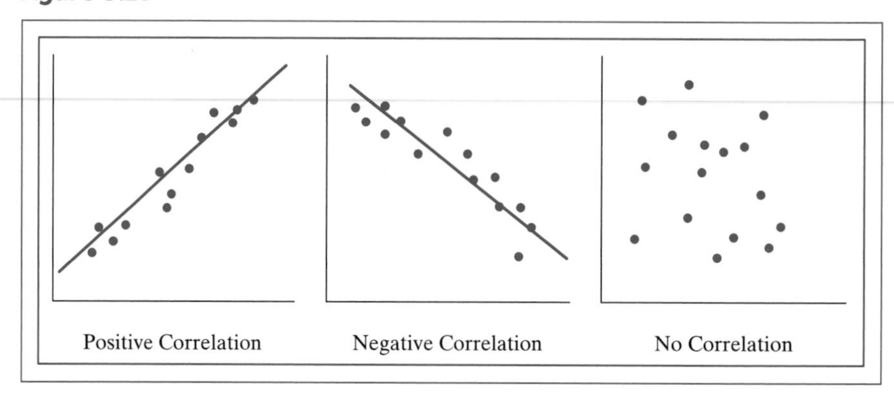

Table 5.16

Months Engaged	Frequency	Relative Frequency
1–4	5	5/30 = 0.17 = 17%
5–8	13	13/30 = 0.43 = 43%
9–12	10	10/30 = 0.33 = 33%
13–16	0	0/30 = 0.00 = 0%
17–21	2	2/30 = 0.07 = 7%
	30	30/30 = 1.00 = 100%

part of the whole set of data each category represents. One way of developing a circle graph is to work from a frequency table (see Table 5.16) and develop a new column called the **relative frequency.**

The relative frequency shows how frequently that category occurs, relative to all of the categories combined. The relative frequency of a category is found by dividing the frequency of that category by the total frequency. Relative frequency can be expressed as a fraction, a decimal number, or a percent. The sum of relative frequencies should be 1 or 100%.

Figure 5.28

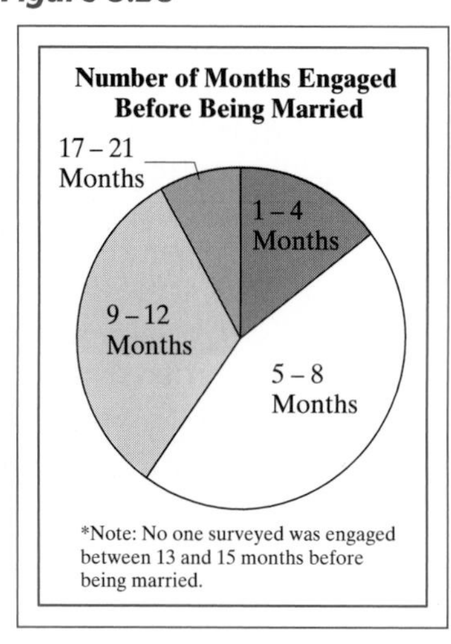

We can use the relative frequencies to form a circle graph by breaking the circle into parts that are defined by the relative frequencies; that is, we want to break the circle into sections so that 33% of the circle is shown to represent the category of engagements from 9 to 12 months. To do so, we have to use the fact that the total number of degrees around the center point of a circle is 360 degrees (360°). Then, to determine the central angle of the section to represent the category 9–12 months, we would have to find 33% of 360°. Computing, we find:

$$33\% \text{ of } 360° = .33 \times 360 = 118.8°.$$

Thus, the central angle of the section representing the category of 9–12 months will be 118.8°.

Table 5.17 shows the computation of the central angles for all of the sections to be shown in the circle graph.

Finally, using the central angle measures from each category, we can form our circle graph with the appropriate central angle measures. We can label each section of the circle graph by its class (see Figure 5.28).

With circle graphs, it is important to indicate that a category whose frequency is 0 is not ignored. Thus, a note indicating why a region for that category is not shown is appropriate.

5.9.8 Stem-and-Leaf Plots

When you have a collection of numerical data, one way to organize the data for examination and interpretation is through a stem-and-leaf plot. A **stem-and-leaf plot** is a chart that shows two columns, the left column representing stems of data

Table 5.17

Months Engaged	Frequency	Relative Frequency	Central Angle Measure (in degrees)
1–4	5	5/30 = 0.17 = 17%	0.17 × 360 = 61.2
5–8	13	13/30 = 0.43 = 43%	0.43 × 360 = 154.8
9–12	10	10/30 = 0.33 = 33%	0.33 × 360 = 118.8
13–16	0	0/30 = 0.00 = 0%	0.00 × 360 = 0
17–21	2	2/30 = 0.07 = 7%	0.07 × 360 = 25.2
	30	30/30 = 1.00 = 100%	360.0

pieces and the right column representing leafs; that is, if a number of numerical data items happen to have a common first digit or common first few digits, we can let the common digits be the stems shown in the left column. Then, the leaves, shown in the right column show the various differentiations among numbers with the same stem.

For example, suppose we have the following pieces of data representing test scores:

35, 79, 66, 80, 90, 91, 92, 88, 86, 46, 77, 76, 65, 70, 59, 92, 83, 61, 62, 71, 85, 89, 75, 84.

We can present these data items in a stem-and-leaf plot as shown:

Stem	Leaf
3	5
4	6
5	9
6	1, 2, 5, 6
7	0, 1, 5, 6, 7, 9
8	0, 3, 4, 5, 6, 8, 9
9	0, 1, 2, 2

In this stem-and-leaf plot, the stems represent the first digit of the test scores and the leaves represent the second digit. From this plot, it is easy to discern how many students scored in the 60s range, in the 70s range, in the 80s range, and so on. We can also notice where there were two similar scores, such as the two scores of 92 as evidenced by the repeating leaves in the 9s stem row.

Stem-and-leaf plots can have more than one digit in the stem column. For example, suppose we did a survey of people at a local park, asking them what year they were born. The stem-and-leaf plot shows those years, with a stem that is three-digits long.

Stem	Leaf
193	0, 1, 5
194	3, 3, 5, 6
195	2, 3, 3, 4, 6, 7, 8
196	0, 1, 2, 2, 3, 6, 7, 7, 8, 8, 9
197	4, 5, 6, 6, 8
198	2, 2, 4, 7, 9
199	0, 0, 1, 1, 2, 3, 3, 3, 4, 4, 5, 6, 6, 6, 7, 8, 8

From the stem-and-leaf plot, we might conclude that at the time we conducted the survey, there were mostly people who were born in the years 1990 to 1998, probably children, and then also people who were born in the years 1960 to 1969, most likely parents of some of the children.

Note that the leaves in stem-and-leaf plots do not have to be a single digit. They may be multidigited, but keep in mind that for ease of quickly reading and interpreting data in a stem-and-leaf plot, it is best not to have leaves that are too lengthy.

Related Topics

Coordinate Graphing
Angle Measure
Fractions
Decimals
Percents

In the Classroom

This chapter focuses on the *Data Analysis and Probability Content Standard*. "The increased curricular emphasis on data analysis proposed . . . is intended to span the grades rather than be reserved for the middle grades and secondary school" (NCTM, 2000, p. 48). Given this emphasis, we believe that you should construct an understanding of some basic (a) principals underlying the concept of probability (Activities 5.1–5.5), (b) counting principals to enable efficient calculation of probability (Activity 5.6), and (c) concepts of statistics and the use of sample data (Activities 5.7–5.11).

Some Expectations of Your Future Students:

Pre-K–2 (NCTM, 2000, p. 108)

- pose questions and gather data about themselves and their surroundings
- sort and classify objects according to their attributes and organize data about the objects
- represent data using concrete objects, pictures, and graphs
- describe parts of the data and the set of data as a whole to determine what the data show
- discuss events related to students' experiences as likely or unlikely

Grades 3–5 (NCTM, 2000, p. 176)

- describe the shape and important features of a set of data and compare related data sets, with an emphasis on how the data are distributed
- propose and justify conclusions and predictions that are based on data and design studies to further investigate the conclusions and predictions
- understand that the measure of the likelihood of an event can be represented by a number from 0 to 1

How Activities 5.1–5.11 Help You Develop an Adult-level Perspective on the Above Expectations:

- By building on your current concepts of probability, you formalize a definition of *probability* and understand why probability values are between 0 and 1.
- By making predictions and conducting experiments to test your predictions, you construct an understanding of theoretical versus experimental probability and of sample space (what could happen) versus observed data.
- By systematizing ways of displaying the sample space, you analyze how multistage/multistep events could occur and explore the concept of independent versus dependent stages. You realize efficient techniques are essential for counting the ways in which such events could occur, as the number of stages and options at each stage increase and you learn such techniques.
- By analyzing sample data taken from experiences to which you can relate, you study important features of a sample and compare samples from the same population. You make predictions about the features of the population by building on your prior understanding of the connections between experimental and theoretical probability.
- By understanding the differences between paired and unpaired sets of data, you continue to build on your concept of dependent versus independent events. You learn graphical representations suitable for displaying different kinds of data and how to interpret them.
- By discussing and studying different statistical reports—questions asked, sampling methods used, and the content of and the way in which statistics are reported—you begin to develop a critical awareness of issues underlying the design of a statistical study. You learn to question and to understand better reported studies, as well as to justify studies you design and the conclusions you draw from sample data.

Bibliography

Bryan, E. H. (1988). Exploring data with box plots. *Mathematics Teacher, 81*, 658–663.

Burrill, G., Burrill, J. C., Coffield, P., Davis, G., de Lange, J., Resnick, D., & Siegel, M. (1992). *Data analysis and statistics across the curriculum: Addenda series, grades 9–12*. Reston, VA: National Council of Teachers of Mathematics.

Corwin, R. B., & Friel, S. N. (1990). *Statistics: Prediction and sampling*. A unit of study for grades 5–6 from *Used numbers: Real data in the classroom*. Palo Alto, CA: Dale Seymour.

Curcio, F. R., Nimerofsky, B., Perez, R., & Yaloz-Femia, S. (1998). Developing concepts in probability: Designing and analyzing games. In L. P. Leutzinger (Ed.), *Mathematics in the middle* (pp. 206–211). Reston, VA: National Council of Teachers of Mathematics.

Dessart, D. J. (1995). Randomness: A connection to reality. In P. A. House (Ed.), *Connecting mathematics across the curriculum* (pp. 177–181). Reston, VA: National Council of Teachers of Mathematics.

Elementary Quantitative Literacy Project (1998). *Exploring statistics in the elementary grades: Book 1 (K–6)*. White Plains, NY: Cuisenaire-Dale Seymour.

Friel, S. H., & Corwin, R. B. (1990). Implementing the *Standards*: The statistics standards in K–8 mathematics. *Arithmetic Teacher 38*, 35–39.

Friel, S. N., Mokros, J. R., & Russell, S. J. (1992). *Middles, means, and in-between*. A unit of study for grades 5–8 from *Used numbers: Real data in the classroom*. Palo Alto, CA: Dale Seymour.

Goldman, P. H. (1990). Teaching arithmetic averaging: An activity approach. *Arithmetic Teacher*, 37(7), 38–43.

Hatfield, L. L. (1992). Activities: Explorations with chance. *Mathematics Teacher, 85*, 280–282, 288–290.

Hollingdale, S. (1994). *Makers of mathematics*. New York: Penguin Books.

Huff, W. D. (1954). *How to lie with statistics*. New York: Norton.

Jones, G. A., Thornton, C. A., Langrall, C. W., & Tarr, J. E. (1999). Understanding students' probabilistic reasoning. In L. V. Stiff (Ed.), *Developing mathematical reasoning in grades K–12* (pp. 146–155). Reston, VA: National Council of Teachers of Mathematics.

Jones, K. S. (1993). The birthday problem again. *Mathematics Teacher, 86*, 373–377.

Landwehr, J. M., & Watkins, A. E. (1987). *Exploring data: Quantitative literacy series*. Palo Alto, CA: Dale Seymour.

Leutzinger, L. P. (1990). Graphical representation and probability. In J. N. Payne (Ed.), *Mathematics for the young child* (pp. 251–263). Reston, VA: National Council of Teachers of Mathematics.

Lindquist, M. M. (1992). *Making sense of data: Addenda series, grades K–6*. Reston, VA: National Council of Teachers of Mathematics.

Manin, H. M., & Zawojewski, J. S. (1993). Dealing with data and chance: An illustration from the middle school addendum to the *Standards. Arithmetic Teacher, 41*, 220–223.

May, E. L., Jr. (1992). Are seven-game baseball playoffs fairer? *Mathematics Teacher, 85*, 528–531.

National Council of Teachers of Mathematics. (1990). Data analysis: Focus issue. *Mathematics Teacher, 83*(2).

Newman, C. M., Obremski, T. E., & Schaeffer, R. L. (1987). *Exploring probability: Quantitative literacy series*. Palo Alto, CA: Dale Seymour.

Orkin, M. (1996). *Can you win?: The real odds for casino gambling, Sport betting, and lotteries*. New York: Freeman.

Parker, J., & Widmer, C. C. (1992). Teaching mathematics with technology: Statistics and graphing. *Arithmetic Teacher, 39*, 48–52.

Phillips, E., Lappan, G., Winter, M. J., & Fitzgerald, W. (1986). *Middle grades mathematics project: Probability*. Menlo Park, CA: Addison-Wesley.

Porter, M. T. (1986). *The rise of statistical thinking 1820–1900*. Princeton, NJ: Princeton University Press.

Richbart, L., & Richbart, C. (1992). Trading Alfs for a Bart: A simulation. *Arithmetic Teacher, 40*, 112–114.

Russell, S. J., & Corwin, R. B. (1989). *Statistics: The shape of the data*. A unit of study for grades 4–6 from *Used numbers: Real data in the classroom*. Palo Alto, CA: Dale Seymour.

Russell, S. J., & Friel, S. N. (1989). Collecting and analyzing real data in the elementary classroom. In P. R. Trafton (Ed.), *New directions for elementary school mathematics* (pp. 134–148). Reston, VA: National Council of Teachers of Mathematics.

Schule, H. S., & Leonard, B. (1989). Probability and intuition. *Mathematics Teacher, 82*, 52–53.

Shaughnessy, J. M., & Dick, T. (1991). Monty's dilemma: Should you stick or switch? *Mathematics Teacher, 84*, 252–256.

Uccellini, J. C. (1996). Teaching the mean meaningfully. *Mathematics Teaching in the Middle Grades, 2*, 112–115.

Zawojewski, J. (1992). *Dealing with data and chance: Addenda series, grades 5–8*. Reston, VA: National Council of Teachers of Mathematics.

CHAPTER 6 SIX

Fraction Models & Operations

CHAPTER OVERVIEW

Many real-world situations require the use of fractions. It is important for anyone who will teach fraction concepts and procedures to children to have a deep understanding of fractions models and operations. In this chapter, you can learn some interesting facts about fractions and will examine some concepts and procedures involving fraction ideas.

BIG MATHEMATICAL IDEAS

Problem-solving strategies, conjecturing, verifying, decomposing, generalizing, using language and symbolism, mathematical structure

NCTM PRINCIPLES & STANDARDS LINKS

Number and Operation; Problem Solving; Reasoning; Communication; Connections; Representation

TOPIC 6.1 **Fractions**
6.2 **Equivalent Fractions**
6.3 **Simplifying Fractions**
6.4 **Improper Fractions and Mixed Numbers**
In the Classroom
Bibliography

TOPIC 6.1 Fractions

cornerstone

Natural Numbers
Whole Numbers
Integers
Whole-Number Division
Integer Division

Introduction

The word *fraction* is commonly used today to denote some part of a whole. Although the idea of fractions has been known for quite awhile, it took a long time to develop the form we use today to write fractions. The word fraction itself comes from a Latin word, *fractus*, which means "to break." In the next few sections, we examine the meanings of fractions and their representations.

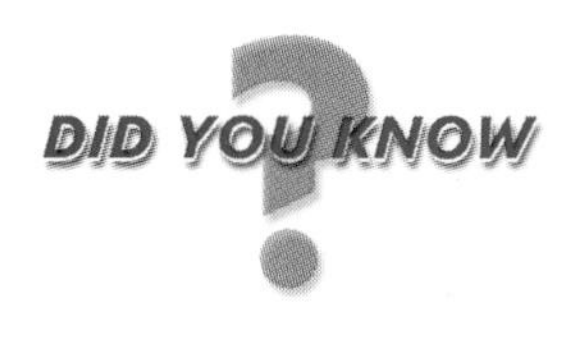

The earliest accounts of the use of fractional quantities on a wide and practical scale come from Egypt (about 1700 B.C.), where they appeared on a roll of leather that dates back to the same time as the Ahmes papyri, the most important source of our knowledge of Egyptian mathematics. This roll of leather is a mathematical table that consists of decompositions of fractions into unit fractions (the single exception was $\frac{2}{3}$). *Unit fractions* are fractions that have 1 in the numerator. The Egyptians developed a means of writing unit fractions, decomposing all other fractions into them, and developing elaborate tables that listed such decompositions.

Egyptian fractions appear in the form of hieroglyphics (pictorial symbols for numerals that we discussed in topics related to numeration) representing unit fractions. For example, we have the following unit fractions (Figure 6.1) written as hieroglyphics.

Figure 6.1

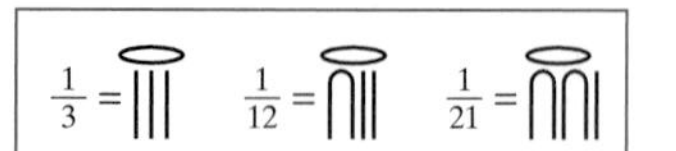

The Egyptians used fractions for very practical purposes. They did not have a monetary system, and trade was based on a barter system of exchange of goods. This probably led them into developing a system of operations with fractions. Although the Egyptians did not work with the modern techniques of operations of fractions, their rather complex system of such operations worked remarkably well for them.

Later, fractions appear in the work of the Indian mathematician, Brahmagupta, who lived in A.D. 598–C. 665. The Hindus wrote fractions by placing integers one on top of another, without a bar between them. The Arabs are credited with the present way of writing down a fraction with a bar placed between the two integers. Sources from these two civilizations show that both of them knew how to divide a fraction by another by inverting the divisor. The techniques somehow mysteriously disappeared until the 17th century but thereafter came to stay as a useful tool in operations with fractions.

For More Information Smith, D. E. (1958). *History of Mathematics, Volume II*. New York: Dover Publications.

A fraction is a representation of a number of the form $\frac{a}{b}$ where a and b can be any numbers. The top number, a, is called the numerator, and the bottom number, b, is the denominator. A fraction can have one of three meanings: (a) a part of a whole, (b) a ratio, or (c) the quotient in a division problem.

6.1.1 Fractions Representing Part of a Whole

When a fraction represents **part of a whole,** the denominator represents the number of equal-sized pieces into which the whole has been divided, whereas the numerator indicates how many of those pieces we are considering at that time. For example, the fraction $\frac{4}{5}$ suggests that some whole, let's say a rectangular region, has been divided into 5 equal-sized pieces. The numerator, 4, suggests that we are interested in only 4 of the 5 pieces. See Figures 6.2 and 6.3 below. Fractions greater than 1 can also represent part of a whole. Figures 6.4 and 6.5 illustrate the fraction $\frac{4}{3}$, still interpreting the denominator as the number of equal-sized pieces into which one whole is broken, while the numerator suggests the number of those pieces to consider. Note that one rectangle represents one whole.

6.1.2 Fractions Representing Ratios

A fraction also can represent a **ratio.** A ratio is a comparison of two or more quantities. In a typical fraction with a numerator over a denominator, the amount in the numerator is being compared to the amount in the denominator. For example, if there are 30 students in a class and 10 of those students are male and 20 are female, the ratio of males to females could be expressed as 10 to 20, 10:20, or $\frac{10}{20}$. Thus, the fraction $\frac{10}{20}$ does not represent part of the whole in this situation. Rather, the ratio of 10 to 20 indicates that there are twice as many women in the class as men.

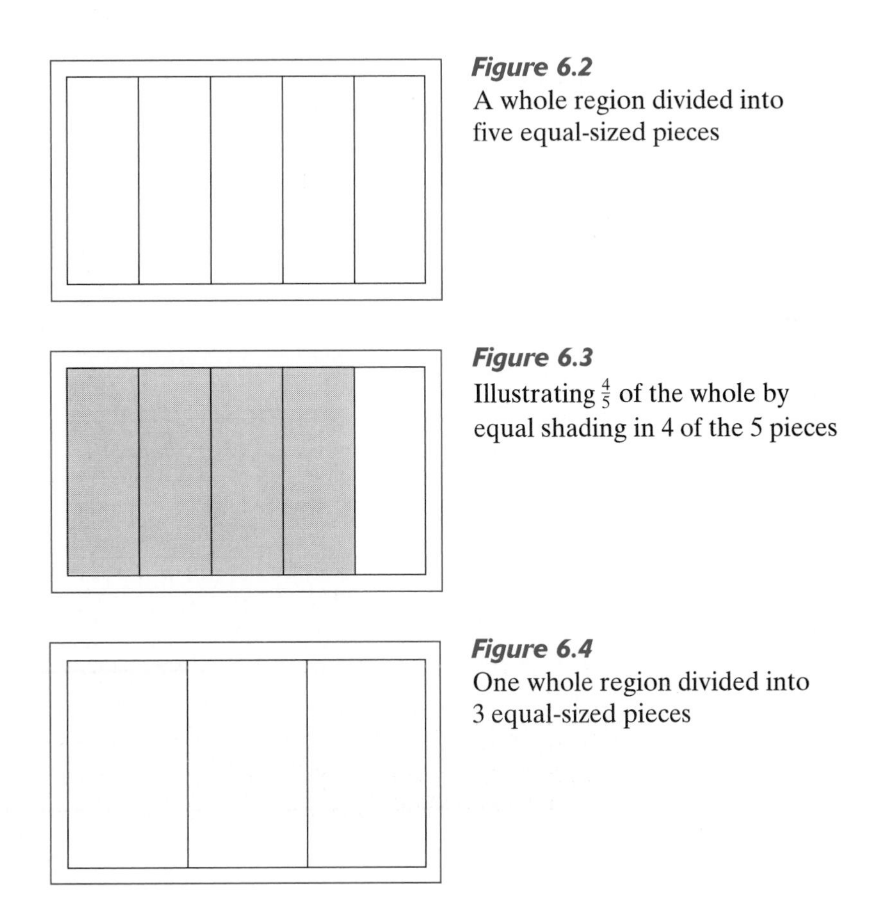

Figure 6.2
A whole region divided into five equal-sized pieces

Figure 6.3
Illustrating $\frac{4}{5}$ of the whole by equal shading in 4 of the 5 pieces

Figure 6.4
One whole region divided into 3 equal-sized pieces

Figure 6.5
Two whole regions broken into 3 equal-sized pieces with 4 of those pieces shaded to represent $\frac{4}{3}$

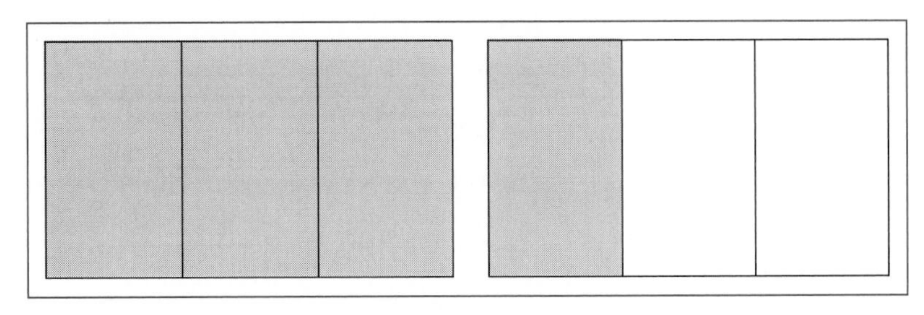

6.1.3 Fractions as Representing a Division Problem

Fractions sometimes simply represent a **quotient.** For example, when computing the division problem $27 \div 4$, the answer can be expressed as $\frac{27}{4}$, or $6\frac{3}{4}$. Because of this interpretation of fractions, students often learn to interpret the bar that separates the numerator from the denominator as a division sign.

Related Topics

Rational Numbers
The Set of Rational Numbers
Equivalent Fractions
Simplifying Fractions
Improper Fractions and Mixed Numbers
Adding Rational Numbers
Subtracting Rational Numbers
Multiplying Rational Numbers
Dividing Rational Numbers
Ratios

TOPIC 6.2 Equivalent Fractions

cornerstone

Fractions
Rational Numbers
Integer Multiplication
Integer Division
Least Common Multiple
Divisibility
Greatest Common Divisor
Prime Factorization

Introduction

Equivalent fractions are fractions that represent the same value. Equivalent fractions are extremely useful when adding and subtracting fractions because we often have to find a common denominator before we can perform the operation. An equivalent fraction, with the correct denominator, is a key to engaging in addition and subtraction. Equivalent fractions also help us solve proportion problems and help us find multiple representations of a particular fraction. In this topic set, we explore equivalent fractions and methods for determining equivalent fractions.

The notion of equivalent fractions as we currently understand and express them took some time to develop. Although the Egyptians had developed a means of representing unit fractions, they had no means of representing other types of fractions, except the fraction $\frac{2}{3}$. Thus, they certainly had no means of expressing a fractional amount in more than one way.

In Renaissance times, Arabs developed a method of representing fractions that occasionally used Roman numerals like $\frac{\text{IX}}{\text{XI}}$ to represent $\frac{9}{11}$. Because the Roman Numeration System does allow for some multiple representation of whole numbers, it is conceivable that Arabian fractions expressed with Roman numerals may have resulted in multiple representations of fractions. For example, because the number 49 can be represented in Roman numerals as IL or XLIX, the fractional representation of $\frac{1}{49}$ could have been either $\frac{\text{I}}{\text{IL}}$ or $\frac{\text{I}}{\text{XLIX}}$. Although this multiple representation of a fraction differs greatly from our current notion of equivalent fractions, it does provide some sense of the evolution of the concept of equivalent fractions.

For More Information Smith, D. E. (1958). *History of Mathematics, Volume II.* New York: Dover Publications.

6.2.1 Finding Equivalent Fractions

Given a fraction $\frac{a}{b}$, we can find fractions equivalent to $\frac{a}{b}$ by multiplying or dividing both the numerator and the denominator by the same nonzero number; that is, fractions of the form $\frac{a \times k}{b \times k}$ or $\frac{a \div k}{b \div k}$, provided $k \neq 0$ are equivalent to $\frac{a}{b}$. To generate a list of fractions equivalent to $\frac{2}{3}$, we could easily multiply both numerator and denominator by the natural numbers, generating the following list of equivalent fractions:

$$\frac{2}{3}, \frac{2 \times 2}{2 \times 3} = \frac{4}{6}, \frac{3 \times 2}{3 \times 3} = \frac{6}{9}, \frac{4 \times 2}{4 \times 3} = \frac{8}{12}, \frac{5 \times 2}{5 \times 3} = \frac{10}{15}, \frac{6 \times 2}{6 \times 3} = \frac{12}{18}, \text{ and so on.}$$

6.2.2 Testing Whether Fractions Are Equivalent

There are a variety of ways to determine whether two given fractions are equivalent. One way would be to **find a common denominator** between the two fractions, rewrite each fraction using the common denominator, and then compare numerators to see if they are also the same. For example, to check whether $\frac{3}{8}$ is equivalent to $\frac{5}{12}$, first rewrite each fraction using the common denominator of 24: $\frac{3}{8} = \frac{9}{24}$ and $\frac{5}{12} = \frac{10}{24}$. Clearly, the numerators differ, thus the two fractions are not equivalent.

Another method of testing whether two fractions are equivalent is **cross multiplication.** That is, $\frac{a}{b} = \frac{c}{d}$ if $ad = bc$. So, given the fractions $\frac{4}{5}$ and $\frac{7}{8}$, we can verify that they are not equivalent because $4 \times 8 \neq 5 \times 7$. Using the cross-multiplication method is essentially a shortcut to the common-denominator method described above. For example, if we were to find a common denominator for $\frac{4}{5}$ and $\frac{7}{8}$, we might choose 5×8 or 40 as the common denominator. To convert these fractions to forms with this common denominator, we would multiply numerator and denominator of $\frac{4}{5}$ by 8 and numerator and denominator of $\frac{7}{8}$ by 5: $\frac{8 \times 4}{8 \times 5} = \frac{32}{40}$ and $\frac{5 \times 7}{5 \times 8} = \frac{35}{40}$.

Notice that when we compare numerators, we are checking whether the products 4×8 and 5×7 are equal. These are the same products we checked when using the cross-multiplication method. Thus, cross-multiplication provides a quick way of checking whether or not fractions would have the same numerator if we were to convert them to fractions having a common denominator.

6.2.3 Ordering Fractions

When placing fractions in order from the smallest to the largest or the largest to the smallest, we can use the same methods we used to compare fractions to check for equivalence; that is, to order a set of fractions from the smallest to the largest, we could first **find a common denominator** that they all share and then compare numerators to place them in order. If there is a large set of fractions to order, this could be a time-consuming, although effective, approach.

Using a variation of the **cross-multiplication method** is a useful means of comparing and ordering fractions. Given two fractions $\frac{a}{b}$ and $\frac{c}{d}$, we say that $\frac{a}{b} < \frac{c}{d}$ if $ad < bc$. Similarly, $\frac{a}{b} > \frac{c}{d}$ if $ad > bc$. Thus, we can verify that $\frac{3}{2} > \frac{5}{4}$ because $3 \times 4 > 2 \times 5$.

Related Topics

The Set of Rational Numbers
Simplifying Fractions
Improper Fractions and Mixed Numbers
Adding Rational Numbers
Subtracting Rational Numbers
Multiplying Rational Numbers
Dividing Rational Numbers
Ratios

TOPIC 6.3 Simplifying Fractions

cornerstone

Fractions
Rational Numbers
Integer Multiplication
Integer Division
Least Common Multiple
Divisibility
Greatest Common Divisor
Prime Factorization

Introduction

Many people believe that simpler is better; consequently, they often try to simplify their lives, frequently by making "big jobs" smaller. This idea of simplification transfers to working with fractions; that is, rather than manipulate fractions that involve large numbers, it is preferable for most people to simplify fractions to equivalent versions using smaller numbers. In the following sections, we examine the process of simplification of fractions.

Until the early 1900s, the simplification (or reduction) of a fraction to lower or lowest terms was commonly known as *abbreviation*. The word *depression* was also used.

Today, we use simplified fractions in our everyday conversations. For example, when you want to order a half-pound of cheese at the deli, it is unconventional to say, "I'll have $\frac{6}{12}$ of a pound of cheese." Simplified fractions are easier for us to work with because we can more readily picture part of a whole if that whole is broken in a small number of equal-sized pieces.

6.3.1 Relatively Prime

A fraction is in **simplest form** when the numerator and the denominator are **relatively prime,** that is, when the numerator and denominator have no factors in common other than 1. Thus, the fraction $\frac{5}{6}$ is in simplest form because 5 and 6 have no factors in common other than 1. On the other hand, the fraction $\frac{8}{12}$ is not in simplest form because 8 and 12 share a common factor of 4, which means this fraction can be simplified further.

6.3.2 Simplifying Fractions by Looking for Common Factors

One means of simplifying a fraction is to look for **factors that the numerator and denominator share.** We can simplify a fraction by dividing numerator and denominator by any factor that they have in common (other than 1). Often, students determine a factor that the numerator and the denominator share either by recognizing something they have in common based on experience or by using divisibility tests to check what factors they might share. In the case of the fraction $\frac{12}{18}$, students might notice that 18 and 12 both share 2 as a factor, so they will divide both numerator and denominator by 2 and reduce the fraction to $\frac{6}{9}$. Students might then notice that 6 and 9 have a common factor of 3 and so divide numerator and denominator by 3, reducing the fraction further to $\frac{2}{3}$. Because 2 and 3 have no factors other than 1 in common, we can conclude that the fraction is now in simplest form.

Rather than reduce a fraction in multiple steps as in the above example, a more efficient method of reducing a fraction to simplest form is to determine the **greatest common factor** that the numerator and the denominator share and to divide both by this factor to find simplest form in one step. Thus, for the fraction $\frac{12}{18}$, the greatest common factor of 12 and 18 is 6. Dividing numerator and denominator by 6 reduces the fraction to its simplest form of $\frac{2}{3}$ in one step.

6.3.3 Simplifying Fractions Using Prime Factorization

Another method of reducing fractions to simplest form is to find the prime factorizations of the numerator and the denominator and factor out all prime factors that the two numbers share. For example, the fraction $\frac{24}{36}$ can be written in prime factorization as $\frac{2 \times 2 \times 2 \times 3}{2 \times 2 \times 3 \times 3}$. Noticing that the numerator and the denominator share two 2s and one 3 as prime factors, one can divide by those factors or cancel out pairs of those prime factors that appear in numerator and denominator to yield a simplified form of $\frac{2}{3}$.

Related Topics

The Set of Rational Numbers
Equivalent Fractions
Improper Fractions and Mixed Numbers
Adding Rational Numbers
Subtracting Rational Numbers
Multiplying Rational Numbers
Dividing Rational Numbers
Ratios

TOPIC 6.4 *Improper Fractions and Mixed Numbers*

cornerstone

Fractions
Rational Numbers
Integer Multiplication
Integer Division
Least Common Multiple
Divisibility
Greatest Common Divisor
Prime Factorization

Introduction

Too often, people restrict their image of fractions to fractions between 0 and 1. There are an infinite number of fractions that can be represented as improper fractions or mixed numbers. Fractions that must be represented by mixed number or improper fractions are those that fall between any two integers other than between 0 and 1 or between 0 and -1. Between any two other integers, say between 3 and 4, the rational numbers on the number line will have a value of 3 plus some fractional part. For example, $3\frac{1}{3}$ or $3\frac{1}{2}$ would be between 3 and 4 on the number line. In the sections that follow, we examine mixed numbers and improper fractions and methods for converting fractions to such forms.

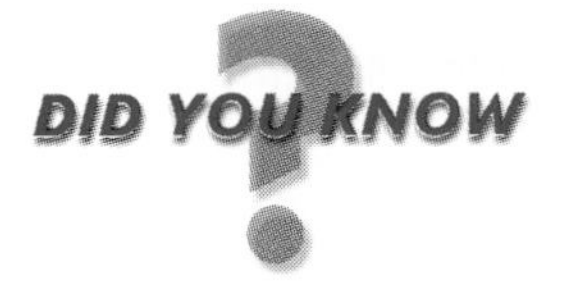

Historically, the idea of mixed numbers took quite a while to develop. At first, cultures were only able to represent unit fractions, those of the form $\frac{1}{n}$, where n was some natural number. In the Middle Ages, these unit fractions were sometimes called simple fractions; the more general form were known as composite fractions. The Babylonians were among the first to represent fractions other than unit fractions, such as the fractions $\frac{2}{3}$, $\frac{2}{18}$, $\frac{4}{18}$, $\frac{5}{6}$, and other cases of a similar level of difficulty.

Fractions within the work of the Indian mathematician, Brahmagupta, who lived in A.D. 598–c. 665, were among the first fractions to appear much like the fractions we write today. The Hindus wrote fractions by placing integers on top of each other, without a line between them. For example, the fraction $\frac{2}{3}$ would have been represented as $\begin{matrix}2\\3\end{matrix}$. The Hindus were also the first culture to represent mixed numbers. They represented mixed numbers by using a type of compound fraction notation. For example, they would write the mixed number $1\frac{1}{3}$ as three integers being placed one on top of the other: $\begin{matrix}1\\1\\3\end{matrix}$

For More Information Smith, D. E. (1958). *History of Mathematics, Volume II*. New York: Dover Publications.

6.4.1 Improper Fractions and Mixed Numbers

Fractions with a numerator that is greater than the denominator are called **improper fractions**. Thus, improper fractions are positive fractions greater than 1 or negative fractions less than -1. Examples of improper fractions include $\frac{9}{7}$ and $-\frac{4}{3}$.

A **mixed number** is a number that has both a whole number and a fractional component. Examples of mixed numbers include $3\frac{1}{2}$ and $-2\frac{2}{5}$. The word *and* is usually used in expressing the mixed number. For example, $3\frac{1}{2}$ would be expressed as "three and one-half," indicating that there are 3 wholes and $\frac{1}{2}$ of another whole.

It is sometimes useful to express an improper fraction as a mixed number or a mixed number as an improper fraction. Converting fractions back and forth from improper-fraction to mixed-number form is particularly helpful when preparing to perform operations with fractions and later simplifying results of operations with fractions.

6.4.2 Changing Mixed Numbers to Improper Fractions

To change a mixed number to an improper fraction, one should first rewrite the whole-number part of the mixed number to an equivalent fraction form. The best choice of denominator would be the denominator of the fractional part of the mixed number. For example, with the mixed number $4\frac{2}{3}$, read as "four and two-thirds," one should first rewrite 4 as $\frac{12}{3}$. Then what we have instead of 4 and $\frac{2}{3}$ is $\frac{12}{3}$ and $\frac{2}{3}$, which can be computed as $\frac{12}{3} + \frac{2}{3} = \frac{14}{3}$.

In the case of a negative mixed number, such as $-2\frac{4}{5}$, one can express this number as negative "2 and $\frac{4}{5}$." Changing the whole number, 2, to $\frac{10}{5}$, we then have negative $\frac{10}{5}$ and $\frac{4}{5}$ or negative $\frac{10}{5} + \frac{4}{5}$ = negative $\frac{14}{5} = -\frac{14}{5}$.

In general, when you have a mixed number of the form $k\frac{a}{b}$, it can be expressed as an improper fraction, using the following formula: $\frac{kb + a}{b}$. Revisiting the example of $4\frac{2}{3}$ used above, one can use the formula finding $4\frac{2}{3} = \frac{4 \cdot 3 + 2}{3} = \frac{14}{3}$. This formula essentially provides a shortcut to rewriting the whole number piece as a fraction, using a common denominator with the fractional piece.

6.4.3 Changing Improper Fractions to Mixed Numbers

To change an improper fraction to a mixed number, one could view the fraction as a division problem. For instance, with the fraction $\frac{13}{4}$, we could divide 13 by 4 to determine how many times the number 4 goes into 13. Solving with long division, we find $13 \div 4 = 3$ with a remainder of 1. Viewing this remainder as a part of the number of times four goes into 13, we can express this remainder as the fraction $\frac{1}{4}$. Thus, we can write the answer to the division problem $13 \div 4$ as $3\frac{1}{4}$. Thus, $\frac{13}{4} = 3\frac{1}{4}$.

In general, we can convert an improper fraction $\frac{k}{b}(k > b)$ to a mixed number using the long-division procedure. Thus, $k \div b = n\frac{a}{b}$, where n is some natural number, a is the remainder, and b is the divisor. For the improper fraction $\frac{16}{15}$, long division of $16 \div 15 = 1$ with a remainder of 1 suggests that $\frac{16}{15} = 1\frac{1}{15}$.

Related Topics

The Set of Rational Numbers
Equivalent Fractions
Simplifying Fractions
Adding Rational Numbers
Subtracting Rational Numbers
Multiplying Rational Numbers
Dividing Rational Numbers

In the Classroom

The fourth of the five chapters whose focus is on the *Number and Operations Content Standard*, this chapter also emphasizes the *Representation Process Standard*. Given that "instructional programs from prekindergarten through grade 12 should enable all students to use representations to model and interpret physical, social, and mathematical phenomena" and that "representations should be treated as essential elements in supporting students' understanding of mathematical concepts and relationships" (NCTM, 2000, p. 67), we believe that you should develop a strong understanding of (a) the concept of fraction as dependent on knowing what constitutes the "one" (Activities 6.1–6.7), and (b) operations with fractions as extending prior concepts of arithmetic operations (Activities 6.8–6.16).

Some Expectations of Your Future Students:

Pre-K–2 (NCTM, 2000, p. 78)

- connect number words and numerals to the quantities they represent, using various physical models and representations
- understand and represent commonly used fractions such as $\frac{1}{4}$, $\frac{1}{3}$, and $\frac{1}{2}$
- understand situations that entail multiplication and division, such as equal groupings of objects and sharing equally

Grades 3–5 (NCTM, 2000, p. 148)

- develop understanding of fractions as parts of unit wholes, as parts of a collection, as locations on number lines, and as divisions of whole numbers
- use models, benchmarks, and equivalent forms to judge the size of fractions
- use visual models, benchmarks, and equivalent forms to add and subtract commonly used fractions

How Activities 6.1–6.16 Help You Develop an Adult-level Perspective on the Above Expectations:

- By examining fractional models and representations and relationships between them, you develop an understanding of fraction as parts of the unit whole. You realize that this understanding depends on understanding division as sharing equally.
- By working with different region models, you develop the concepts of unit fractions and equivalent fractions and construct a deeper understanding of the relationship between the numerical representations of a fraction and the concept of the fraction.
- By using models appropriate to the context, you associate meaning with each and become confident in making sense of problems that involve fractions. For example, you may use the set (discrete) model in the context of a number (discrete) of people, but the linear (continuous) model in the context of time (continuous) elapsed.
- By identifying the arithmetic operations involved and modeling them using different fractional models, you reinforce your understanding of the models of operations learned in Chapter 3. Misconceptions such as "multiplication always makes bigger" or "division always makes smaller" are thus challenged and replaced by models of operations that will encompass all real numbers.

Bibliography

Behr, M. J., Post, T. R., & Wachsmuth, L. (1986). Estimation and children's concept of rational number size. In H. L. Schoen (Ed.), *Estimation and mental computation* (pp. 103–111). Reston, VA: National Council of Teachers of Mathematics.

Bezuk, N. S. (1988). Fractions in the early childhood mathematics curriculum. *Arithmetic Teacher, 35*, 56–60.

Bezuk, N. S., & Bieck, M. (1993). Current research on rational numbers and common fractions: Summary and implications for teachers. In D. T. Owens (Ed.), *Research ideas for the classroom: Middle grades mathematics* (pp. 118–136). Old Tappan, NJ; Macmillan.

Bezuk, N. S., & Cramer, K. (1989). Teaching about fractions: What, when, and how? In P R. Trafton (Ed.), *New directions for elementary school mathematics* (pp. 156–161). Reston, VA: National Council of Teachers of Mathematics.

Cramer, K., & Bezuk, N. (1991). Multiplication of fractions: Teaching for understanding. *Arithmetic Teacher, 39*, 34–37.

Dorgan, K. (1994). What textbooks offer for instruction in fraction concepts. *Teaching Children Mathematics, 1*, 150–155.

Edge, D. (1987). Fractions and panes. *Arithmetic Teacher, 34*, 13–17.

Ettline, J. F. (1985). A uniform approach to fractions. *Arithmetic Teacher, 32*, 42–43.

Greenwood, J. (1989). Problem solving with fractions. *Mathematics Teacher, 82*, 44–50.

Hollis, L. Y. (1984). Teaching rational numbers—Primary grades. *Arithmetic Teacher, 31*, 36–39.

Howard, A. C. (1991). Addition of fractions: The unrecognized problem. *Mathematics Teacher, 84*, 710–713.

Kieren, T. E. (1984). Helping children understand rational numbers. *Arithmetic Teacher, 31*, 3.

Kieren, T., Davis, B., & Mason, R. (1996). Fraction flags: Learning from children to help children learn. *Mathematics Teaching in the Middle School, 2*, 14–19.

Lester, F. K. (1984). Preparing teachers to teach rational numbers. *Arithmetic Teacher 31*, 54–56.

Mack, N. K. (1993). Making connections to understand fractions. *Arithmetic Teacher, 40*, 362–364.

Moss, J., & Case, R. (1999). Developing children's understanding of the rational numbers: A new model and an experimental curriculum. *Journal for Research in Mathematics Education, 30*, 122–147.

Ockenga, E. (1984). Chalk up some calculator activities for rational numbers. *Arithmetic Teacher, 31*, 51–53.

Ott, J. M., Snook, D. L., & Gibson, D. L. (1991). Understanding partitive division of fractions. *Arithmetic Teacher, 39*, 7–11.

Post, T. R., & Cramer, K. (1987). Children's strategies in ordering rational numbers. *Arithmetic Teacher, 35*, 33–35.

Pothier, Y., & Sawada, D. (1990). Partitioning: An approach to fractions. *Arithmetic Teacher, 38*, 12–16.

Sinicrope, R., & Mick, H. W. (1992). Multiplication of fractions through paper folding. *Arithmetic Teacher, 40*, 116–121.

Skypek, D. H. B. (1984). Special characteristics of rational numbers. *Arithmetic Teacher, 31*, 10–12.

Steffe, L. P., & Olive, J. (1991). Research into practice: The problem of fractions in the elementary school. *Arithmetic Teacher, 38*, 22–24.

Trafton, P. R., & Zawojewski, J. S. (1984). Teaching rational number division: A special problem. *Arithmetic Teacher, 31(6)*, 20–22.

Witherspoon, M. L. (1993). Fractions: In search of meaning. *Arithmetic Teacher, 10*, 482–485.

CHAPTER 7 SEVEN

Real Numbers: Rationals & Irrationals

CHAPTER OVERVIEW

As technology (especially calculators) becomes more and more prevalent for everyday use, as well as for scientific purposes, so does the use of decimals (and percent). At the same time, ratio and proportion are also quite useful to solve a variety of real-world problems. In this chapter, you can learn some interesting facts about rational and irrational numbers and will examine some concepts and procedures involving real-number ideas.

BIG MATHEMATICAL IDEAS

Problem-solving strategies, conjecturing, verifying, decomposing, generalizing, representation, using language and symbolism, limit, mathematical structure

NCTM PRINCIPLES & STANDARDS LINKS

Number and Operation; Problem Solving; Reasoning; Communication; Connections; Representation

TOPIC ***7.1*** Ratio
7.2 Proportion
7.3 Introduction to Decimals
7.4 Decimals and Fractions
7.5 Terminating and Repeating Decimals
7.6 Decimal Addition and Subtraction
7.7 Decimal Multiplication
7.8 Decimal Division
7.9 Percent
7.10 Interest
7.11 Rational Numbers
7.12 Looking at Rational Numbers as the Set of Rational Numbers
7.13 Adding Rational Numbers
7.14 Subtracting Rational Numbers
7.15 Multiplying Rational Numbers
7.16 Dividing Rational Numbers
7.17 Irrational Numbers
7.18 Real Numbers
In the Classroom
Bibliography

Ratio

cornerstone **Fractions**
Rational Numbers

Introduction

Whether you want to compute the speed with which we are driving or whether we need to figure out how much it would cost to purchase three sets of an item, given that we know the cost of one set, we are working with ratios. In the following sections, you will learn the definition of a ratio and examine different types of ratios. You will also learn of some of the relationships between ratios and fractions.

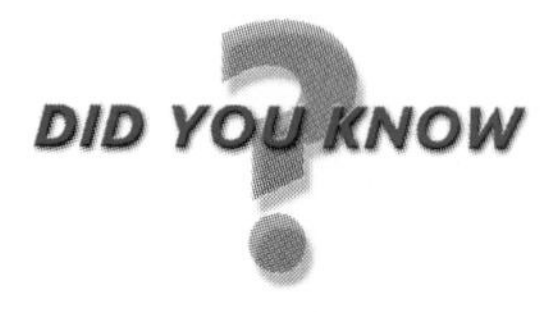

The **Golden Ratio** is a ratio that describes the relationship between the lengths of line segments formed when we divide a line segment *AB* into two segments at an interior point, *P*, between *A* and *B* so that the ratio of the length of *AB* to *AP* is equal to the ratio of the length of *AP* to *PB*. Dividing a line segment according to this ratio is said to create a division that the ancient Greeks thought was pleasing to the eye. Related to the Golden Ratio is the **Golden Rectangle,** which essentially is a rectangle divided into two sections, one a square and one a rectangle, so that the ratio of the sides of the rectangle is the Golden Ratio. Rectangular shapes that are Golden Rectangles are also said to be shapes that are pleasing to the eye. Many architectural designs are built according to the Golden Ratio. One such famous landmark of Greek architecture is the Parthenon.

For More Information Bunt, L. N. H., Jones, P. S., & Bedient, J. D. (1988). *The Historical Roots of Elementary Mathematics*. New York: Dover Publications.

7.1.1 Definition of Ratio

A **ratio** is a comparison of two or more quantities. In most day-to-day situations, we generally compare two quantities. For example, in a classroom we might compare the number of boys in the classroom (which is 15) to the number of girls in a classroom (which is 10). We would say, the ratio of boys to girls is 15 to 10.

7.1.2 Ratio Notation

Ratios can be expressed several ways. One way to express the comparison of 15 boys in a classroom to 10 girls in a classroom is by expressing the ratio using the word *to*, saying the ratio is **15 to 10.** Instead of using the word *to*, we often represent ratios using a colon symbol ":". Thus, we can say the ratio is **15 : 10.** A final way to express a ratio is by writing the ratio as a fraction where the numerator is one quantity to be compared and the denominator is the other quantity. Thus, the ratio can be expressed $\frac{15}{10}$.

7.1.3 Types of Ratios

When comparing two quantities, we can create ratios of three different varieties. We can compare (1) a part to a part, (2) a part to a whole, or (3) a whole to a part.

Comparing **a part to a part** takes place when we take a whole set and break it into two subsets to compare. For example, of the 500 employees at Mathematics Incorporated, 400 of the employees own foreign cars and 100 own domestic cars. Thus, we can talk about the ratio of foreign car owners to domestic car owners as 400:100.

Sometimes, we compare a **part to a whole.** In the example of the employees at Mathematics Incorporated, we could say the ratio of foreign-car owners to total employees is 400:500. We can also compare **a whole to a part.** For example, comparing the total number of employees to the number of domestic-car owners is 500:100.

7.1.4 Rates

We often express ratios using the term ***per*** in the comparison. For example, when comparing how far we have traveled in a certain period of time, we often describe our speed in terms of miles per hour. Whenever we compare a given quantity to time, we are describing a **rate.** Some frequently used rates are comparisons of distance:time (speed), growth : time (growth rate), money earned : time (interest rate).

To compute a rate, such as speed, we can express the comparison of quantities as a ratio in fraction form and perform the division suggested by the fraction representation to simplify the rate. For example, if I know I traveled 165 miles in 3 hours, I can compute my speed in terms of miles per hour (mph) as: $\frac{165}{3} = 55$ mph.

7.1.5 Comparing Ratios and Fractions

Because one representation of a ratio is to express a ratio as a fraction, we should explore when we can manipulate ratios like fractions and when they should be treated differently. A key similarity between fractions and ratios is the notion of equivalence. Like finding equivalent fractions, we can find **equivalent ratios,** and we find them by multiplying or dividing numerator and denominator by the same nonzero amount; that is, given the ratio $\frac{a}{b}$, then for any real number, k, $k \neq 0$, $\frac{a}{b} = \frac{a \times k}{b \times k} = \frac{a \div k}{b \div k}$.

For example, if there are 200 men and 100 women in the movie theater, then the ratio of men to women in the movie theater is $\frac{200}{100}$. By dividing numerator and denominator by 100, we can simply say that the ratio of men to women in the movie theater is $\frac{2}{1}$, indicating that there are twice as many men as there are women.

A key difference between ratios and fractions is how we add them. For example, when adding fractions, we find a common denominator and then add numerators, placing the sum over the common denominator (see Adding Rational Numbers for more details and examples). For example, $\frac{2}{3} + \frac{1}{4} = \frac{8}{12} + \frac{3}{12} = \frac{11}{12}$.

On the other hand, in the **addition of ratios,** we sometimes add numerators and add denominators to create a new ratio. For example, suppose on a first quiz that a student earned 6 points out of 8 possible points. Thus, the ratio of points earned to total possible points is $\frac{6}{8}$. Then suppose on a second quiz that this student earned 8 points out of 10 possible points. Thus, the ratio of points earned to total possible points on the second quiz is $\frac{8}{10}$. Now if we want to compute the ratio of the total number of points earned on quizzes to total possible points, we can find the total number of points earned by adding the numerators of the ratios for each quiz and then find the total number of points possible for the two quizzes by adding the denominators. We then find that the ratio of total points earned to total possible points on the two quizzes is: $\frac{6}{8} + \frac{8}{10} = \frac{14}{18}$.

Related Topics

Proportion
Percent

TOPIC 7.2 Proportion

Introduction

We encounter proportional relationships regularly. When we view models of the floor plans of a house or read maps, we have to understand the scale used to depict these models. The scale defines a proportional relationship between the model and the actual dimensions of the house or the actual distance between locations on a map. In the following sections, we learn what proportions are and how to solve proportions where some piece of information in a proportion is unknown.

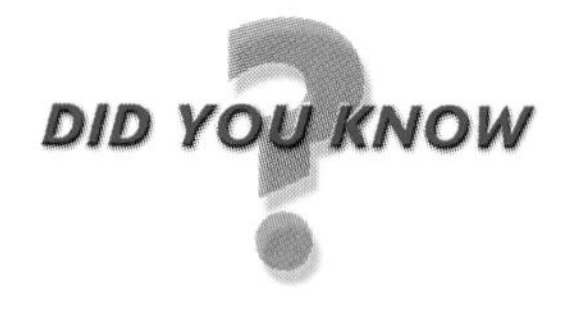

When solving a proportion, there are three knowns and one unknown. Historically, the three knowns were given many names. However, the process for using the three terms and arithmetically solving for the fourth term was known as the **Rule of Three.** This expression originated with the Hindus around the year A.D. 620. The Rule of Three was so important for solving mercantile problems that among merchants this rule of proportions became known as the **Golden Rule.**

For More Information Smith, D. E. (1958). *History of Mathematics, Volume II.* New York: Dover Publications.

7.2.1 Definition of Proportion

A **proportion** is an equation where both expressions on either side of the equals sign are ratios; that is, if $\frac{a}{b}$ and $\frac{c}{d}$ are two ratios such that $\frac{a}{b} = \frac{c}{d}$, then this equality statement is called a proportion. Examples of proportions are: $\frac{1}{2} = \frac{3}{6}$, $\frac{2}{3} = \frac{10}{15}$, and $\frac{x}{x+1} = \frac{2x}{2x+2}$.

7.2.2 Solving Proportions Using Equivalent Ratios

Proportions are useful in solving problems involving ratios where we are missing one piece of information. For example, if eggs cost 90 cents per one dozen, how much will it cost to buy 4 dozen eggs? In this situation, we can set up a proportion comparing number of dozens to cost per dozen:

$$\frac{1}{90} = \frac{4}{x}$$

Given this situation, we need to solve for x. Because we know we are seeking an equivalent ratio, we can create a series of ratios equivalent to $\frac{1}{90}$ until we find the one we need:

$$\frac{1}{90} = \frac{2}{180} = \frac{3}{270} = \frac{4}{360} = \frac{5}{450} \dots .$$

From this list of equivalent ratios, we conclude that it will cost 360 cents, or \$3.60, for 4 dozen eggs.

In a second example, if we want to solve the proportion $\frac{3}{5} = \frac{x}{25}$, we can again generate a list of ratios equivalent to $\frac{3 \cdot 3}{5 \cdot 5} = \frac{6}{10} = \frac{9}{15} = \frac{12}{20} = \frac{15}{25} = \frac{18}{30} = \frac{21}{35} \dots$. We conclude, then that x must be 15.

7.2.3 Solving Proportions Using Cross Multiplication

Proportions are frequently solved using **cross multiplication;** that is, given a proportion $\frac{a}{b} = \frac{c}{d}$, then $ad = bc$. Knowing this fact of cross multiplication, we can solve any proportion involving an unknown by setting up the cross multiplication and solving for x.

For example, given the proportion: $\frac{4}{10} = \frac{9}{x}$, we know that

$$4x = 9 \times 10$$
$$4x = 90$$
$$x = \frac{90}{4} = \frac{45}{2} = 22\frac{1}{2} \text{ or } 22.5.$$

In a second example, suppose that the ratio of men to women at State University is 4 to 5. If there are 2,200 women at State University, how many men are there? We solve by setting up a proportion, cross multiplying, and solving for x.

$$\frac{4}{5} = \frac{x}{2200}$$

Then,

$$4 \times 2200 = 5x$$
$$8800 = 5x$$
$$1760 = x.$$

7.2.4 Directly and Indirectly Proportional

If two items x and y are proportional, that is if x is **proportional** to y, then there exists some **constant of proportionality,** k, such that $x = ky$ or $\frac{x}{y} = k$.

For example, if we want to show that 40 is proportional to 8, we try to find a value, k, such that $40 = 8k$. Solving for 5, we identify the constant of proportionality is 5. Therefore, 40 is proportional to 8.

Most of the time when the claim is made that two numbers are proportional, the intended meaning is that the two numbers are directly proportional. So when we have the relationship where $x = ky$, we can say that x is **directly proportional** to y. On the other hand, when the relationship between x and y is such that $x = \frac{1}{k} \cdot y$ or $x = \frac{y}{k}$, we claim that x is **indirectly proportional** to y.

For example, we can show that 6 is indirectly proportional to 30. Setting up the equation, we have $6 = \frac{1}{k} \cdot 30$; solving for k, we find that $k = 5$. Thus, because we could find a value for k that satisfies the equation, we verified that 6 is indeed indirectly proportional to 30.

Percent

TOPIC 7.3 Introduction to Decimals

Base Ten
Place Value
Integers

Introduction

Decimals are a natural part of our base-ten system. We usually think of fractional amounts when we hear the term *decimal* because we envision a decimal point separating a whole number part from a fractional part. In this section, we explore the definition of a decimal, where a decimal point features in the determination of place value, and how to compare and order decimals.

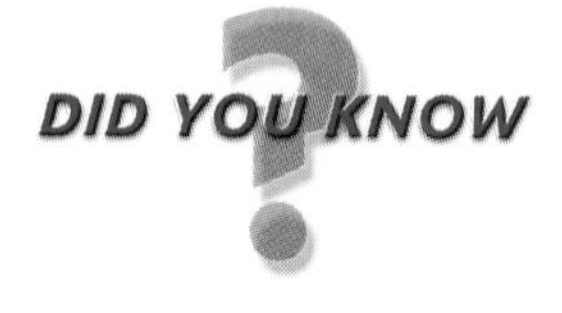

Several mathematicians are credited with having done work leading to the invention of decimal notation of fractions. One man, Jean de Meurs, who wrote in the early fourteenth century, brought intuition to the benefits of decimal notation in his work with the extraction of roots associated with $\sqrt[n]{a}$. Another influence leading to the invention of the decimal fraction was the rule for dividing by numbers of the form $a \times 10^n$. Work in this area is attributed to mathematicians Piero Borghi and Francesco Pellos in the mid-1400s. The first mathematician, however, to comprehend fully the significance of decimals is Christoff Rudolff, whose work with decimals appeared in Augsburg in 1530. Notably, he used a decimal bar, "|", instead of a decimal point, but his concept of place values was as it is today.

For More Information Smith, D.E. (1958). *History of Mathematics, Volume II.* New York: Dover Publications.

7.3.1 Definition of Decimal

When most people use the term **decimal,** they refer to a number that has some **decimal fraction** component. A **decimal point,** represented by the symbol ".", separates integer value parts of a decimal number (comprised of symbols to the left of the decimal point) from fractional value parts of a decimal number (comprised of symbols to the right of the decimal point.). Some examples of decimal numbers are: 10.34, −3.67421, and 0.9. When a number is expressed as a decimal fraction plus an integer, it is often referred to as a **mixed decimal.**

7.3.2 Decimals, Base Ten, & Place Value

The decimal point in a decimal number helps to indicate the place value of the digits in the number. The places to the left of the decimal point are the places we use to express whole number and integer amounts: ones place, tens place, hundreds place, and so on. The places to the right of the decimal point are used to express fractional amounts: tenths place, hundredths place, thousandths place, and so on. The place value chart in Figure 7.1 illustrates places to the right and the left of the decimal point in the decimal system.

Figure 7.1

	10^3	10^2	10^1	10^0	10^{-1}	10^{-2}	10^{-3}	
	1000	100	10	1	1/10	1/100	1/1000	
etc.	thousands	hundreds	tenths	ones	tenths	hundreds	thousands	etc.

Decimal point (between ones and tenths)

Notice that, consistent with a base-ten system, each place value represents a power of 10. Places to the right of the decimal point can be expressed as 10 raised to a negative integer power.

7.3.3 Expressing a Decimal Number

Using the chart in Figure 7.1, we can interpret decimal numbers and their values. For example, the number 34.8 can be interpreted as representing "thirty-four and eight tenths." Note that the decimal point is read as the word *and*, indicating that the integer amount is combined with the fraction amount. The number 2.01 is read as "two and one hundredth."

When reading decimals that have more than one significant digit to the right of the decimal point, it is proper to express the string of digits to the right of the decimal point as a fraction where the numerator is the whole number represented by those digits and the denominator is the place where the right-most significant (nonzero) digit sits. For example, the decimal 12.345 would be read as "twelve and three hundred forty-five thousandths." Similarly, the decimal 0.99 would be read as "ninety-nine hundredths."

7.3.4 Zeros in Decimals

Decimal numbers can include zeros to hold place value. Thus, the decimal 1.504 represents 1 one, 5 tenths, and 4 thousandths. The number 1.504 would be read "one and five hundred four thousandths." Although we could add an infinite string of zeros to the right of the last nonzero digit in a decimal number and not affect the value of that decimal, it is a universally accepted practice to end a decimal number with a nonzero, right-most digit. Thus, although 15.670 is technically equivalent in value to 15.67, it is best to write the decimal without additional zeros. Exceptions to this rule include situations where scientific or statistical measurements require answers expressed to certain degrees of accuracy. For example, it might be necessary to provide a scientific measurement rounded to the nearest thousandth. If the actual measurement is 0.1495773 centimeters (cm), the appropriate measurement rounded to the nearest thousandth would be 0.150 cm.

7.3.5 Base-Ten Blocks as a Model for Decimals

Base-ten blocks provide an excellent tool for modeling decimal fractions. Letting the flat block represent 1 whole, the long block then represents one tenth, and the unit block represents one hundredth. With these values defined, we can represent the decimal number 3.25 using base-ten blocks as illustrated in Figure 7.2.

Modeling decimals using base-ten blocks can be extremely helpful when trying to visualize the addition and subtraction of decimal numbers.

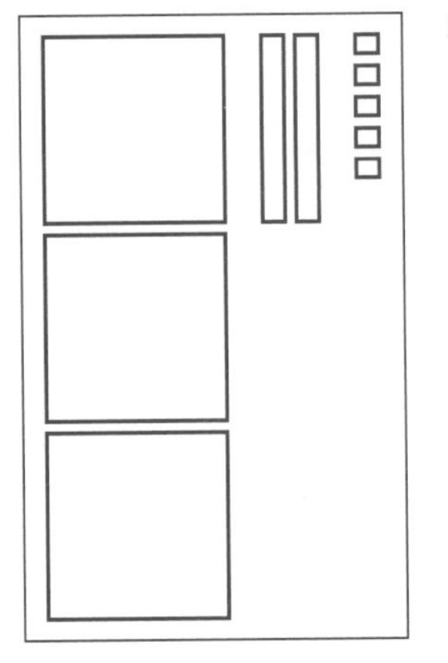

Figure 7.2

7.3.6 Rounding of Decimals

In many practical problems involving decimals, it is necessary to round a decimal number to a particular decimal place. The place to which we round is determined by the degree of accuracy we want in our solution. The rules for rounding decimal fractions are similar to the rules for rounding whole number amounts.

To round a decimal number to a particular place, check the digit to the right of that place to determine whether you round up or down. If the digit to the right of the place to which you are rounding is 5 or more, then round up. If the digit is less than 5, round down. Unlike when rounding whole numbers where all digits to the right of the place to which you are rounding become zeros, when rounding to a fractional decimal place, we truncate, or cut off, the rounded number at the place to which we are rounding, showing no digits to the right of that place.

For example, to round 78.569 to the nearest hundredth, we check the digit in the thousandths place. Because that digit is a 9, we will round the 6 in the tenths place up to a 7 and then truncate the number at that point. Thus, 78.569 rounded to the hundredths place is 78.57. Rounding 0.78927 to the nearest thousandth, we get the number 0.789.

Related Topics

Decimals and Fractions
Adding Decimals
Subtracting Decimals
Exponents

TOPIC 7.4 *Decimals and Fractions*

cornerstone

Decimals
Fractions
Rational Numbers

Introduction

When we think of fractions, we generally picture a numerator over a denominator. However, fractional amounts can also be expressed as decimals. There is a connection between decimal and common fractions. In the following sections, we examine a method for converting certain fractions to terminating decimal form and a means of converting terminating decimals to common fraction form.

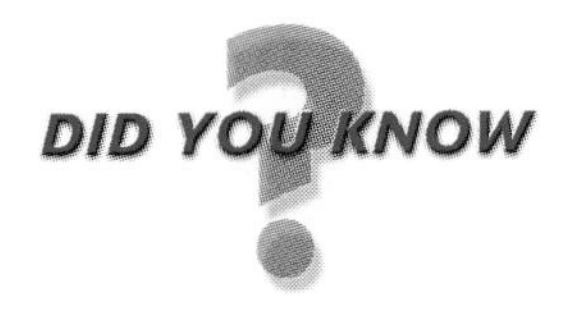

Fractional amounts were originally expressed in the form of a numerator over a denominator. When computing business transactions, performing long-division problems, fractions began to take on very complicated forms. By the advent of printing capabilities in the mid-1400s, a need for a more varied method of expressing fractional amounts was clearly necessary. Decimal representation helped to alleviate the problem of expressing very complicated fractions.

For More Information Smith, D. E. (1958). *History of Mathematics, Volume II.* New York: Dover Publications.

7.4.1 Decimals and Common Fractions

Decimal fractions and **common fractions** (a fraction written as a numerator and a denominator) both represent part of a whole. All decimal fractions with a right-most significant (nonzero) digit can be expressed as a common fraction with a numerator and a denominator. Similarly, common fractions can be written in one of several decimal forms.

7.4.2 Converting Decimals to Common-Fraction Form

Decimals that have a right-most significant (nonzero) digit are called terminating decimals. All terminating decimals can be expressed as common fraction form. For example, $0.25 = \frac{25}{100}$ and $0.374 = \frac{374}{1000}$. A procedure to follow to convert terminating decimals to common fraction form is as follows:

1. Determine the place value of the right-most significant digit. This place will be the denominator of the common fraction form.
2. Write the digits of the decimal number in order without the decimal point. This number will be the numerator of the common fraction form.

For example, to convert the decimal 0.1254 to a fraction, we observe that the place value of the right-most digit is the ten thousandths place. Thus, the denominator will be 10,000. Next, writing the digits without the decimal point, we have the number 1254, which will be the numerator of the fraction. Thus, $0.1254 = \frac{1254}{10,000}$.

Mixed decimals are converted in the same manner. For example, the decimal 65.2 is converted to fraction form by determining the denominator to be 10 because the right-most digit is in the tenths place and then writing the digits 652 in the numerator. Therefore, $65.2 = \frac{652}{10}$.

7.4.3 Converting Certain Common Fractions to Decimals

Common fractions that have a denominator that is a power of 10, such as 10, 100, 1,000, and so on, can easily be converted to decimal form. A set of procedures that guides the conversion is the following:

1. The denominator of the common fraction indicates the place value of the right-most digit in the decimal to which we are converting the fraction.
2. The numerator will be written as is, and the decimal point will be placed so that the last digit ends in the place value indicated by the denominator.

For example, the fraction $\frac{87}{100}$ has a denominator of 100, so when converting to decimal form, the decimal's right-most digit will be in the hundredths place. The number 87 is written as is, and a decimal point is inserted so that the digit 7 is in the hundredths place. Consequently, $\frac{87}{100} = 0.87$.

Converting the number $\frac{2189}{100}$ to decimal form, note that we write the number 2189 and place the decimal point so that the digit 9 is in the hundredths place. Thus, $\frac{2189}{100} = 21.89$.

In a situation where you want to write $\frac{85}{1000}$ as a decimal, we write the number 85 and next need to place the decimal point so that the digit 5 is in the thousandths place. To do so, we need to add a place-holding zero in front of the 85. Thus, $\frac{85}{1000} = 0.085$.

7.4.4 Fractions That Do Not Have Power-of-10 Denominators

For common fractions that do not have a denominator that is a power of 10, many can still be converted to terminating decimal form. Those fractions that can be converted to terminating decimals have the following characteristic: *If the prime factorization of the denominator of a common fraction, when in simplest form, involves only powers of 2, powers of 5, or combinations of powers of 2 and 5, then the fraction can be rewritten with a denominator that is a power of 10.* For example, given the fraction $\frac{3}{20}$, we observe that the prime factorization of the denominator, $20 = 2 \times 2 \times 5$, consists of only twos and fives. Therefore, we know that we can find a fraction equivalent to $\frac{3}{20}$ that has a denominator that is a power of 10. Multiplying the numerator and the denominator of $\frac{3}{20}$ by $\frac{5}{5}$, we find that $\frac{3}{20}$ can be written in the equivalent form $\frac{15}{100} = 0.15$.

7.4.5 Using a Calculator for Converting

Another method for converting common fractions to decimal form is to recall that fractions can represent division problems. Thus, dividing the numerator by the denominator by long division or by using a calculator will yield the decimal form of the fraction. Use caution when using the calculator when converting fractions to decimal form. Calculators will round or truncate decimals after they have exceeded their digit capacity. This is particularly true for fractions that cannot be rewritten with a denominator that is a power of 10. These fractions will convert to nonterminating decimals (described in a section entitled Terminating and Repeating Decimals). Calculators will present nonterminating decimals as if they were terminating.

7.4.6 Comparing and Ordering Decimals Using Base-Ten Blocks

To determine which of the decimals 0.35 and 0.4 is larger, it is sometimes useful to model these decimals using base-ten blocks and then compare. Figure 7.3 illustrates the base-ten block representations of 0.35 and 0.4.

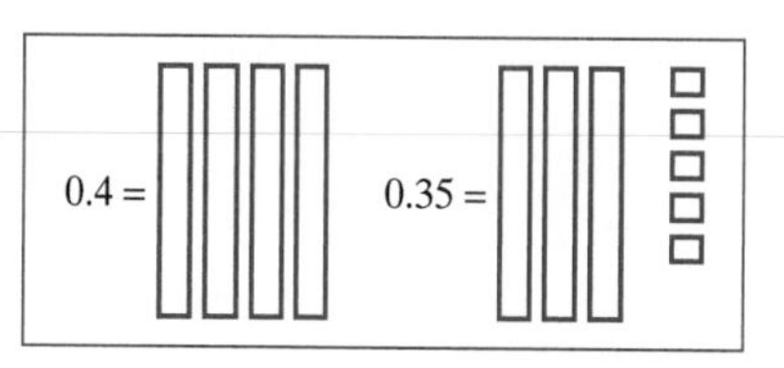

Figure 7.3

Letting one flat base-ten block represent one whole, the longs represent tenths and the units represent hundredths. Thus, the decimal 0.4 represents 4 tenths, or 4 longs, and the decimal 0.35 represents 35 hundredths, or 3 tenths and 5 hundredths. From this representation, it is clear that $0.4 > 0.35$.

7.4.7 Ordering Decimals By Lining Up Decimal Points

To compare decimals, one method involves lining up decimal points and comparing place value by place value. For example, to order the decimals 2.3333, 2.3434, 2.3, 2.4343, and 2.4435, begin by lining up the decimal points of each decimal number, adding zeros to the right of the right-most digits so that we have the same number of digits to compare in each number.

(1) 2.3333
(2) 2.3434
(3) 2.3000
(4) 2.4343
(5) 2.4435

Let us order these decimals from the smallest to the largest. Start by comparing the left-most digit of all the numbers. Because all five numbers have a two in the ones place, we cannot distinguish between them, so we move to the right to the tenths place. In this place, the smallest-valued digit is a 3. Three appears in the first three numbers, so one of these three will be the smallest decimal. To further compare these three numbers, we move to the hundredths place. The first has a 3 in the hundredths place, the second has a 4 in the hundredths place, and the third has a 0 in the hundredths place. Thus, the third number is the smallest of the three, the second number is next smallest, and the first number is the largest of the three. Therefore, the three smallest decimals in order are: 2.3, 2.3333, and 2.3434.

To determine which of the last two decimal numbers is the larger, compare the hundredths places. The last decimal number has a 4, whereas the fourth decimal number has a 3 in the hundredths place. Therefore $2.4343 < 2.4435$. So the order of the five decimal numbers from the smallest to the largest is: 2.3, 2.3333, 2.3434, 2.4343, 2.4435.

7.4.8 Ordering Decimals On a Number Line

Decimals can be ordered on a number line. We can show tenths by breaking a unit length into ten equal-sized pieces, each worth one tenth. See Figure 7.4. To show 0.2, 0.3, 0.25, and 0.28 on a number line, we see, that 0.25 and 0.28 are between 0.2 and 0.3, with $0.25 < 0.28$. Imagining breaking the length between 0.2 and 0.3 into ten equal-sized parts, we can estimate the placement of 0.25 and 0.28 on the number line.

From the number-line representation of these decimals, it is clear that $0.2 < 0.25 < 0.28 < 0.3$.

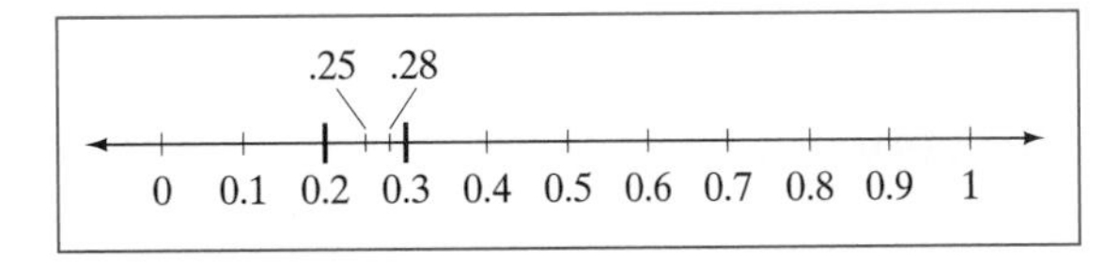

Figure 7.4

Related Topics **Terminating and Repeating Decimals**

TOPIC 7.5 Terminating and Repeating Decimals

Decimals
Decimals and Fractions
Rational Numbers

Introduction

Rational numbers can be expressed in decimal form as either terminating decimals or repeating decimals. In the following sections, we examine these types of decimals through definitions and examples. We also provide methods for converting fractions to repeating decimal form and for converting repeating decimals to common fraction form.

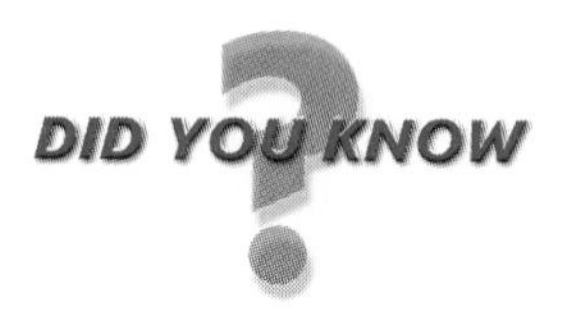

Probably, the most popular use for decimal fractions is to represent dollars and cents. Decimals that represent dollar and cents amounts are terminating decimals. Oftentimes, money amounts must be rounded to the nearest hundredth so that a comprehensible money amount can be established.

7.5.1 Terminating Decimals

Terminating decimals are decimal numbers that have a right-most significant (nonzero) digit. Examples of terminating decimals are: 10.234, −567.2, 0.336677.

7.5.2 Repeating Decimals

Some decimals do not terminate; rather, they continue on in an identifiable, repeating pattern. Thus, with these **repeating decimals,** there is no right-most digit. Examples of repeating decimals are: 0.33333 ... and 0.16666 ....The ellipses—that is, the symbols "..."—indicates that the digits continue in the same pattern. In the first example, the digit 3 keeps repeating over and over again. In the second example, the digit 6 repeats continuously.

Rather than writing a repeating decimal with an ellipsis, you can use what is called a **repeating bar** to indicate the **repeatand,** or the string of digits that repeats. For example, we can write 0.33333 ... as $0.\overline{3}$. Similarly, the decimal 0.16666 ... can be written as $0.1\overline{6}$. Notice on this second example that only the digit 6 appears under the repeating bar, indicating that only the digit 6 repeats.

In another example, given the repeating decimal $0.\overline{257}$, we can write this in ellipsis form as 0.257257257....Note that in this example, all three digits are under the repeating bar; thus all three digits repeat continuously in the same order.

7.5.3 Rational Numbers as Repeating Decimals

A rational number is any number that can be expressed as $\frac{a}{b}$ where a and b are both integers and $b \neq 0$. Any rational number can be converted to decimal form using long division. Terminating decimals ultimately get to a point where the remainder is zero in the long division. Repeating decimals, on the other hand, get to a point where a quotient begins to repeat. For example, to convert the rational number $\frac{4}{9}$ to decimal form, we divide 9 into 4:

$$\begin{array}{r} .44 \\ 9\overline{)4.00} \\ -3\,6 \\ \hline 40 \\ -36 \\ \hline 4 \end{array}$$

We observe that because the remainder continues to be 4, we are caught in a loop that will continue to repeat itself. Thus, we can conclude that the quotient in the repeating decimal $0.\overline{4}$.

As a second example, let us verify that $\frac{1}{7}$ can be written in decimal form as $0.\overline{142857}$.

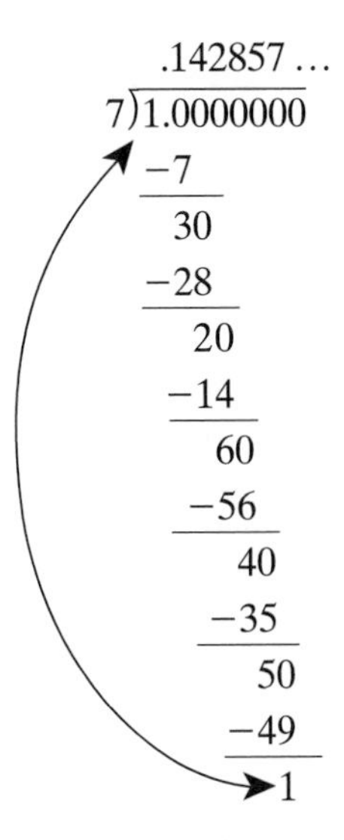

Notice that once we reach a point where the remainder is 1, we are essentially in the same division situation we were in during the first step of our division process; that is, when we initially divided 7 into 1, the first step was to bring down a zero and note that 7 goes into 10 one time with a remainder of 3. If we were to continue with the long-division problem, we would bring down a zero, divide 7 into 10, and note that it goes in one time with a remainder of 3. Thus, we would begin to repeat what has been done before. So, we know to stop and to indicate that the string $0.\overline{142857}$ repeats infinitely.

Checking the results of our long division using a calculator, note that the calculator rounds the repeating decimal to the closest terminating decimal it can fit into its view screen. Thus, a reminder: Take caution when using a calculator.

7.5.4 Converting Repeating Decimals to Rational-Number Form

All repeating decimals are rational numbers; that is, they can be written in the form $\frac{a}{b}$ where a and b are integers, with $b \neq 0$. Thus, there is an established procedure for converting repeating decimals to rational-number form.

The procedure is algebraic in nature. We will illustrate the procedure by converting the decimal $0.\overline{56}$ to rational-number form. First, be clear that $0.\overline{56} \neq 0.56$. Therefore, $0.\overline{56} \neq \frac{56}{100}$. Also, because $0.\overline{56} = 0.565656\ldots$ then $100 \times 0.\overline{56} = 56.565656\ldots = 56.\overline{56}$. Finally, note that the overall goal of the algorithm is to create two decimals with the same repeating part so that we can subtract away the repeating part.

The algorithm for converting repeating decimals to rational number form is as follows:

Example 1: Let $x = 0.\overline{56}$. Then $100x = 56.\overline{56}$. Since $x = 0.\overline{56}$, subtracting x from the left side of the equation $100x = 56.\overline{56}$ and subtracting $0.\overline{56}$ from the right-hand side of the equation should still maintain the equality of the equation. Thus, we find:

$$\begin{aligned} x &= 0.\overline{56} \\ 100x &= 56.\overline{56} \\ -x &\quad -0.\overline{56} \\ \hline 99x &= 56 \end{aligned}$$

Solving for x by dividing both sides of the equation by 99 we find:

$$x = \frac{56}{99}.$$

Example 2: Suppose we want to convert the decimal $5.\overline{142}$. Letting $x = 5.\overline{142}$, we want to create a new decimal that is a multiple of $5.\overline{142}$ so that the repeating strand of the new decimal "lines up with" the repeating strand of $5.\overline{142}$. This time, because the repeating strand is three digits in length, we multiply $5.\overline{142} \times 1000 = 5142.\overline{142}$. Thus, performing the algorithm we find:

$$\begin{aligned} x &= 5.\overline{142} \\ 1000x &= 5142.\overline{142} \\ -x &\quad -5.\overline{142} \\ \hline 999x &= 5137 \end{aligned}$$

Solving for x we find:

$$x = \frac{5137}{999}$$

Example 3: To convert the decimal $0.1\overline{4}$, we can follow the same procedure, adding an extra step at the end.

$$\begin{aligned} x &= 0.1\overline{4} \\ 10x &= \ \ 1.4\overline{4} \\ -x &\quad -0.1\overline{4} \\ \hline 9x &= 1.3 \end{aligned}$$

Note that at this point, we have eliminated the repeating part. Now, when solving for x we find:

$$x = \frac{1.3}{9}.$$

At this point, the solution is not in rational-number form because the numerator is not an integer. We can multiply numerator and denominator by 10 to find an equivalent fraction that is in rational-number form:

$$x = \frac{13}{90}.$$

Related Topics **Irrational Numbers**

TOPIC 7.6 *Decimal Addition and Subtraction*

cornerstone **Decimals**
Decimals and Fractions

Introduction

The ability to add and subtract decimals is a valuable tool, particularly when it comes to working with money and balancing financial accounts. In the following sections, the rules for decimal addition and decimal subtraction are introduced and modeled using base-ten blocks.

Simon Stevin (1548–1620) is considered to be among the most influential Dutch mathematicians. He is best noted for his contributions in numerical computation, including rational and real numbers. In particular, his numerical computations extended the use of the Hindu-Arabic base-ten system, representing decimal amounts in his numeric computations.

For More Information Smith, D. E. (1958). *History of Mathematics, Volume II*. New York: Dover Publications.

7.6.1 Adding Decimals

The rule for adding decimals says to (1) add decimals vertically, (2) line up decimal points in the addends, and (3) add digits in the same place-value positions together, beginning from the right-most digits, carrying sets of 10 to the next column when necessary. Once the decimal addends are lined up, any place where there is a digit missing in one of the addends can be replaced with a zero so that both addends have the same number of digits to the right of the decimal point.

For example, to add 345.7256 + 16725.89, we first write the addends vertically, making sure to line up decimal points and to add two zeros on to the end of the second addend:

$$\begin{array}{r} 345.7256 \\ +\ 16725.8900 \\ \hline \end{array}$$

Then, adding column by column, carrying when appropriate, we find the sum:

$$\begin{array}{r} 345.7256 \\ +\ 16725.8900 \\ \hline 17071.6156 \end{array}$$

7.6.2 Modeling the Addition of Decimals

Addition of decimals can be modeled using base-ten blocks. Exchanges made when carrying can be clearly visualized with this model. For example, modeling the addition problem 0.77 + 1.56, we first build each addend with base-ten blocks as shown in Figure 7.5.

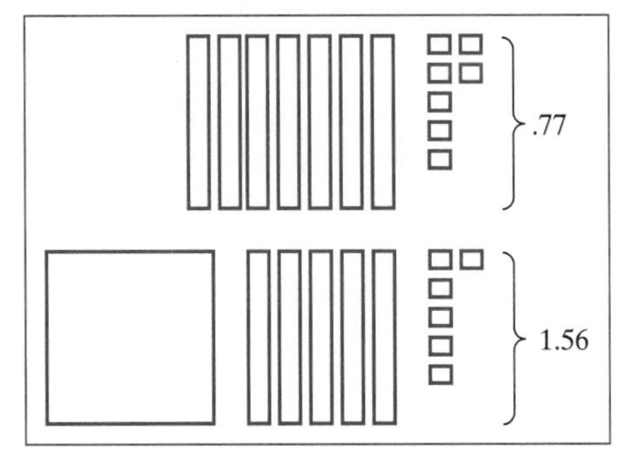

Figure 7.5

Next, we combine units (hundredths) and exchange one set of 10 units for a long (tenth), as seen in Figure 7.6.

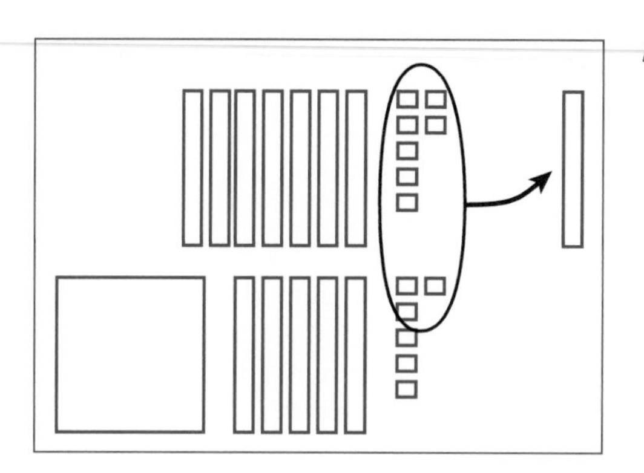

Figure 7.6

Then, the longs are combined, and 10 of the longs (tenths) are exchanged for 1 flat (1 whole) as shown in Figure 7.7.

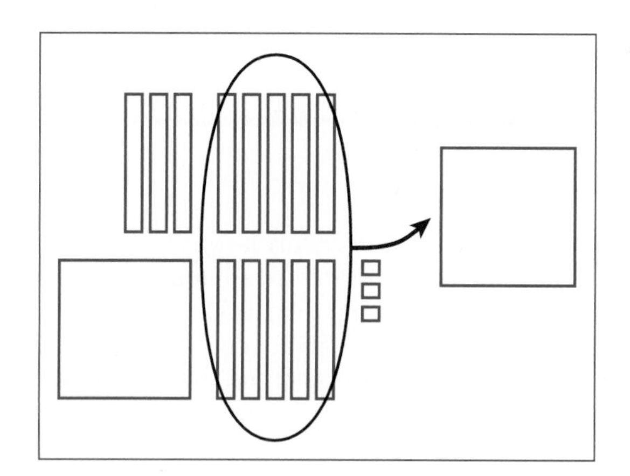

Figure 7.7

Once this exchange is done, we combine flats (one wholes) and see that our sum consists of 2 flats, 3 longs, and 3 units, which represents a sum of 2.33 (see Figure 7.8).

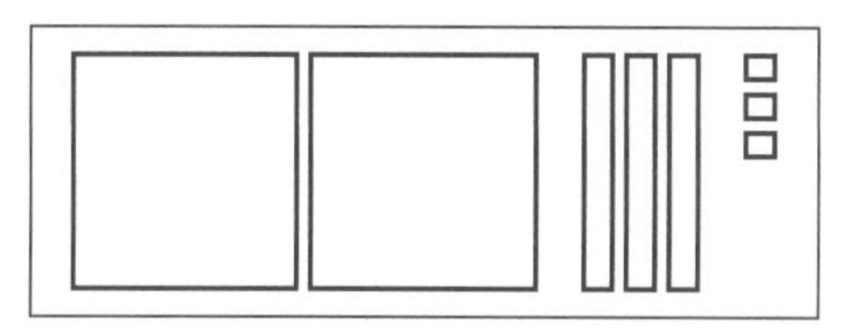

Figure 7.8

7.6.3 Subtracting Decimals

The rules for subtracting decimals are similar to those for addition: to (1) subtract the decimals vertically, (2) line up decimal points in the minuend and subtrahend, and (3) subtract digits in the same place-value positions, beginning from the rightmost digits, exchanging 1 from a column to a set of 10 for the next column when necessary. Once the decimals are lined up, any place in the minuend or the subtrahend where there is a digit missing can be replaced with a zero so that both decimals have the same number of digits to the right of the decimal point.

For example, to subtract 235.98 from 300, we first write the minuend and the subtrahend vertically, making sure to line up decimal points and to add two zeros on to the end of the minuend:

$$\begin{array}{r} 300.00 \\ -\ 235.98 \\ \hline \end{array}$$

Now, subtract column by column, exchanging when necessary to find the difference:

$$\begin{array}{r} 300.00 \\ -\ 235.98 \\ \hline 64.02 \end{array}$$

7.6.4 Modeling the Subtraction of Decimals

Subtraction of decimals can be modeled using base-ten blocks. Exchanges made when borrowing can be clearly visualized with this model. For example, modeling the subtraction problem 3.23 − 1.56, we first build the minuend, 3.23, with base-ten blocks as shown in Figure 7.9.

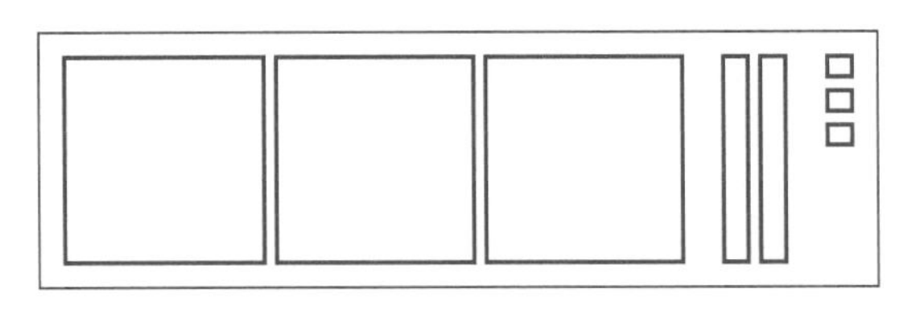

Figure 7.9

Starting with the units (one-hundredths), we need to subtract 6 units as indicated in the digit 6 in the decimal 1.56. To subtract 6 units, we exchange one long (one-tenth) for 10 units (hundredths) and then take away 6 units as seen in Figure 7.10.

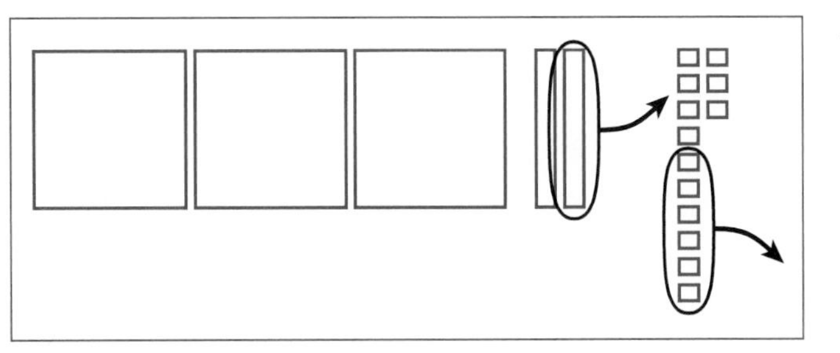

Figure 7.10

Then, we are left with 1 long, but need to subtract 5 longs, as indicated by the digit 5 in the subtrahend 1.56. To do so, we exchange 1 flat (one whole) for 10 longs (tenths). From the 11 longs (tenths), we can remove 5 longs (tenths). See Figure 7.11.

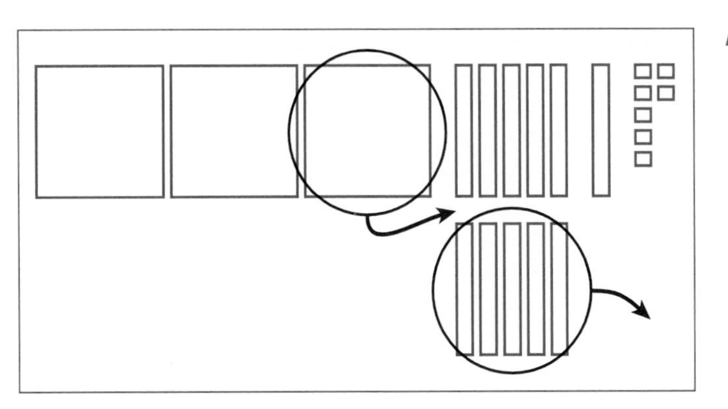

Figure 7.11

Finally, left with 2 flats (wholes), 6 longs (tenths), and 7 units (hundredths), we can subtract one whole (as seen in Figure 7.12) as indicated by the digit 1 in the subtrahend 1.56.

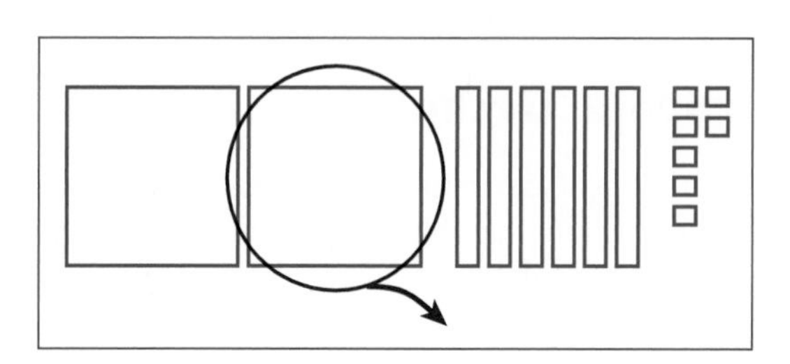

Figure 7.12

After all exchanges and subtractions, the base-ten blocks (see Figure 7.13) reveal that the difference found in the problem 3.23 − 1.56 is 1.67.

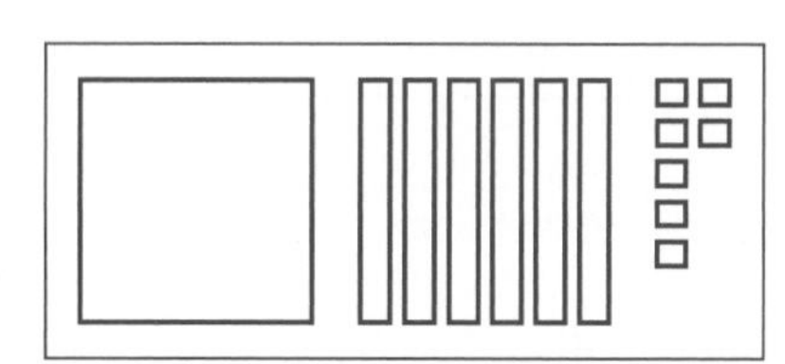

Figure 7.13

Related Topics — **Decimal Multiplication**

TOPIC 7.7 *Decimal Multiplication*

Decimals
Decimals and Fractions
Decimal Addition and Subtraction

Introduction

The multiplication of decimals involves a very interesting rule. In this section, the rule is explained and examined as to why the rule makes sense. Examples of decimal multiplication are provided as well.

Mathematician Leonard of Pisa, better known as Fibonacci (1180–1250), was an Italian mathematician who helped to make the use of Hindu-Arabic numerals common throughout Europe. He is best known for the Fibonacci Sequence, which is an infinite, iterative sequence of numbers where the next number in the sequence is found by determining the sum of the two prior numbers in the sequence. The basic Fibonacci Sequence is: 0, 1, 1, 2, 3, 5, 8, 13, 21, 34, 55, 89, 144, ... Fibonacci sequences have been found to be related to decimal representations of certain rational numbers. For example, the rational number $\frac{1}{89} = 0.01123595506\ldots$. Notice, interestingly, that the first six digits to the right of the decimal point mirror the first six numbers in the Fibonacci Sequence. Also note that $\frac{1}{89} = \frac{0}{10^1} + \frac{1}{10^2} + \frac{1}{10^3} + \frac{2}{10^4} + \frac{3}{10^5} + \frac{5}{10^6} + \frac{8}{10^7} + \frac{13}{10^8} + \ldots.$ Notice in this representation that the string of numerators exactly matches the Fibonacci Sequence.

For More Information Reimer, L., & Reimer, W. (1995). *Mathematicians Are People Too: Stories from the Lives of Great Mathematicians, Volume 2*. Palo Alto, CA: Dale Seymour Publications.

7.7.1 Rule for Decimal Multiplication

When multiplying decimals, the following steps apply: (1) Multiply the decimals as if they were integers, (2) find the sum of the number of digits to the right of the decimal point of each factor, (3) place the decimal point in the product so that the number of decimal points to the right of the decimal point is equal to the sum of digits found in step (2).

Example 1: When multiplying the decimals 12.64×3.8, we first find the product as if the factors were integers: $1264 \times 38 = 48032$. Then, because the number of digits to the right of the decimal in 12.64 is two and the number of digits to the right of the decimal point of 3.8 is one, we find the sum $2 + 1 = 3$ and place a decimal point in the number 48032 so that there are 3 digits to the right of the decimal point. Thus, $12.64 \times 3.8 = 48.032$.

Example 2: To find the product 100.335×2.55, we find the product $100335 \times 255 = 25585425$ and then place the decimal point so that there are five digits to the right of the decimal point. Thus, $100.335 \times 2.55 = 255.85425$.

7.7.2 Connection Between Decimal and Fraction Multiplication

To understand why the rule for multiplication of decimals makes sense, it helps to see the connection between decimal and fraction multiplication. When we change decimals to their equivalent common fraction forms, we can multiply the fractions and gain insight into why the rule for decimal multiplication works.

For example, multiplying 2.5×3.01 using the rules for decimal multiplication, we find a product of 7.525. Notice that the number of digits to the right of the decimal point in the product is three, which is the sum of the number of digits to the right of the decimal points of the two factors. Now, writing these two factors as fractions, we have the multiplication problem $\frac{25}{10} \times \frac{301}{100}$. Using rules for multiplying fractions (see Rational-Number Multiplication) by multiplying numerators and denominators, we find the product $\frac{25}{10} \times \frac{301}{100} = \frac{7525}{1000} = 7.525$.

Notice that the denominator in the first factor is 10 or 10^1 and the denominator in the second factor is 100 or 10^2. The denominator in the product is 1,000 or 10^3. Multiplying $10^1 \times 10^2 = 10^{1+2} = 10^3$ (see Exponents section). The exponent indicates the number of zeros in the power of 10. Note that the number of zeros in the denominator of the product is equal to the sum of the number of zeros in the denominators of the two factors. Thus, the rule for decimal multiplication where we add the number of digits in the decimal factors is akin to adding the number of zeros in the denominators of the fractional factors.

7.7.3 Area Model of Decimal Multiplication

The area model is a visual means of representing multiplication of common fractions (see Rational-Number Multiplication). The area model can also illustrate multiplication of decimal fractions. To illustrate 0.4×0.6 with the area model, we break the region into 10 rows, each row representing one tenth, and 10 columns, each column representing one tenth (see Figure 7.14).

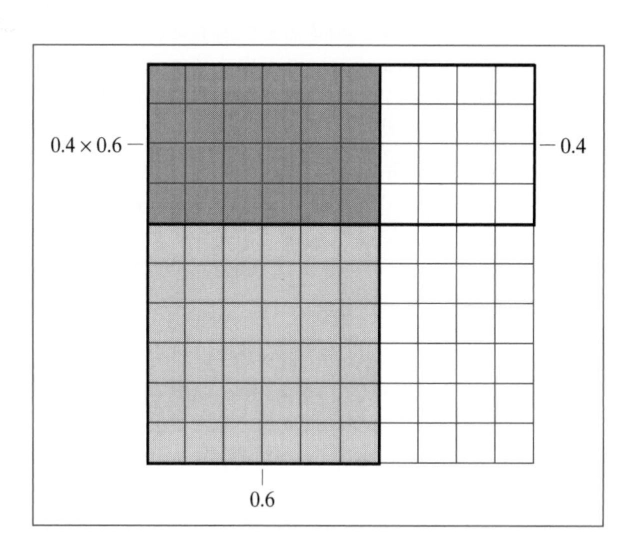

Figure 7.14

The area where the region of 4 rows representing 0.4 overlaps the region of 6 columns representing 0.6 is a region representing the product 0.4×0.6. This product region covers 24 small units, each worth 1 hundredth. Thus, the product is 24 hundredths, or 0.24.

Rational-Number Multiplication
Exponents

TOPIC 7.8 *Decimal Division*

Decimals
Decimals and Fractions
Integer Division

Introduction

Decimal division may seem to be a complicated process to many. However, described in the following sections is a set of procedures to follow for successful decimal division. Further, a means of explaining why decimal division works the way it does is included.

Archimedes (287–212 B.C.) was the greatest mathematician of ancient times. Archimedes loved challenges. Once, he set out to determine how much sand it would take to fill the universe. To begin, he counted the number of grains of sand that made up a cluster the size of a poppy seed. Then, he counted the number of poppy seeds that would fit on a human's finger. Then, he counted the number of fingers it would take to fill a stadium. He kept going until he came up with 10^{63} grains. Thus, he was the first to develop a system of writing large numbers with exponents. In another challenge to himself, Archimedes was determined to find the exact ratio between the circumference of a circle and its diameter. After much experimentation, he concluded that the ratio was somewhere between $3\frac{1}{7}$ and $3\frac{10}{71}$. This ratio, called pi (π), has now been calculated to many decimal places, but is most often approximated as 3.14.

For More Information Reimer, L., & Reimer, W. (1990). *Mathematicians Are People Too: Stories from the Lives of Great Mathematicians, Volume 1*. Palo Alto, CA: Dale Seymour Publications.

7.8.1 Rule for Decimal Division

When the divisor in a division problem is a decimal fraction, the steps for long division are to (1) move the decimal point in the divisor the number of places to the right until the decimal point is to the right of the last digit, (2) simultaneously move the decimal point in the dividend the same number of places that we moved the decimal point in the divisor, (3) place a decimal point above the dividend where the quotient will be written so that the decimal point in the quotient lines up with where the decimal point in the dividend is, and finally, (4) divide the numbers as if they were integers, paying attention to keep the decimal point in the final quotient where indicated in step (3).

Example 1: To divide 507.45 ÷ 0.15, we first write the problem in long division form:

$$.15\overline{)507.45}$$

Then, noting that we must move the decimal point in the divisor two places to the right, we simultaneously move the decimal point in the dividend two places to the right. Thus the division problem becomes:

$$\begin{array}{r} . \\ 15.\overline{)50745.0} \end{array}$$

Note that the decimal point in the quotient corresponds to the placement of the decimal point in the dividend. Now, completing the long division, we find the quotient to be 3,383:

$$\begin{array}{r} 3383.0 \\ 15.\overline{)50745.0} \end{array}$$

Example 2: To solve the problem 10.68 ÷ 3.2, we set up the long division problem

$$3.2\overline{)10.68}$$

and then move the decimal point in both the dividend and the divisor one place to the right:

$$32\overline{)106.8}$$

Performing the long division of 106.8 ÷ 32, we find the quotient to be 3.3375:

$$\begin{array}{r} 3.3375 \\ 32\overline{)106.8000} \end{array}$$

7.8.2 A Key Fact About Division

In any division problem, if we multiply both dividend and the divisor by the same nonzero amount, the quotient is unaffected. For example, $10 \div 5 = 2$. Similarly, $(3 \times 10) \div (3 \times 5) = 30 \div 15 = 2$. Essentially, the factor introduced in the dividend is divided by itself in the divisor, thus essentially forming a name for 1. Therefore, a quotient is unaffected when both dividend and divisor are multiplied by the same nonzero amount.

From another perspective, viewing the division problem $10 \div 5$ written in fraction form, $\frac{10}{5}$, we can think of multiplying numerator and denominator by the same number (a name for one) to form an equivalent fraction. That is, $\frac{10}{5} = \frac{3 \times 10}{3 \times 5} = \frac{30}{15} = 30 \div 15 = 2$. Again, the value of the quotient in the division problem is unaffected.

7.8.3 Connection Between Decimal and Fraction Division

To understand why the rule for decimal division works, we can look to the connection between decimal and fraction division. For example, consider the problem 30.45 ÷ 0.6:

$$0.6\overline{)30.45} \quad \longrightarrow \quad \begin{array}{r} 50.75 \\ 6\overline{)304.50} \end{array}$$

To verify that these two division problems are equivalent, rewrite the division problem 30.45 ÷ 0.6 in fraction form as $\frac{3045}{100} \div \frac{6}{10}$. Multiplying both quotient and divisor by the 10, we produce a new division problem that will have the same quotient as the original, as explained in the previous section. Therefore,

$$30.45 \div 0.6 = \frac{3045}{100} \div \frac{6}{10} = \frac{30450}{100} \div \frac{60}{10} = 304.5 \div 6.$$

Thus, the rule for decimal division essentially leads the problem solver to multiply both dividend and divisor by the appropriate power of 10 that creates a whole-number divisor. No matter what power of 10 both dividend and divisor are multiplied by, the quotient is unaffected, and the division problem is made easier.

Decimal Multiplication

TOPIC 7.9 Percent

Introduction

Whether we are computing a tip or determining what our salary will be after a certain cost-of-living increase, percents are a part of the calculations. Percents also provide a means of describing how likely it is to rain on a given day or how much of a chance we have at winning the lottery. In the following sections, we define percent and illustrate a process of solving simple percent problems.

▸ The word *percent* comes from the Latin phrase *per centum*, meaning "per hundred."

7.9.1 Definition of Percent

A **percent** is a ratio where one of the items being compared is the amount 100. In general, given any nonnegative real number, r, we say r percent, denoted $r\%$, is the ratio $\frac{r}{100}$. For example, we say $45\% = \frac{45}{100}$. Likewise, $250\% = \frac{250}{100}$.

7.9.2 Percents and Decimals

Because percents can be written as fractions with a denominator of 100, they can also easily be written as decimal numbers. To translate a percent to its decimal equivalent, first, write the percent in fraction form. Then, convert the fraction to decimal form (see Decimals and Fractions for more details of converting fractions to decimals). Some examples of equivalent percent, fraction, and decimal representations are provided in Table 7.1.

Table 7.1

Percent	Fraction Form	Decimal Equivalent
20%	$\frac{20}{100}$	.20
35%	$\frac{35}{100}$	.35
150%	$\frac{150}{100}$	1.50
1%	$\frac{1}{100}$	0.01
0.5%	$\frac{.5}{100}$	0.005
3.5%	$\frac{3.5}{100}$	0.035

7.9.3 Solving Basic Percent Problems

To solve basic percent problems, we can set up a proportion where one of the ratios is a fractional representation of a percent. Thus, to solve a percent problem, we can set up a ratio: $\frac{A}{B} = \frac{C}{100}$, where A, B, and C are real numbers, $B \neq 0$. Most percent problems involve knowing two of the three unknowns A, B and C and having to solve for the third, most likely using the cross multiplication method of solving proportions.

Example 1: What is 30% of 65? In this situation, we know the value for $C = 30$. We also know the value for $B = 65$. What we do not know is the value of A. Thus, we set up the proportion: $\frac{A}{65} = \frac{30}{100}$. Solving, using cross multiplication, we find:

$$100 \times A = 30 \times 65$$
$$100 \times A = 1950$$
$$A = \frac{1950}{100} = 19.5$$

Thus, we conclude that 19.5 is 30% of 65.

Example 2: Fifty is 40% of what number? In this case, we know the value for $C = 40$. We also know the value for A. We need to find the value for B. We set up the proportion: $\frac{50}{B} = \frac{40}{100}$. Solving with cross multiplication, we determine:

$$50 \times 100 = 40 \times B$$
$$5000 = 40 \times B$$
$$200 = B.$$

Therefore, 50 is 40% of 200.

Example 3: What percent of 80 is 20? In this example, we know $A = 20$ and $B = 80$, but we do not know the value for C. The proportion we need to solve is: $\frac{20}{80} = \frac{C}{100}$. Solving we find:

$$20 \times 100 = 80 \times C$$
$$2000 = 80 \times C$$
$$25 = C.$$

So, 25% of 80 is 20.

7.9.4 Percent Change

Sometimes, we want to determine the percent by which an amount increased or decreased. In these cases, we need to compute a percent change. To compute a **percent change,** we set up and solve the proportion: $\frac{\text{Actual Change}}{\text{Original Amount}} = \frac{x}{100}$.

Example 1: If your salary increased from $20,000 to $24,000, by what percent did your salary increase? First, we compute the actual change in salary to be $4,000: $24{,}000 - 20{,}000 = 4{,}000$. Noting that the original salary is $20,000, we set up the proportion: $\frac{4,000}{20,000} = \frac{x}{100}$. Solving, we find that the percent increase was 20%.

Example 2: Suppose that enrollment at the local college dropped from 8,000 students to 7,500 students. What was the percent decrease? Using the actual change in student population of 500 and the original student enrollment of 8,000, we compute the percent decrease with the proportion: $\frac{500}{8,000} = \frac{x}{100}$. Solving for x, we find the percent decrease was 6.25%.

Example 3: If at the beginning of the day the temperature was 30°, and during the day the temperature rose 10%, what was the high temperature of the day? In this case, we know the original temperature, and we know the percent change, but we do not know the actual change. We can set up the proportion: $\frac{x}{30} = \frac{10}{100}$. Solving for x, we find that the actual change in temperature was 3°. Therefore, the high temperature for the day must have been $3° + 30° = 33°$.

Related Topics
Interest
Probability

TOPIC 7.10 Interest

Introduction

Interest is a concept that is involved in such financial transactions as computing a mortgage on a house, determining the monthly cost of a car loan, or predicting how money in a savings account will grow over time. In the following sections, we define terms associated with financial transactions that entail interest and learn how to compute the value of transactions that involve simple interest and compound interest.

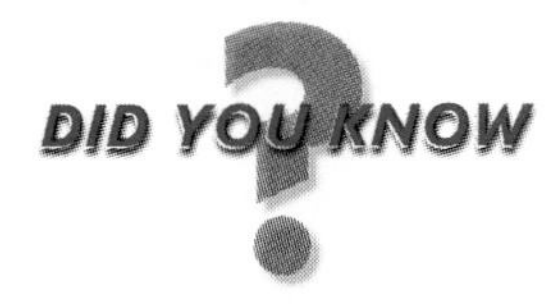

To solve most problems that involve interest, we rely on computers or calculators because we need to compute exponential values, often with very large exponents. Calculator keys that allow us to input a base amount and raise that amount to a given exponent are usually labeled with symbols such as: x^y or $x \wedge y$ or simply the symbol, ^.

7.10.1 Basic Terms in an Interest Problem

When we make a financial investment, the initial amount we invest is called the **principal.** When we invest the principal at a certain interest rate, we earn additional money called **interest. Interest rates** are generally expressed as percents, such as $r\%$. Most interest rates in investments are **annual percentage rates** (APR), implying that at the end of a year, the principal amount you invested will earn the interest computed by the APR.

For example, if you invest \$1,000 for one year at an annual percentage rate of 6%, then at the end of the year, you have earned an amount of interest that is 6% of \$1,000, which is \$60. Thus, the total value of your investment at the end of the year is equal to your initial principal investment plus your interest earned, which equals \$1,000 + \$60 = \$1,060.

7.10.2 Computing Simple Interest

If you invest your money and you earn interest **once** at the end of each year, then this is a **simple interest** situation. If you invest a principal amount, P, and earn $r\%$ interest at the end of one year, then the value of your investment at the end of one year, denoted A, is $A = P + \frac{r}{100}P$. By factoring out P, A can be expressed as $A = P(1 + \frac{r}{100})$.

Now, suppose you keep your money invested for another year. The amount invested at the beginning of the second year includes the original principal plus the interest earned from the first year. So, at the end of the second year, you have earned interest on the entire amount $P(1 + \frac{r}{100})$. Thus, at the end of the second year, your investment is worth: $A = P(1 + \frac{r}{100})[1 + \frac{r}{100}]$ or $A = P(1 + \frac{r}{100})^2$.

Continuing in this manner, we can find that at the end of the third year of investing, this same principal amount that the value at the end of the third year is

$$A = P\left(1 + \frac{r}{100}\right)\left[1 + \frac{r}{100}\right]\left[1 + \frac{r}{100}\right] \text{or } A = P\left(1 + \frac{r}{100}\right)^3.$$

In general, when computing the value of an investment involving simple interest, that is, interest that is computed once at the end of a year, the value of a principal investment, *P*, at the end of *t* years, at an APR of *r* is $A = P(1 + \frac{r}{100})^t$.

For example, if you invest \$600 for 8 years at an annual interest rate of 5%, the value of your investment at the end of the 8 years is:

$$A = 600\left(1 + \frac{6}{100}\right)^8 = 600\,(1 + .06)^8 = 600(1.06)^8 = 1110.56, \text{ or } \$1{,}110.56.$$

7.10.3 Computing Compound Interest

When interest is computed for an investment more than once during a year, we say that **interest is compounded** during the year. Given an annual interest rate of $r\%$, we can have interest compounded n times during a year; each time the current value of the account, including any previously earned interest, is increased by $r\%$. To compute the value of an account where interest is compounded n times during a year at an annual interest rate of $r\%$, we use the formula:

$$A = P\left(1 + \frac{r/100}{n}\right)^{nt}.$$

Example 1: Suppose you invest \$400 for 5 years at an annual percentage rate of 7% compounded quarterly (that is 4 times during the year). The value of your investment at the end of the 5 years is:

$$A = 400\left(1 + \frac{7/100}{4}\right)^{4(5)} = 400(1.0175)^{20} = 565.91 \text{ or } \$565.91.$$

Example 2: Suppose you borrow \$100,000 for 20 years and your debt grows at an annual interest rate of 4% compounded monthly (that is, 12 times during the year). How much will you have to pay back to the lender at the end of the 20 years? To solve, compute:

$$A = 100{,}000\left(1 + \frac{4/100}{12}\right)^{20(12)} = 100{,}000\,(1.0033)^{240} = 220{,}493.07, \text{ or } \$220{,}493.07.$$

Related Topics

Ratio
Proportion
Percent
Exponents

TOPIC 7.11 Rational Numbers

cornerstone

Natural Numbers
Whole Numbers
Integers
Whole-Number Division
Integer Division
Fractions

Introduction

The word *rational* stems from *ration,* which refers to an allotment or a portioning out of a fixed amount. Thus, the action of division is linked to rationing. The term *rational number* usually causes people to think of a fraction of some sort. However, as you will see in the sections that follow, a rational number is a special kind of fraction and can be represented by three distinct models.

DID YOU KNOW

One model for representing the whole in the part-whole definition of rational numbers is a linear model. Rational numbers are often understood as points on a number line. For example, a twelve-inch ruler provides an image of a total length being divided into equal-sized sections representing 1 inch, $\frac{1}{2}$ inch, $\frac{1}{4}$ inch, $\frac{1}{8}$ inch, and $\frac{1}{16}$ inch.

7.11.1 Definition of a Rational Number

A **rational number** is a number that can be expressed as a fraction of the form $\frac{a}{b}$ where both a and b are integers (provided that $b \neq 0$). The number $\frac{6}{7}$, for example, is a rational number because both 6 and 7 are integers. It can also be said that the number 0.4 is a rational number because it can be expressed in an equivalent form of $\frac{4}{10}$, which is a fraction with an integer numerator, 4, and an integer denominator, 10.

7.11.2 Rational Numbers Versus Fractions

The terms *fractions* and *rational numbers* are often used interchangeably. However, we should be clear about the relationship between the two expressions. In comparing the definitions of the two terms, we see that the set of rational numbers can almost be viewed as a subset of the set of fractions.

Fractions are any number expressions written as a numerator over a denominator. Rational numbers, on the other hand, are those numbers that can be written as fractions whose numerators and denominators are both integers. Thus, although $\frac{\pi}{2}$ is a fraction, it is not a rational number because the number π is not an integer.

It is tempting to conclude that the set of rational numbers is necessarily a subset of the set of fractions; however, we have to acknowledge that rational numbers can be expressed in forms other than fraction form. For example, the rational number $\frac{1}{2}$ can be expressed as 0.5. Thus it seems inappropriate to state that the set of rational numbers, which includes numbers that may be written in decimal form, is a subset of the set of fractions.

7.11.3 Three Models of Rational Numbers

There are three models that are used to illustrate rational numbers: (a) the region (or area) model, (b) the linear (or number line) model, and the (c) set (or discrete) model. Each of these models is most easily demonstrated using the part of a whole interpretation of a rational-number fraction.

The Region Model. The model selected to represent a rational number is generally determined by the whole being described in a given context. In the **region**

model, for example, the whole represents some enclosed two-dimensional area that is being divided into equal-sized parts. Suppose that when you left a cake to cool on the counter, your dog ate $\frac{3}{4}$ of the cake. The pictorial image that comes to mind for representing a cake is a rectangular region. Thus, $\frac{3}{4}$ of the cake can be represented as in Figure 7.15:

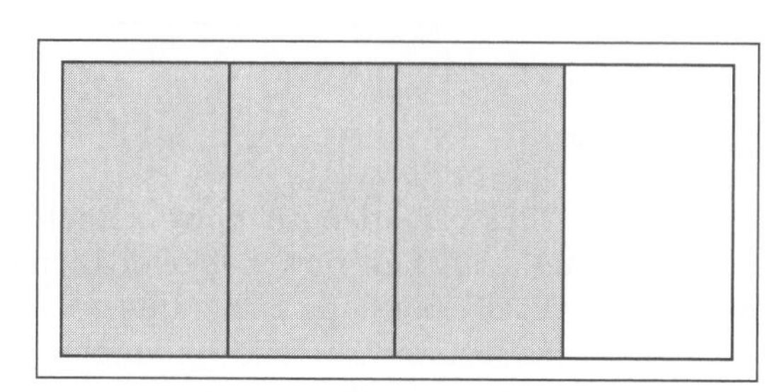

Figure 7.15

The Linear Model. A second model of rational numbers is a **linear model.** Linear representations make sense when the whole being divided into equal-sized parts is some one-dimensional measure, such as distance (e.g., length, width), time, temperature, or money. Thus, if one were to illustrate that Grandpa had walked $\frac{2}{5}$ of the way home before stopping to rest, $\frac{2}{5}$ could be represented as in Figure 7.16. As in the region model, the whole, in this case a line segment, was broken into 5 equal-sized pieces as indicated by the denominator, and 2 of those 5 pieces, as suggested by the numerator, were highlighted to represent the rational number $\frac{2}{5}$ in this context.

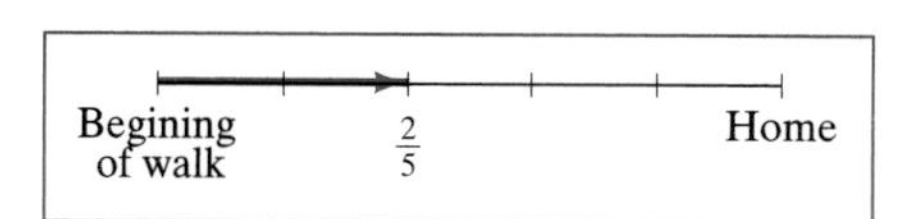

Figure 7.16

The Set Model. The third model of rational numbers is the **set model.** A set model is called for when the whole being described is a collection of discrete objects. For example, if one were to make the statement that $\frac{3}{8}$ of the class was wearing blue jeans on a given day, the whole is the class, which is made up of a group of individual people. Often, in a set model, we can refer to the whole, such as the class, without knowing how many objects are actually in the set. In these cases, drawing an accurate image of $\frac{3}{8}$ of the class using the set model might be difficult.

A context in which the set model is the best means of representing a rational number and can easily be done is in the situation where a student claims that in her bag of 15 gum drops, $\frac{3}{5}$ of them are licorice. The image that comes to mind is see in Figure 7.17:

Figure 7.17

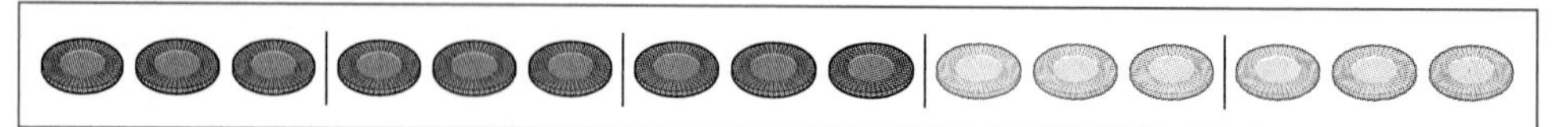

In the above set model, the 15 gum drops are represented by 15 individual circles. The collection of circles is divided into 5 same-sized groups (3 in each group). Then, as the numerator suggests, 3 of those 5 groups (or 9 circles altogether) are shaded in to represent $\frac{3}{5}$ of the set.

Related Topics

The Set of Rational Numbers
Equivalent Fractions
Simplifying Fractions
Improper Fractions and Mixed Numbers
Adding Rational Numbers
Subtracting Rational Numbers
Multiplying Rational Numbers
Dividing Rational Numbers
Ratios

TOPIC 7.12 Looking at Rational Numbers as the Set of Rational Numbers

Sets
The Set of Natural Numbers
The Set of Whole Numbers
The Set of Integers
Fractions

Rational Numbers
Whole-Number Division
Integer Division
Fractions

Introduction

The set of rational numbers is related to other sets of numbers such as the set of integers, the set of whole numbers, and the set of natural numbers. These sets of numbers share some common properties. However, the set of rational numbers introduces new properties unique to this set.

Try to visualize the set of points on a number line corresponding to all the rational numbers. If we concentrate only on those rational numbers between 0 and 1, this task might seem a bit more manageable. Note that on a number line, it is easy to determine that the next whole number after 0 is 1. (See Figure 7.18.) However, can we say what the next rational number after 0 is in the same sense? Pursuing an answer to this simple question may lead one to the very interesting discovery of the density property of rational numbers that is discussed in detail, along with the definition of and properties of rational numbers, in the sections that follow.

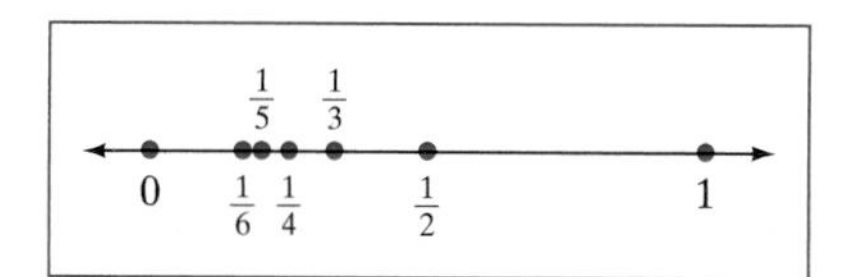

Figure 7.18

The set of natural numbers seem to have served the purposes of most civilized societies until about the sixteenth century. Until then, people broke items into pieces and spoke of the parts, but even after weights of measure came into use, it was not the custom to speak of such a fraction as, for example, $\frac{3}{4}$ of a pound. The world avoided fractions by creating smaller units. For example, the Romans used the following: A pound of copper was referred to as a *denarius*. To talk about less than one denarius, they created the term *as* to refer to $\frac{1}{16}$ of a denarius or an ounce. So, to talk about an amount equivalent to what we would express as $\frac{1}{8}$ of a pound, the Romans would call it *2 asses* instead.

As the need for representing rational number amounts emerged more distinctly, cultures slowly began to represent unit fractions (those with a 1 in the numerator). As late as the seventeenth century, Russian manuscripts on surveying spoke of a "half-half-half-half-half-third" of a certain measure instead of $\frac{1}{96}$ of the measure. Even today the unit fraction is used to some extent in the diamond trade in speaking of parts of a carat. Over time, however, most enduring cultures developed a means of expressing not only the set of unit fractions, but also the larger set of the rational numbers written in fraction form.

For More Information Smith, D. E. (1958). *History of Mathematics, Volume II*. New York: Dover Publications.

7.12.1 Comparing Sets of Numbers

The **set of rational numbers** refers to the entire collection of numbers that can be expressed in the form $\frac{a}{b}$ where both a and b are integers (provided that $b \neq 0$). Consequently, the set of rational numbers (denoted Q) includes all natural numbers (denoted N), Whole numbers (denoted W), and integers (denoted I), as well as noninteger fractions of the form $\frac{a}{b}$ where both a and b are integers. Thus, the set of rational numbers can be viewed as a set containing very important subsets as suggested by the following:

$$N \subset W \subset I \subset Q^{\cdot}$$

This chain of subsets indicates that any natural number is also a rational number. Thus, any natural number can be expressed in fraction form where numerator and denominator are both integers. For example, the number 3 can be expressed as $\frac{3}{1}$, which meets the criteria of a rational number because 3 and 1 are both integers. The same can be said of any whole number or integer in that any whole number or integer can be expressed as itself in the numerator over a denominator of 1.

7.12.2 Infinite Sets

A **finite set** is a set whose cardinal number is a whole number. An **infinite set** is one that is not finite. Each of the sets of numbers N, W, I, and Q are infinite sets. If each set is infinite, it might seem difficult to compare the sizes of the sets. For example, the set of whole numbers can be defined as the set of natural numbers plus the number zero. With this definition of the set of whole numbers, it is tempting to say that the set of whole numbers is "bigger" than the set of natural numbers because we know it contains one more element, zero. However, because both sets are infinite, and because we can show a one-to-one correspondence between the two sets, that is, match each element in the set of natural numbers to exactly one element in the set of whole numbers as seen in Figure 7.19, we say that sets N and W are the same size.

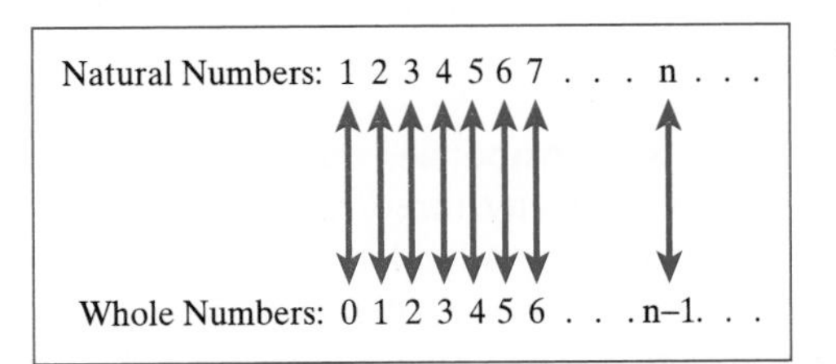

Figure 7.19

7.12.3 The Density of the Set of Rational Numbers

The set of rational numbers is an infinite set, and because a one-to-one correspondence can be shown to exist between the set of whole numbers and the set of rational numbers, one can conclude that the sets Q and W are the same size. However, although the set of rational numbers is technically the same size as the other infinite sets mentioned above, it has a property that distinguishes it from the other sets. This property is called the Density Property of rational numbers.

The Density Property of Rational Numbers says that between any two rational numbers, there is an infinite number of rational numbers. This property suggests that a number line would be "dense" with rational numbers. Thus, given any two rational numbers, say $\frac{7}{9}$ and $\frac{8}{9}$, we should be able to find an infinite number of rational numbers between them.

One way to find at least one rational number between $\frac{7}{9}$ and $\frac{8}{9}$ is to write equivalent forms of each fraction and search to find an "obvious" rational number between them. For example, rewrite $\frac{7}{9}$ and $\frac{8}{9}$ as $\frac{14}{18}$ and $\frac{16}{18}$, respectively, and note that $\frac{15}{18}$ clearly falls between them. This process can be carried further to find other rational numbers between $\frac{7}{9}$ and $\frac{8}{9}$.

Another strategy for finding at least one number between $\frac{7}{9}$ and $\frac{8}{9}$ would be to find the midpoint between them by averaging them, that is, add the two rational numbers together and divide by 2: $(\frac{7}{9} + \frac{8}{9}) \div 2 = \frac{15}{9} \div 2 = \frac{15}{9} \times \frac{1}{2} = \frac{15}{18}$. Because of the Density Property, this strategy can be repeated endlessly, finding other numbers between $\frac{7}{9}$ and $\frac{8}{9}$ by next finding the midpoint between $\frac{7}{9}$ and $\frac{15}{18}$, and the midpoint between $\frac{8}{9}$ and $\frac{15}{18}$, and so on.

7.12.4 Other Properties of Rational Numbers

The set of rational numbers has properties, other than the Density Property, that describe how numbers within the set relate to one another. For example, like the sets of natural numbers, whole numbers, and integers, the set of rational numbers is closed under addition and closed under multiplication.

Closure Properties for Addition and Multiplication of Rational Numbers. Given any two rational numbers a and b, the sum $a + b$ is a rational number. Therefore, the **set of rational numbers is closed under addition.** For example, choose $a = \frac{1}{3}$ and $b = \frac{2}{5}$. Then $a + b = \frac{1}{3} + \frac{2}{5} = \frac{11}{15}$, and $\frac{11}{15}$ is a rational number.

Similarly, given any two rational numbers a and b, the product $a \cdot b$ is a rational number. Therefore, the **set of rational numbers is closed under multiplication.** For example, choose $a = \frac{1}{4}$ and $b = \frac{3}{7}$. Then $a \times b = \frac{1}{4} \times \frac{3}{7} = \frac{3}{28}$, and $\frac{3}{28}$ is a rational number.

Commutative Properties of the Addition and Multiplication of Rational Numbers. For any two rational numbers a and b, $a + b = b + a$. Thus, the **set of rational numbers is commutative under addition.** Similarly for any two rational numbers a and b, $a \times b = b \times a$, illustrating that the **set of rational numbers is commutative under multiplication.** Therefore, with either operation, the order of addends or factors does not affect the sum or product, respectively, of the rational numbers. For example, $\frac{1}{5} + \frac{3}{4} = \frac{19}{20}$ and $\frac{3}{4} + \frac{1}{5} = \frac{19}{20}$. Likewise, $\frac{3}{8} \times \frac{2}{7} = \frac{6}{56}$ and $\frac{2}{7} \times \frac{3}{8} = \frac{6}{56}$.

The Associative Properties for the Addition and Multiplication of Rational Numbers. Given any three rational numbers a, b, and c, the sum of the three numbers, even when added using different groupings, does not change; that is, $(a + b) + c = a + (b + c)$, meaning that the **set of rational numbers is associative under addition.** For example, when adding $\frac{1}{2}$, $\frac{1}{3}$, and $\frac{1}{4}$, we find:

$$\left(\frac{1}{2} + \frac{1}{3}\right) + \frac{1}{4} = \frac{5}{6} + \frac{1}{4} = \frac{13}{12}$$
$$\frac{1}{2} + \left(\frac{1}{3} + \frac{1}{4}\right) = \frac{1}{2} + \frac{7}{12} = \frac{13}{12}.$$

Further, given any three rational numbers a, b, and c, the product of the three numbers, even when regrouped, is not affected; that is, $(a \times b) \times c = a \times (b \times c)$, meaning that the **set of rational numbers is associative under multiplication.** For example, when multiplying $\frac{1}{2}$, $\frac{1}{3}$, and $\frac{1}{4}$, we find:

$$\left(\frac{1}{2} \times \frac{1}{3}\right) \times \frac{1}{4} = \frac{1}{6} \times \frac{1}{4} = \frac{1}{24}$$
$$\frac{1}{2} \times \left(\frac{1}{3} \times \frac{1}{4}\right) = \frac{1}{2} \times \frac{1}{12} = \frac{1}{24}.$$

Identity Properties for the Addition and Multiplication of Rational Numbers. Like for the addition and multiplication of integers, the set of rational numbers has additive and multiplicative identities. **The additive identify for the set of rational numbers is 0** because for any rational number, a, $a + 0 = 0 + a = a$. For example $\frac{1}{6} + 0 = 0 + \frac{1}{6} = \frac{1}{6}$. Further, **the multiplicative identity for the set of rational numbers is 1** because for any rational number, a, $a \times 1 = 1 \times a = a$. For example, $\frac{1}{6} \times 1 = 1 \times \frac{1}{6} = \frac{1}{6}$.

Multiplication by Zero for the Set of Rational Numbers. As with the set of integers, the set of rational numbers has a special property associated with **multiplication by zero;** that is, for any rational number, a, $a \times 0 = 0 \times a = 0$. Thus, any rational number times zero is zero.

The Distributive Property of Multiplication over Addition for the Set of Rational Numbers. For any three rational numbers a, b, and c, the **Distributive Property of multiplication over addition** holds in that $a \times (b + c) = (a \times b) + (a \times c)$. For example, choosing three rational numbers $\frac{2}{3}$, $\frac{4}{5}$, and $\frac{5}{2}$:

$$\frac{2}{3} \times \left(\frac{4}{5} + \frac{5}{2}\right) = \frac{2}{3} \times \frac{33}{10} = \frac{66}{30} = \frac{11}{5}$$

$$\left(\frac{2}{3} \times \frac{4}{5}\right) + \left(\frac{2}{3} \times \frac{5}{2}\right) = \frac{8}{15} + \frac{10}{6} = \frac{66}{30} = \frac{11}{5}.$$

The Additive Inverse Property for the Set of Rational Numbers. Like the set of integers, the set of rational numbers has an **Additive Inverse Property** that states that for any rational number, a, there exists some other rational number, b, such that $a + b = 0 = b + a$. The additive inverse of a usually takes the form $-a$. Thus, the Additive Inverse Property can be stated as, for any rational number, a, there exists some other rational number, $-a$, such that $a + -a = 0 = -a + a$.

For example, given the rational number $\frac{5}{6}$, there exists the rational number $-\frac{5}{6}$ so that $\frac{5}{6} + -\frac{5}{6} = 0$ and $-\frac{5}{6} + \frac{5}{6} = 0$.

The Multiplicative Inverse Property for the Set of Rational Numbers. A new property that holds for the set of rational numbers that does not hold for the set of integers is the **Multiplicative Inverse Property.** According to this property, given any rational number, a, there exists some other rational number, b, such that $a \times b = 1$ and $b \times a = 1$. We say that b is the multiplicative inverse of a. Generally, b takes the form of $\frac{1}{a}$. Thus, the multiplicative inverse property can be stated as, for every rational number, a, there exists some other rational number, $\frac{1}{a}$, such that $a \times \frac{1}{a} = 1$ and $\frac{1}{a} \times a = 1$. This multiplicative inverse is also referred to as the reciprocal.

For example, for the rational number $\frac{4}{3}$ there exists the rational number $\frac{1}{\frac{4}{3}}$ or $\frac{3}{4}$, such that $\frac{4}{3} \times \frac{3}{4} = \frac{3}{4} \times \frac{4}{3} = 1$.

Related Topics

Equivalent Fractions
Simplifying Fractions
Improper Fractions and Mixed Numbers
Adding Rational Numbers
Subtracting Rational Numbers
Multiplying Rational Numbers
Dividing Rational Numbers
Properties of Integers
Ratios

TOPIC 7.13 Adding Rational Numbers

Fractions
Rational Numbers
Equivalent Fractions
Simplifying Fractions
Improper Fractions and Mixed Numbers
Integer Addition
Integer Multiplication
Integer Division
Least Common Multiple

Introduction

Addition of rational numbers is a key concept in the school mathematics curriculum. From young students to adults of all ages, people use the addition of rational numbers to solve real-life problems. There are many situations where you have two fractional amounts and want to find out how much you have altogether. In many cases, adding two or more rational numbers involves finding a common denominator before the addition can take place. Within this topic, we will demonstrate methods for adding rational numbers.

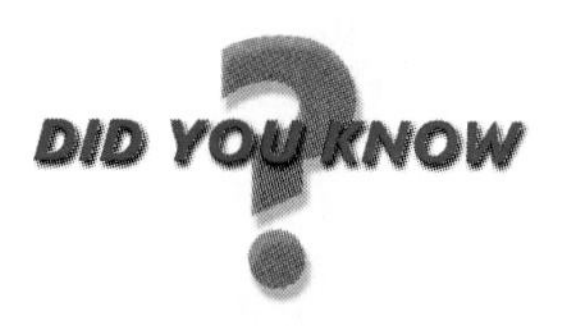

Early mathematicians dealt with the addition of fractions. The Egyptians, for example, preferred to express ordinary proper fractions as the sum of two or more "subfractions." To indicate that a particular fraction was a subfraction, the Egyptians used an accent symbol. So using the accented symbol for one-half, $\angle'$, followed by the accented symbol for one-fourth, δ', the Egyptian system represents $\angle'\delta' = \frac{1}{2}\frac{1}{4}$ or $\frac{1}{2} + \frac{1}{4} = \frac{3}{4}$.

Later mathematicians, using fraction representation that we use today, added fractions much as we do currently. They usually chose a common denominator by finding the product of the denominators of the two addends.

For More Information Smith, D. E. (1958). *History of Mathematics, Volume II.* New York: Dover Publications.

7.13.1 Addition of Rational Numbers with Common Denominators

When adding two rational numbers that have a common denominator, we simply add the numerators and put the sum over the common denominator. For example, $\frac{3}{8} + \frac{4}{8} = \frac{7}{8}$. The number-line model clearly illustrates the addition. Breaking one whole unit into 8 equal-sized pieces, each piece representing $\frac{1}{8}$, we can see the addition of $\frac{3}{8} + \frac{4}{8}$. See Figure 7.20:

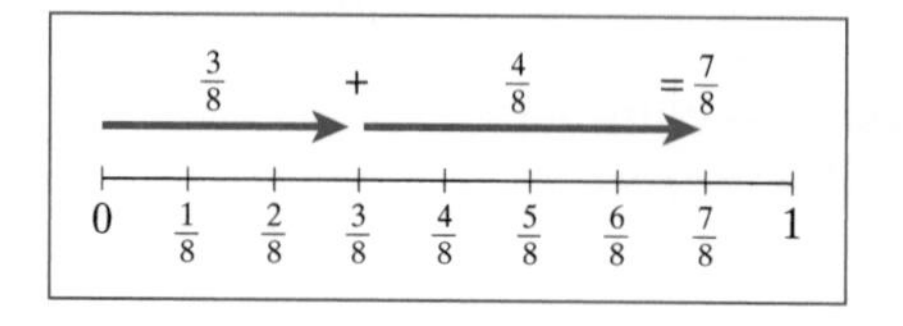

Figure 7.20

7.13.2 Formal Definition of the Addition of Rational Numbers

In most cases, when adding two rational numbers, the rational numbers do not have a common denominator, so the first step is to find a common denominator. A simple choice for the common denominator is the product of the two denominators. In the case of $\frac{1}{4} + \frac{2}{3}$, one could choose 12 as the common denominator, rewriting the problem as $\frac{3}{12} + \frac{8}{12}$ and then adding $\frac{3+8}{12} = \frac{11}{12}$.

The process of finding the common denominator before adding fractions can be nicely illustrated on the number-line model. For example, when adding $\frac{1}{4} + \frac{2}{3}$, we can break one whole unit on the number line into 4 equal-sized pieces as suggested by the 4 in $\frac{1}{4}$ (see Figure 7.21) and then break each of those four pieces into 3 equal-sized pieces as suggested by the 3 in $\frac{2}{3}$. The result in Figure 7.22 is a whole unit broken into 12 equal-sized pieces, each representing $\frac{1}{12}$. Thus, the number line in Figure 7.22 could be used to illustrate the addition of $\frac{1}{4} + \frac{2}{3}$ as $\frac{3}{12} + \frac{8}{12}$ rather easily. Notice that $\frac{1}{4} + \frac{2}{3} = \frac{1 \times 3 + 4 \times 2}{4 \times 3} = \frac{3 + 8}{12} = \frac{11}{12}$.

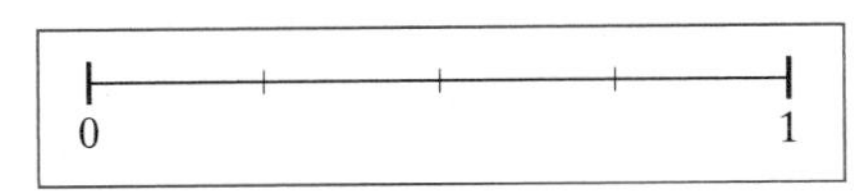

Figure 7.21
Number line divided into fourths

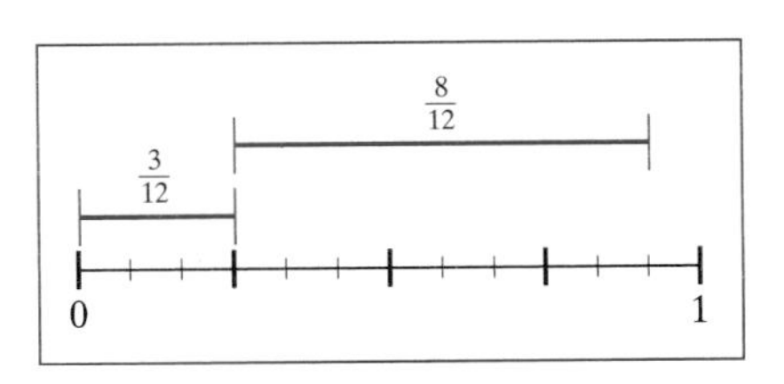

Figure 7.22
Number line divided into twelfths

The addition of two rational numbers can be expressed formally in this way: $\frac{a}{b} + \frac{c}{d} = \frac{ad + bc}{bd}$, which combines the processes of finding a common denominator, adding numerators, and placing the sum of the numerators over the common denominator. Thus, $\frac{2}{5} + \frac{4}{9}$ can be found by $\frac{2 \times 9 + 4 \times 5}{5 \times 9} = \frac{18 + 20}{45} = \frac{38}{45}$.

7.13.3 Addition of Mixed Numbers

When adding two rational numbers where one or more of the addends is a mixed number, you can use two strategies.

Method 1. Add the whole-number parts and the fractional parts separately, and then find the total sum by combining the two parts together. For example, adding $1\frac{1}{3} + 4\frac{2}{5}$, you can rewrite the problem and solve as follows:

$$1\frac{1}{3} + 4\frac{2}{5} = 1 + \frac{1}{3} + 4 + \frac{2}{5} = (1 + 4) + \left(\frac{1}{3} + \frac{2}{5}\right) = 5 + \left(\frac{5}{15} + \frac{6}{15}\right) = 5 + \frac{11}{15} = 5\frac{11}{15}.$$

Using this method, you may encounter problems where the sum of the fractional parts is greater than 1. For example, in the problem $2\frac{2}{3} + 4\frac{3}{4}$, the above method yields the following sum:

$$2\frac{2}{3} + 4\frac{3}{4} = 2 + \frac{2}{3} + 4 + \frac{3}{4} = (2 + 4) + \left(\frac{2}{3} + \frac{3}{4}\right) = 6 + \left(\frac{8}{12} + \frac{9}{12}\right) = 6 + \frac{17}{12} = 6\frac{17}{12}.$$

Because $\frac{17}{12}$ is an improper fraction, we need to change this to a mixed number so that the total sum can be expressed simply as a mixed number. The $\frac{17}{12}$ can be expressed as $1\frac{5}{12}$. Therefore, the sum is $6 + 1\frac{5}{12} = 7\frac{5}{12}$.

Method 2. A second method for adding rational numbers where one or more of the addends is a mixed number is to first change the mixed numbers to improper fraction form. Then add the fractions, finding a common denominator. Finally, convert your answer back to mixed number form.

For example, $3\frac{3}{7} + 10\frac{7}{12}$ can be rewritten as $\frac{24}{7} + \frac{127}{12}$. Finding a common denominator of 84, the problem can again be rewritten as $\frac{288}{84} + \frac{889}{84} = \frac{288 + 889}{84} = \frac{1177}{84} = 14\frac{1}{84}$.

Related Topics

The Set of Rational Numbers
Subtracting Rational Numbers
Multiplying Rational Numbers
Dividing Rational Numbers

TOPIC 7.14 Subtracting Rational Numbers

cornerstone

Fractions
Rational Numbers
Equivalent Fractions
Simplifying Fractions
Improper Fractions and Mixed Numbers
Integer Addition
Integer Subtraction
Integer Multiplication
Integer Division
Least Common Multiple
Adding Rational Numbers

Introduction

Subtraction of rational numbers is much like the addition of rational numbers. As with addition of rational numbers, there are many real-life problems that can be solved by subtracting two rational numbers. There are many situations where subtraction makes sense. Whether you have a situation where subtraction stems from a take-away context, a missing addend context, or a comparison context, rational numbers are often involved in the problem. In many cases, subtracting one rational number from another involves finding a common denominator before the subtraction can take place.

DID YOU KNOW

Early mathematicians, using fraction representation that we use today, subtracted fractions much like our current practices. They found common denominators, using the product of the denominators of the two fractions involved in the subtraction problem.

For More Information Bunt, L.N., Jones, P. S., & Bedient, J.D. (1998). *The Historical Roots of Elementary Mathematics*. New York: Dover Publications

7.14.1 The Subtraction of Rational Numbers with Common Denominators

Much like the addition of rational numbers, subtraction of two rational numbers that share a common denominator can be expressed as finding the difference between the numerators and placing that difference over the common denominator. For example, $\frac{5}{6} - \frac{3}{6} = \frac{5-3}{6} = \frac{2}{6}$.

On a number line, the subtraction of rational numbers can be seen when one whole unit is broken into 6 equal-sized parts, each representing $\frac{1}{6}$. See Figure 7.23.

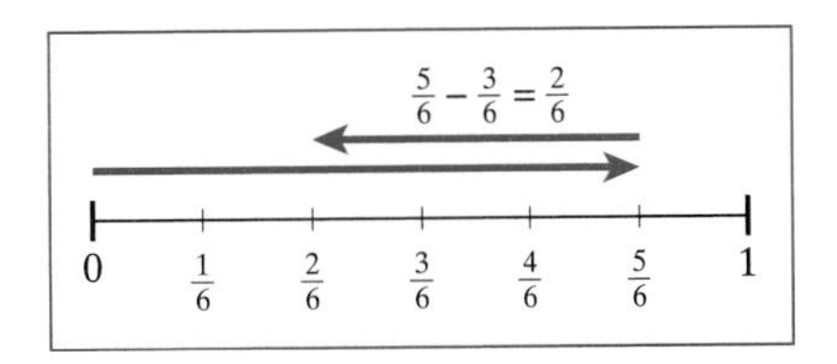

Figure 7.23

7.14.2 Formal Definition of the Subtraction of Rational Numbers

When subtracting two rational numbers that do not have a common denominator, we can find a common denominator by multiplying the two denominators together. In the example $\frac{6}{7} - \frac{3}{5}$, the common denominator of 35 ($7 \cdot 5 = 35$) yields the equivalent problem $\frac{30}{35} - \frac{21}{35} = \frac{9}{35}$.

In general, the definition for the subtraction of any two rational numbers can be expressed as follows: $\frac{a}{b} - \frac{c}{d} = \frac{ad - bc}{bd}$. Using this definition, we can determine a common denominator and find the difference between the numerators, placing the difference over the common denominator. For example, $\frac{6}{9} - \frac{3}{5} = \frac{6 \times 5 - 3 \times 9}{9 \times 5} = \frac{30 - 27}{45} = \frac{3}{45}$.

In addition to the number-line model, the region model can illustrate the addition and subtraction of rational numbers. For the subtraction problem $\frac{3}{4} - \frac{2}{6}$, we can begin by breaking the region into 4 equal-sized rows (suggested by the 4 in the denominator in $\frac{3}{4}$), each row representing $\frac{1}{4}$ (see Figure 7.24), and 6 equal-sized columns (suggested by the 6 in the denominator in $\frac{2}{6}$), each column representing $\frac{1}{6}$ (see Figure 7.25). Therefore, the region is broken into 24 equal-sized rectangles, each $\frac{1}{24}$ in size, illustrating one such common denominator, 24. See Figure 7.26.

Figure 7.24

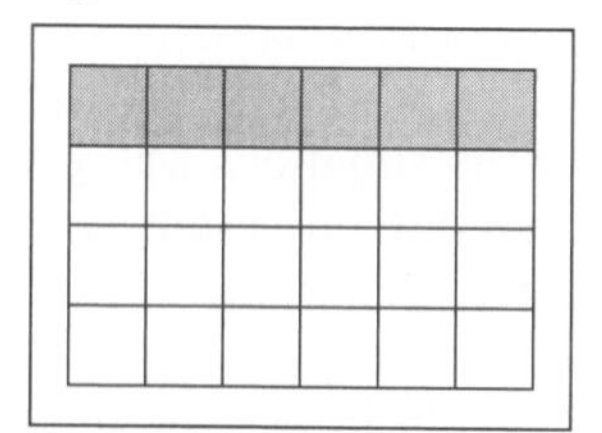

Figure 7.25

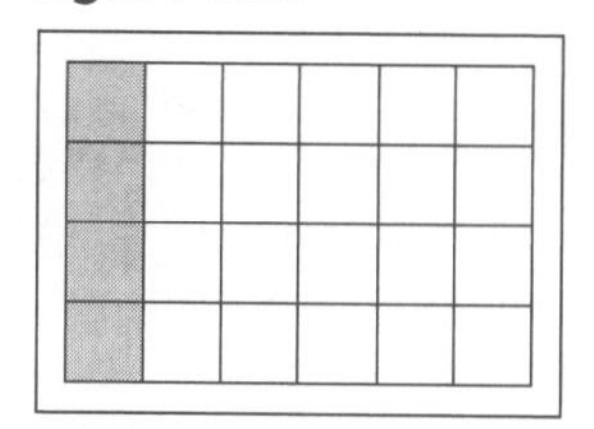

Figure 7.26

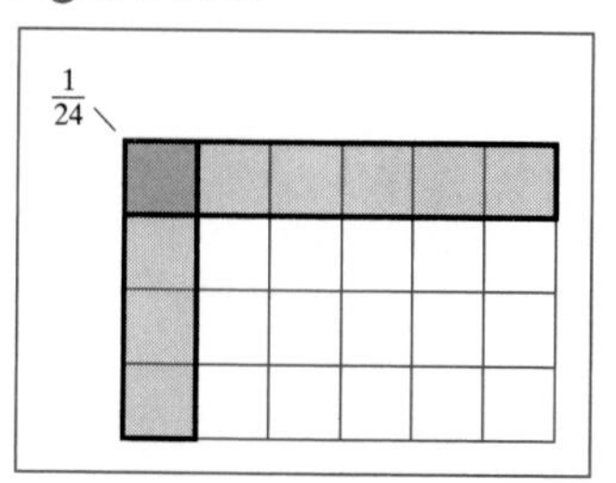

The subtraction of $\frac{3}{4} - \frac{2}{6}$ can be explained using this region model. First, illustrate that $\frac{3}{4}$ can be represented by shading in three rows or a total of 18 rectangles. See Figure 7.27. Because each column, or 4 rectangles, is $\frac{1}{6}$ of the whole, then $\frac{2}{6}$ is represented by 8 rectangles. So, when subtracting $\frac{2}{6}$ from $\frac{3}{4}$, this can be illustrated by removing 8 rectangles from the shaded 18 rectangles. (See Figure 7.28.) The difference is 10 rectangles or $\frac{10}{24}$.

Figure 7.27

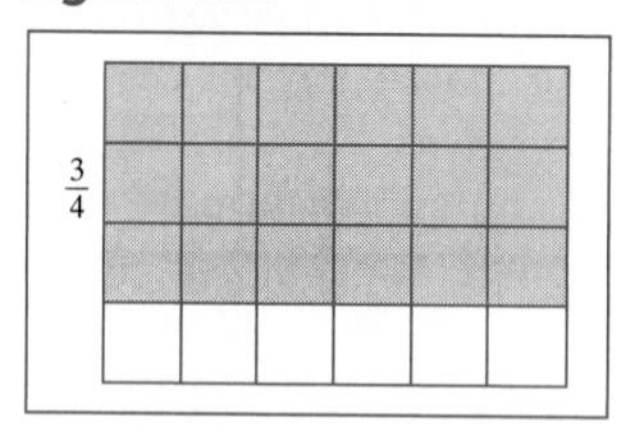

Figure 7.28

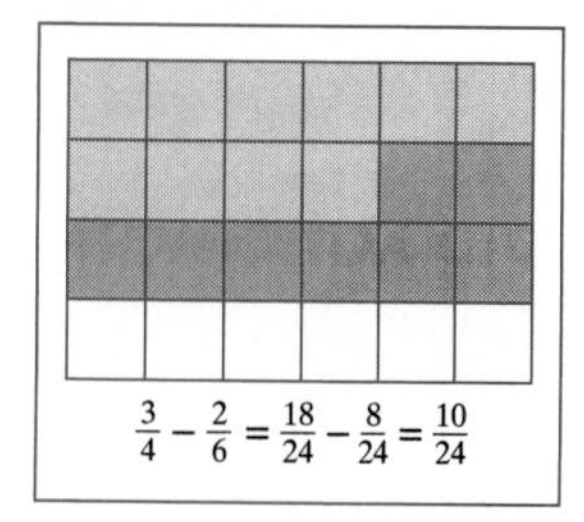

7.14.3 Subtraction of Mixed Numbers

The subtraction of rational numbers, when one or more of the elements of the subtraction problem is a mixed number, can be done using two different methods.

Method 1. Keep the rational numbers written as mixed numbers. Subtract the whole-number part from the whole-number part and the fractional part from the fractional part. For example, in the problem $5\frac{4}{5} - 2\frac{1}{2}$, you can think of the problem as follows:

$$5\frac{4}{5} - 2\frac{1}{2} = (5 - 2) + \left(\frac{4}{5} - \frac{1}{2}\right) = 3 + \left(\frac{8}{10} - \frac{5}{10}\right) = 3 + \frac{3}{10} = 3\frac{3}{10}$$

This method works easily unless the fractional part of the number being subtracted is greater than the fractional part of the minuend. For example, the problem $6\frac{1}{3} - 1\frac{3}{4}$ can be rewritten as $(6 - 1) + (\frac{1}{3} - \frac{3}{4})$. However, we now have a problem with subtracting $\frac{1}{3} - \frac{3}{4}$. What we must do is to rethink the problem in a way so that we "borrow" one from the whole part of the minuend and rewrite the fractional part as an improper fraction. For example, rewrite $6\frac{1}{3} - 1\frac{3}{4}$ as $5\frac{4}{3} - 1\frac{3}{4}$. Then, we can break the subtraction of the mixed number into the whole and fractional pieces:

$$5\frac{4}{3} - 1\frac{3}{4} = (5 - 1) + \left(\frac{4}{3} - \frac{3}{4}\right) = 4 + \left(\frac{16}{12} - \frac{9}{12}\right) = 4 + \frac{7}{12} = 4\frac{7}{12}.$$

Method 2. A second method for subtracting rational numbers when mixed numbers are involved is to first convert mixed numbers to improper fractions and then subtract using common denominators. For example, we can rewrite the problem $10\frac{1}{6} - 3\frac{2}{3}$ as $\frac{61}{6} - \frac{11}{3}$. Then, finding common denominators, the problem is solved by $\frac{61}{6} - \frac{11}{3} = \frac{61}{6} - \frac{22}{6} = \frac{61 - 22}{6} = \frac{39}{6} = 6\frac{3}{6} = 6\frac{1}{2}$, converting the final answer to a mixed number.

Related Topics

The Set of Rational Numbers
Multiplying Rational Numbers
Dividing Rational Numbers

TOPIC 7.15 Multiplying Rational Numbers

cornerstone

Fractions	**Integer Subtraction**
Rational Numbers	**Integer Multiplication**
Equivalent Fractions	**Integer Division**
Simplifying Fractions	**Least Common Multiple**
Improper Fractions and Mixed Numbers	**Adding Rational Numbers**
Integer Additio.n	**Subtracting Rational Numbers**

Introduction

Multiplication of rational numbers is helpful in solving problems that involve finding "part of a part;" that is, if you want to know what one-half of one-third is, the operation to choose to solve this problem is multiplication. We could solve by finding $\frac{1}{2} \times \frac{1}{3} = \frac{1}{6}$. In the sections that follow, we examine the algorithm for multiplying rational numbers, illustrating the multiplication using various models.

Early mathematicians in the Egyptian and Greek cultures encountered difficulties with the multiplication of fractions because of the nature of the symbols used to represent fractions. Other early mathematicians struggled with the concept that the product of two proper fractions, that is, two fractions between 0 and 1, would be less than the factors themselves.

In general, the process of multiplication of simple fractions has not changed dramatically during the last few centuries. One notable difference, however, is that early mathematicians did not cancel common factors between the numerators and the denominators of the factors prior to the multiplication of numerator times numerator and denominator times denominator. Today, we readily recognize the advantages canceling common factors before multiplying.

For More Information Bunt, L.N., Jones, P. S., & Bedient, J.D. (1998). *The Historical Roots of Elementary Mathematics*. New York: Dover Publications.

7.15.1 Understanding Multiplication of Two Rational Numbers

To understand what happens when multiplying two rational numbers, it is easiest to start with the multiplication of two rational numbers between zero and one. When multiplying $\frac{1}{3}$ and $\frac{1}{2}$, you could ask the question, "What is $\frac{1}{3}$ of $\frac{1}{2}$?" Asking the multiplication problem this way, we could imagine having a picture of $\frac{1}{2}$ and then determining what $\frac{1}{3}$ of this amount is. Using a number line, as in Figure 7.29, we can show $\frac{1}{2}$ of a whole length and then break this half into three equal-sized pieces to examine $\frac{1}{3}$ of this half.

Figure 7.29

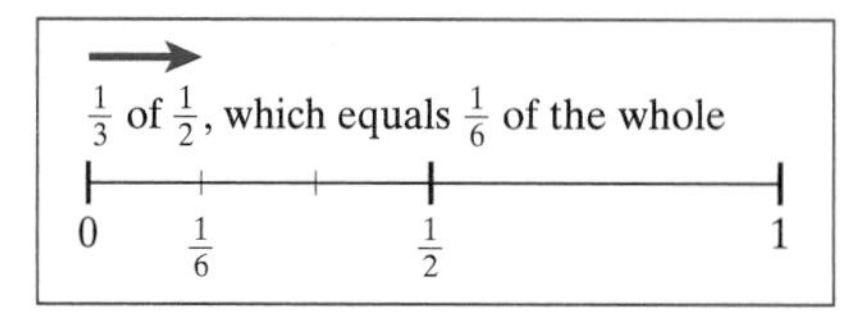

Figure 7.30

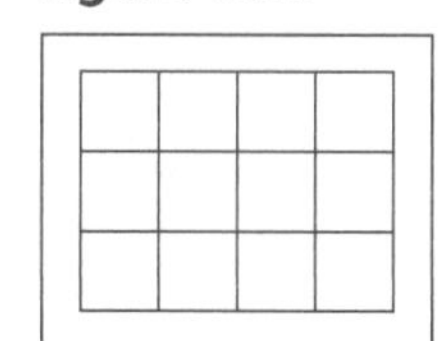

We can further illustrate this idea of taking "part of a part" using the region model. To show $\frac{2}{3} \times \frac{3}{4}$ using the region model, we can break the region into 3 equal-sized rows and 4 equal-sized columns, breaking the region into 12 equal-sized rectangles. See Figure 7.30.

In Figure 7.31, we see that to find $\frac{2}{3}$ of $\frac{3}{4}$, we first illustrate $\frac{3}{4}$ by shading in 3 columns, and then viewing the $\frac{3}{4}$ as divided into 3 equal-sized groups, we can show $\frac{2}{3}$ of $\frac{3}{4}$ by shading 6 of the 9 rectangles. Thus, the solution is $\frac{6}{12}$ or $\frac{1}{2}$.

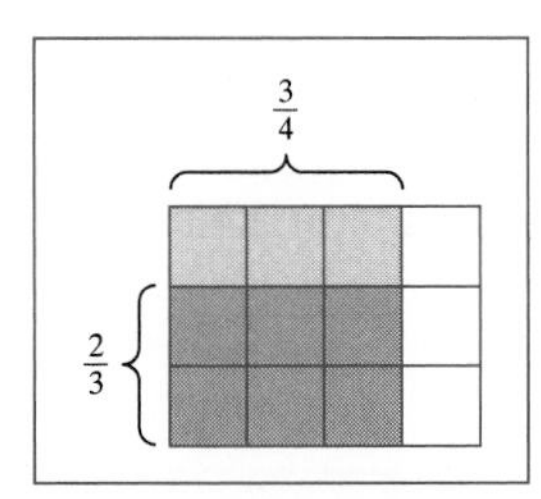

Figure 7.31

7.15.2 Definition of Multiplication of Rational Numbers

The product of two rational numbers can be found by multiplying numerators and multiplying denominators to find the numerator and denominator, respectively, of the product of the two rational numbers. Formally, we say, $\frac{a}{b} \times \frac{c}{d} = \frac{a \times c}{b \times d}$. For example, $\frac{2}{7} \times \frac{3}{4} = \frac{2 \times 3}{7 \times 4} = \frac{6}{28}$.

7.15.3 Multiplication of Mixed Numbers

When multiplying rational numbers where one or more of the factors are mixed numbers, there are two approaches.

Method 1. One way of multiplying mixed numbers is to first convert the mixed numbers to improper fraction form. Once in improper fraction form, the formal definition can be used to find the product. The product can then be converted back to mixed number form. For example,

$$2\frac{2}{5} \times 5\frac{2}{3} = \frac{12}{5} \times \frac{17}{3} = \frac{12 \times 17}{5 \times 3} = \frac{204}{15} = 13\frac{9}{15} = 13\frac{3}{5}.$$

Method 2. Mixed numbers can be multiplied in mixed number form. However, it requires using the Distributive Property of multiplication over addition. For example, in the problem above $2\frac{2}{5} \times 5\frac{2}{3}$, we can rewrite the problem as $(2 + \frac{2}{5}) \times (5 + \frac{2}{3})$, and then use the Distributive Property of multiplication over addition in two stages as seen below:

$$\left(2+\frac{2}{5}\right)\times\left(5+\frac{2}{3}\right)=2\times\left(5+\frac{2}{3}\right)+\frac{2}{5}\times\left(5+\frac{2}{3}\right)=(2\times 5)+\left(2\times\frac{2}{3}\right)+\left(\frac{2}{5}\times 5\right)+\left(\frac{2}{5}\times\frac{2}{3}\right).$$

Multiplying to find these products we finish the problem, finding the same solution in Method 1:

$$= 10 + \frac{4}{3} + 2 + \frac{4}{15} = 12 + \left(\frac{20}{15} + \frac{4}{15}\right) = 12 + \frac{24}{15} = 12 + 1\frac{9}{15} = 13\frac{3}{5}.$$

Related Topics

The Set of Rational Numbers
Dividing Rational Numbers

TOPIC 7.16 Dividing Rational Numbers

cornerstone

Fractions
Rational Numbers
Equivalent Fractions
Simplifying Fractions
Improper Fractions and Mixed Numbers
Integer Addition
Integer Subtraction
Integer Multiplication
Integer Division
Least Common Multiple
Adding Rational Numbers
Subtracting Rational Numbers
Multiplying Rational Numbers

Introduction

The operation of division, whether dividing whole numbers, integers, or rational numbers, involves answering the question, "How many of one set fits inside another set?" For example, when dividing $20 \div 4$, one might ask the question, "How many fours are in twenty?" Likewise, when dividing $\frac{3}{5} \div \frac{1}{2}$, one might translate the problem into answering the question, "How many one-halves are there in three-fifths?" Within this topic, we illustrate the process of rational-number division using several models and discuss the relationship between rational-number multiplication and division.

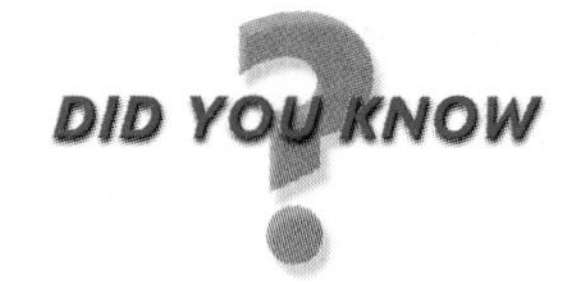

Historically, the division of rational numbers was the most difficult operation to develop. Even though we regularly accept division by a rational number as equivalent to multiplication by the reciprocal of the rational number, this process only came into general use in the early 1900s. Nevertheless, early versions of this process were used in the Middle Ages by both the Hindus and the Arabs. The roots of the process stemmed from the fact that it seemed natural to substitute division by an integer with multiplication by its reciprocal. For example, $\frac{2}{3} \div 4 = \frac{2}{3} \times \frac{1}{4}$. The idea that dividing by 4 is akin to finding one-fourth of a number, that is, multiplying by one-fourth, led to further explorations into division of rational numbers.

Early methods to expand the notion of division by rational numbers included finding a common denominator for the two fractions in the division problem; that is, for the problem $\frac{2}{3} \div \frac{3}{4}$, mathematicians would rewrite the problem with common denominators as $\frac{8}{12} \div \frac{9}{12}$. Then, they would find their solution by taking the quotient of the numerators: $\frac{8}{12} \div \frac{9}{12} = \frac{8 \div 9}{12 \div 12} = \frac{8/9}{1} = \frac{8}{9}$.

Today, division by a rational number helps to solve problems that involve finding how many times a part of something fits into another part. As with the other operations, division of rational numbers can be situated within the context of a linear model, a region model, or a set model. However, division in the context of a set model is difficult to illustrate because with division of rational numbers, the quotient is often much larger than the fractions representing the dividend or the divisor in the problem.

For More Information Smith, D. E. (1958). *History of Mathematics, Volume II*. New York: Dover Publications.

7.16.1 Understanding Division of Rational Numbers

It is essential to understand what question is really being asked when we divide a number by a rational number. In the problem $\frac{1}{3} \div \frac{1}{6}$, the question being asked is, "How many $\frac{1}{6}$s are there in $\frac{1}{3}$? Illustrating this question on a number line helps provide some understanding.

In Figure 7.32, we have broken one whole unit into 3 equal-sized parts to get a picture of $\frac{1}{3}$. Then, knowing that we want to know how many $\frac{1}{6}$s are in $\frac{1}{3}$, we illustrate $\frac{1}{6}$ of the whole, noting that two $\frac{1}{6}$s fit into $\frac{1}{3}$.

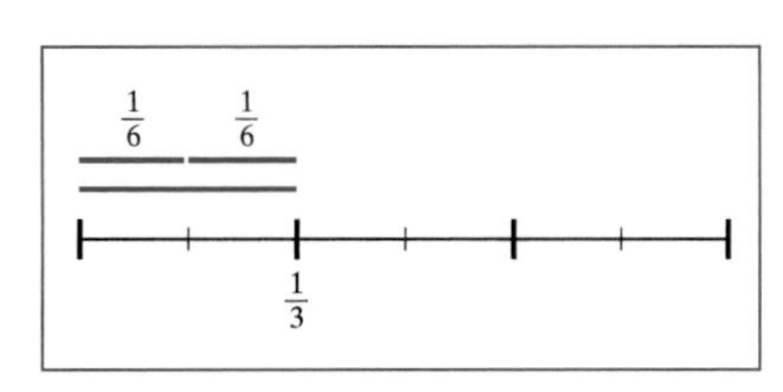

Figure 7.32

7.16.2 Definition of Reciprocal

When working with rational numbers of the form $\frac{a}{b}$, where a and b are integers, we say that the **reciprocal** of $\frac{a}{b}$ is the fraction $\frac{b}{a}$. This reciprocal is the multiplicative inverse of the number; that is, $\frac{a}{b} \times \frac{b}{a} = 1$.

7.16.3 Definition of Dividing Rational Numbers

When dividing one rational number by another, we use the definition $\frac{a}{b} \div \frac{c}{d} = \frac{a}{b} \times \frac{d}{c} = \frac{a \times d}{b \times c}$. Many learn this definition by saying, "When dividing by a rational number $\frac{a}{b}$, simply multiply by the reciprocal $\frac{b}{a}$ instead." For example, in the problem $\frac{1}{3} \div \frac{1}{6}$, we can use the definition to find that $\frac{1}{3} \div \frac{1}{6} = \frac{1}{3} \times \frac{6}{1} = \frac{6}{3} = 2$.

7.16.4 Why Can We Multiply by the Reciprocal?

There are several ways to explain why multiplying by the reciprocal yields the same answer as dividing by a rational number. Consider the problem $\frac{1}{2} \div \frac{1}{4}$. The definition for division by rational numbers says that the answer to this problem is the same as the answer to the problem $\frac{1}{2} \times \frac{4}{1}$. Let us illustrate the two problems, using a region model for fractions.

Figure 7.33 illustrates the problem $\frac{1}{2} \div \frac{1}{4}$, asking the question, "How many $\frac{1}{4}$s are in $\frac{1}{2}$?" The answer is that there are 2 one-fourths in one-half. Thus, the numeric answer is 2.

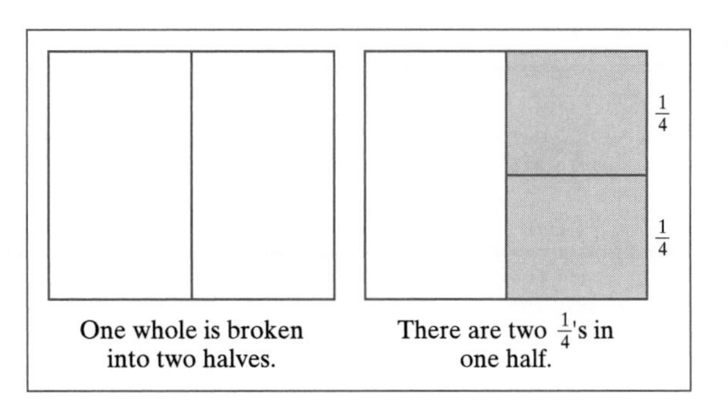

Figure 7.33

When we illustrate the problem $\frac{1}{2} \times \frac{4}{1}$ in Figure 7.34, we see that when we have one-half times 4 or $4\frac{1}{2}$s. When we put the four $\frac{1}{2}$s together, they make 2 wholes. Thus, the numeric answer is, again, 2.

Figure 7.34

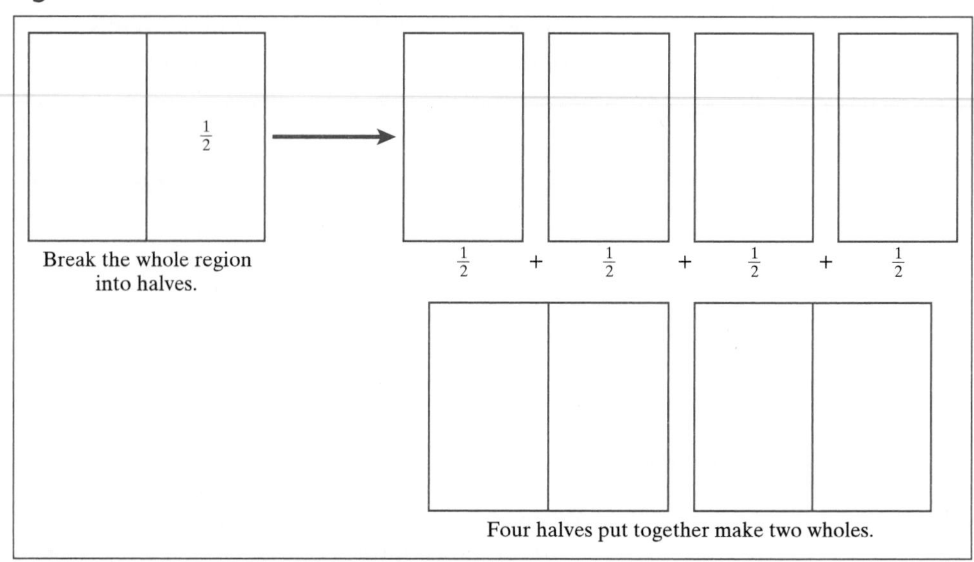

7.16.5 Dividing Mixed Numbers

When dividing rational numbers that are mixed numbers, the best method is to convert mixed numbers to improper-fraction form first. Then, we can easily use the definition of division to find the quotient. For example, in the problem $2\frac{1}{2} \div 3\frac{3}{4}$, we can rewrite these mixed numbers as improper fractions and solve using the definition: $\frac{5}{2} \div \frac{15}{4} = \frac{5}{2} \times \frac{4}{15} = \frac{20}{30} = \frac{2}{3}$.

Related Topics
The Set of Rational Numbers
Decimal Fractions

TOPIC 7.17 *Irrational Numbers*

cornerstone **Terminating and Repeating Decimals**

Introduction

Irrational numbers are most easily described through examples. Herein, we will define irrational numbers and discuss the decimal representation of irrational numbers. We also highlight the concept of roots and other notable examples of irrational numbers.

The discovery of irrational numbers was marked by a great deal of secrecy. The Pythagoreans, who first discovered such numbers, were unable to accept their existence. The Pythagoreans were a society of Greek scholars that was started by Pythagoras (c. 540 B.C.). The society became so secretive about its work that it is believed that Hippasus, one of its members, was drowned after he shared the secret of irrational numbers with an outsider! By 300 B.C., many irrational numbers were known.

Examples of such numbers include $\sqrt{2}$, $\sqrt{3}$, and $\sqrt{6}$. By the first century A.D., the Hindus were performing operations on irrational numbers. Among the most interesting irrational numbers is the number denoted π and defined as the ratio of the circumference of a circle to its diameter. The value of pi is most often represented by the truncated decimal 3.14, even though its true value is an infinite string of decimals with no identifiable pattern.

For More Information Reimer, L., & Reimer, W. (1990). *Mathematicians Are People, Too: Stories from the Lives of Great Mathematicians, Volume 1*. Palo Alto, CA: Dale Seymour Publications.

7.17.1 Nonterminating, Nonrepeating Decimals

Some decimals are **nonterminating, nonrepeating decimals.** This type of decimal has an infinite string of digits to the right of the decimal point, and the string of digits has no repeating pattern. Examples of nonterminating, nonrepeating decimals are: 0.12112111211112 ..., −23.034213531 ..., 3.141592654 ..., and 1.414213562 ....

7.17.2 Definition of Irrational Numbers

Numbers that have a decimal representation as nonterminating, nonrepeating decimals are **irrational numbers.** These are numbers that cannot be represented in the form $\frac{a}{b}$, where a and b are integers, $b \neq 0$; that is, irrational numbers are numbers that are not rational numbers. Many irrational numbers can be expressed in forms other than nonterminating, nonrepeating decimals. Two examples of irrational numbers expressed in a form other than decimal form are π (pronounced "pi") and $\sqrt{2}$ (the square root of 2). Note, $\pi = 3.141592654\ldots$ and $\sqrt{2} = 1.414213562\ldots$ in decimal form.

7.17.3 Square Roots

The **square root** of a number is a number that, when multiplied by itself, produces the given number. A positive (real) number has two square roots (a positive and a negative real number). A negative (real) number does not have any real-number square roots. The notation for square root is the symbol $\sqrt{\ }$. Thus, $\sqrt{x}$ is read as "the square root of x." Positive numbers that are perfect squares have whole-number square roots.

For example, $\sqrt{16} = 4$ and $\sqrt{16} = -4$ because $4 \times 4 = 16$ and $(-4) \times (-4) = 16$. To indicate that the square root includes both the positive and negative root value, we often use the symbol $\pm$. Thus, we can write, $\sqrt{16} = \pm 4$.

Positive numbers that are not perfect squares still have square roots; however, the square roots will not be whole numbers. In fact, in many cases the square roots will be irrational numbers. There are methods for approximating square roots of numbers; however, the calculator is an available and sufficient tool for completing this task. Just note that calculators often do not indicate both the positive and the negative square root values; rather, they simply display the positive square root. Also, because many square roots are irrational numbers, calculators will not fully represent the nonterminating, nonrepeating nature of the decimal. Therefore, take caution when using calculators for finding square roots.

7.17.4 Roots

Just as we can raise numbers to exponential powers other than 2, we can take roots of numbers other than square roots. For example, we can find the cube root of a number. A cube root of a number is a number that, when multiplied by itself twice, produces the given number. The notation for a cube root is the symbol $\sqrt[3]{}$. For example, $\sqrt[3]{8} = 2$ because $2^3 = 2 \times 2 \times 2 = 8$.

We can find any numbered root of a number; that is, we can find the "fourth root," written, $\sqrt[4]{}$, the "sixth root," written, $\sqrt[6]{}$, and so on. In general, *the "nth root," written,* $\sqrt[n]{}$ *is a number that, when multiplied by itself n − 1 times, produces the number itself.*

7.17.5 Rational Exponents

Roots of numbers can be expressed using **rational exponents,** that is, exponents that are rational numbers. Symbolically, we have $\sqrt[n]{x} = x^{1/n}$. For example, $100^{1/2} = \sqrt{100} = 10$. Likewise, $81^{1/4} = \sqrt[4]{81} = 3$.

Related Topics Real Numbers

TOPIC 7.18 *Real Numbers*

Whole Numbers
Integers
Rational Numbers
Irrational Numbers

Introduction

The set of real numbers includes whole numbers, negative numbers, fractional numbers, and beyond. In the following sections, we will examine the definition of a real number and how the set of real numbers relates to other numbers systems.

An important and dramatic area of mathematics is the theory of infinity. The pioneer in this area was German mathematician named Georg Cantor (1845–1918). Cantor was able to prove that there are an overwhelming number of irrational numbers among the real numbers. Relatedly, he also proved that two line segments, regardless of their lengths, have the same number of points.

For More Information Bunt, L.N., Jones, P. S., & Bedient, J.D. (1998). *The Historical Roots of Elementary Mathematics*. New York: Dover Publications.

7.18.1 The Set of Real Numbers

A **real number** is any rational or irrational number. The **set of real numbers** is the collection of all rational and irrational numbers. The set of real numbers can also be described as the infinite set of all points on a number line. See Figure 7.35.

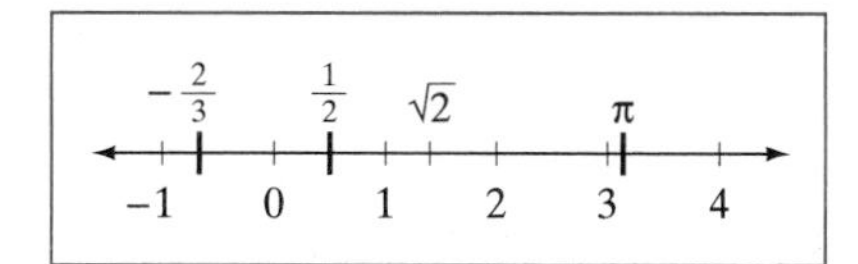

Figure 7.35

7.18.2 Relationship Between the Reals and Other Number Systems

The set of real numbers (the reals) is the largest set of numbers that many people have the opportunity to explore. The reals contains many other sets of numbers. The relationship between the reals and other number systems is expressed using subset notation in Figure 7.36.

Figure 7.36

Natural Numbers ⊂ *Whole Numbers* ⊂ *Integers* ⊂ *Rational Numbers* ⊂ *Real Numbers*

↓ Rational + Irrationals

The notation indicates that the set of real numbers contains the rational numbers, which, in turn, contain all the integers, which contain the whole numbers, which contain the natural numbers.

7.18.3 Properties of the Real Numbers

Because the real numbers contains all rational numbers, the reals have the same properties that the rational numbers have. The addition of the irrational numbers into the set of reals does not add any new properties. The properties for rational numbers are described in detail in the section entitled, "Properties of Rational Numbers." A list and an abbreviated description of the properties of real numbers is provided below:

1. The set of real numbers is closed under addition, subtraction, multiplication, and division; that is, the sum, the difference, the product, or the quotient of any two real numbers is also a real number.
2. The set of real numbers is Commutative under addition and multiplication. This says that the order of addends or the order of factors does not affect the sum or product, respectively.
3. The set of real numbers is Associative under addition and multiplication. This property says that when adding or multiplying three real numbers together, the way we choose to group the addends or factors together when adding or multiplying two at a time will not affect the sum or product.
4. The set of real numbers has an Additive Identity of zero. When you add any real number, a, to zero, the sum is always a.
5. The set of real numbers has a Multiplicative Identity of 1. When you multiply any real number, a, by 1, the product is always a.
6. The set of real numbers has additive inverses. For any real number a, there exists another real number, $-a$, so that $a + -a = 0$.
7. The set of real numbers has multiplicative inverses. For any real number a, there exists another real number, $\frac{1}{a}$, so that $a \times \frac{1}{a} = 1$.
8. The set of real numbers has the Distributive Property of multiplication over addition; that is, for any real numbers a, b, and c, $a \times (b + c) = a \times b + a \times c$.

Related Topics

Properties of Rational Numbers

In the Classroom

The last chapter to focus on the *Number and Operations Content Standard,* this chapter also emphasizes the *Connections Process Standard:* "Throughout the pre-K–12 span, students should routinely ask themselves, "How is this problem or mathematical topic like things I have studied before?" (NCTM, 2000, p. 65). Given that these students should be enabled to "recognize and use connections among mathematical ideas" and "understand and apply mathematics in contexts outside of mathematics" (NCTM, 2000, p. 64), we believe it essential for you to have an understanding of (a) the concepts of ratio and proportion in relationship to fraction concepts, and a meaningful interpretation of different graphs used to represent these ideas (Activities 7.1–7.3); (b) the concepts of decimal and percent in relationship to fraction concepts, and a meaningful extension of the base-ten numeration system for decimal representation (Activities 7.4–7.11); and (c) how decimal representations of real numbers indicate the existence of numbers that are not rational, and how the latter can be conceptualized geometrically (Activities 7.12–7.14).

Some Expectations of Your Future Students:

Pre-K–2 (NCTM, 2000, p. 78)

- use multiple models to develop initial understandings of place values and the base-ten number system
- develop a sense of whole numbers, and represent and use them in flexible ways, including relating, composing, and decomposing numbers

Grades 3–5 (NCTM, 2000, p. 148)

- understand the place-value structure of the base-ten number system, and be able to represent and compare whole numbers and decimals
- recognize and generate equivalent forms of commonly used fractions, decimals, and percents
- develop and use strategies to estimate computations involving fractions and decimals in situations relevant to students' experience

How Activities 7.1–7.14 Help You Develop an Adult-level Perspective on the Above Expectations:

- By exploring the concept of ratios in relation to that of fractions, you study the concept of proportion and construct a deeper understanding of equivalent fractions.
- By using proportional thinking, you apply the concepts of ratios and equivalent fractions in solving real-life problems that model linear relationships or mathematical similarity.
- By building on your understanding of fractions and of how base-ten blocks represent our numeration system, you extend the latter to construct a model of the decimal system. You then extend prior models of operations to explain why algorithms using decimals work.
- By understanding percent representation in terms of decimals and fractions, you develop multiple strategies for solving common problems involving percentages.
- By analyzing patterns in the decimal representations of real numbers, you realize the existence of numbers that are not rational. You learn to construct models of some rational numbers by using concepts studied in geometry and measurement. You thus learn to represent and use the real numbers in flexible ways, including relating, composing, and decomposing them.

Bibliography

Ben-Chaim, D., Fey, J. T., Fitzgerald, W. M., Benedetto, C., & Miller, J. L. (1998). Proportional reasoning among 7th grade students with different curricular experiences. *Educational Studies in Mathematics, 36,* 247–273.

Bennett, A. B., & Nelson, L. T. (1994). A conceptual model for solving percent problems. *Mathematics Teaching in the Middle School, 1,* 20–25.

Bezuk, N. S., & Armstrong, B. E. (1992). Activities: Understanding fraction multiplication. *Mathematics Teacher, 85,* 729–733, 739–744.

Bezuk, N. S., & Armstrong, B. E. (1993). Activities: Understanding division of fractions. *Mathematics Teacher, 86,* 43–46, 56–60.

Boling, B. A. (1985). A different method for solving percentage problems. Mathematics *Teacher, 78,* 523–524.

Chappell, M. F., & Thompson, D. R. (1999). Modifying our questions to assess students' thinking. *Mathematics Teaching in the Middle School, 4,* 470–474.

Coburn, T. G. (1986). Percentage and the hand calculator. *Mathematics Teacher, 79,* 361–367.

Cramer, K., & Bezuk, N. (1991). Multiplication of fractions: Teaching for understanding. *Arithmetic Teacher, 39* (3), 34–37.

Cramer, K., & Karnowski, L. (1995). The importance of informal language in representing mathematical ideas. *Teaching Children Mathematics, 1,* 332–335.

Cramer, K. A., Post, T. R., & Behr, M. J. (1989). Interpreting proportional relationships. *Mathematics Teacher 82,* 445–452.

Cramer, K., & Post, T. R. (1993). Making connections: A case for proportionality. *Arithmetic Teacher 40,* 342–346.

Cramer, K., Post, T. R., & Currier, S. (1993). Learning and teaching ratio and proportion: Research implications. In D. T. Owens (Ed.), *Research ideas for the classroom: Middle grades mathematics* (pp. 159–178). Old Tappan, NJ: Macmillan.

Dewar, J. M. (1984). Another look at the teaching of percent. *Arithmetic Teacher, 31,* 48–49.

Edwards, F. M. (1987). Geometric figures make the LCM obvious. *Arithmetic Teacher, 34,* 17–18.

Erickson, D. K. (1990). Activities: Percentages and Cuisenaire rods. *Mathematics Teacher, 83,* 648–654.

Glatzer, D. J. (1984). Teaching percentage: Ideas and suggestions. *Arithmetic Teacher, 31,* 24–26.

Grossman, A. S. (1983). Decimal notation: An important research finding. *Arithmetic Teacher, 30,* 32–33.

Haubner, M. A. (1992). Percents: Developing meaning through models. *Arithmetic Teacher, 40,* 232–234.

Hiebert, J. (1987). Research report: Decimal fractions. *Arithmetic Teacher, 34,* 22–23.

Huinker, D. (1992). Decimals and calculators make sense. In J. T. Fey (Ed.), *Calculators in mathematics education* (pp. 56–64). Reston, VA: National Council of Teachers of Mathematics.

Hurd, S. P. (1991). Egyptian fractions: Ahmes to Fibonacci to today. *Mathematics Teacher, 84,* 561–568.

Langford, K., & Sarullo, A. (1993). Introductory common and decimal fraction concepts. In R. J. Jensen (Ed.), *Research ideas for the classroom: Early childhood mathematics* (pp. 223–247). Old Tappan. NJ: Macmillan.

Mack, N. K. (1998). Building a foundation for understanding the multiplication of fractions. *Teaching Children Mathematics, 5,* 34–38.

Meeks, K. I. (1992). Decimals, rounding, and apportionment. *Mathematics Teacher, 85,* 523–525.

Nowlin, D. (1996). Division with fractions. *Mathematics Teaching in the Middle School, 2,* 116–119.

Oppenheimer, L., & Hunting, R. P. (1999). Relating fractions and decimals: Listening to students talk. *Mathematics Teaching in the Middle School, 4,* 318–321.

Ott, J. M. (1990). A unified approach to multiplying fractions. *Arithmetic Teacher, 37*(7), 47–49.

Owens, D. T., & Super, D. B. (1993). Teaching and reaming decimal fractions. In D T. Owens (Ed.), *Research ideas for the classroom: Middle grades mathematics* (pp. 137–158). Old Tappan, NJ: Macmillan.

Payne, J. N., & Towsley, A. E. (1990). Implementing the *Standards*: Implications of NCTM's *Standards* for teaching fractions and decimals. *Arithmetic Teacher, 37,* 23–26.

Payne, J. N., Towsley, A. E., & Huinker, D. M. (1990). Fractions and decimals. In J. N. Payne (Ed.), *Mathematics for the young child* (pp. 175–200). Reston, VA: National Council of Teachers of Mathematics.

Prevost, F. J. (1984). Teaching rational numbers: Junior high school. *Arithmetic Teacher, 31*(6), 43–46.

Quintero, A. H. (1987). Helping children understand ratios. *Arithmetic Teacher, 34,* 17–21.

Rees, J. M. (1987). Two-sided pies: Help for improper fractions and mixed numbers. *Arithmetic Teacher, 35,* 28–32.

Sarver, V. T., Jr. (1986). Why does a negative times a negative produce a positive? *Mathematics Teacher, 79,* 178–180.

Steiner, E. E. (1987). Division of fractions: Developing conceptual sense with dollars and cents. *Arithmetic Teacher, 34,* 36–42.

Thompson, C. S., & Walker, V. (1996). Connecting decimals and other mathematical content *Teaching Children Mathematics, 2,* 496–502.

Vance, J. M. (1986). Estimating decimal products: An instructional sequence. In H. L. Schoen (Ed.), *Estimation and mental computation* (pp. 127–134). Reston, VA: National Council of Teachers of Mathematics.

Vance, J. M. (1986). Ordering decimals and fractions: A diagnostic study. *Focus on Learning Problems in Mathematics, 8*(2), 51–59.

Warrington, M. A., & Kamii, C. K. (1998). Multiplication with fractions: A Piagetian, constructivist approach. *Mathematics Teaching in the Middle School, 3,* 339–343.

Wiebe, J. H. (1986). Manipulating percentages. *Arithmetic Teacher, 33*(5), 23–26.

Williams, S. E., & Copley, J. V. (1994). Promoting classroom dialogue: Using calculators to discover patterns in dividing decimals. *Mathematics Teaching in the Middle School, 1,* 72–75.

Zawojewski, J. (1983). Initial decimal concepts: Are they really so easy? *Arithmetic Teacher, 30*(7), 52–56.

CHAPTER 8 EIGHT

Patterns & Functions

CHAPTER OVERVIEW

The concept of function is a central theme, a big idea, running through many areas of mathematics. In this chapter, you can learn some interesting facts about functions and will examine some concepts and procedures involving function ideas.

BIG MATHEMATICAL IDEAS

Problem-solving strategies, functions and relations, representation, conjecturing, verifying, mathematical structure

NCTM PRINCIPLES & STANDARDS LINKS

Patterns, Functions, and Algebra; Problem Solving; Reasoning; Communication; Connections; Representation

TOPIC **8.1** **Functions**
8.2 **Equations and Inequalities**
8.3 **Cartesian Coordinate System**
8.4 **Graphs and Equations of Lines**
8.5 **Systems of Linear Equations**
8.6 **Solving Systems of Linear Equations Symbolically**
8.7 **Exponents**
8.8 **Scientific Notation**
In the Classroom
Bibliography

TOPIC 8.1 Functions

Set Operations

Introduction

Functions describe relationships. We often hear expressions such as "Worker productivity is a function of employee satisfaction" and "Wisdom is a function of time and experience." The message in these expressions is that functions provide a means of showing how one item affects or is related to another. In this section, we examine terminology and notation associated with functions, as well as a variety of ways of representing functions.

Russian mathematician Sonya Kovalevsky (1850–1891) was inspired to study mathematics by an unusual source. Her father, General Krukovsky, had retired from the military and moved his family into an ancient castle near the Lithuanian border. When her parents ordered wallpaper for all the rooms, they made a measuring error and did not order enough. When papering Sonya's room, instead of using wallpaper, her father covered her walls with a set of lithographed lectures on differential and integral calculus. These were topics he intended to study himself. Sonya became extremely interested in the writings on her walls and worked to decipher the formulas. From this experience, Sonya developed the basis of her mathematical understanding. When she grew up, she went on to win a renowned competition through the Paris Academy of Sciences for original work in mathematics and contributed greatly to the field of mathematics.

For More Information Reimer, L., & Reimer, W. (1995). *Mathematicians Are People Too: Stories from the Lives of Great Mathematicians, Volume 2*. Palo Alto, CA: Dale Seymour Publications.

8.1.1 Definition of Functions and Function Notation

Figure 8.1

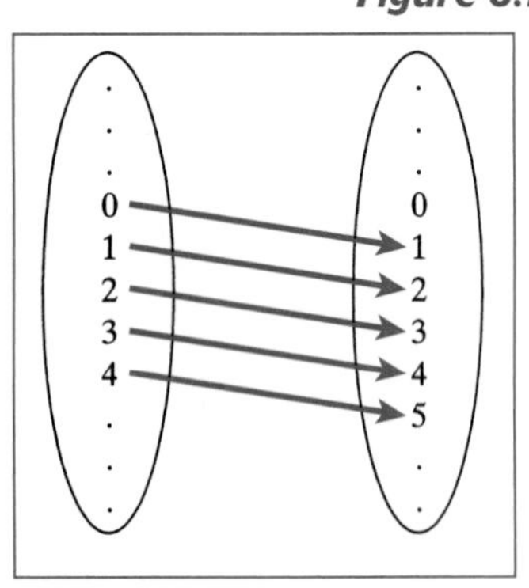

A **function** is a mapping, or association, of elements of one set to elements of another set. The mapping is such that each element of the first set is associated with **exactly one** element in the second set. For example, consider the function that maps the set of integers to the set of integers by the mapping $y = x + 1$. For any integer, x, in the first set, we associate it with the integer y in the second set, which can be determined by the mapping of adding 1 to x. Thus, with the function $y = x + 1$, we associate the x-value, 1, with the y-value of $1 + 1$ or 2. Likewise, the x-value -2 is associated with the y-value $-2 + 1$ or -1. Figure 8.1 illustrates this mapping.

We often write functions in terms of x and y, in that we describe how a value for y can be generated given a value for x. We say that y is a function of x, denoted $y = f(x)$. Thus, we can express the function $y = x + 1$ as $f(x) = x + 1$. Thus, to determine the y-value or function value associated with the x-value $x = 5$, we find $f(5) = 5 + 1 = 6$. We have associated $x = 5$ with the y-value, or $f(x)$ value, 6.

Not all mappings are functions. For example, the mapping $y = \sqrt{x}$ is not a function. A function associates exactly one y-value with each x-value. If we try to evaluate $y = \sqrt{x}$ at the x-value $x = 16$, we find that $\sqrt{16}$ has two values, 4 and -4. Therefore, $y = \sqrt{x}$ is not a function.

8.1.2 Domain and Range

The **domain** of a function is the set of all x-values in a set that can be mapped to another set, according to the association described by the function. The **range** is the set of all y-values that can be generated from mapping x-values in the domain, according to the association described by the function.

For example, given the function $y = \frac{x}{x+1}$, we describe the domain by all the x-values that can be mapped through this function to produce a corresponding y-value. Note, though, that fractions cannot have a zero in the denominator. Therefore, if we try to insert the x-value $x = -1$ into the function, we are in the sit-uation of having the fraction $\frac{-1}{0}$. Because this fraction makes no sense and is not defined, we cannot include the value $x = -1$ in the domain. Therefore, we can say that the domain of the function $y = \frac{x}{x+1}$ is the set of real numbers, except where $x = -1$.

To determine the range of this function, we can try to solve the function for x algebraically to see if there are any restrictions on what values y can take on. Rewriting $y = \frac{x}{x+1}$ so we express x as a function of y, we find:

$$\begin{aligned} y &= \frac{x}{x+1} \\ y(x+1) &= x \\ yx + y &= x \\ y &= x - yx \\ y &= x(1-y) \\ \frac{y}{1-y} &= x. \end{aligned}$$

When we examine the equation $y = \frac{y}{1-y}$ we find that y cannot take on the value $y = 1$ because this value would create a nonsensical fraction with zero in the denominator. Thus, we conclude that $y = 1$ will not be an element in the range. Therefore, the range of the function $y = \frac{x}{x+1}$ is the set of all real numbers except for $y = 1$.

8.1.3 Ways to Describe Functions

There are a variety of ways to describe and represent functions, including: equations, graphs, function machines, tables, mappings, and sets of ordered pairs.

A common means of representing functions is through **equations.** Functions such as $y = 2x + 3$ show an algebraic relationship between two variables x and y. Given various values of x, we can use the equations to find corresponding values for y (or vice versa).

Graphs can also represent functions. From graphs, such as the one shown in Figure 8.2, we can infer a relationship between corresponding x- and y-values. From the graph in Figure 8.2, we can tell that there is a relationship between corresponding x- and y-values; that is, as x-values grow larger, y-values grow larger at a faster rate.

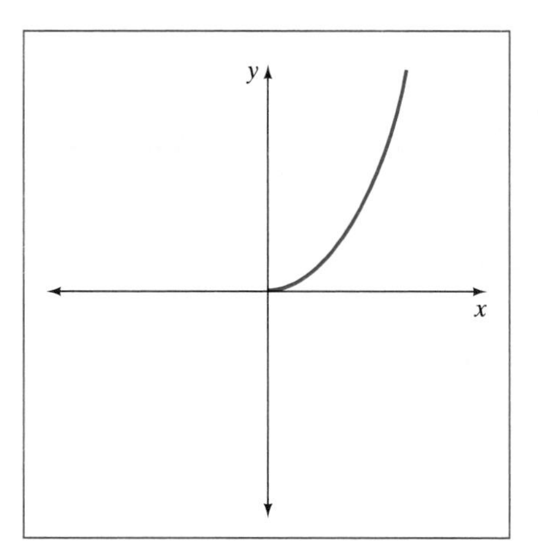

Figure 8.2

Functions can also be described through function machines. A **function machine** is like a "mystery box" that applies some rule to the x-value that is put into the function machine to produce a corresponding y-value that exits the function machine. Figure 8.3 illustrates a function machine.

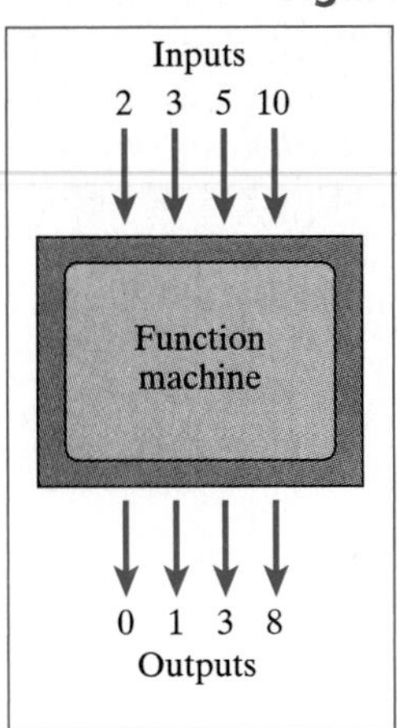

Figure 8.3

With a function machine, we can "guess" the function or rule that is applied to input values to produce output values. In the example of Figure 8.3, we can verify that the function machine takes an input value and subtracts 2 from it to produce the corresponding output values.

Tables provide a means of describing a function. Consider the table of x- and y-values as shown in Table 8.1.

Table 8.1

x	y
−2	−20
−1	−10
0	0
1	10
2	20
3	30
4	40
5	50

We can understand the relationship between the x- and the y- values shown to be that the y-value is 10 times the x-value. This relationship could be represented in equation form as $y = 10x$ or $f(x) = 10x$.

Sometimes a **mapping** is an effective way to illustrate a function. Figure 8.4 shows a mapping of the function that can be written in equation form as $f(x) = 3x + 2$.

A second example, seen in Figure 8.5, depicts a function that maps the value 1 in the domain to the value 4 in the range, maps the value 3 in the domain to the value 6 in the range, and so on.

Note that both the values 7 and 11 in the domain are mapped to the value 10 in the range. It is acceptable for a function to have two different x-values associated with the same y-value. However, it would not be a function if there was a single x-value mapped to two different y-values.

Another way of representing a function is through **ordered pairs.** An ordered pair is a pairing of an x-value and a y-value, written parenthetically as (x,y). Given a collection of related ordered pairs, we can understand the function that relates each x- value to its partnered y-value.

For example, as we examine the ordered pairs $(1,3), (2,6), (3,9), (4,12), (10,30)$, we can uncover the relationship between x and y as the function of the y-coordinate in the ordered pair is three times the x-coordinate in the ordered pair.

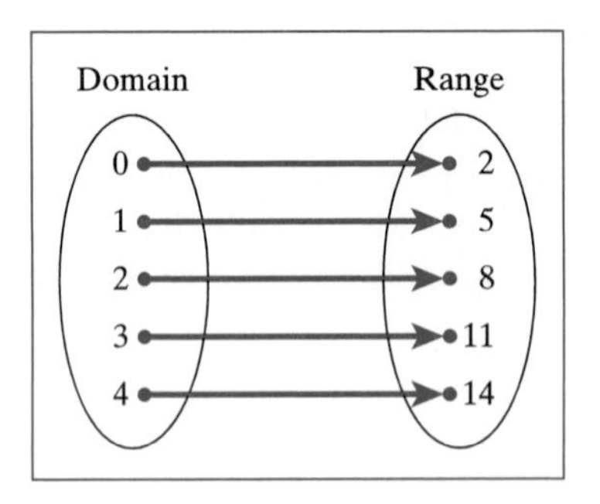

Figure 8.4

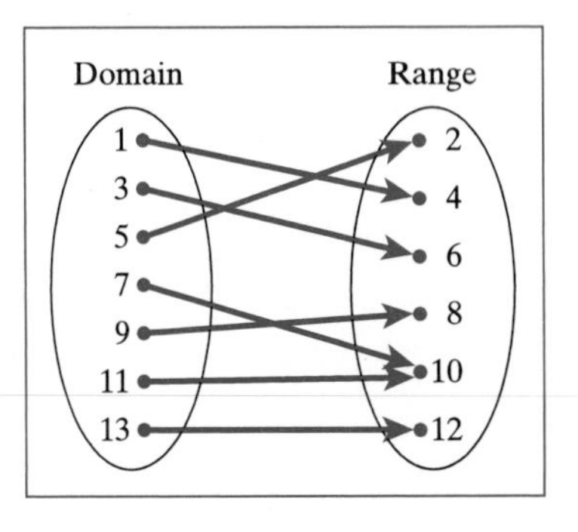

Figure 8.5

8.1.4 Vertical-Line Test

One way to determine if a mapping is a function is to use the **vertical-line test.** To use this test, we do the following:

1. Draw a graph of the mapping.
2. Conceptualize drawing all possible vertical lines (i.e., lines parallel to the vertical axis) through the graph.
3. Determine if any of the vertical lines intersect the graph at more than one point.
4. If no line touches the graph at more than one point, the mapping is a function.
5. If one or more lines touches the graph at more than one point, the mapping is not a function.

For example, consider the graphs in Figure 8.6. The first graph passes the vertical-line test, that is, represents a function, because any vertical line we can draw through the graph will only touch the graph at exactly one point. On the other hand, the second graph is not a function because we can find at least one vertical line that passes through two points of the graph.

The vertical-line test makes sense because what the test detects are points where there are two y-values associated with a single x-value. Thus, failing the vertical-line tests means that the graph does not match the fundamental definition of a function.

Figure 8.6

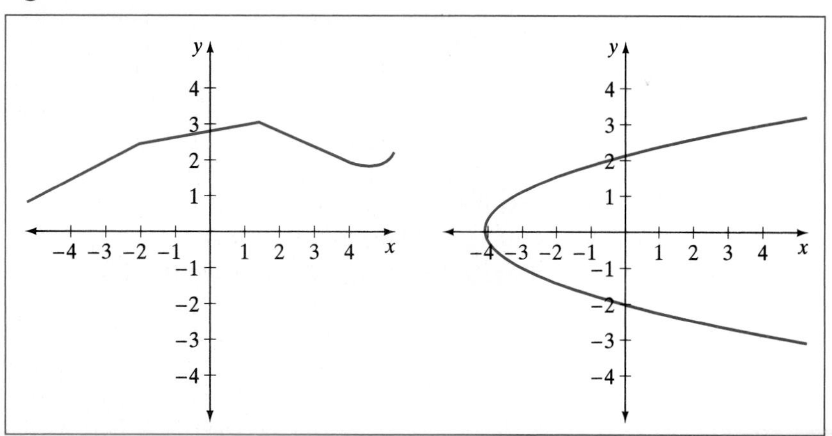

8.1.5 Composition of Functions

When we have two functions, say $f(x)$ and $g(x)$, we can find the **composition** of $f(x)$ with $g(x)$, denoted $(f \circ g)(x)$ or $f(g(x))$ by evaluating the function $f(x)$ using the entire function $g(x)$ as the independent variable.

For example, given the functions $f(x) = 2x + 5$ and $g(x) = 10x - 1$, we can find $f(g(x))$ by substituting the function $10x - 1$ into the x-value in the function $f(x)$. That is, $f(g(x)) = 2(10x - 1) + 5 = 20x - 2 + 5 = 20x + 3$. We can evaluate the new function formed from the composition of f and g. For example, with the function $f(g(x)) = 20x + 3$, we can find $f(g(4)) = 20(4) + 3 = 27$.

As a second example, suppose we have the functions $f(x) = x^2 + 3x - 9$ and $g(x) = x + 2$, we find $f(g(x)) = (x + 2)^2 + 3(x + 2) - 9 = x^2 + 4x + 4 + 3x + 6 - 9 = x^2 + 7x + 1$. Evaluating $f(g(-1) = (-1)^2 + 7(-1) + 1 = 1 - 7 + 1 = -5$. The function machine shown in Figure 8.7 shows the composition of these functions.

The composition of functions might be further clarified by picturing $f(g(x))$ as a mapping as depicted in Figure 8.8.

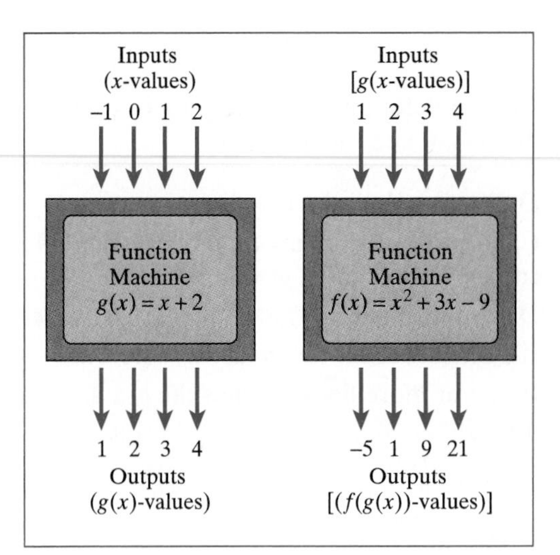

Figure 8.7

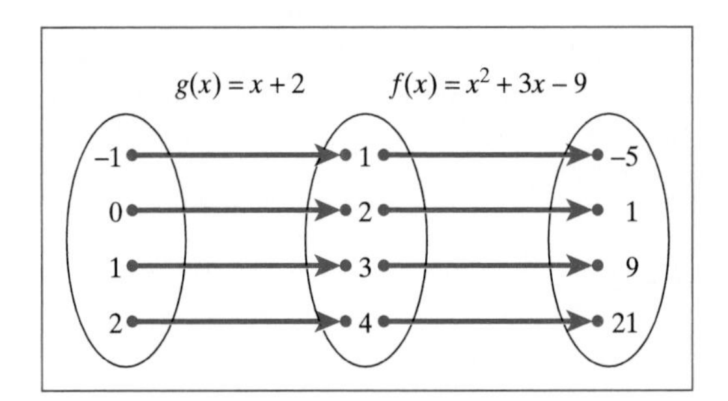

Figure 8.8

Related Topics

Patterns
Cartesian Coordinate System
Measurement and Coordinate Geometry

TOPIC 8.2 Equations and Inequalities

Multiplying Rational Numbers
Dividing Rational Numbers
Adding Rational Numbers
Subtracting Rational Numbers

Introduction

We make comparisons in almost everything we do. When those comparisons seek to find whether two things are numerically equal or if one is greater than the other, we must engage in solving an equation or an inequality. In the following sections, we examine the definitions of equations and inequalities and the properties associated with them.

Historically, equations were not classified clearly by type. Today, we have a system of describing polynomial equations by the highest degree, that is, the largest exponent, that appears in the polynomial expression. For example, when the highest exponential power in an equation is 1, the equation is called a linear equation. However, when the highest exponential power in an equation is 2, the equation is classified as a quadratic equation.

8.2.1 Definition of Equation

An **equation** is a statement of equality between two expressions. The equality between the expressions is denoted by an **equals sign,** " = ". Equations can relate any variety of expressions. Examples of equations include $4 = 2 + 2$, $7 = x + 3$, $y = 8x - 5$.

8.2.2 Properties of Equations

The beauty of equations and the key behind solving algebraic equations is that they have a number of properties that allow us to maintain the equality as we alter the expressions simultaneously.

Given any real number, a, and expressions x and y:

Addition Property of Equations. If $x = y$, then $x + a = y + a$; that is, adding the same amount to the expressions on both sides of an equation creates two new expressions that are also equal. For example, given that $4 = 2 + 2$, then $4 + 8 = 2 + 2 + 8$ (that is, $12 = 12$).

Subtraction Property of Equations. If $x = y$, then $x - a = y - a$; that is, subtracting the same amount from the expressions on both sides of an equation creates two new expressions that are also equal. For example, given that $10 = 10$, then $10 - 3 = 10 - 3$ (that is, $7 = 7$).

Multiplication Property of Equations. If $x = y$, then $a \times x = a \times y$; that is, multiplying the same amount times the expressions on both sides of an equation creates two new expressions that are also equal. For example, given that $y = 5$, then $2 \times y = 2 \times 5$ (that is, $2y = 10$).

Division Property of Equations. If $x = y$, then $x \div a = y \div a (a \neq 0)$; that is, dividing the expressions on both sides of an equation by the same amount creates two new expressions that are also equal. For example, given that $x = 6$, then $x \div 2 = 6 \div 2$ (that is, $\frac{x}{2} = 3$).

8.2.3 Definition and Notation of Inequalities

An **inequality** is a statement that one quantity is less than (or greater than) another. For real numbers a and b, if a is less than b, their relationship is denoted, $a < b$. Likewise, if a is greater than b, their relationship is denoted $a > b$. In some cases, we have the situation where a is either less than b or equal to b. In this case we say a is less than or equal to b, and denote the relationship by $a \leq b$. Similarly, when a is greater than or equal to b, we denote this with the symbols, $a \geq b$. Examples of inequality statements include: $5 < 13$, $7 > 2 + 3$, $x + 8 \leq 8$, and $3 \geq x$.

8.2.4 Properties of Inequalities

There are properties associated with inequalities that allow us to alter both sides of the inequalities simultaneously and still maintain the less-than (or greater-than) relationship. The following properties are stated with regard to the less-than relationship but also hold true for greater-than situations, less-than or equal-to situations, and greater-than or equal-to situations.

Given any real number, a, and quantities x and y:

Addition Property of Inequalities. If $x < y$, then $x + a < y + a$; that is, adding the same number to both sides of an inequality maintains the inequality. For example, given $2 < 7$, then $2 + 5 < 7 + 5$ (that is, $7 < 12$).

Subtraction Property of Inequalities. If $x < y$, then $x - a < y - a$; that is, subtracting the same number from both sides of an inequality maintains the inequality. For example, given $8 < 10$, $8 - 1 < 10 - 1$ (that is, $7 < 9$).

Multiplication Property of Inequalities. If $x < y$, then (a) $a \times x < a \times y$ when $a > 0$, and (b) $a \times x > a \times y$ when $a < 0$; that is, multiplying both sides of an inequality by the same amount affects an inequality differently, depending on whether the factor is positive or negative.

When multiplying by a positive factor, the inequality stays the same. For example, given $6 < 9$, then $10 \times 6 < 10 \times 9$, (that is, $60 < 90$). However, when multiplying by a negative factor, the inequality changes directions (that is, a less than becomes a greater than and vice versa). For example, given $2 < 5$, then $-8 \times 2 > -8 \times 5$ (that is, $-16 > -40$).

Division Property of Inequalities. If $x < y$, then $x \div a < y \div a$ when $a > 0$, and $x \div a > y \div a$ when $a < 0$; that is, dividing both sides of an inequality by the same amount affects an inequality differently, depending on whether the divisor is positive or negative. (Note that division by 0 is undefined.)

When dividing by a positive factor, the inequality stays the same. For example, given $2 < 9$, then $2 \div 5 < 9 \div 5$ (that is, $\frac{2}{5} < \frac{9}{5}$). However, when dividing by a negative divisor, the inequality changes directions (that is, a less than becomes a greater than and vice versa). For example, given $6 < 20$, $6 \div (-2) > 20 \div (-2)$ (that is, $-3 > -10$).

Properties of Rational-Number Addition
Properties of Rational-Number Multiplication

TOPIC 8.3 Cartesian Coordinate System

Introduction

We often have to give directions to a particular location to others. Those directions often involve pinpointing a position on a map according to certain distances from one point to another. Understanding the location of points on a map is similar to understanding the Cartesian Coordinate System. In this section, we will examine key concepts related to the Cartesian Coordinate System and how to identify the location of points within the system.

René Descartes (1596–1650) was a French mathematician who is credited with being the first to combine the study of algebra and geometry, creating coordinate geometry. In fact, the Cartesian Coordinate System is named in his honor. It is said that Descartes slept a lot and that his insight into the field of coordinate geometry came to him in a dream.

For More Information Reimer, L., & Reimer, W. (1995). *Mathematicians Are People Too: Stories from the Lives of Great Mathematicians, Volume 2*. Palo Alto, CA: Dale Seymour Publications.

8.3.1 Definition of the Cartesian Coordinate System and Related Terms

The **Cartesian Coordinate System** is the set of all points in a plane that are identified by a pair of x- and y- coordinates. **Coordinates,** denoted together as the **ordered pair** (x, y), are numbers that designate the position of a point in the plane where the first number indicates the position along the x-axis and the second number indicates the position along the y-axis. The coordinate plane is determined by two intersecting, perpendicular lines called **the x-axis and the y-axis.** Note that the x- and y-axes each contain an infinite set of points equivalent to the real-number line. The point where these axes intersect is called the **origin** and is denoted $(0, 0)$, indicating that the location of the point along the x-axes is at zero and the location of the point along the y-axes is zero (see Figure 8.9).

The coordinate plane is broken into four sections, called **quadrants,** by the x- and y-axes. (See Figure 8.9). The first quadrant, denoted Q1, is the section where both the x- and y-coordinates are positive. Note that the point $(3, 1)$ in Figure 8.9 lies in the first quadrant. The second quadrant, Q2, is the section where the x-coordinate is negative and the y-coordinate is positive. The third quadrant, Q3, is the section where both the x- and the y-coordinates are negative. The point $(-1, -3)$ as seen in Figure 8.9 is in Q3. Finally, the fourth quadrant, Q4, is the section where the x-coordinate is positive and the y-coordinate is negative.

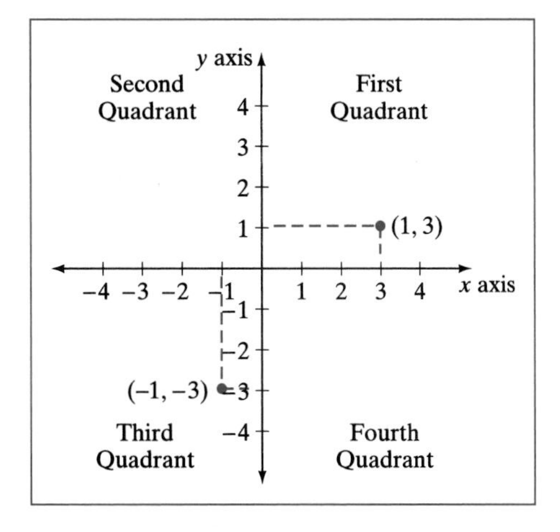

Figure 8.9

8.3.2 Equations of Horizontal and Vertical Lines

In the coordinate plane, we can represent any collection of points. It is often valuable to indicate a set of points that all lie on the same line. When those lines are vertical or horizontal lines, they can be identified by x- and y-coordinates, respectively. Examine the vertical line, m, in Figure 8.10. Notice that for each point that can be identified along line m, the x-coordinate is always 2. So, for any choice of y-coordinate, the x-coordinate is fixed at 2. Therefore, we can refer to vertical line m as the line $x = 2$. In general, a vertical line where the x-coordinate is always the same number, a, can be referred to as the line $x = a$.

Now, in examining the horizontal line, n, in Figure 8.10, we make a different observation. In this case, the points that are identified as being on line n all have the same y-coordinate of 3. Thus, given any x-coordinate, the y-coordinate is always the same number, 3. Therefore, we can refer to line n as the line $y = 3$. In general, horizontal lines where the y-coordinate is always the same number, b, can be referred to as $y = b$.

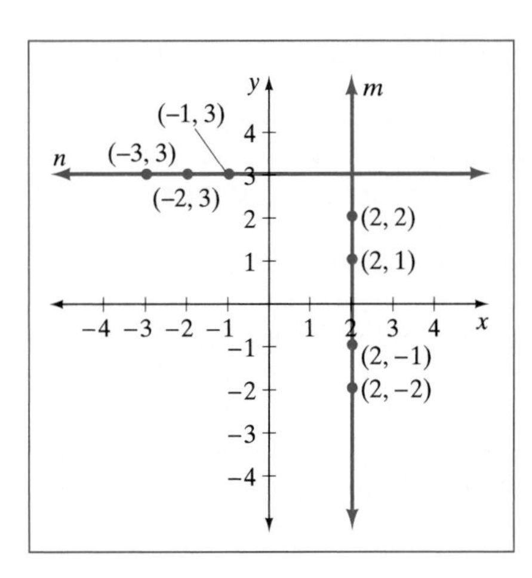

Figure 8.10

Related Topics Graphs and Equations of Lines

TOPIC 8.4 Graphs and Equations of Lines

Introduction

Many situations in life involve relationships that are considered linear. Some might say that there is a linear relationship between the amount of time spent studying and the grade you earn on an exam. In this section, we examine lines and related concepts such as slopes and points on a line, particularly points that intersect the x-axis and the y-axis, which are called intercepts.

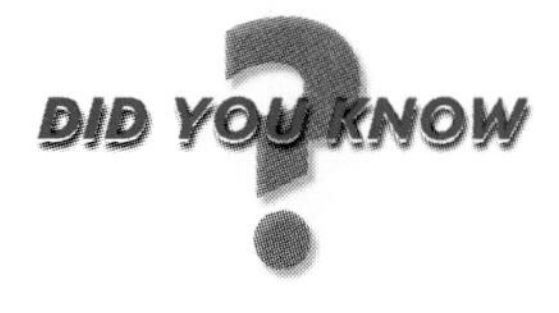

Maria Agnesi (1718–1799), the daughter of an Italian mathematics professor, was said to be a child prodigy. She is one of the most celebrated Italian woman of the scientific revolution. Her most noted work, the book entitled *Foundations of Analysis*, includes a systematic presentation of algebra, analytic geometry, calculus, and differential equations. She made great strides in work related to the graphing of equations, particularly the "bell-shaped curve," often used to describe a normal distribution of data graphically.

For More Information Reimer, L., & Reimer, W. (1995). *Mathematicians Are People Too: Stories from the Lives of Great Mathematicians, Volume 2.* Palo Alto, CA: Dale Seymour Publications.

8.4.1 Slopes of Lines

Slope refers to the inclination of a line. A line that rises vertically as we trace along the line from left to right is said to have a **positive slope.** For example, in Figure 8.11, lines a, b, and c have positive slopes. The slope of line b, which runs exactly between the x-axis and the y-axis, is a slope of 1. The slope of line c, which is less steep than line b, will be a positive fraction that is less than 1—approximately $\frac{1}{2}$. The slope of line a, which is steeper than line b, is a positive number that is greater than 1—approximately a slope of 3 as pictured in Figure 8.11.

A line that lowers vertically as we trace along the line from left to right is said to have a **negative slope.** For example, lines d, e, and f in Figure 8.12 have negative slopes. Line e has a slope of approximately -1. Line d is steeper than line e and has a slope of approximately -2. Finally, line f is less steep than line e and has a slope of approximately $-\frac{1}{2}$.

We should note that all lines are characterized by slope, even horizontal and vertical lines. Horizontal lines have a slope of zero; however, the slope of vertical lines are undefined. Vertical lines are said to have no slope.

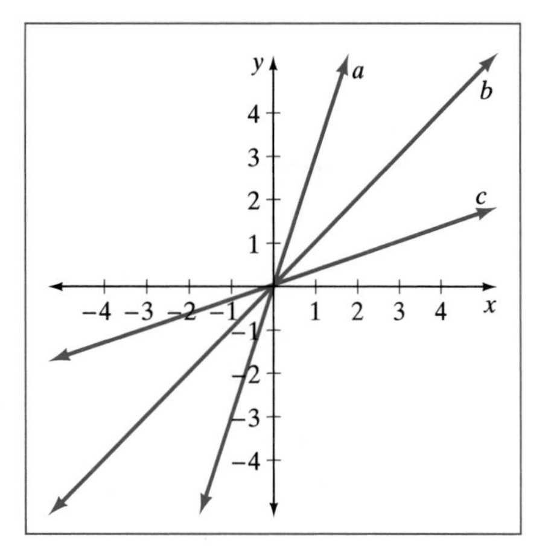

Figure 8.11

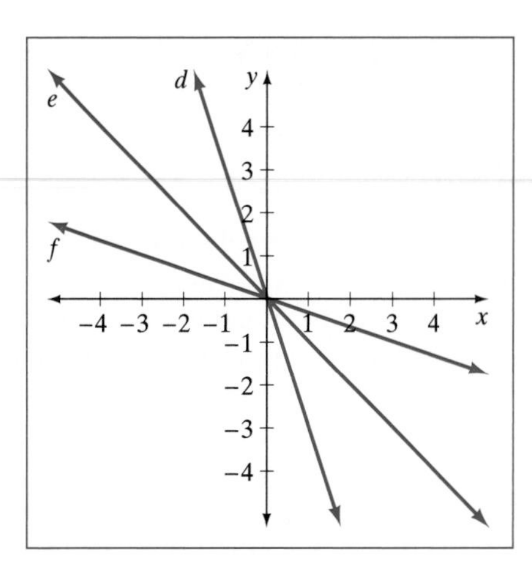

Figure 8.12

8.4.2 Computing the Slope of a Line

To determine the slope of a line, you need to know two points on the line. Given two points on a line, (x_1, y_1) and (x_2, y_2), we can determine the slope, denoted m, of the line between the points by computing $m = \frac{y_2 - y_1}{x_2 - x_1}$. For example, given points (4, 3) and (−1, 6), we compute the slope of the line between the points $m = \frac{6 - 3}{-1 - 4} = \frac{3}{-5} = -\frac{3}{5}$. Plotting the points and drawing the line that connects them, we verify that the slope of the line is, in fact, negative and that the line appears to have a slope of $-\frac{3}{5}$ (see Figure 8.13).

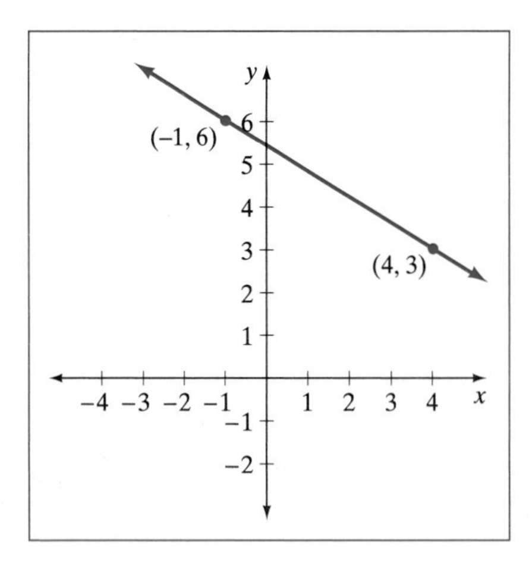

Figure 8.13

8.4.3 Lines of the Form $y = mx$

Lines in the plane that pass through the origin can be written in equation form by $y = mx$. The slope of the line through the origin is designated by m. For example, the line $y = 3x$ is a line that passes through the origin and has a slope of 3. We can approximate what the line looks like, or we can use the slope, 3, and the fact that the line goes through the point (0, 0), and we can determine a second point on the line.

Slope is determined by computing the change in y-values divided by the change in x-values, that is, $y_2 - y_1$ divided by $x_2 - x_1$. Many describe slope as "rise over run" because the numerator describes how far (up or down) we rise from one point and how far we run (right or left) to locate another point on a line.

For example, we can view the slope of 3 as $\frac{3}{1}$ where the numerator, 3, denotes the change in y-values and the denominator, 1, denotes the change in x-values. Thus, starting at the origin (0, 0), we can determine another point on the line by moving

3 units upward from (0, 0) for the change in y direction and moving 1 unit to the right for the change in the x direction (see Figure 8.14).

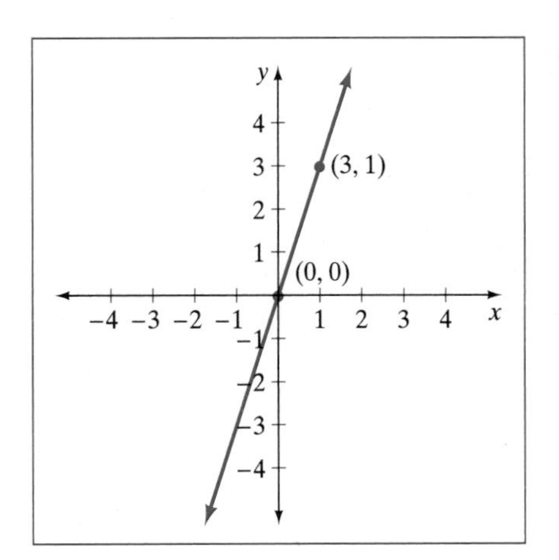

Figure 8.14

Moving up 3 units and to the right 1 unit, we identify another point on the line $y = 3x$ as the point (1,3). Connecting the two points with a line, we observe that the slope appears to be a slope of 3 as planned.

In a second example, given the line $y = \frac{-1}{2}x$, we can determine a point other than the origin on the line. Starting at the origin, we move down 1 unit (down because it is a negative number) and to the right 2 units (to the right because the 2 is positive). Thus, we identify that the point $(2, -1)$ is a point on the line $y = \frac{-1}{2}x$, and we can now graph the line by joining the two points $(0, 0)$ and $(2, -1)$. See Figure 8.15.

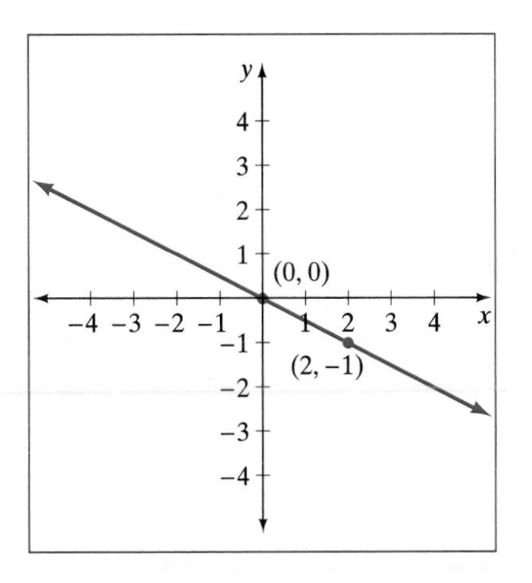

Figure 8.15

8.4.4 X- and Y-Intercepts

An intercept is a point that intersects one of the axes. An ***x*-intercept** is a point where a line intersects the x-axis. Because the x-intercept is a point on the x-axis, the corresponding y-value for any x-intercept is $y = 0$. Thus, given an equation of a line, for example, $y = 5x + 3$, we can determine any x-intercepts by setting $y = 0$ and solving for x:

$$y = 5x + 3$$

Setting $y = 0$, we find:

$$0 = 5x + 3$$
$$-3 = 5x$$
$$-\frac{3}{5} = x.$$

Thus, the x-intercept for the line $y = 5x + 3$ is the point $(-\frac{3}{5}, 0)$.

A **y-intercept** is a point that intersects the y-axis. For any point that intersects the y-axis, the x-value is zero. Thus, to find a y-intercept, we can set $x = 0$ and solve for y. For example, with the equation $y = 5x + 3$, we can determine any y-intercepts by setting $x = 0$:

$$y = 5x + 3.$$

Setting $x = 0$ we find:

$$y = 5 \times 0 + 3$$
$$y = 3.$$

Thus, the y-intercept for the line $y = 5x + 3$ is the point $(0, 3)$. Plotting the x- and y-intercepts for the line $y = 5x + 3$, we can then graph the line by connecting the intercepts. See Figure 8.16.

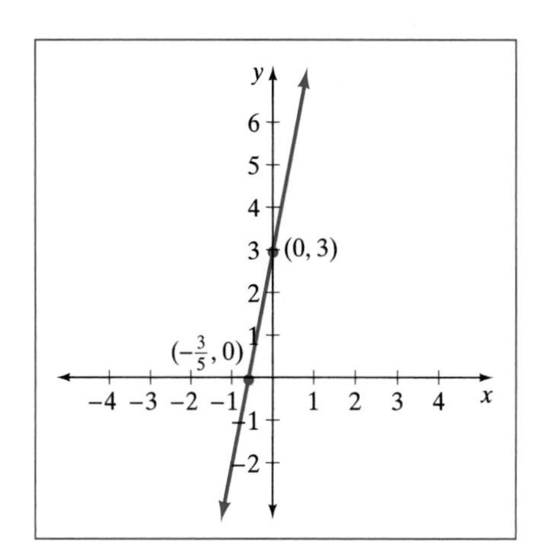

Figure 8.16

8.4.5 Slope-Intercept Form of a Line $y = mx + b$

We call the equation of a line in the form $y = mx+b$ the **slope-intercept form.** In this form, m is the slope, and b is the y-intercept. Thus, in the example $y = 2x + 6$, the slope of the line is 2 and the y-intercept is the point $(0, 6)$. In the example $y = 4x - 5$, the slope is 4 and the y-intercept is the point $(0, -5)$.

8.4.6 Point-Slope Formula for the Equation of a Line

Given the slope, m, of a line and a point (x_1, y_1) on the line, we can determine the equation of the line using the point-slope formula. The point-slope formula is: $y - y_1 = m(x - x_1)$. For example, given the point $(x_1, y_1) = (2, 5)$ on a line with slope $m = -2$, we can determine the equation of the line by $y - 5 = -2(x - 2)$. Distributing the -2 across the parenthetical expression and adding 5 to both sides of the equation, we simplify the equation to slope-intercept form:

$$y - 5 = -2(x - 2)$$
$$y - 5 = -2x + 4$$
$$y = -2x + 9.$$

Related Topics

Systems of Linear Equations
Solving Systems of Linear Equations

TOPIC 8.5 Systems of Linear Equations

cornerstone **The Cartesian Coordinate System**
Graphs and Equations of Lines

Introduction

When we try to determine a relationship between two variables, such as a relationship between smoking and lung cancer, we often have to work within certain constraints. The relationship, as well as these constraints, can be expressed as equations. To understand thoroughly the relationship between the variables at hand, we have to consider all equations simultaneously. In this section, we describe a process of considering linear equations simultaneously and a method of solving those equations.

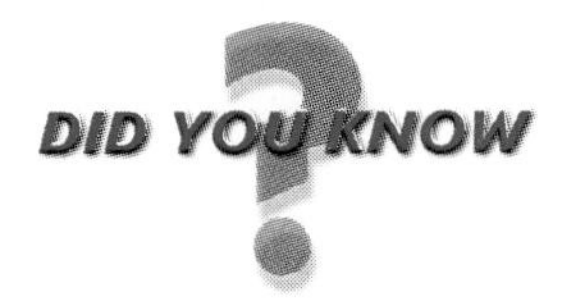

Professor George Dantzig from Stanford University is known for creating a technique for solving linear programming problems called the simplex method. The need for such a method for solving simultaneous linear equations was motivated by the problem of moving personnel and materials more efficiently during World War II. Dantzig's work provided the solution to this very immediate problem and aided the efforts of the Allies in the war.

For More Information Smith, D. E. (1958). *History of Mathematics, Volume II.* New York: Dover Publications.

8.5.1 Simultaneous Equations

A **system of equations** is a collection of equations for which there is some relationship or for which together they set conditions on a mathematical situation. Two or more equations can have different sets of solutions. However, *sometimes we wish to consider the equations simultaneously* and determine what solutions solve all of the equations at the same time. For example, consider the equations $x + y = 2$ and $3x + y = 5$. Considered individually, the equations have many solutions. For example, the equation $x + y = 2$ has such solutions as $x = 3$ and $y = -1$, $x = 0$ and $y = 2$, $x = -4$ and $y = 6$, and so on. The equation $3x + y = 5$ has such solutions as $x = 0$ and $y = 5$, $x = 2$ and $y = -1$, $x = -3$ and $y = 4$, and so on. However, if we consider the two equations simultaneously and try to determine values for x and y that solve both equations at the same time, we would find that only the solution $x = 1$ and $y = 1$ solves the equations simultaneously.

8.5.2 Geometrical Relationships of the Graphs of Two Lines

When we examine the relationship between two linear equations, we find that one of three relationships exists: (1) The lines intersect at some singular point, (2) the lines are parallel to each other and never intersect, or (3) the lines are coincidental; that is, they lie one on top of the other. Let us explore each of these relationships and their geometrical representations (see Figure 8.17).

In any situation where we have two linear equations with different slopes, the two lines will intersect at one singular point. For example, the lines $y = 2x + 6$ and $y = 5x - 2$ have slopes 2 and 5, respectively. Therefore, the two lines will, at some point, intersect. See Figure 8.18.

When two lines have the same slope, they will be parallel or coincidental. If two lines are coincidental, they are essentially the same line but might be written in alternative forms. For example, the lines $y = x - 5$ and $2y = 2x - 10$ are essentially the same lines. If we divide both sides of the linear equation $2y = 2x - 10$, we have

Figure 8.17

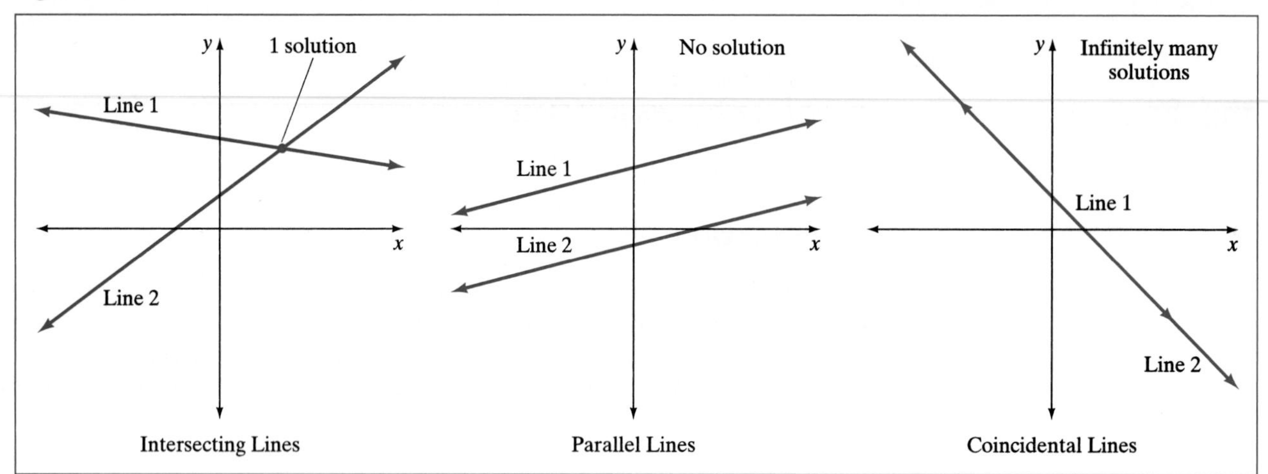

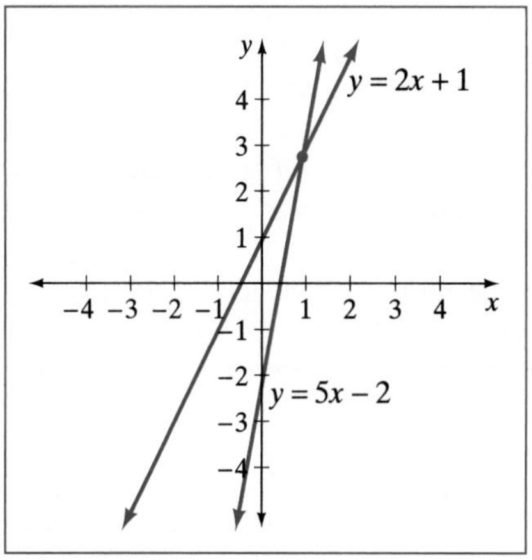

Figure 8.18

the equation $y = x - 5$. Therefore, the graphs of the two lines basically have the same slope and the same y-intercept. Thus, the lines lie one on top of each other (see Figure 8.19).

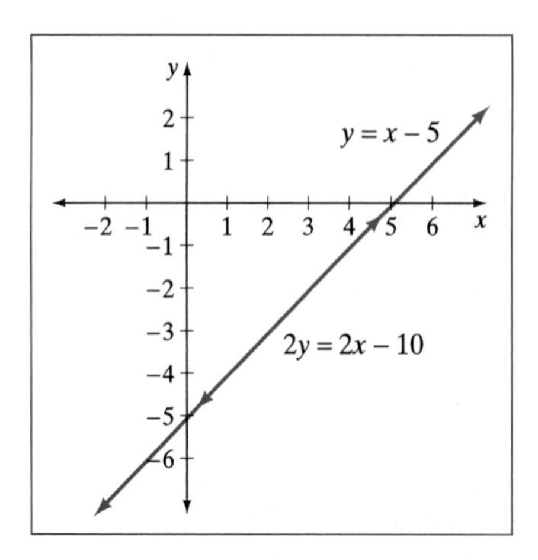

Figure 8.19

If two lines are parallel, they have the same slope but different y-intercepts. For example, the lines $y = 3x + 1$, $y = 3x - 2$, $y = 3x$, and $y = 3x + 3$ are parallel because they all have a slope of 3 and different y-intercepts. See Figure 8.20 for a graphical representation of these lines.

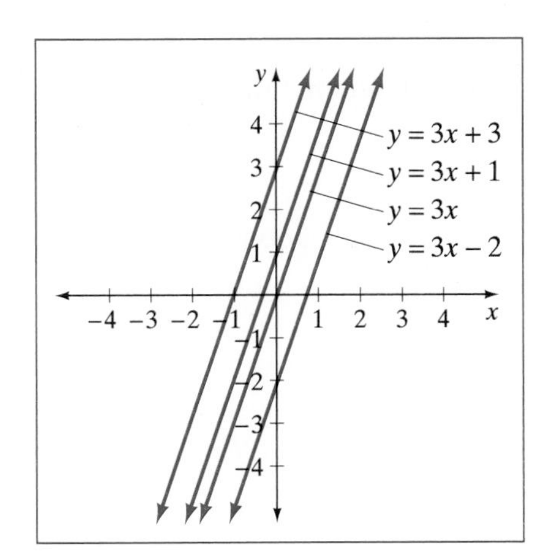

Figure 8.20

When solving simultaneous linear equations, we will find that those lines that intersect at a single point will have a single simultaneous solution. On the other hand, two lines that are parallel will represent linear equations that have no simultaneous solutions. Finally, when two lines are coincidental, they represent linear equations that have an infinite number of simultaneous solutions.

8.5.3 Solving Simultaneous Linear Equations

Given two linear equations, we can find simultaneous solutions by finding the point (or points) where they intersect. Consider, for example, the equations $y = 4x + 2$ and $y = -x + 3$. Graphically, we can see that the two lines intersect at one point (see Figure 8.21).

To determine the precise point of intersection, we use an algebraic process. Because we want to find when the equations take on the same value, let us set the two equations equal to each other and solve for x. So, given $y = 4x + 2$ and $y = -x + 3$, we form the equation:

$$4x + 2 = -x + 3.$$

Now, solving for x:

$$\begin{aligned} 4x + 2 &= -x + 3 \\ 5x &= 1 \\ x &= \frac{1}{5}. \end{aligned}$$

Then, because the two equations are equal at the point where $x = \frac{1}{5}$, we can find the corresponding y-value at the point by evaluating either of the equations at the x-value

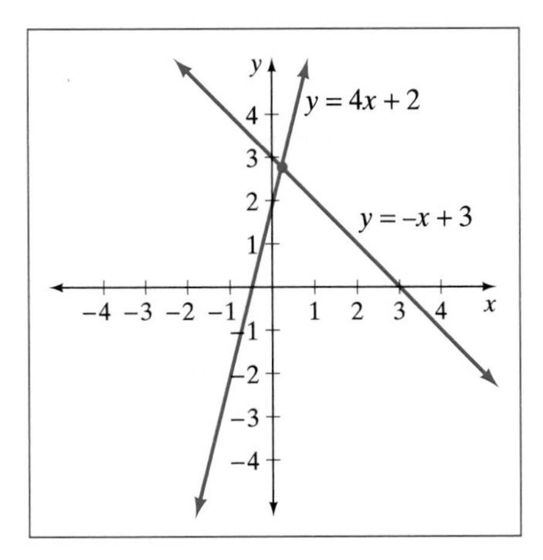

Figure 8.21

$x = \frac{1}{5}$. Solving, we find that $y = \frac{14}{5}$. Referring back to the graph of the two equations in Figure 8.21, we can see that the point of intersection appears to have the coordinates $(\frac{1}{5}, \frac{14}{5})$.

Related Topics

Solving Systems of Linear Equations Symbolically

TOPIC 8.6 Solving Systems of Linear Equations Symbolically

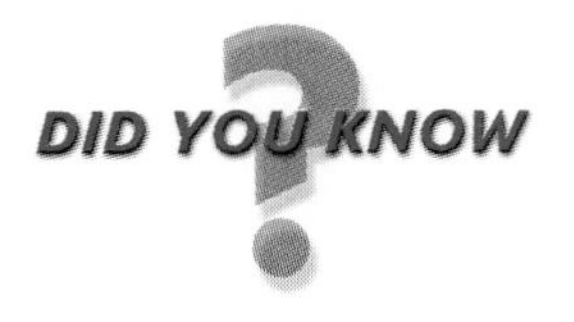

cornerstone

Cartesian Coordinate System
Graphs of Linear Equations
Systems of Linear Equations

Introduction

Solving systems of linear equations can be done in a variety of ways. In addition to solving systems of linear equations by looking for points of intersection, we can also use methods of substitution and elimination. In this section, we examine these methods and also develop a process for determining the equation of the line between two points.

DID YOU KNOW?

Narendra Karmarkar, a linear programmer with AT&T Bell Laboratories in 1984, invented a method for solving linear programming problems quickly. In fact, his method is so fast that it can be used to provide immediate answers to problems that arise requiring an efficient and accurate response.

For More Information Smith, D. E. (1958). *History of Mathematics, Volume II*. New York: Dover Publications.

8.6.1 Solving Systems of Equations by Substitution

Given a set of linear equations, we can sometimes solve them through basic **substitution.** For example, given the equations: $y = x + 3$ and $y - 5 = 4x - 4$, we can substitute the expression $x + 3$ for y in the equation $y - 5 = 4x - 4$, getting: $(x + 3) - 5 = 4x - 4$. Solving for x algebraically we find: $x = \frac{2}{3}$.

8.6.2 Solving by Systems of Equations by Elimination

In some cases, solving systems of equations involves **eliminating** one of the variables so that we can solve for the other. For example, given the equations: $y = -3x + 7$ and $2y = 2x + 1$, we can set the equations up vertically, lining up like variables and constants:

$$\begin{aligned} y &= -3x + 7 \\ 2y &= 2x + 1 \end{aligned}$$

First, to create a situation where we have the same number of x or y in each equation, we can multiply both sides of one equation by the appropriate number to create an equivalent equation. In this case, if we multiply both sides of the top equation by a factor of 2, we have the equations:

$$\begin{aligned} 2y &= -6x + 14 \\ 2y &= 2x + 1 \end{aligned}$$

Next, we can view the top equation as the minuend in a subtraction problem and the bottom number as the subtrahend. Subtracting like terms from like terms, we find the difference between the two equations:

$$\begin{aligned} 2y &= -6x + 14 \\ -\ 2y &= 2x + 1 \\ \hline 0 &= -8x + 13 \end{aligned}$$

Finally, because we have successfully eliminated the variable y from the equations, we can solve for x. We find that $x = \frac{13}{8}$. With this value of x, we can determine the corresponding value of y by evaluating one of the original equations at $x = \frac{13}{8}$, finding that $x = \frac{17}{8}$.

8.6.3 Determining the Equation of a Line Through Two Points

Given a point and the slope of a line, we can use the point-slope formula to determine the equation of a line. However, if all we know are two points on a line, we can still determine the equation of the line through those points. Using the two points, we can find the slope between the two points using the slope formula. Then, using this slope and one of the points, we can use the point-slope formula to determine the equation of the line.

For example, let us determine the equation of the line that passes through the points (2, 3) and (4, −3). First, we find the slope, $m = \frac{y_2 - y_1}{x_2 - x_1} = \frac{-3 - 3}{4 - 2} = \frac{-6}{2} = -3$. Then, choosing one of the points, say (2, 3), we use the point-slope formula to find the equation of the line:

$$\begin{aligned} y - y_1 &= m(x - x_1) \\ y - 3 &= -3(x - 2) \\ y &= -3x + 6 + 3 \\ y &= -3x + 9 \end{aligned}$$

TOPIC 8.7 Exponents

cornerstone **Base Ten**
Order of Operations

Introduction

Exponents provide a means of expressing many large amounts concisely. In this section, we define exponents and related terms and introduce exponential notation. We also highlight well-known exponential numbers, including square and cubic numbers.

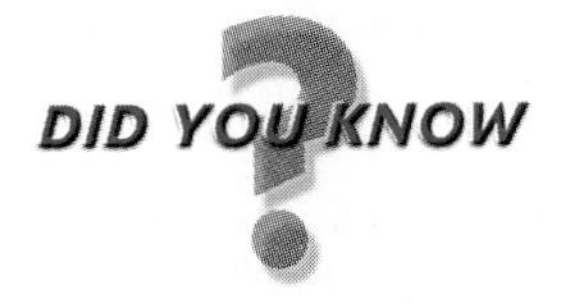

DID YOU KNOW

The concept of an exponent was known in theory well before practical uses for the exponent were developed. However, it was mathematician, John Wallis (c. 1650), who showed that x^0 should represent 1. Wallis was also the person who established notation for fractional and negative exponents.

For More Information Smith, D. E. (1958). *History of Mathematics, Volume II*. New York: Dover Publications.

8.7.1 Definitions of Terms Associated with Exponents

An **exponent** is a symbol placed above and after a number or expression, referred to as the **base,** that raises the base to the **power** designated by the symbol. In the example x^n, the expression x is the base, and the exponent is n.

The operation that an exponential symbol represents is the multiplication of the designated base times itself the number of times indicated by the exponent. Thus, in the exponential expression 10^3, we interpret the symbols to mean $10^3 = 10 \times 10 \times 10 = 1,000$. Because 1,000 can be expressed as 10^3, we can refer to 1,000 as *a power of 10*, indicating that the value 1,000 can be generated by multiplying 10 times itself a given number of times.

People often confuse the phrases *a power of* and *a multiple of*. To clarify, consider the expression *a power of 10*, for example. This refers to numbers generated by raising 10 to an exponential power such as $10^1 = 10$, $10^2 = 100$, $10^3 = 1{,}000$, $10^4 = 10{,}000$, $10^5 = 100{,}000$, and so on. On the other hand, *multiples of 10* refer to the set of numbers generated by multiplying 10 times an integer value, such as $1 \times 10 = 10$, $2 \times 10 = 20$, $3 \times 10 = 30$, $4 \times 10 = 40$, $5 \times 10 = 50$, and so on. Note that a number that is a power of 10, or of any base, is also a multiple of 10, or that base, but the reverse does not hold true.

8.7.2 Squares and Cubes

We refer to the exponential expression of a base raised to the power of 2 as the **square** of that number. Similarly, we refer to the exponential expression of a base raised to the power of 3 as the **cube** of that number. For example, 4^2 is read, "four squared," and the number 16 is referred to as "the square of 4." Likewise, 5^3 is read as "five cubed," and the number 125 is referred to as "the cube of 5."

For powers other than 2 or 3, we generally refer to the exponential expression by the ordinal number represented by the exponent; that is, 2^4 is read, "two to the fourth power," and the number 16 is referred to as "the fourth power of two."

8.7.3 Zero as an Exponent

When an exponent is zero, we say the value of that exponential expression is 1; that is, given any base, x, we say $x^0 = 1$. Thus, $10^0 = 1$, $2^0 = 1$, $(a + b)^0 = 1$, and so on.

8.7.4 Negative Exponents

Exponents can be negative. Given any base, x, and positive number, n, the value of $x^{-n} = \frac{1}{x^n}$. That is, the value of an exponential expression with a negative exponent is equal to the reciprocal of the exponential expression raised, instead, to the positive value (or absolute value) of the negative exponent. For example, $10^{-2} = \frac{1}{10^2} = \frac{1}{100}$, $3^{-2} = \frac{1}{3^2} = \frac{1}{9}$, $2^{-3} = \frac{1}{2^3} = \frac{1}{8}$, and so on.

8.7.5 Properties of Exponents

There are several properties of exponents that allow us to combine and simplify exponential expressions. Given rational numbers a, b, and c:

1. The product of two exponential expressions with the same base is equal to that base raised to the sum of the exponents in each factor; that is, $a^b \times a^c = a^{b+c}$. For example, $10^2 \times 10^3 = 10^{2+3} = 10^5$. This makes sense because in writing 10^2 as 10×10 and 10^3 as $10 \times 10 \times 10$, we verify that $10^2 \times 10^3 = 10 \times 10 \times 10 \times 10 \times 10 = 10^5$.
2. The quotient of two exponential expressions with the same base is equal to that based raised to the difference of the exponent in the dividend minus the exponent in the divisor; that is, $a^b \div a^c = a^{b-c}$ or, written in fraction form: $\frac{a^b}{a^c} = a^{b-c}$. For example, $10^5 \div 10^3 = 10^{5-3} = 10^2$. Writing the division problem in fraction form, we can verify that the property makes sense:

$$\frac{10^5}{10^3} = \frac{10 \times 10 \times 10 \times 10 \times 10}{10 \times 10 \times 10} = \frac{10 \times 10}{1} = 10^2.$$

3. The value found when an entire exponential expression with base a and exponential power b is further raised to another power, c, is equal to the base, a, raised to the product of the exponents $b \times c$; that is $(a^b)^c = a^{b \times c}$.

   For example, $(2^2)^3 = 2^{2\times 3} = 2^6 = 64$. Verifying, we notice that $(2^2)^3$ means, $(2^2) \times (2^2) \times (2^2) = (2 \times 2) \times (2 \times 2) \times (2 \times 2) = 2^6$.

Related Topics **Scientific Notation**

TOPIC 8.8 Scientific Notation

cornerstone
Decimals
Exponents

Introduction

Scientific notation is a precise means of representing decimal numbers. In this section, we discuss the definition of scientific notation. We also explore a means of interpreting decimal numbers written in scientific notation, as well as a means of converting decimal expression to scientific notation.

If a calculation on a calculator produces a very large number with many digits, the answer in the display window of the calculator is often given in scientific notation. The presentation may take different forms, but, generally, large amounts are presented as: 7.132507 representing the scientific notation: 7.1325×10^7. Notice that the calculator leaves a space between the factor, 7.1325, and the power of 10, 07. Some calculators use "E" instead of a space, as in 7.1325 E 07.

8.8.1 Definition of Scientific Notation

Numbers can be written in what is called scientific notation. **Scientific notation** is the expression of a number as the product of (1) a number, x, such that $1 \le |x| < 10$, and (2) a power of 10 that would position the decimal point to represent the number at hand correctly. The digits we write to express the factor, x, in the scientific notation are called **significant digits.**

Examples of numbers written in scientific notation are the following:

Example 1: 5.782×10^4

Example 2: 1.24409×10^{-3}

Example 3: -4.77×10^2

Example 1 above is a number written in scientific notation to 4 significant digits because there are 4 digits in the factor being multiplied by a power of 10. Likewise, in Example 2, the number is written to 6 significant digits, and in Example 3, the number is written to 3 significant digits.

8.8.2 Writing Numbers Expressed in Scientific Notation as Decimals

Each number written in scientific notation can be expressed more simply as a decimal. To do so, we use the fact that when we multiply a number by 10, the resulting product is the same digits in the original number with the decimal point moved one position to the right. Similarly, multiplying a number by 10^2 yields a product involving the same digits but with the decimal point moved two positions to the right.

In general, when multiplying a number by a positive power of 10—that is, by 10^n, where $n > 0$—the resulting product is the same set of digits in the original number with the decimal point moved n positions to the right.

Example 1: The product of $4.98 \times 10^2 = 498$ because the decimal point in 4.98 was moved two positions to the right.

Example 2: The product of $1.398 \times 10^6 = 1,398,000$ because the decimal point in 1.398 is moved six positions to the right. Notice that zeros needed to be added to indicate that the decimal point was moving beyond the significant digits in 1.398.

When we multiply a number by 10^{-1}, it is akin to multiplying by $\frac{1}{10}$, which is the same as dividing the number by 10. Therefore, the result of multiplying a number by 10^{-1} is moving the decimal point in the original number one position to the left. Similarly, multiplying a number by 10^{-2} results in a product that has the digits in the original number with the decimal point moved two positions to the left.

Thus, in general, when multiplying a number by a negative power of 10—that is, by 10^{-n}, where $n > 0$—the resulting product is the same set of digits in the original number with the decimal point moved n positions to the left.

Example 1: The product of $7.8855 \times 10^{-2} = 0.078855$ because the decimal point was moved two positions to the left. Notice that a zero needed to be added to indicate that the decimal point had moved beyond the significant digits in 7.8855.

Example 2: The product of $4.12 \times 10^{-8} = 0.0000000412$ because the decimal point was moved eight positions to the left.

8.8.3 Writing Decimal Numbers in Scientific Notation

Given a decimal number, we can write that decimal number in scientific notation. To do so, we need to develop two factors: one that is a number greater than or equal to 1 and less than 10, and one that is the appropriate power of 10. The first factor is created by repositioning the decimal point in the given decimal number so that there is one nonzero digit to the left of the decimal point. The second factor is a power of 10 so that the exponent indicates the number of positions, to the left if negative and to the right if positive, that the decimal point in the first factor needs to be moved to produce the original decimal number.

Example 1: To write the number 745,300 in scientific notation, we create the first factor by moving the decimal point in 745,300 so that only the digit 7 is to the left of the decimal point. Thus, the first factor is 7.453. Then the second factor is determined by finding how many positions the decimal point in 7.453 needs to be moved (and in which direction) to get back to the decimal number 745,300. Clearly, the decimal point needs to be 5 positions to the right, so the second factor must be 10^5. Therefore, $745{,}300 = 7.453 \times 10^5$ in scientific notation.

Other examples of decimal numbers written in scientific notation are:

Example 2: $0.00234 = 2.34 \times 10^{-3}$.

Example 3: $79512.14 = 7.951214 \times 10^4$.

Example 4: $-0.00000589 = -5.89 \times 10^{-6}$.

Related Topics Irrational Numbers

In the Classroom

This chapter focuses on the *Algebra Content Standard* with emphasis on the *Representation Process Standard*: "By viewing algebra as a strand in the curriculum from prekindergarten on, teachers can help students build a solid foundation of understanding and experience as a preparation for more-sophisticated work in algebra in the middle grades and high school" (NCTM, 2000, p. 37); all students should be able to "select, apply, and translate among mathematical representations to solve problems" (NCTM, 2000, p. 67). Given that algebra has often been associated with symbolic manipulation, we believe you should develop a solid conceptual understanding of (a) variables and their multiple representations (Activities 8.1–8.4), and (b) patterns, functions, and equations (Activities 8.5–8.14).

Some Expectations of Your Future Students:

Pre-K–2 (NCTM, 2000, p. 90)

- recognize, describe, and extend patterns such as sequences of sounds and shapes or simple numeric patterns, and translate from one representation to another
- analyze how both repeating and growing patterns are generated
- describe quantitative change, such as a student growing 2 inches in one year

Grades 3–5 (NCTM, 2000, p. 148)

- represent the idea of a variable as an unknown quantity, using a letter or a symbol
- model problem situations with objects, and use representations such as graphs, tables, and equations to draw conclusions
- investigate how a change in one variable relates to a change in a second variable
- identify and describe situations with constant or varying rates of change, and compare them

How Activities 8.1–8.14 Help You Develop an Adult-level Perspective on the Above Expectations:

- By exploring the idea of variable in familiar contexts, you develop an understanding of their representation in symbolic, tabular, and graphical forms and of the relationship between these representations.
- By investigating and describing numerical patterns, you learn to represent them in tabular, graphical, and symbolic form and begin to understand the limitations of each representation or of how they are to be interpreted in a given context.
- By analyzing the relationships that underlie familiar situations that are described verbally, modeled geometrically, or demonstrated physically, you investigate how a change in one variable relates to a change in a second and discuss the concept of a function.
- By working with different functions, you discuss the concept of and develop multiple ways of understanding the rate of change of a function. These include functions that are represented graphically and those that you have to model kinesthetically, as well as those represented in tabular form.

Bibliography

Ajose, S. (1991). Activities: Patterns in the hundred chart. *Mathematics Teacher, 84*, 43–48, 118–124.

Bidwell, J. K. (1987). Using reflections to find symmetric and asymmetric patterns. *Arithmetic Teacher*, 34(7), 10–15.

Coburn, T. G. (1993). *Patterns: Addenda series, grades K–4*. Reston, VA: National Council of Teachers of Mathematics.

Coxford, A. F. (Ed.). (1988). *The ideas of algebra*. Reston, VA: National Council of Teachers of Mathematics.

Day, R. P. (1995). Using functions to make mathematical connections. In P. A. House (Ed.), *Connecting mathematics across the curriculum* (pp. 54–64). Reston, VA: National Council of Teachers of Mathematics.

Driscoll, M. J. (1999). *Fostering algebraic thinking: A guide for teachers, grades 6–10*. Portsmouth, NH: Heinemann.

Edwards, E. L. (Ed.). (1990). *Algebra for everyone*. Reston, VA: National Council of Teachers of Mathematics.

Emie, K. T. (1995). Mathematics and quilting. In P. A. House (Ed), *Connecting mathematics across the curriculum* (pp. 170–176) Reston, VA: National Council of Teachers of Mathematics.

English, L. D., & Warren, E. A. (1998). Introducing the variable through pattern exploration. *Mathematics Teacher, 91*, 166–172.

Falkner, K. P., Levi, L., & Carpenter, T. P. (1999). Children's understanding of equality: A foundation for algebra. *Teaching Children Mathematics, 6*, 232–236.

Fey, J. T., & Heid, M. K. (1995). *Concepts in algebra: A technological approach*. Dedham, MA: Janson.

Geer, C. P. (1992). Exploring patterns, relations, and functions. *Arithmetic Teacher, 39*(9), 19–21.

Giambrone, T. M. (1983). Challenges for enriching the curriculum: Algebra. *Mathematics Teacher, 76*, 262–263.

Greenes, C., & Findell, C. (1999). Developing students' algebraic reasoning abilities. In L. V. Stiff (Ed.), *Developing mathematical reasoning in grades K–12* (pp. 127–137). Reston, VA: National Council of Teachers of Mathematics.

Greenes, C., & Findell, C. (1999). *Groundworks: Algebraic thinking* (3 vols: grades 1, 2, and 3). Chicago, IL. Creative Publications.

Greenes, C., & Findell, C. (1999). *Groundworks: Algebra puzzles and problems* (4 vols: grades 4, 5, 6, and 7). Chicago, IL: Creative Publications.

Hastings, E. H., & Yates, D. S. (1983). Microcomputer unit: Graphing straight lines. *Mathematics Teacher, 76*, 181–186.

Kieran. C. (1991). Research into practice: Helping to make the transition to algebra. *Arithmetic Teacher, 38*, 49–51.

Litwiller, B. H., & Duncan, D. R. (1985). Pentagonal patterns in the addition table. *Arithmetic Teacher, 32*, 36–38.

Loewen, A. C. (1991). Lima beans, paper cups, and algebra. *Arithmetic Teacher, 38*, 34–37.

MacGregor, M., & Stacey, K. (1999). A flying start to algebra. *Teaching Children Mathematics, 6*, 78–85.

Morelli, L. (1992). A visual approach to algebra concepts. *Mathematics Teacher, 85*, 434–437.

National Council of Teachers of Mathematics & Mathematical Sciences Education Board (1998). *The nature and role of algebra in the K–14 curriculum: Proceedings of a national symposium, May 27 and 28, 1997*. Washington, DC: National Academy Press.

Nibbelink, W. H. (1990). Teaching equations. *Arithmetic Teacher, 38*, 48–51.

Onslow, B. (1990). Pentominoes revisited. *Arithmetic Teacher, 37*, 5–9.

Osbome, A., & Wilson, P. S. (1988). Moving to algebraic thought. In T. R. Post (Ed.), *Teaching mathematics in grades K–8: Research-based methods* (pp. 421–442). Needham Heights, MA: Allyn & Bacon.

Patterson, A. C. (1999). Grasping graphing. *Mathematics Teacher, 92*, 758–762.

Peitgen, H., Jürgens, H., & Saupe, D. (1992). *Fractals for the classroom part one: Introduction to fractals and chaos*. New York: Springer-Verlag.

Peitgen, H., Jügrens, H., Saupe, D., Maletsky, E., Perciante, T., & Yunker, L. (1991). *Fractals for the classroom: Strategic activities (Vol. 1)*. New York: Springer-Verlag.

Phillips, E. (1991). *Patterns and functions: Addenda series, grades 5–8*. Reston, VA: National Council of Teachers of Mathematics.

Quinn, A. L., Koca, R. M., Jr., & Weening, F. (1999). Developing mathematical reasoning using attribute games. *Mathematics Teacher, 92*, 768–775.

Schifter, D. (1999). Reasoning about operations: Early algebraic thinking in grades K–6. In L. V. Stiff (Ed.), *Developing mathematical reasoning in grades K–12* (pp. 62– 81). Reston, VA: National Council of Teachers of Mathematics.

Schoenfeld, A. H., & Arcavi, A. (1988). On the meaning of variable. *Mathematics Teacher, 81*, 424–427.

Schultz, J. E. (1991). Implementing the *Standards*: Teaching informal algebra. *Arithmetic Teacher, 38*, 34–37.

Specht, J. (2000). *More than graphs: Mathematics explorations for T1-73, T1-82, T1-83, and T1-83 Plus graphing calculators*. Emeryville, CA: Key Curriculum Press.

Thompson, A. G. (1985). On patterns, conjectures, and proof: Developing students' mathematical thinking. *Arithmetic Teacher, 33*(1), 20–23.

Vance, J. H. (1998). Number operations from an algebraic perspective. *Teaching Children Mathematics, 4*, 282–285.

Van de Walle, J. A., & Holbrook, H. (1987). Patterns, thinking, and problem solving. *Arithmetic Teacher, 31*(8), 9–12.

Van Dyke, F. (1998). Visualizing cost, revenue, and profit. *Mathematics Teacher, 91*, 488–493, 500–503.

Wagner, S. (1983). What are these things called variables? *Mathematics Teacher, 76*, 474–479.

Wagner, S., & Parker, S. (1993). Advancing algebra. In S. Wagner (Ed.), *Research ideas for the classroom: High school mathematics* (pp. 119–139). Old Tappan, NJ: Macmillan.

Wilson, M. R., & Shealy, B. E. (1995). Experiencing function relationships with a viewing tube. In P. A. House (Ed.), *Connecting mathematics across the curriculum* (pp. 219–224) Reston, VA: National Council of Teachers of Mathematics.

CHAPTER 9 NINE

Geometry

CHAPTER OVERVIEW

Geometry is among the richest and oldest branches of mathematics. We think of geometry as the study of space experiences. This study focuses mainly on shapes as abstractions from the environment, which can be informally investigated and analyzed. In this chapter, you can learn some interesting facts about geometry and will examine some concepts and procedures involving geometry ideas.

BIG MATHEMATICAL IDEAS

Problem-solving strategies, shape and space, congruence, similarity, verifying, conjecturing, generalizing, decomposing

NCTM PRINCIPLES & STANDARDS LINKS

Geometry and Spatial Sense; Problem Solving; Reasoning; Communication; Connections; Representation

TOPIC 9.1 Two-dimensional Geometry Basics
9.2 Planes
9.3 Line Relationships
9.4 Line Segments
9.5 Rays and Angles
9.6 Angle Relationships
9.7 Polygons
9.8 Regular Polygons
9.9 Triangles
9.10 Quadrilaterals
9.11 Circles
9.12 Congruence
9.13 Symmetry
9.14 Line and Angles Constructions
9.15 Angle-Bisector Constructions
9.16 Perpendicular-Line-Through-a-Point Constructions
9.17 Parallel-Line Construction
9.18 Circumscribed-Circle Constructions
9.19 Inscribed-Circle Constructions
9.20 Triangle Congruence
9.21 Triangle-Congruence Properties
9.22 Quadrilateral Properties
9.23 Similarity
9.24 Triangle-Proportion Properties
9.25 Fractals and Rep-tiles
9.26 Three-dimensional Geometry Basics
9.27 Polyhedra, Prisms, and Pyramids
9.28 Regular Polyhedra
9.29 Cylinders, Cones, and Spheres
9.30 van Hiele Levels
In the Classroom
Bibliography

TOPIC 9.1 Two-Dimensional Geometry Basics

cornerstone **Cartesian Coordinate System**

Introduction

Most ideas, whether in mathematics or not, have foundational ideas upon which other ideas are built. In geometry, three foundational ideas are point, line and plane. These ideas are taken as *undefined*, meaning that no formal definition is given for them. But as foundational ideas in geometry, they are used to define and describe many other ideas in geometry and, thus, if they *were* defined, we would run into the problem of having circular definitions (e.g., idea #1 is defined in terms of idea #2, but idea #2 is defined in terms of idea #1). In this section, we discuss points and lines, as well as relationships between points and lines.

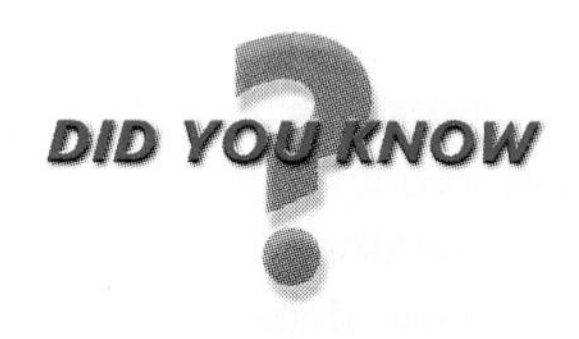

The word *geometry* comes from two Greek words, *ge* meaning "earth" and *metria* meaning "measure," suggesting that, initially, this field of study may have been related to and used for measuring land. Today, geometry is much more than measurement. It has evolved into the study of figures and shapes in two and three dimensions.

9.1.1 Points and Lines

We can describe a **point** as a location. Points do not have any dimension (i.e., no length or width), but they represent a location within the set of all points (called space). We talk about driving from point A to point B, the point of a pencil, or the sharp point of a corner. All of these bring to mind a location in space. We usually represent a point by a dot, and we use a capital letter to refer to it, as shown in Figure 9.1.

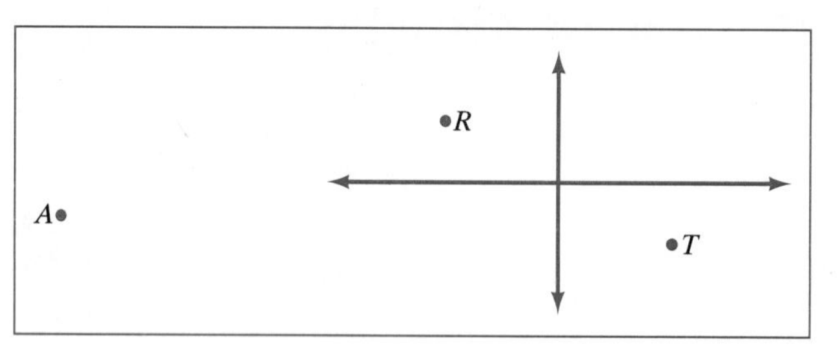

Figure 9.1

We can describe a **line** as a one-dimensional set of an infinite number of points with direction. A line has no thickness, but it has infinite length because it extends in opposite directions forever. In our everyday talk, we refer to walking in a straight line from here to there, to not stepping over the line, or to lining up. All of these bring to mind a collection of points that stretch out in opposite directions. We usually represent a line, as shown in Figure 9.2, with the arrows at the ends of the line indicating that the set of points goes on indefinitely in opposite directions. A line can be named by using a single lowercase letter (e.g., line m) or by using two points that are on the line (e.g., line XY or $\overleftrightarrow{XY}$).

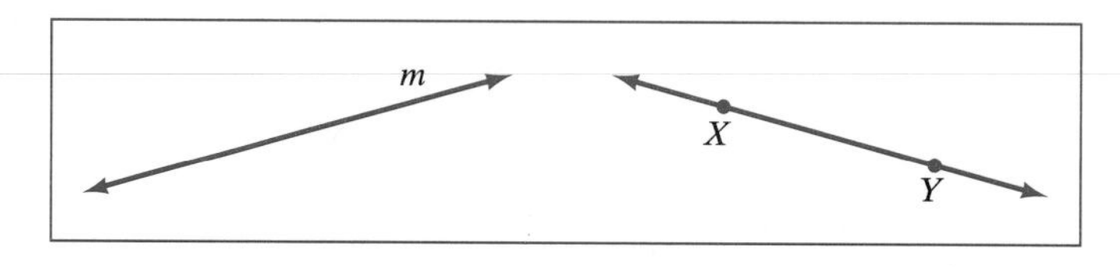

Figure 9.2

9.1.2 Relationships Among Points and Lines

Although an infinite number of lines can be drawn through a single point, only one line can be drawn through any two given points. This is an important relationship among two points and a line. Another important relationship is collinearity. Points that are **collinear** are points that lie on the same line. Two points are always collinear, but three or more points are not necessarily collinear. Points that do not lie on the same line are called **noncollinear.**

Points S, T, and R in Figure 9.3 are collinear, but points S, T, and P are noncollinear. We say that point T is **between** points S and R because they are collinear and that if we moved along $\overleftrightarrow{SR}$ from point S to point R, we would come to point T before coming to point R.

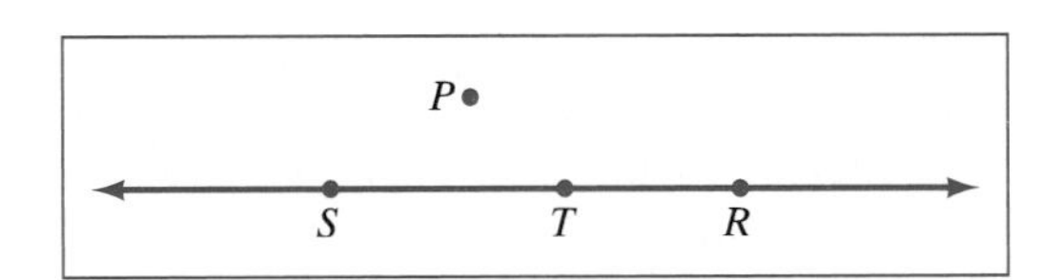

Figure 9.3

Related Topics

Planes
Line Relationships
Line Segments
Rays and Angles
Cartesian Coordinate System

TOPIC 9.2 Planes

cornerstone **Two-dimensional Geometry Basics**

Introduction

Together with point and line, plane is one of the three foundational undefined ideas in geometry. We use point, line, and plane as building blocks on which to build the rest of geometry. In this section, we discuss planes and relationships among points, lines, and planes.

Euclid was a Greek mathematician who lived about 300 B.C. Although no great mathematical discoveries are attributed to him, he is credited with pulling together geometry ideas and putting them together in the thirteen books called *The Elements*. The first six books discuss plane geometry.

For More Information Bunt, L. N. H., Jones, P. S., & Bedient, J. D. (1988). *The Historical Roots of Elementary Mathematics*. New York: Dover Publications.

9.2.1 Planes

Just as we can describe a line as a one-dimensional set of an infinite number of points with direction, we can describe a **plane** as a two-dimensional set of an infinite number of points extending forever in all directions. A plane has no thickness but it has infinite length and width because it extends indefinitely as a surface. We can think of a plane as an infinite sheet of paper that is perfectly flat and has no thickness. We usually represent a plane, as shown in Figure 9.4, with a four-sided figure, and we name the plane by using three points in the plane (e.g., Plane JKL) or by using a lowercase Greek letter (e.g., Plane α).

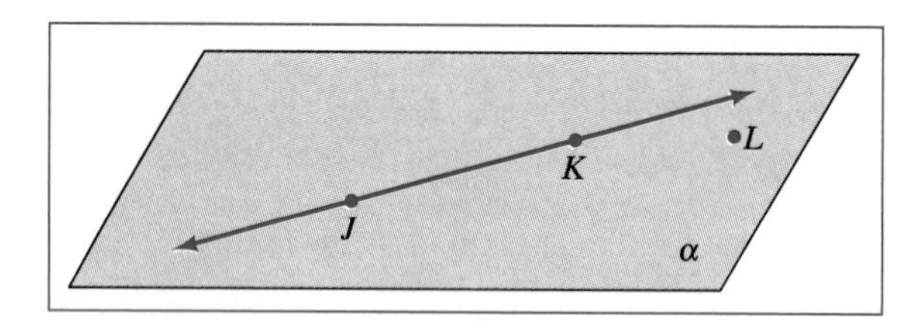

Figure 9.4

Points that are **coplanar** are points that lie in the same plane. Two points or three points are always coplanar. Four or more points are not necessarily coplanar. Points that do not all lie in the same plane are called **noncoplanar.**

Points T, U, V, and W in Figure 9.5 are coplanar, but points S, T, U, and V are noncoplanar.

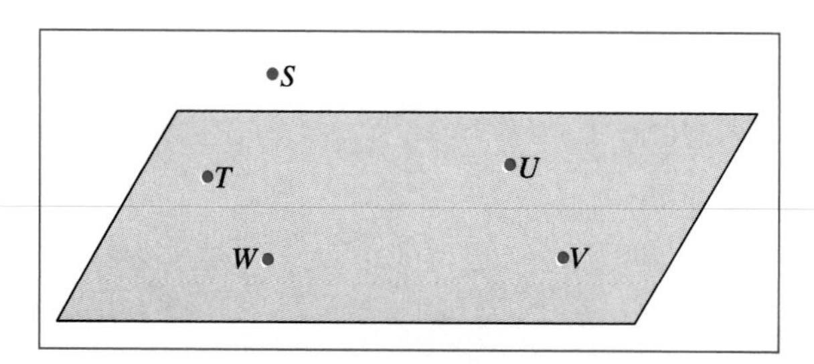

Figure 9.5

9.2.2 Relationships Among Points, Lines and Planes

An infinite number of planes can be drawn through one point and even two points, as shown in Figure 9.6. But for any three points, exactly one plane can contain them. This is an important relationship that can also be stated as: Three noncollinear points determine exactly one plane. Another relationship is that a line and a point not on the line determine exactly one plane.

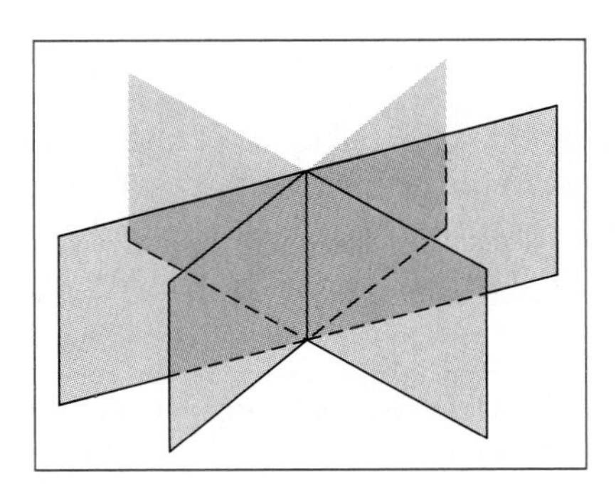

Figure 9.6

A third important relationship among points, lines, and planes is that if a line contains two points that are in a plane, then the line is also in the plane.

Line Relationships
Line Segments
Rays and Angles
Cartesian Coordinate System

TOPIC 9.3 Line Relationships

cornerstone

Two-dimensional Geometry Basics
Planes

Introduction

In geometry, there are special kinds of lines, as well as special relationships between lines and planes. In this section, we discuss coplanar, noncoplanar, and skew lines, intersecting and concurrent lines, parallel and perpendicular lines, and relationships among lines and planes.

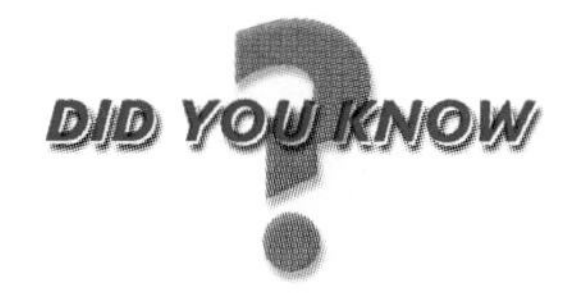

In *The Elements*, Euclid's systematic account of all the geometry that was known at his time, he based geometry on five postulates. The parallel postulate, as his fifth postulate came to be called, acquired an independent status among the postulates. This postulate stated that two straight lines that meet certain conditions have a point of intersection. One person who challenged this fifth postulate was Nicholai Lobachevsky, a Russian mathematician who lived from 1792–1856. By challenging and rejecting this postulate, Lobachevsky created another set of postulates that were consistent and led to an entirely new kind of geometry. Unlike Euclidean geometry, this geometry was not on a flat surface; in fact, it was based on a spherical surface and thus could be used to model geometry on the surface of Earth. By the creation of this geometry, Lobachevsky had succeeded in overturning a belief that had been held since the time of Euclid, nearly 2,000 years before Lobachevsky—the belief that the fifth postulate was necessary in a geometry system. His success later prompted the creation of other kinds of non-Euclidean geometries that have proved very useful in studying different phenomena in modern times.

For More Information Sibley, T. Q. (1998). *The Geometric Viewpoint: A Survey of Geometries*. Reading, MA: Addison-Wesley.

9.3.1 Coplanar, Noncoplanar, and Skew Lines

Two lines that can be contained in the same plane are called **coplanar** lines. Two lines that cannot be contained in the same plane are **noncoplanar** lines. In Figure 9.7, lines m and n are coplanar, and lines n and ℓ are coplanar, but lines m and ℓ are noncoplanar.

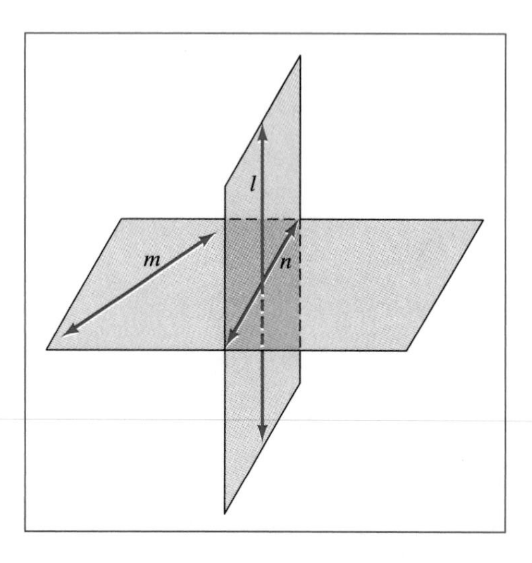

Figure 9.7

Another name for noncoplanar lines is **skew** lines. Any two lines that cannot lie in the same plane are called skew.

A line in a plane divides the plane into three regions: two half-planes and the line itself. We name a half-plane by using the line that determines the half-plane and a point not on the line that lies in the half-plane. For example, in Figure 9.8, four half-planes are shown: the half-plane determined by $\overleftrightarrow{NO}$ and point D, the half-plane determined by $\overleftrightarrow{NO}$ and point E, the half-plane determined by $\overleftrightarrow{PQ}$ and point D, and the half-plane determined by $\overleftrightarrow{PQ}$ and point E.

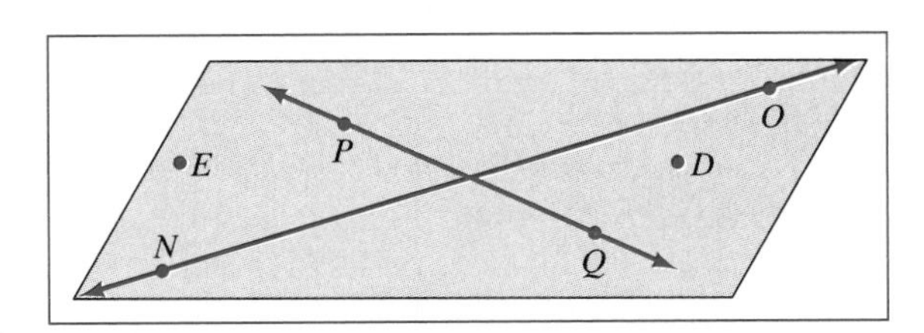

Figure 9.8

9.3.2 Intersecting and Concurrent Lines

Lines are called **intersecting** lines if they share exactly one point. If two lines intersect, they determine a plane. To see why this is true consider Figure 9.9. We know that two points determine a line, so suppose that points A and B determine $\overleftrightarrow{AB}$ and points C and B determine $\overleftrightarrow{CB}$. The two lines are intersecting lines because they have point B in common. Using the fact that three noncollinear points determine a plane, we can see that $\overleftrightarrow{AB}$ and point C determine a plane. Likewise, the same plane is also determined by $\overleftrightarrow{CB}$ and point A.

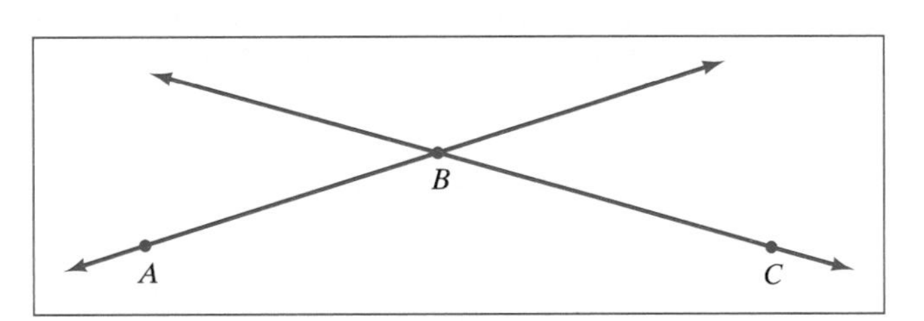

Figure 9.9

Two or more lines that share the same point are called **concurrent** lines. These lines can be coplanar or noncoplanar, as shown in Figure 9.10.

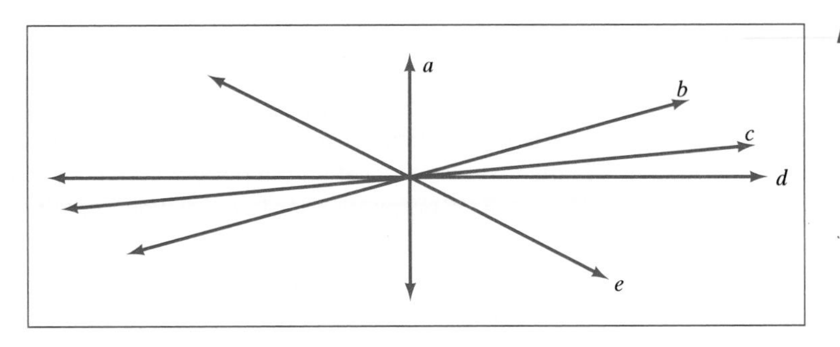

Figure 9.10

9.3.3 Parallel and Perpendicular Lines

Parallel lines are lines that are distinct from each other, are coplanar and share no common points. Lines r and s in Figure 9.11 are parallel lines: We use the symbol $\|$ to say that $r \| s$. If two lines are parallel, they determine a plane.

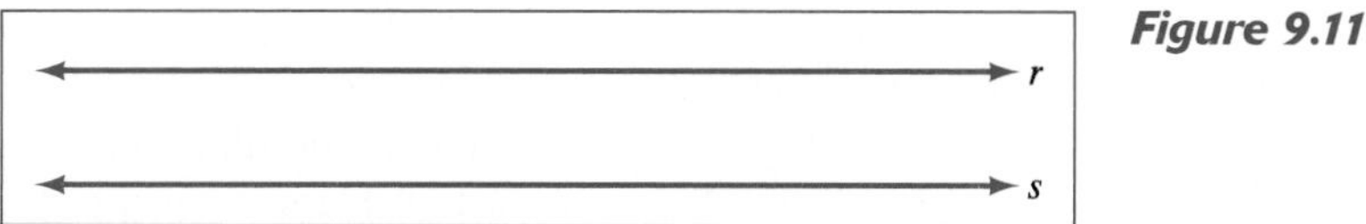

Figure 9.11

Lines are said to be **perpendicular** if they intersect to form right angles. Lines k and ℓ in Figure 9.12 are perpendicular because they form right angles when they intersect. We use the symbol $\perp$ to say that k and ℓ are perpendicular, and we write this relationship, $k \perp \ell$. We also use the symbol ⌝ to show that an angle in a figure is a right angle.

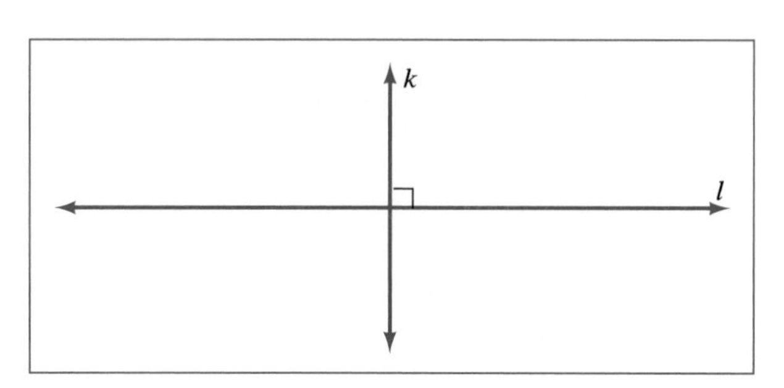

Figure 9.12

9.3.4 Relationships among Lines and Planes

There are three possible relationships between a line and a plane: (1) A line and a plane can have no points in common. In Figure 9.13, line m and plane β have no points in common; thus, line m is parallel to plane β. (2) A line and a plane can have one point in common. Line n and plane β have one point in common; we say that line n intersects plane β. Because line n does not lie in the plane, it must intersect the plane at exactly one point. (3) If a line and a plane have more than one point in common, the line must lie in the plane. Line p lies in plane β.

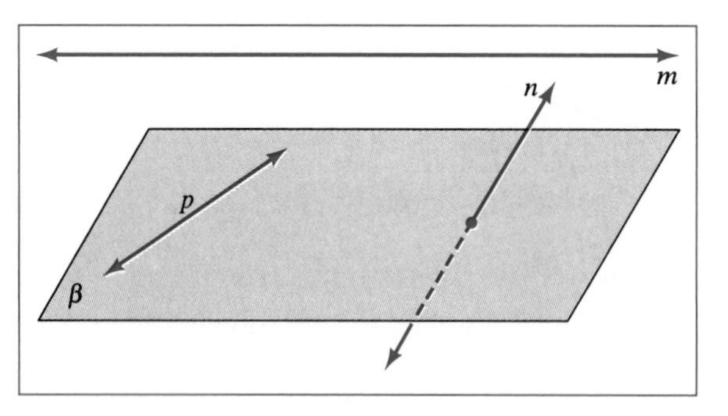

Figure 9.13

If a line intersects a plane in such a way that a right angle is formed, then the line is perpendicular to the plane; line t is perpendicular to plane λ in Figure 9.14. When a line is perpendicular to a plane, that line is also perpendicular to every line in the plane that passes through the point where the perpendicular line intersects the plane. Thus, line t is perpendicular to every line in plane λ that passes through point C.

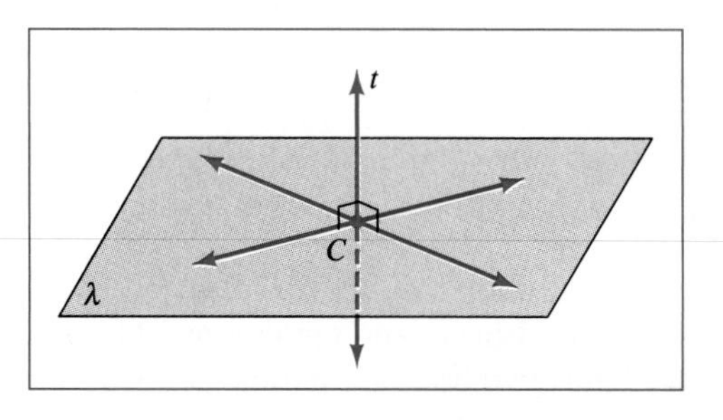

Figure 9.14

Related Topics

- Line Segments
- Rays and Angles
- Slope
- Cartesian Coordinate System

TOPIC 9.4 Line Segments

cornerstone

Two-dimensional Geometry Basics
Planes
Line Relationships

Introduction

Line is one of the three foundational ideas on which geometry is built. Along with point and plane, line is one of the undefined terms in geometry. In this section, we discuss line segments, endpoints, and midpoints.

Euclid's first postulate in *The Elements* states that any point can be connected to any other point by drawing exactly one line segment. The idea of a line segment, as opposed to a line which continues in two directions indefinitely, is very useful because we live in a world where we deal with finite lengths.

For More Information Bunt, L. N. H., Jones, P. S., & Bedient, J. D. (1988). *The Historical Roots of Elementary Mathematics.* New York: Dover Publications.

An important part of the concept of a line is that a line is a set of an infinite number of points. A **line segment** is a subset of a line and consists of two points on the line, called the **endpoints** of the line segment, and all the points on the line that are between the two endpoints. A line segment is named by its two endpoints, and we use a bar over the two capital letters representing the points to show that it is a line segment (e.g., $\overline{AB}$, $\overline{BC}$, and $\overline{AC}$ are line segments that are subsets of $\overleftrightarrow{AC}$ in Figure 9.15).

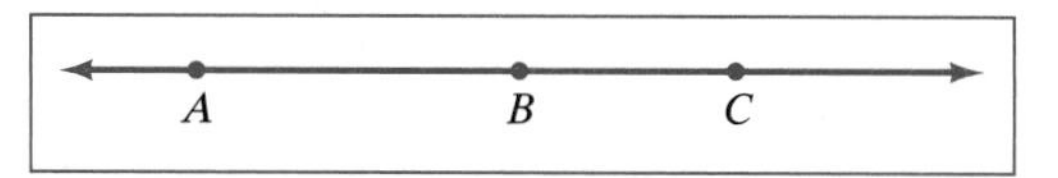

Figure 9.15

Because a line segment has endpoints, it has a finite length. Thus, every line segment also has a **midpoint,** which is the point on the line segment that divides it into two segments of the same length. Point M in Figure 9.16 is called the midpoint of $\overline{SR}$ because M is between S and R and the length of $\overline{SM}$ is equal to the length of $\overline{MR}$. Every line segment has exactly one midpoint, and we say that the midpoint bisects the line segment.

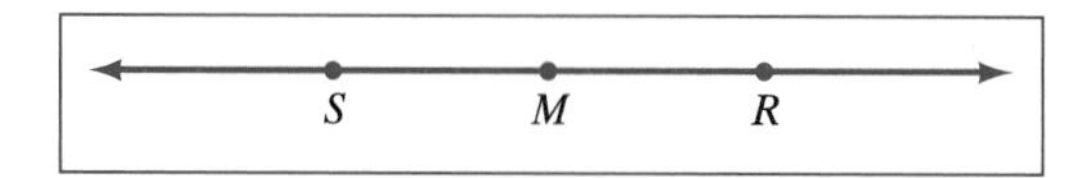

Figure 9.16

Symmetry
Distance Formula
Cartesian Coordinate System

TOPIC 9.5 Rays and Angles

cornerstone **Two-dimensional Geometry Basics**

Introduction

The ideas of rays and angles have many uses, not only in developing further geometrical relationships, but especially in practical matters of constructing and measuring objects. In this section, we discuss rays and angles, including the interior and exterior of an angle.

Although we have evidence that early civilizations used the concepts of rays and angles, there is reason to believe that formal discussion and definition of these ideas did not occur until Euclid, a Greek mathematician living around 300 B.C., compiled *The Elements*. These thirteen books pulled together most of the geometry ideas known in Greece at that time.

For More Information Bunt, L. N. H., Jones, P. S., & Bedient, J. D. (1988). *The Historical Roots of Elementary Mathematics*. New York: Dover Publications.

9.5.1 Rays

Like a line segment, a **ray** is a subset of a line. A ray consists of a point on a line, called the endpoint, and all the points on that line that are on one side of the endpoint. A ray is named by two points: The endpoint is listed first, followed by another point on the ray. It is important that the endpoint is listed first to show in which direction the ray extends. For example, in Figure 9.17, $\overrightarrow{EF}$ and $\overrightarrow{FE}$ are not the same ray.

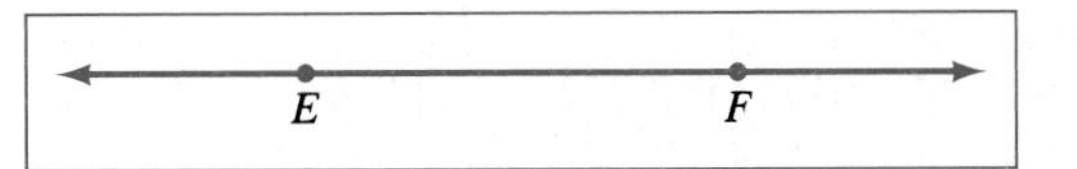

Figure 9.17

9.5.2 Angles

An **angle** is formed by two rays that have a common endpoint, such as $\overrightarrow{XY}$ and $\overrightarrow{XZ}$ in Figure 9.18. The common endpoint of the rays is called the **vertex** of the angle. An angle is named by a point on one ray, the vertex, and a point on the other ray; thus, the angle in Figure 9.18 can be called $\angle YXZ$ or $\angle ZXY$. If there is only one angle at a vertex, the angle can also be named by the vertex point, such as $\angle X$. However, if there are several angles at a vertex point, three points must be used to name the angle. The rays form the **sides** of an angle, so $\overrightarrow{XY}$ and $\overrightarrow{XZ}$ are the sides of $\angle X$.

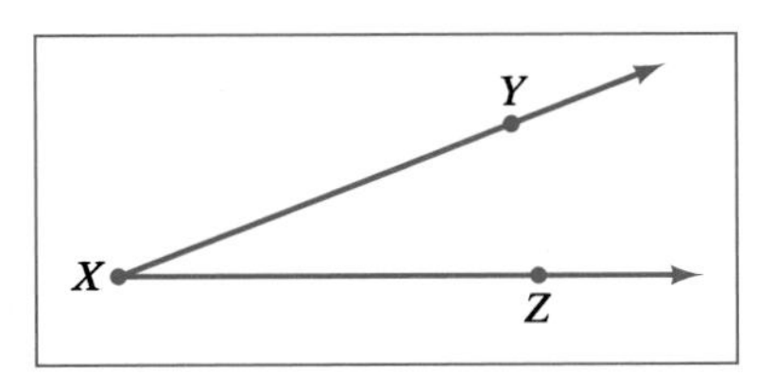

Figure 9.18

Because three noncollinear points determine a plane, every angle lies in a plane. An angle separates the plane into three disjoint sets of points: the set of points that makes up the interior of the angle, the set of points that makes up the angle, and the set of points that makes up the exterior of the angle. The **interior** of $\angle JKL$ in Figure 9.19 is the set of all points P in the plane that contains $\angle JKL$ such that P and J are on the same side of $\overleftrightarrow{KL}$ and P and L are on the same side of $\overleftrightarrow{JK}$. The **exterior** of $\angle JKL$ is the set of all points in the plane that are neither on the angle nor in its interior.

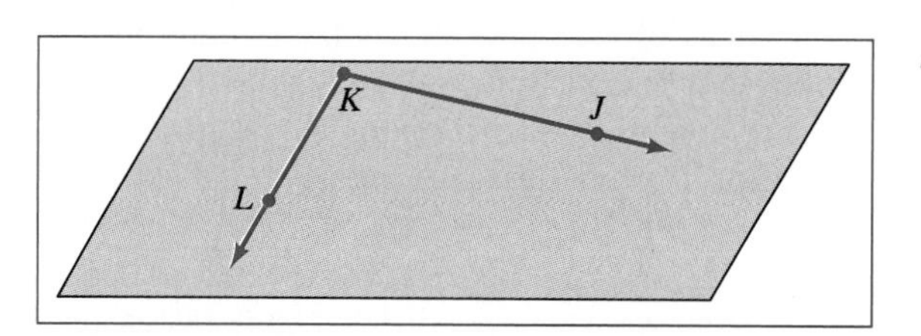

Figure 9.19

Related Topics
Angle Measurement
Angle Relationships

Angle Relationships

cornerstone

Rays and Angles
Angle Measurement
Congruence
Line Relationships

Introduction

There are a variety of angle relationships that are important to understand because they often play a role in various geometric relationships (e.g., triangle congruence). In this section, we discuss adjacent angles, linear pair angles, vertical angles, complementary and supplementary angles, angles formed by a transversal and two lines, and angles formed by a transversal and two parallel lines.

Some of the relationships that Euclid derived in *The Elements* dealt with angle relationships. These same angle relationships are part of our geometry curriculum today and continue to be useful in proving other relationships and in practical matters, such as cutting congruent shapes out of metal for use in manufacturing.

For More Information Bunt, L. N. H., Jones, P. S., & Bedient, J.D. (1988). *The Historical Roots of Elementary Mathematics.* New York: Dover Publications.

9.6.1 Adjacent Angles, Linear Pair Angles, and Vertical Angles

Two angles are said to be **adjacent** if they lie in the same plane, share a common side and vertex, and have interiors that do not overlap. Angles ABC and CBD are adjacent in Figure 9.20, whereas $\angle ABC$ and $\angle ABD$ are not adjacent.

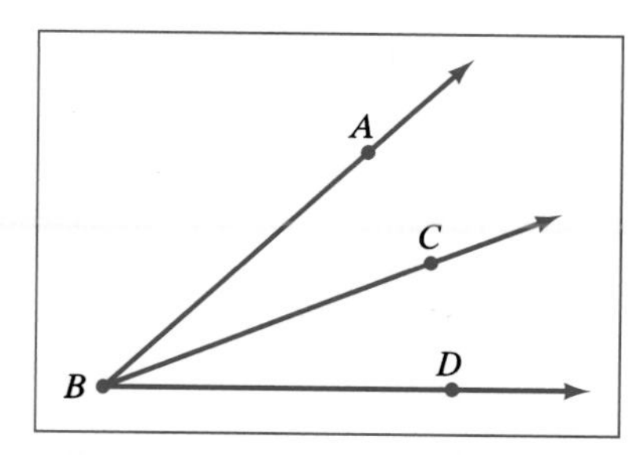

Figure 9.20

If two adjacent angles have their noncommon sides form a straight angle, then the angles are called **linear pair angles.** In Figure 9.21, angles UVW and WVX are linear pair angles, whereas $\angle WVX$ and $\angle XVY$ are not. Because linear pair angles form a straight angle, the sum of the two angle measures is 180°.

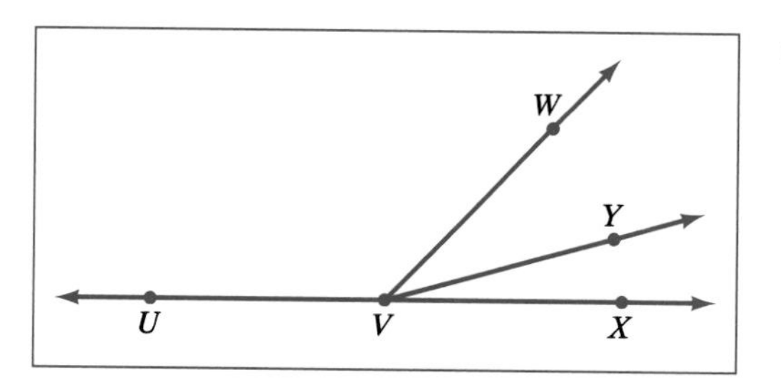

Figure 9.21

When two lines intersect, they form four angles. Two angles are called **vertical angles** if their sides form two straight angles. In Figure 9.22, $\angle 1$ and $\angle 3$ are vertical angles and $\angle 2$ and $\angle 4$ are vertical angles. An important relationship between vertical angles is that vertical angles are always congruent.

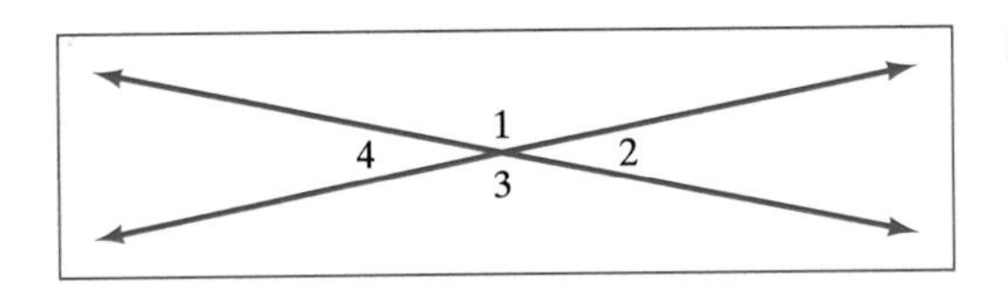

Figure 9.22

9.6.2 Complementary and Supplementary Angles

If the sum of the measures of two angles is 90°, then the angles are called **complementary angles.** Each angle is said to be a complement of the other. If the sum of the measures of two angles is 180°, then the angles are called **supplementary angles,** and the angles are supplements of each other (see Figure 9.23).

Figure 9.23

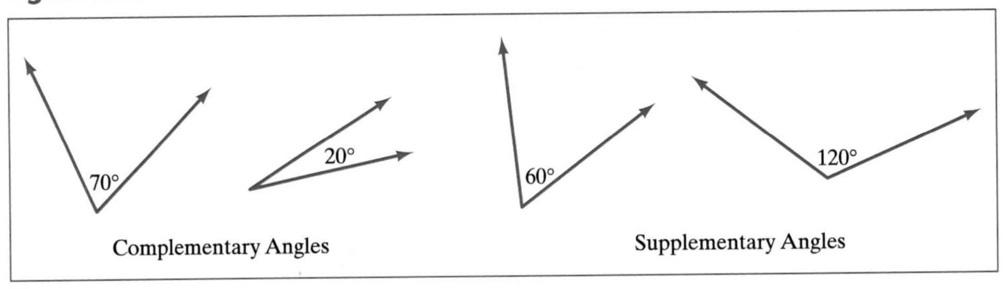

9.6.3 Angles Formed by a Transversal and Two Lines

A **transversal** of two coplanar lines is a line that intersects the two coplanar lines in two different points. When two coplanar lines are cut by a transversal, eight angles are formed as shown in Figure 9.24. We have special names for these angles, according to their location in relation to the two lines and the transversal. $\angle 3$, $\angle 4$, $\angle 5$, and $\angle 6$ are called **interior** angles because they are contained between the two coplanar lines; $\angle 1$, $\angle 2$, $\angle 7$, and $\angle 8$ are called **exterior** angles. **Alternate interior angles** are pairs of interior angles on opposite sides of the transversal; $\angle 3$ and $\angle 6$, $\angle 4$ and $\angle 5$ are pairs of alternate interior angles. Likewise, $\angle 1$ and $\angle 8$ as well as $\angle 2$ and $\angle 7$ are pairs of **alternate exterior angles. Corresponding angles** are angles that are located in the same place on each line in relation to the transversal; $\angle 1$ and $\angle 5$, $\angle 2$ and $\angle 6$, $\angle 3$ and $\angle 7$, and $\angle 4$ and $\angle 8$ are corresponding angles.

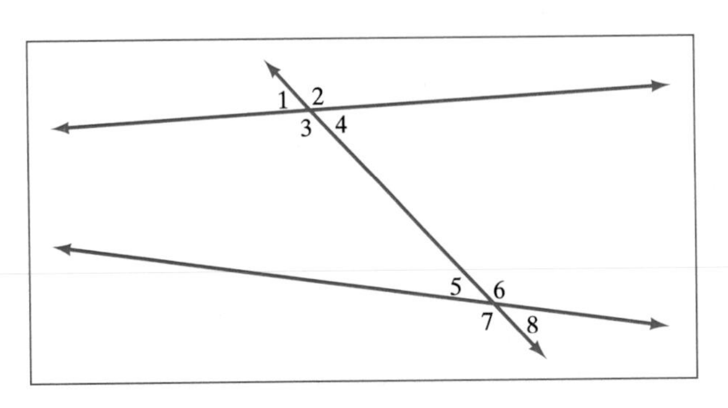

Figure 9.24

9.6.4 Angles Formed by a Transversal and Two Parallel Lines

If two parallel lines are cut by a transversal, then a pair of (a) alternate interior angles, (b) exterior angles, or (c) corresponding angles must be congruent. The converse is also true. In other words, if two lines are cut by a transversal and a pair of (a) alternate interior angles, (b) exterior angles, or (c) corresponding angles are congruent, then the two lines must be parallel (see Figure 9.25).

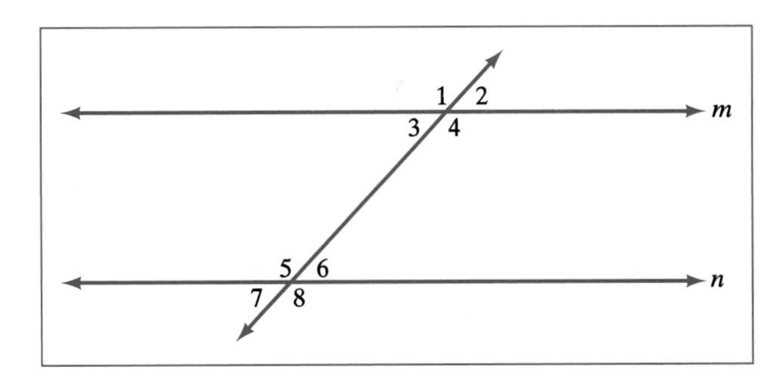

Figure 9.25

Related Topics

Line and Angle Constructions
Angle Bisector Construction
Triangle Congruence
Triangle Congruence Properties

TOPIC 9.7 Polygons

cornerstone

Line Segments
Rays and Angles
Congruence

Introduction

All around us, in our everyday lives, there are geometrical figures. Many of these are known to us as triangles, rectangles, squares, pentagons, octagons, among other names. These are all types of polygons. In this section, we discuss a polygon and its features, convex and concave polygons, and names of polygons.

The word *polygon* comes from Greek: *poly* meaning "many," and *gon* meaning "sides." Polygons are often thought of as many-sided figures.

9.7.1 A Polygon and Its Features

A **polygon** is a curve composed entirely of line segments that is simple (the curve does not cross itself) and closed. Figure 9.26 shows some examples of polygons and nonpolygons.

Figure 9.26

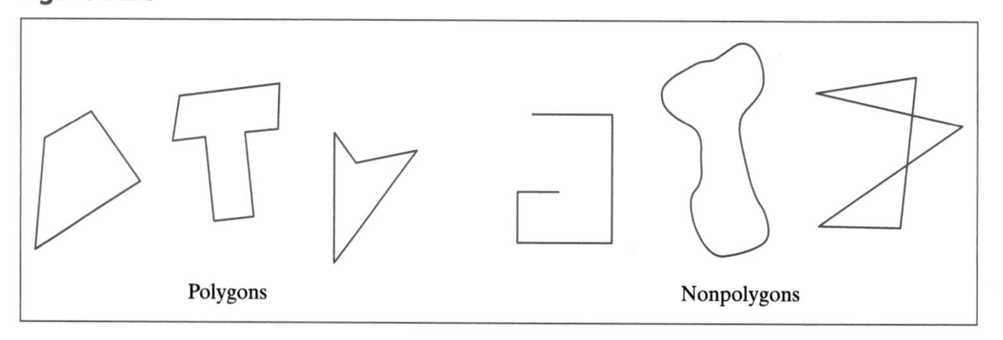

The line segments that comprise a polygon are called the **sides,** and the point where two sides meet is called a vertex (plural **vertices**). A **diagonal** is a line segment that is drawn between any two nonconsecutive vertices in a polygon. A polygon separates the plane into three disjoint subsets. These are the interior of the polygon, the polygon itself, and the exterior of the polygon (see Figure 9.27). A polygon and its interior together make up the set of points known as a **polygonal region.**

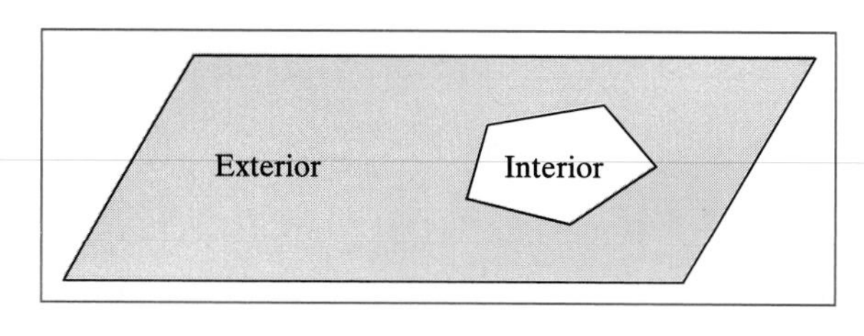

Figure 9.27

At each vertex, there is an interior angle (or vertex angle) that is formed by the vertex and the two sides that meet at the vertex. There are also exterior angles that can be formed at each vertex by extending sides of the polygon, as shown in Figure 9.28.

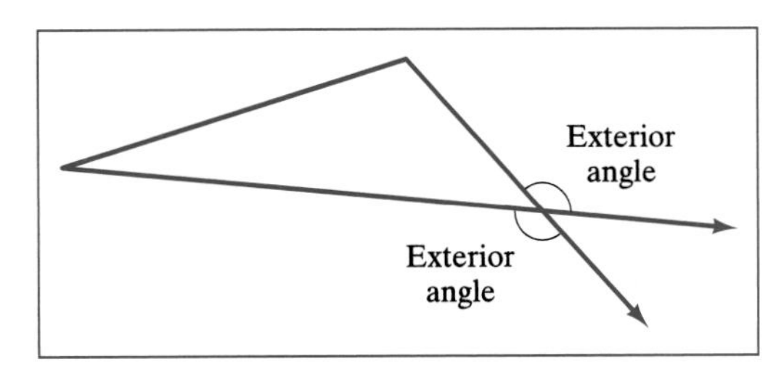

Figure 9.28

9.7.2 Convex and Concave Polygons

A polygon is **convex** if and only if for two distinct points in the polygonal region, the segment that joins those points is contained entirely in the polygonal region. If the segment that joins these two points does not lie in the polygonal region, then the polygon is **concave.** Figure 9.29 shows convex and concave polygons.

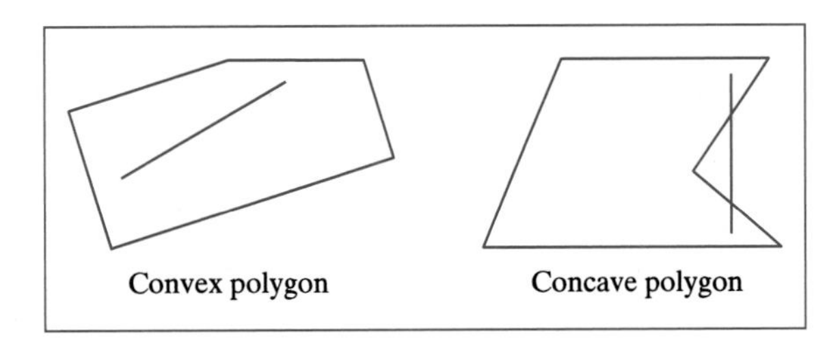

Figure 9.29

9.7.3 Names of Polygons

Polygons are named by the number of sides or vertices that they have. For example:

Polygon	Number of Sides or Vertices
Triangle	3
Quadrilateral	4
Pentagon	5
Hexagon	6
Heptagon	7
Octagon	8
Nonagon	9
Decagon	10
n-gon	n

Related Topics

Regular polygons
Triangles
Quadrilaterals
Area of Regular Polygons and Circles
Symmetry

TOPIC 9.8 Regular Polygons

Polygons
Congruence
Angle Measurement
Triangles

Introduction

In the previous section, we discussed polygons, their features, and particular kinds of polygons. In this section, we discuss what it means for a polygon to be called regular, the sum of the measures of the angles in any convex polygon, and the sum of the measures of the exterior angles in any convex polygon.

In 1801, Carl F. Gauss proved which regular polygons could be constructed using only a straightedge and a compass. He found that a regular polygon can be constructed if the number of sides can be expressed as $2^n(2^{2^k} + 1)$, where $2^{2^k} + 1$ is prime. The only known primes of this form are 3, 5, 17, 257, and 65,537. These are called Fermat primes. In 1796, Gauss constructed a regular 17-gon, and in 1837, he and Pierre Wantzel proved that no other regular polygons can be constructed using only a straightedge and compass.

For More Information Sibley, T. Q. (1998). *The Geometric Viewpoint: A Survey of Geometries*. Reading, MA: Addison-Wesley.

9.8.1 Regular Polygons

A polygon is called **regular** if all of its angles are congruent and all of its sides are congruent. Another way to say this is to say that a regular polygon is one that is **equiangular** (all angles congruent) and **equilateral** (all sides congruent.) The polygons in Figure 9.30 are regular. We use the congruence marks, as shown, to indicate that angles and sides are congruent.

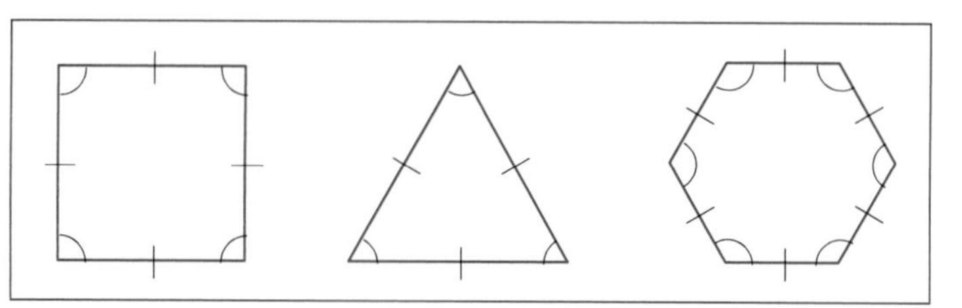

Figure 9.30

9.8.2 Sum of the Measures of Angles in a Convex Polygon

For any convex polygon with n sides, the sum of the measures of the interior angles is $(n - 2)180°$. The $(n - 2)$ represents the number of triangles that can be formed in the polygon by drawing diagonals from one vertex, as shown in Figure 9.31. Because the sum of the measures of the angles in a triangle is 180°, multiplying the number of triangles formed in a polygon in the manner described previously by 180° gives the sum of the measure of the interior angles in the polygon. For example:

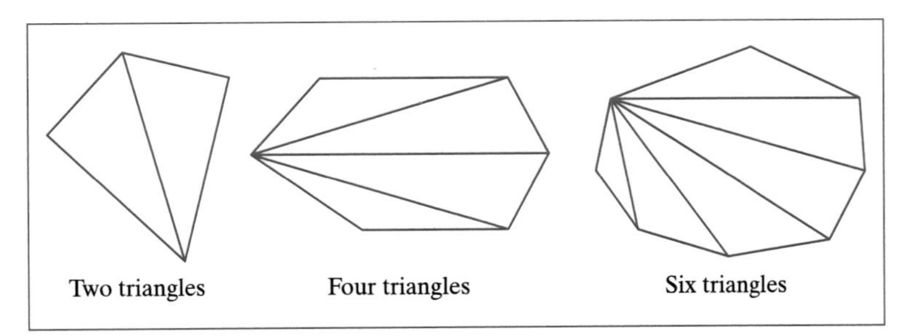

Figure 9.31

Polygon	Sum of the Measures of the Interior Angles
Triangle	180°
Quadrilateral	360°
Pentagon	540°
Hexagon	720°
Heptagon	900°
Octagon	1080°
Nonagon	1260°
Decagon	1440°

Once you find the sum of the measures of the interior angles in a regular polygon, you can find the measure of each angle by dividing the sum by the number of sides (or angles) because all the angles are congruent. Thus, $\frac{(n-2)180°}{n}$ gives the measure of each angle in a regular n-gon. Some regular polygons and the measure of each of their interior angles is as follows:

Regular Polygon	Measure of Each Interior Angle
Equilateral Triangle	60°
Square	90°
Regular Pentagon	108°
Regular Hexagon	120°
Regular Octagon	135°

9.8.3 Sum of the Measures of the Exterior Angles in a Convex Polygon

As we discussed in the previous section, exterior angles can be formed at each vertex in a polygon. If we form just one of these exterior angles at each vertex, we can find the sum of these exterior angles no matter how many sides a convex polygon has. Consider the pentagon in Figure 9.32 with one exterior angle at each vertex.

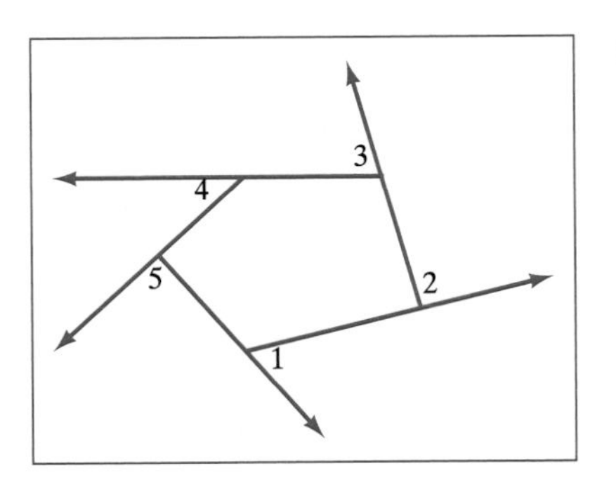

Figure 9.32

Using what we know about linear pair angles, we can see that at each vertex the interior angle and the exterior angle form linear pair angles and, thus, the sum of their measures is 180°. Because a pentagon has five vertices, there are five linear pair angles formed, and so the sum of the pairs of interior and exterior angles is 5(180°). Because we already know that the sum of the interior angles in a convex pentagon is 540°, to find the sum of the measures of just the exterior angles, we can subtract the sum of the measures of the interior angles from the sum of the measures of the interior and exterior angles—$5(180°) - 3(180°) = 360°$. This discussion can be generalized to fit any convex polygon: $n(180°) - (n - 2)180° = 2(180°) = 360°$. Thus, in any convex polygon, the sum of the exterior angles (one at each vertex) is 360°.

Related Topics
Quadrilaterals
Area of Regular Polygons and Circles

TOPIC 9.9 Triangles

cornerstone

Polygons
Angle Relationships
Congruence
Parallel-Line Construction

Introduction

Polygons are named by the number of sides they have. Thus, a three-sided polygon is called a triangle. In this section, we discuss features of a triangle, special kinds of triangles, and the sum of the measures of the interior angles in a triangle.

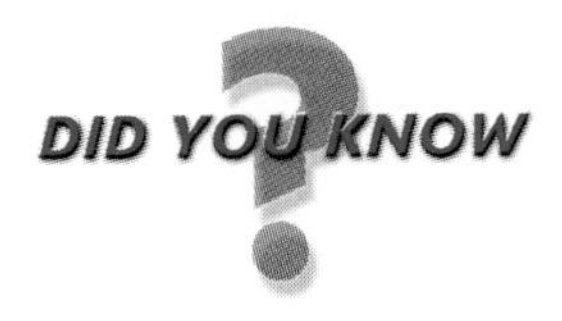

DID YOU KNOW

A triangle is stable in structure. If you form a polygon, with four or more sides, out of cardboard strips and fasten the strips at the vertices, this polygon can be moved so that its shape changes. A triangle, on the other hand, will not change shape once the vertices are fastened. This is why triangles are often used as braces in building structures.

9.9.1 Features of a Triangle

A **triangle** is a polygon that has three sides. An **altitude** in a triangle is a perpendicular segment, drawn from a vertex to the line containing the opposite side. Three altitudes can be drawn in any triangle, as shown in Figure 9.33. Any side of a triangle can be the **base** of the triangle; for that base, the **height** of the triangle is the altitude, drawn to the line containing that side.

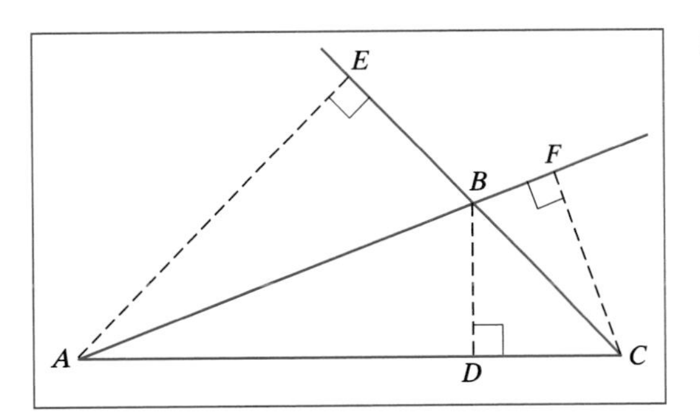

Figure 9.33

9.9.2 Special Kinds of Triangles

We have special names for triangles when classifying them by their angles or by their sides (see Figure 9.34). A triangle with one right angle is called a **right triangle.** A triangle with only acute angles is called an **acute triangle.** A triangle with one obtuse angle is called an **obtuse triangle.**

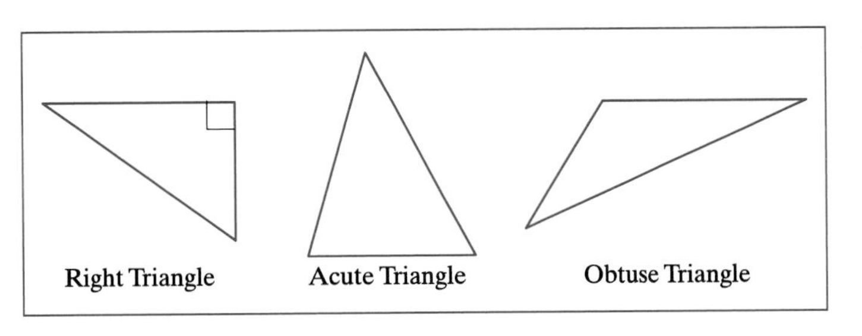

Figure 9.34

A triangle with no sides congruent is called a **scalene triangle.** A triangle with at least two sides congruent is called an **isosceles triangle.** A triangle with all three sides congruent is called an **equilateral triangle** (see Figure 9.35).

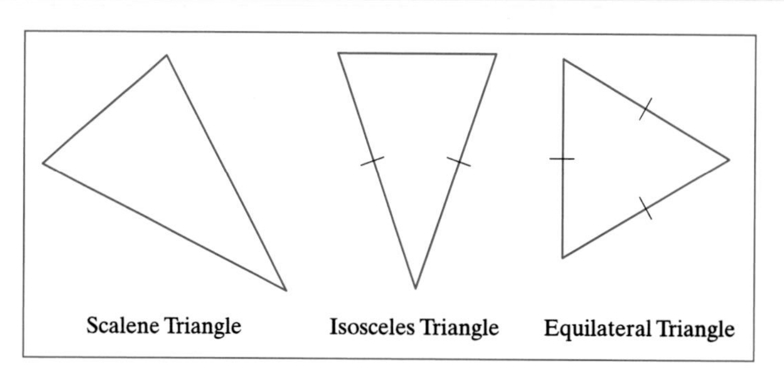

Figure 9.35

The relationships among these special kinds of triangles can be illustrated in a hierarchy, as shown in Figure 9.36.

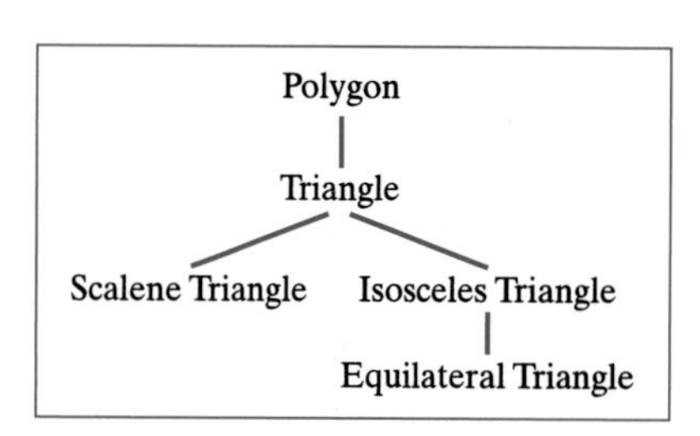

Figure 9.36

9.9.3 Sum of the Measures of the Interior Angles in a Triangle

The sum of the measures of the interior angles in a triangle, in a plane, is 180°. To understand why this is true, consider the following proof using triangle XYZ with angles 1, 2, and 3, in Figure 9.37. Draw a line m through Y that is parallel to $\overleftrightarrow{XZ}$. As discussed in the section called Parallel-Line Construction, we know that we can construct a line parallel to another line through a point that is not on that line. Now, consider the angles labeled $\angle 4$ and $\angle 5$ in the figure. Because line $m \parallel \overleftrightarrow{XZ}$, and $\overleftrightarrow{XY}$ and $\overleftrightarrow{ZY}$ are transversals, $\angle 5 \cong \angle 1$ and $\angle 4 \cong \angle 3$ because they are alternate interior angles. Thus, $m(\angle 5) = m(\angle 1)$ and $m(\angle 4) = m(\angle 3)$, and so because $m(\angle 5) + m(\angle 2) + m(\angle 4) = 180°$ (together they form a line), then $m(\angle 1) + m(\angle 2) + m(\angle 3) = 180°$.

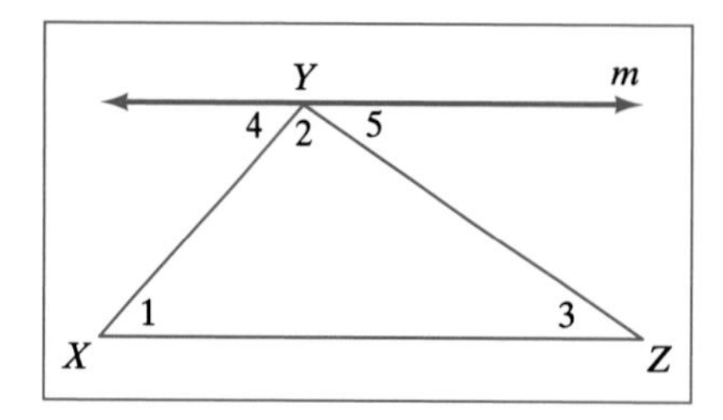

Figure 9.37

Related Topics
Angle Measurement
Symmetry
Area of Triangles
Quadrilaterals

TOPIC 9.10 Quadrilaterals

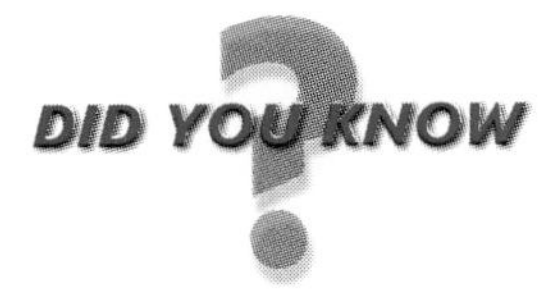

cornerstone

Polygons
Triangles
Congruence

Introduction

Quadrilaterals are a specific type of polygon. In this section, we discuss features of a quadrilateral and the special kinds of quadrilaterals.

DID YOU KNOW

In this section, we use the most common definition of a *trapezoid*. Sometimes, another definition is used—a *trapezoid* is a quadrilateral with exactly one pair of parallel sides. If we use this definition, this changes the relationship among the rest of the special quadrilaterals. For example, a parallelogram would no longer be a trapezoid.

For More Information Moise, E. E., & Downs, Jr., F. L. (1975). *Geometry*. Menlo Park, CA: Addison-Wesley.

9.10.1 Features of a Quadrilateral

A **quadrilateral** is a polygon that has four sides. Any side of a quadrilateral can be the **base** of the quadrilateral; for that base, the **height** of the quadrilateral is the altitude drawn from the opposite vertex to the line containing that side, as shown in Figure 9.38.

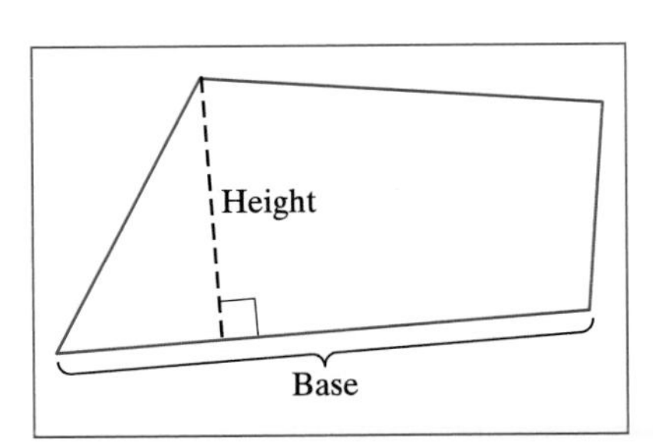

Figure 9.38

9.10.2 Special Kinds of Quadrilaterals

We have special names for quadrilaterals that have certain properties (see Figure 9.39). A quadrilateral with at least two distinct pairs of consecutive sides congruent is called a **kite.** A quadrilateral with at least one pair of parallel sides is called a **trapezoid.** An

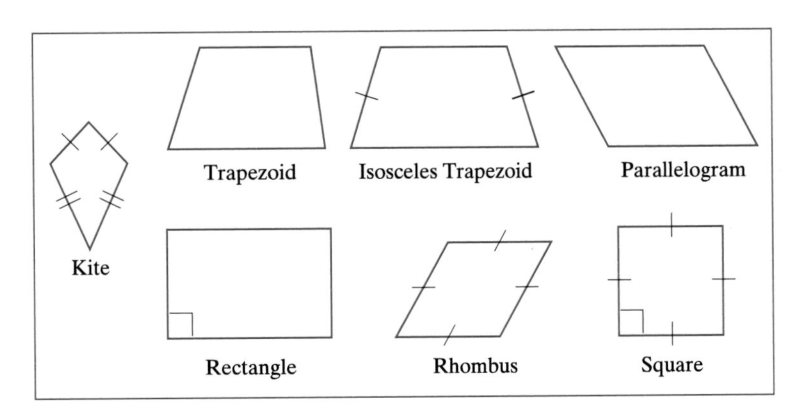

Figure 9.39

isosceles trapezoid is a trapezoid with exactly one pair of opposite sides congruent. A quadrilateral with both pairs of opposite sides parallel is called a **parallelogram.** A **rectangle** is a parallelogram with a right angle. A **rhombus** is a parallelogram with all sides congruent. A **square** is a rectangle with all sides congruent. The relationships among these special kinds of quadrilaterals can be illustrated in a hierarchy, as shown in Figure 9.40.

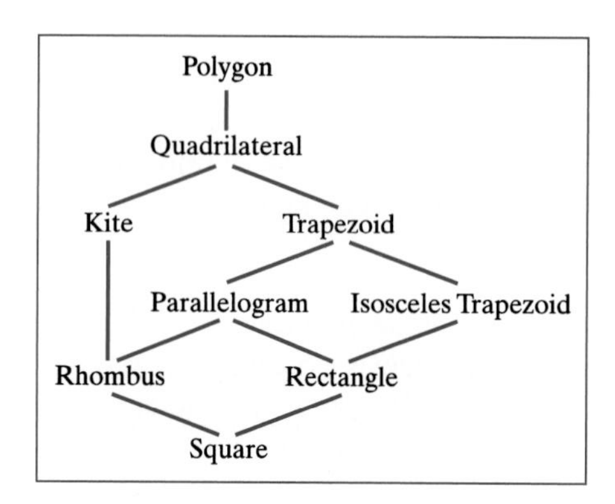

Figure 9.40

Related Topics

Angle Measurement
Symmetry
Area of Quadrilaterals

TOPIC 9.11 Circles

cornerstone **Polygons**
Regular Polygons

Introduction

As the number of sides that a regular polygon has increases, the polygon becomes closer and closer to the shape of a circle. In this section, we discuss features of a circle and the compass.

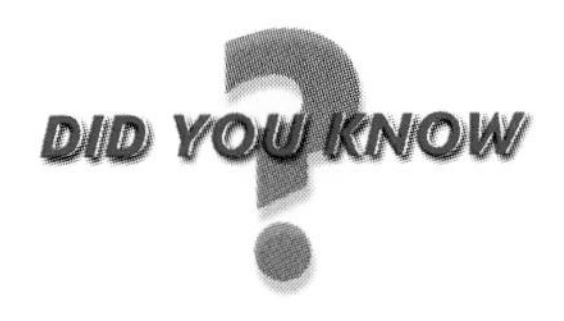

▶ The Babylonians estimated that it takes about 360 days for the sun to travel completely around a circular path. They divided the path into 360 congruent parts. We now call each of these parts a degree, and we say that the measure of a circle is 360°.

For More Information Bunt, L. N. H., Jones, P. S., & Bedient, J. D. (1988). *The Historical Roots of Elementary Mathematics*. New York: Dover Publications.

9.11.1 Features of a Circle

A **circle** is the set of points in a plane that are the same distance from a given point, called the **center.** The **radius** (plural, **radii**) of a circle is the length of a segment with an endpoint on the circle and the center as the other endpoint. The segment is also called a radius. The **diameter** of a circle is the length of a segment that has its endpoints on the circle and passes through the center. The segment is also called a diameter. A **chord** is any line segment that has its endpoints on the circle. Thus, a diameter is a chord (see Figure 9.41).

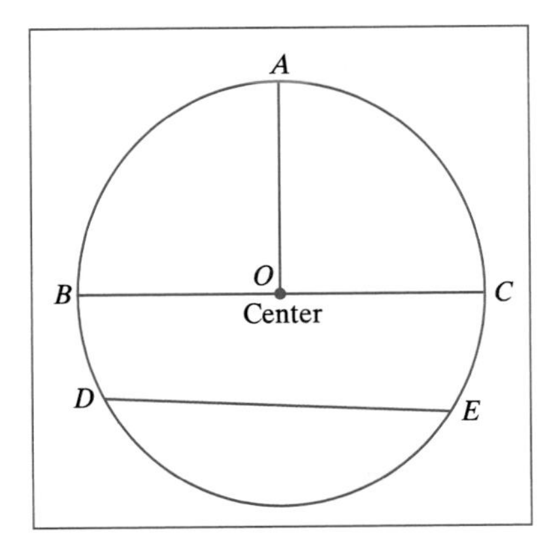

Figure 9.41

An **arc** of a circle is a connected subset of points of the circle. An arc that consists of half of the circle is called a **semicircle**. An arc is named by two points on the circle if the arc is less than half of the circle. If the arc is greater than or equal to half of the circle, we name the arc using three letters, as shown in Figure 9.42.

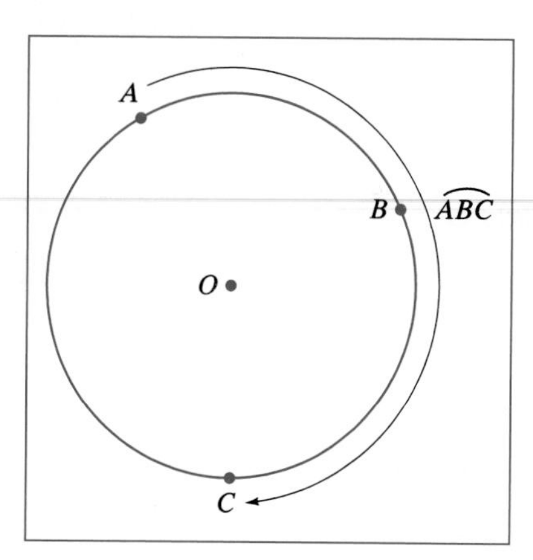

Figure 9.42

9.11.2 The Compass

The ancient Greeks used a straightedge (with no markings) and a collapsible **compass** to construct geometric figures and to demonstrate geometric relationships. The modern compass, shown in Figure 9.43, can be used to draw a circle of a given radius, AB. Open the compass legs to a length of AB, designate a center point, O, put the pointed leg of the compass at O, and move the other leg around O to draw the circle. Two circles are congruent if their radii are congruent.

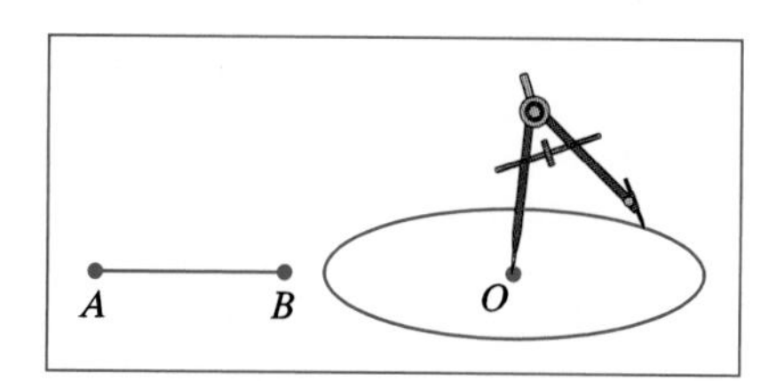

Figure 9.43

Related Topics

Symmetry
Linear Measurement
Perimeter
Circumference
Area of Regular Polygons and Circles

TOPIC 9.12 Congruence

cornerstone

Line Segments
Rays and Angles
Polygons

Introduction

When two geometric objects have the same shape and size, we say that they are **congruent.** Congruent objects are very common in our world because manufacturing of many objects is done by mass-producing objects made out of the same mold. In this section, we discuss congruent segments, angles, and polygons.

The Greek mathematician Euclid did not use the term *congruent* in his famous volumes, *The Elements*; rather, he used the idea of equality to describe how equal line segments and equal angles can be moved in the plane until they coincide.

For More Information Bunt, L. N. H., Jones, P. S., & Bedient, J. D. (1988). *The Historical Roots of Elementary Mathematics.* New York: Dover Publications.

9.12.1 Congruent Segments

Two line segments that have equal lengths are called **congruent segments.** We use the symbol $\cong$ to mean congruent. We say that $\overline{AB} \cong \overline{CD}$ in Figure 9.44 because they are both 5 cm in length.

Figure 9.44

A B C D

9.12.2 Congruent Angles

Two angles are called **congruent angles** if they have the same measures. In Figure 9.45, $\angle RST \cong \angle JLK$ because they both have measures of 70°.

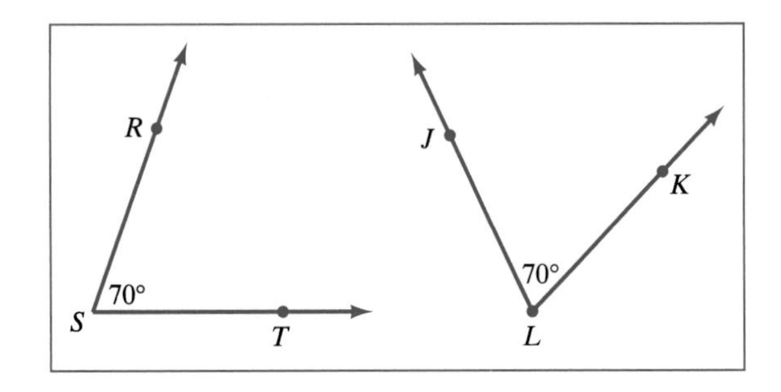

Figure 9.45

9.12.3 Congruent Polygons

Two polygons that are congruent have the same shape and the same size; they can be placed so that they exactly fit one on top of the other. In Figure 9.46, $ABCDE \cong STUVW$.

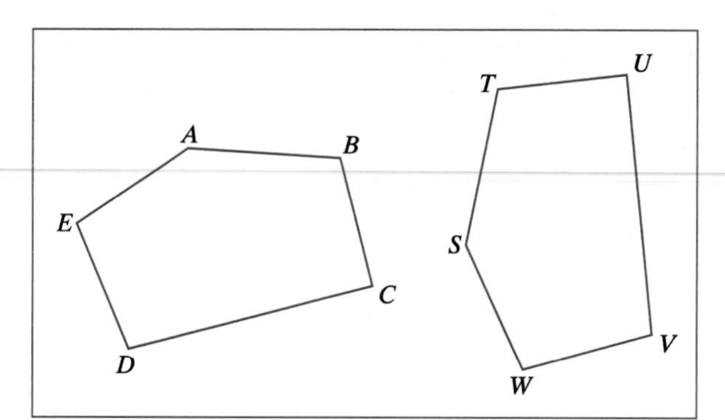

Figure 9.46

Related Topics
Triangles
Quadrilaterals
Angle Measurement

TOPIC 9.13 Symmetry

cornerstone
Polygons
Triangles

Introduction

Repeated patterns occur throughout nature and human-made objects. It is these repeated patterns that underline symmetry. In this section, we discuss reflection, rotational, and plane symmetry.

Hermann Weyl, in his classic book *Symmetry*, wrote: "Symmetric means something like well-proportioned, well-balanced, and symmetry denotes that sort of concordance of several parts by which they integrate into a whole. Beauty is bound up with symmetry" (1952, p. 3).

For More Information Weyl, H. (1952). *Symmetry*. Princeton, NJ: Princeton University Press.

9.13.1 Reflection Symmetry

A figure has **reflection symmetry** (also called **bilateral symmetry** or **line symmetry**) if there is a line (called the **line of symmetry**) such that the reflection of the figure across the line of symmetry is its own image. For example, the figures in Figure 9.47 have reflection symmetry. The second and third figures have more than one line of symmetry.

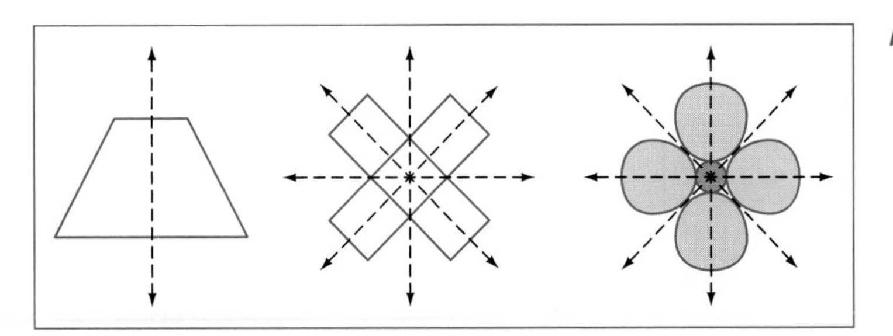

Figure 9.47

9.13.2 Rotational Symmetry

A figure has **rotational symmetry** if the figure can be rotated less than 360° about some point so that the rotated figure coincides with the original figure. (The reason for the "less than 360°" criterion is that any rotated figure will coincide with its original figure when rotated 360°.) Figure 9.48 shows figures that have 60° rotational symmetry and 90° rotational symmetry, respectively. The equilateral triangle also has 120° rotational symmetry, and the square has both 180° rotational symmetry and 270° rotational symmetry. A figure that has 180° rotational symmetry is said to have **point symmetry.**

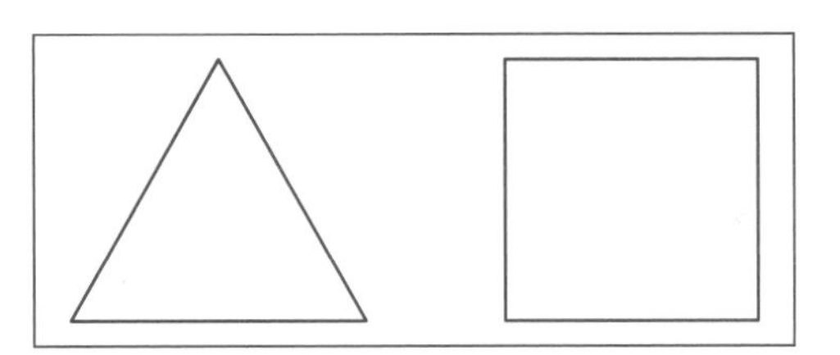

Figure 9.48

9.13.3 Plane Symmetry

Just as it is possible for a two-dimensional figure to have one or more lines of symmetry, a three-dimensional figure can have one or more **planes of symmetry.** A figure has a plane of symmetry when there is a plane such that the reflection of the figure across the plane of symmetry is its own image. Some three-dimensional figures with plane symmetry are shown in Figure 9.49.

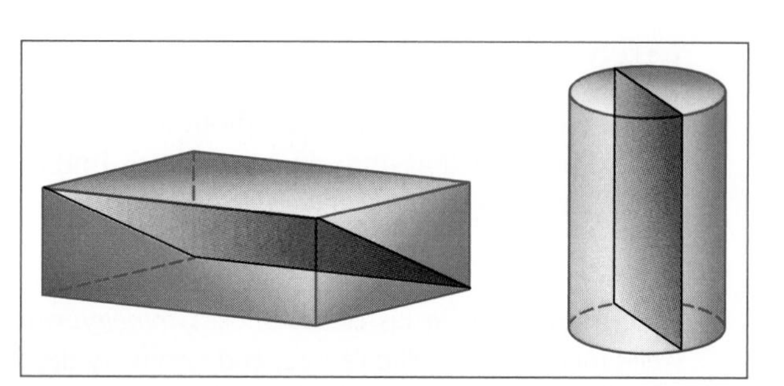

Figure 9.49

Related Topics

Reflections
Rotations

TOPIC 9.14 Line and Angle Constructions

cornerstone **Circles**
Line Segments

Introduction

A geometric construction is a mathematical activity in which we make geometric figures and relationships using only a compass, a straightedge, and given geometric figures. In this section, we discuss geometric tools, how to construct a line segment congruent to a given line segment, and how to construct an angle congruent to a given angle.

The Greek mathematician Euclid used constructions in three of his five postulates that are the basis of the geometric ideas laid out in *The Elements*. The first three postulates are as follows.

Let the following be postulated:

1. To draw a straight line from any point to any point.
2. To produce a finite straight line continuously in a straight line.
3. To describe a circle with any center and distance.

For More Information Sibley, T. Q. (1998). *The Geometric Viewpoint: A Survey of Geometries*. Reading, MA: Addison-Wesley.

9.14.1 Geometric Tools

The ancient Greeks used two tools to construct geometric objects and relationships: a compass and a straightedge. A **compass** is a tool for constructing circles. The compasses used by the ancient Greeks were collapsible. This means that given two points A and B, the collapsible compass could be used to draw a circle with center A passing through B, but the compass could not be moved to another center C to draw a circle of the same radius because the opening of the compass could not be fixed. The type of compass we use today is shown in Figure 9.50. A **straightedge** is a tool for constructing straight lines. It has no markings on it as rulers do.

Figure 9.50

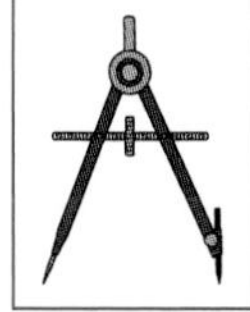

Other geometric tools that we use today are **rulers** (for measuring lengths of line segments) and **protractors** (for measuring an angle in degrees or radians). Although both of these tools are very useful, we typically use only a compass and a straightedge in making geometric constructions in the study of what is referred to as Euclidean geometry.

9.14.2 Constructing a Line Segment Congruent to a Given Line Segment

To construct a line segment on line m congruent to a given line segment, $\overline{FG}$, open the compass so that the pointed end is on F and the pencil end is on G (or vice versa), as in Figure 9.51.

Figure 9.51

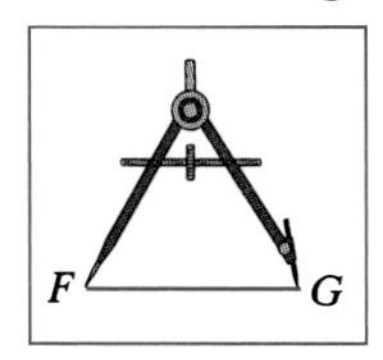

Label a point J on line m and place the pointed end of the compass on J. Keeping the width of the compass open to a length of FG, mark off an arc that intersects line m, and label the point of intersection point K, as in Figure 9.52. Thus, $\overline{JK} \cong \overline{FG}$.

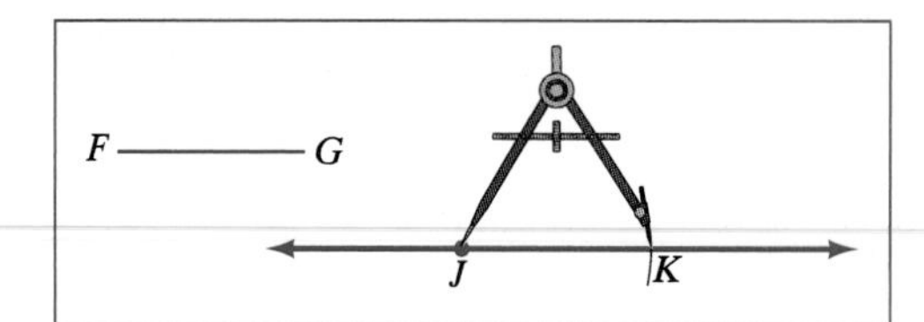

Figure 9.52

9.14.3 Constructing an Angle Congruent to a Given Angle

To construct an angle, with one side on $\overrightarrow{MX}$, congruent to a given angle $\angle P$, place the pointed end of the compass at P, mark off an arc $\overset{\frown}{OQ}$ and then mark off an arc having an equal radius with the center at M (see Figure 9.53). Label the point of the intersection of the arc with $\overrightarrow{MX}$ point N.

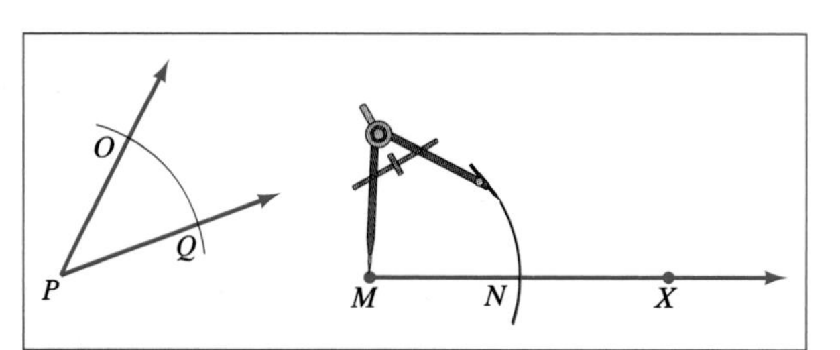

Figure 9.53

Now place the pointed end at Q and open the compass so that the pencil end is at O. Place the pointed end at N and mark off an arc with an equal radius so that this arc intersects the arc that was drawn with center M. Label this point of intersection point L (see Figure 9.54). Draw a ray from M through L. $\angle LMN \cong \angle OPQ$.

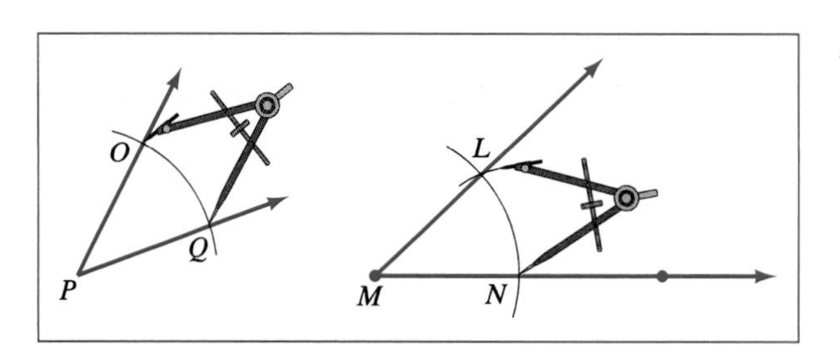

Figure 9.54

Related Topics

Triangle Congruence
Triangle-Congruence Properties
Angle-Bisector Construction

TOPIC 9.15 Angle-Bisector Constructions

cornerstone **Line and Angle Constructions**

Introduction

Sometimes in a geometric construction, we need to separate an angle into two congruent angles. This is known as bisecting an angle. In this section, we discuss what an angle bisector is and how to bisect an angle.

The ancient Greeks were intrigued by challenging geometric constructions. One problem they worked on was trisecting an angle, that is, dividing an angle into three congruent angles. The challenge was to construct one-third of a given angle, using only compass and straightedge. The Greeks were not able to complete this construction using only these tools, although they did make exact constructions using other methods. In 1837, a mathematician named Pierre Wantzel proved, using algebraic means, that trisecting an angle is an impossible task using only compass and straightedge.

For More Information Bunt, L. N. H., Jones, P. S., & Bedient, J. D. (1988). *The Historical Roots of Elementary Mathematics*. New York: Dover Publications.

9.15.1 Angle Bisector

An **angle bisector** is a ray that separates an angle into two congruent angles. In Figure 9.55, two angles with their angle bisectors are shown.

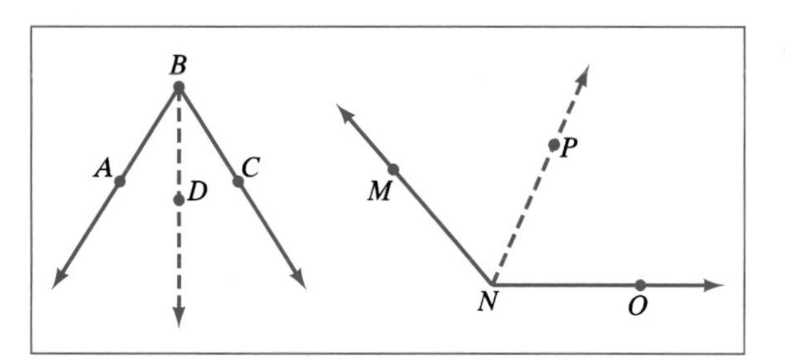

Figure 9.55

9.15.2 Constructing an Angle Bisector

To bisect a given angle $\angle K$, place the pointed end of the compass at K, and mark off an arc $\widehat{JL}$, as shown in the first drawing in Figure 9.56. Place the pointed end at L, and draw an arc as shown in the second drawing in Figure 9.56. Place the pointed end at J, and draw an arc that has the same radius as shown in the third drawing in Figure 9.56. Label the point of intersection of the two arcs point M, and draw $\overrightarrow{KM}$ as shown in the fourth drawing in Figure 9.56. $\overrightarrow{KM}$ is the bisector of $\angle K$, and $\angle JKM \cong \angle MKL$.

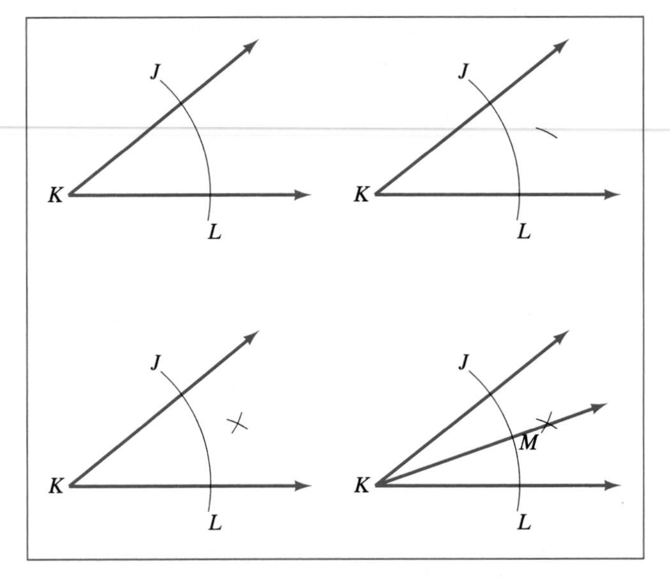

Figure 9.56

TOPIC 9.16 Perpendicular-Line-Through-a-Point Constructions

cornerstone **Line and Angle Constructions**

Introduction

The distance between a line and a point not on the line is the length of the line segment drawn from the point that is perpendicular to the line. In this section, we discuss how to construct a line perpendicular to a line through a point on the line, how to construct a line perpendicular to a line through a point not on the line, and how to construct a perpendicular bisector of a line segment.

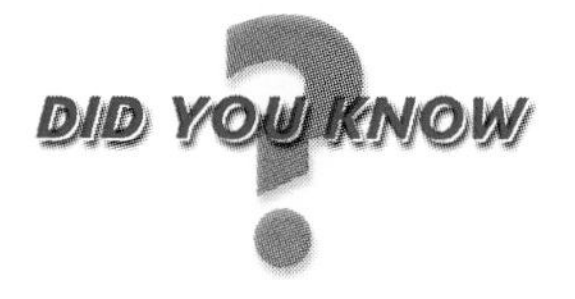

Following the definitions, five postulates, and common notions that Euclid laid out in his geometric work, *The Elements*, he listed propositions (relationships that could be demonstrated using ones already established). The eleventh and twelfth propositions of Book I state:

> *To draw a straight line at right angles to a given straight line from a given point on it. To a given infinite straight line, from a given point which is not on it, to draw a perpendicular straight line.*

For More Information Sibley, T. Q. (1998). *The Geometric Viewpoint: A Survey of Geometries.* Reading, MA: Addison-Wesley.

9.16.1 Constructing a Line Perpendicular to a Line Through a Point on the Line

To construct a line perpendicular to a given line n, through a point C on line n, place the pointed end of the compass at C, and mark off an arc that intersects line n in two points, as shown in the first drawing in Figure 9.57. Label these two points B and D. Open the compass to a wider setting, and draw intersecting arcs with centers at B and D, as shown in the second drawing in Figure 9.57. Label the two points of intersection points E and F. Draw the line through E and F. $\overleftrightarrow{EF}$ is perpendicular to line n through point C.

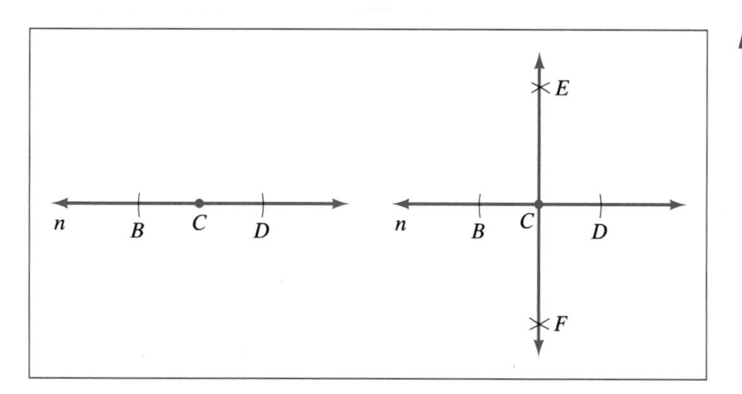

Figure 9.57

9.16.2 Constructing a Line Perpendicular to a Line Through a Point Not on the Line

To construct a line perpendicular to a given line r, through a point Q not on line r, place the pointed end of the compass at Q, and draw an arc that intersects line r in two points, as shown in the first drawing in Figure 9.58. Label these two points R and S.

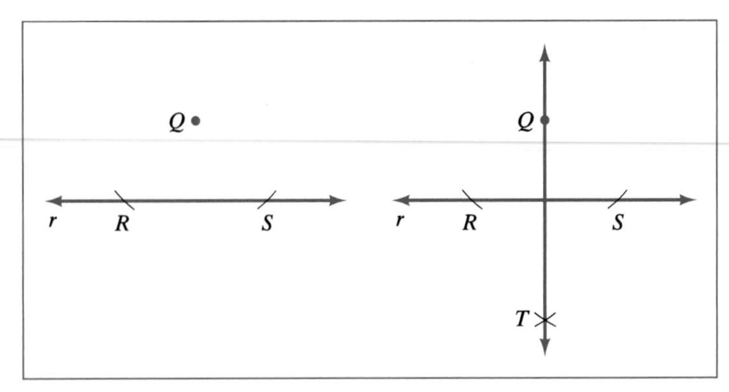

Keeping the compass open to the same width, draw intersecting arcs with centers at R and S as shown in the second drawing in Figure 9.58. Label the point of intersection T. Draw the line through T and Q. $\overleftrightarrow{TQ}$ is perpendicular to line r through point Q.

9.16.3 Constructing a Perpendicular Bisector of a Line Segment

A **perpendicular bisector** of a line segment is a line that is perpendicular to the line segment at its midpoint. To construct a perpendicular bisector of a given line segment, $\overline{FG}$, draw intersecting arcs with centers at F and G, as shown in the first drawing in Figure 9.59. Label the points of intersection points H and I. Draw the line through H and I as shown in the second drawing in Figure 9.59. Label the point of its intersection with $\overline{FG}$ point J. $\overleftrightarrow{HI}$ is the perpendicular bisector of $\overline{FG}$, and J is the midpoint of $\overline{FG}$.

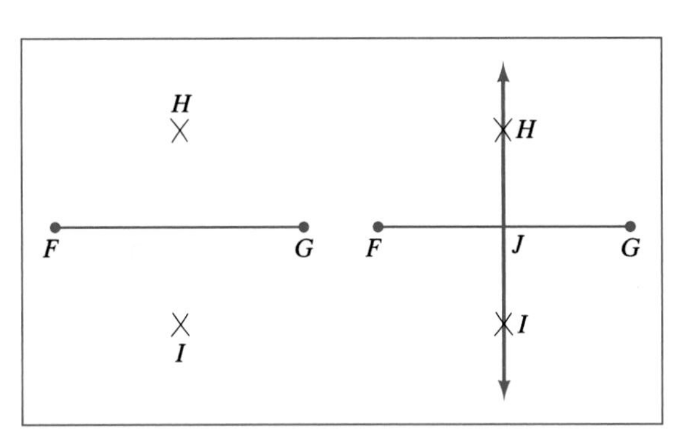

Figure 9.59

Related Topics Parallel-Line Construction

Parallel-Line Construction

cornerstone **Line Relationships**

Introduction

Many geometric relationships use parallel lines as a basis for the relationship. In this section, we discuss how to construct a line parallel to a line through a point not on the line.

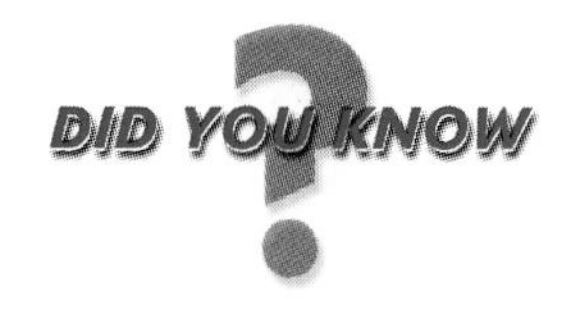

Using Euclid's first three postulates (listed in Section 9.14) and his fourth postulate—All right angles are equal to one another—allows for the construction of parallel lines.

For More Information Sibley, T. Q. (1998). *The Geometric Viewpoint: A Survey of Geometries.* Reading, MA: Addison-Wesley.

To construct a line parallel to a given line s, through a point G not on line s, draw any line t through G that intersects line s, as shown in the first drawing in Figure 9.60. Label the point of intersection point F. Place the pointed end of the compass at F, and draw an arc with radius FG that intersects line s, as shown in the second drawing in Figure 9.60. Label the point of intersection point E. Keeping the compass open to the same width, draw intersecting arcs with centers at E and G, as shown in the third drawing in Figure 9.60. Label the point of intersection D. Draw the line through G and D. $\overleftrightarrow{GD}$ is parallel to line s through point G.

Figure 9.60

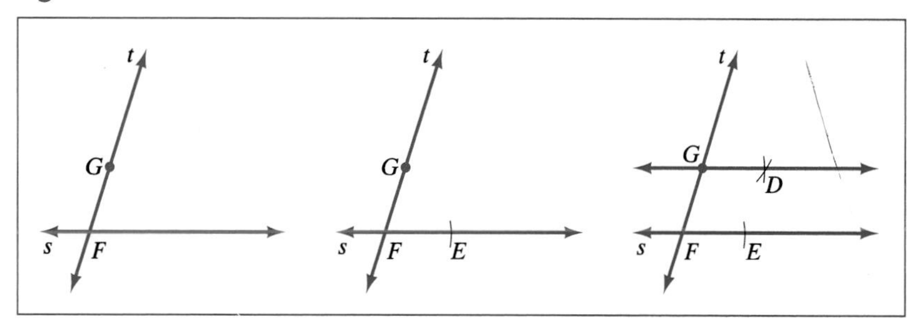

Related Topics **Perpendicular-Line-Through-a-Point Constructions**

TOPIC 9.18 Circumscribed-Circle Constructions

Circles
Triangles
Perpendicular-Line-Through-a-Point Constructions

Introduction

There are some special relationships between circles and triangles. One of these relationships is that every triangle can have a circle circumscribed around it. In this section, we discuss circumscribed circles and how to circumscribe a circle around a triangle.

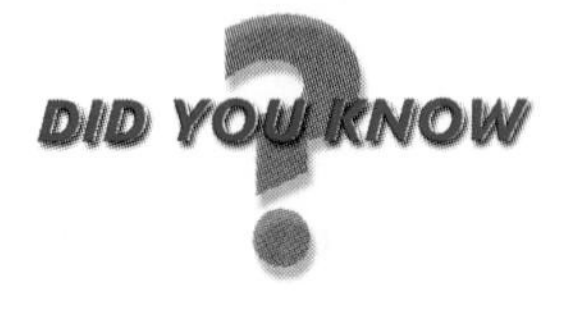

Another of the famous construction problems from ancient Greece was the task of constructing a square whose area is equal to the area of a given circle, using only compass and straightedge. As with the angle-trisection problem, the Greeks were able to construct this using other methods but were not able to do the construction using only compass and straightedge. In 1882, the German mathematician Ferdinand Lindemann proved that this task is impossible using only compass and straightedge by proving that the number π is transcendental.

For More Information Bunt, L. N. H., Jones, P. S., & Bedient, J. D. (1988). *The Historical Roots of Elementary Mathematics*. New York: Dover Publications.

9.18.1 Circumscribed Circles

A **circumscribed circle** is a circle that surrounds a polygon such that the vertices of the polygon lie on the circle. The polygons shown in Figure 9.61 are circumscribed by circles.

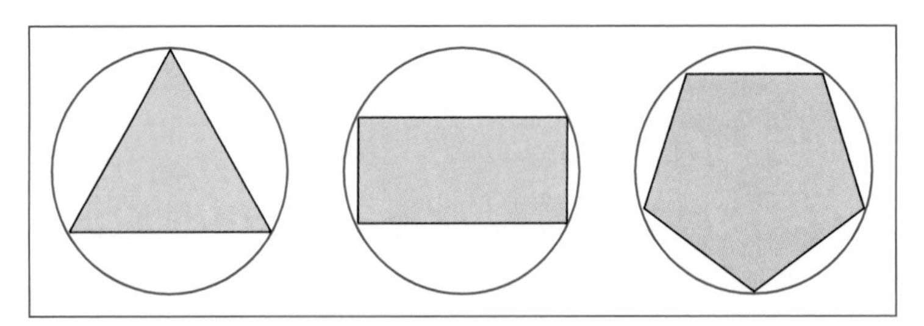

Figure 9.61

9.18.2 Constructing a Circumscribed Circle About a Triangle

To construct the circumscribed circle about a given triangle XYZ, we want to construct a point T that is equidistant from points X, Y, and Z. To accomplish this, construct the perpendicular bisector of each side of triangle XYZ, as shown in Figure 9.62.

The intersection of these perpendicular bisectors is a point called the **circumcenter** of the triangle. Label this point T. Construct the circle with center T and radius TX, as shown in Figure 9.63. This circle circumscribes triangle XYZ.

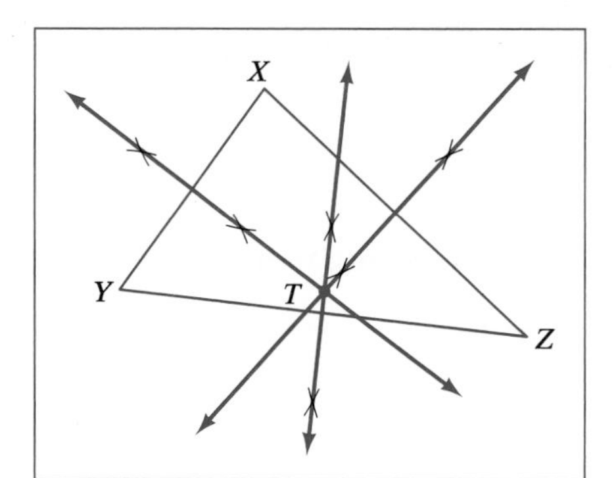

Figure 9.62

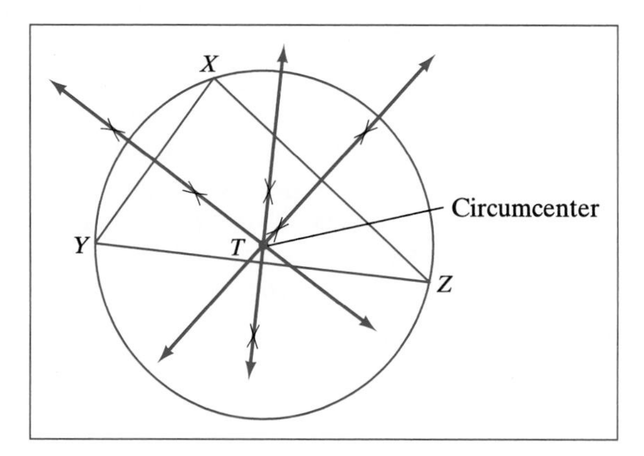

Figure 9.63

Related Topics

Inscribed-Circle Constructions

TOPIC 9.19 Inscribed-Circle Constructions

Circles
Triangles
Perpendicular-Line-Through-a-Point Constructions Circles

Introduction

Another special relationship between triangles and circles is that every triangle can have a circle inscribed within it. In this section, we discuss inscribed circles, how to inscribe a circle within a triangle, and orthocenters and centroids.

The third famous construction problem from ancient Greece was the task of constructing the face of a cube whose volume is twice the volume of another cube, using only compass and straightedge. As with the angle-trisection problem and the squaring-the-circle problem, the Greeks were able to construct this using other methods but were not able to do the construction using only compass and straightedge. In 1837, the mathematician Pierre Wantzel proved that this task, along with the angle-trisection problem, is impossible using only compass and straightedge.

For More Information Bunt, L. N. H., Jones, P. S., & Bedient, J. D. (1988). *The Historical Roots of Elementary Mathematics*. New York: Dover Publications.

9.19.1 Inscribed Circles

An **inscribed circle** is a circle that is surrounded by a polygon such that the inscribed circle intersects each side of the polygon at exactly one point. The polygons shown in Figure 9.64 have circles inscribed within them.

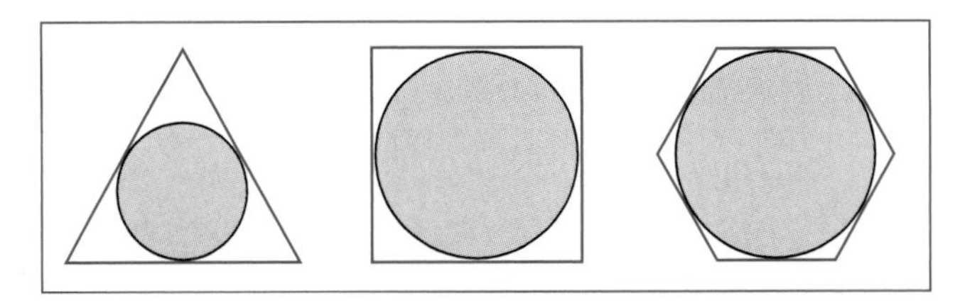

Figure 9.64

9.19.2 Constructing an Inscribed Circle Within a Triangle

To construct the circle inscribed within a given triangle JKL, we want to construct a point M that is equidistant from $\overline{JK}$, $\overline{JL}$, and $\overline{KL}$. To accomplish this, construct the angle bisector of each angle of triangle JKL, as shown in Figure 9.65.

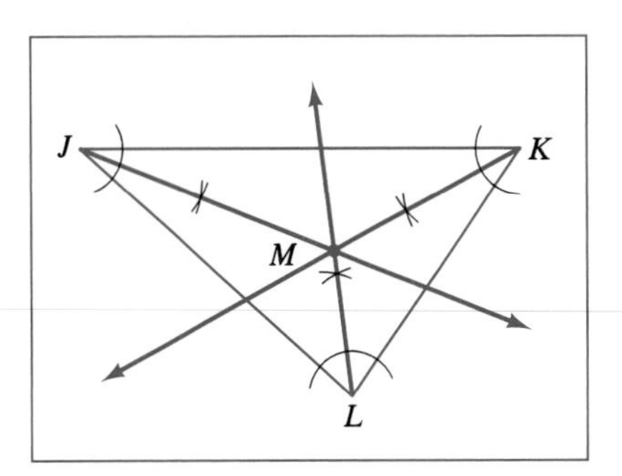

Figure 9.65

The intersection of these angle bisectors is a point called the **incenter** of the triangle. Label this point M. Construct a line perpendicular to $\overline{JK}$ that passes through M. Label the point of intersection of this line with $\overline{JK}$ point N. Construct the circle with center M and radius MN, as shown in Figure 9.66. This circle is inscribed in triangle JKL.

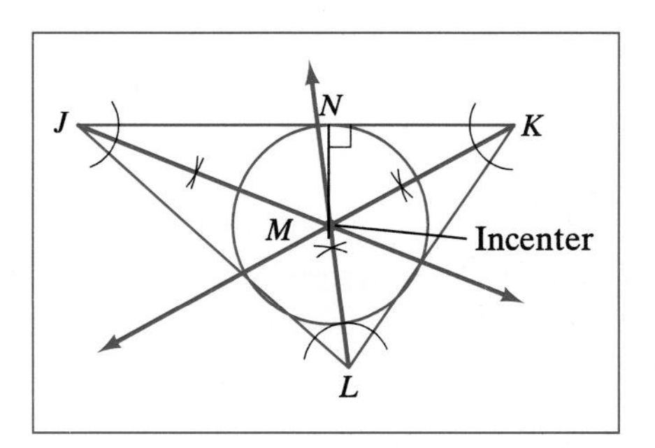

Figure 9.66

9.19.3 Orthocenters and Centroids

Besides circumcenters and incenters, there are two other centers connected with triangles. The **orthocenter** of a triangle is the point of intersection of the triangle's altitudes. The **centroid** of a triangle is the point of intersection of the triangle's medians. The orthocenter and centroid of a triangle are shown in Figure 9.67.

Figure 9.67

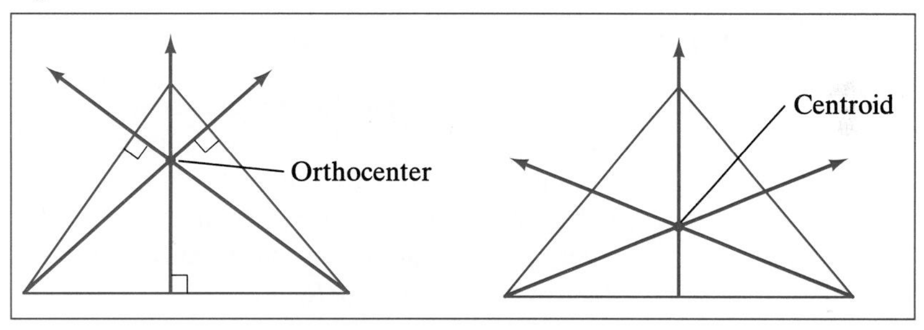

Related Topics **Triangle-Congruence Properties**

TOPIC 9.20 Triangle Congruence

cornerstone

Line and Angle Constructions
Parallel-Line Construction
Perpendicular-Line-Through-a-Point Constructions

Introduction

When two figures are congruent, they must have the same size and shape. In this section, we discuss congruent triangles and their features and how two triangles may or may not be congruent.

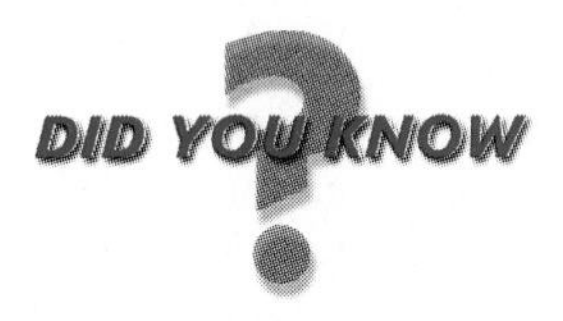

Euclid used two senses of equal—equal in measure (angle, length, area, or volume) and congruent. He showed that two figures are congruent from the equal measures of corresponding parts.

For More Information Sibley, T. Q. (1998). *The Geometric Viewpoint: A Survey of Geometries.* Reading, MA: Addison-Wesley.

9.20.1 Congruent Triangles and Their Features

Two triangles are **congruent** if they coincide when one is placed on the other. In other words, their corresponding angles and corresponding sides are congruent. Corresponding angles and sides are the ones that match when one triangle is placed on a congruent triangle. Two congruent triangles, ΔSTU and ΔPQR, are shown in Figure 9.68. These triangles are congruent because $\angle S \cong \angle P$, $\angle T \cong \angle Q$, $\angle U \cong \angle R$, $\overline{ST} \cong \overline{PQ}$, $\overline{TU} \cong \overline{QR}$, and $\overline{SU} \cong \overline{PR}$.

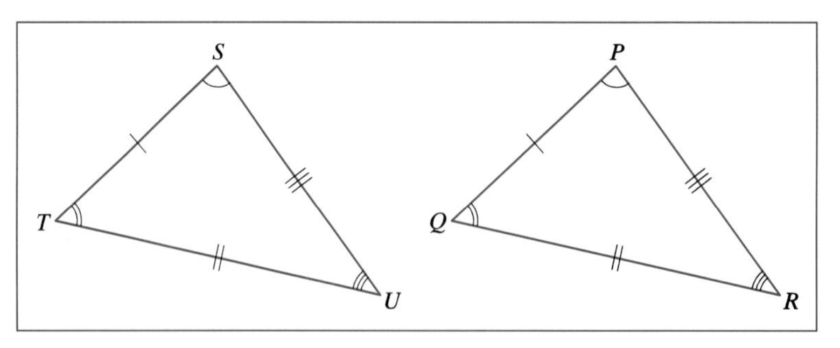

Figure 9.68

9.20.2 How Two Triangles May or May Not be Congruent

Suppose two triangles, ΔABC and ΔDEF, are congruent to each other, as shown in Figure 9.69. The expression $\Delta ABC \cong \Delta DEF$ says not merely that ΔABC and ΔDEF are congruent, but also that they are congruent in a particular way (i.e., under a particular one-to-one correspondence between vertices). Thus, although $\Delta ABC \cong \Delta DEF$, ΔABC is not necessarily congruent to this triangle with the vertices listed in another order (e.g., ΔDFE).

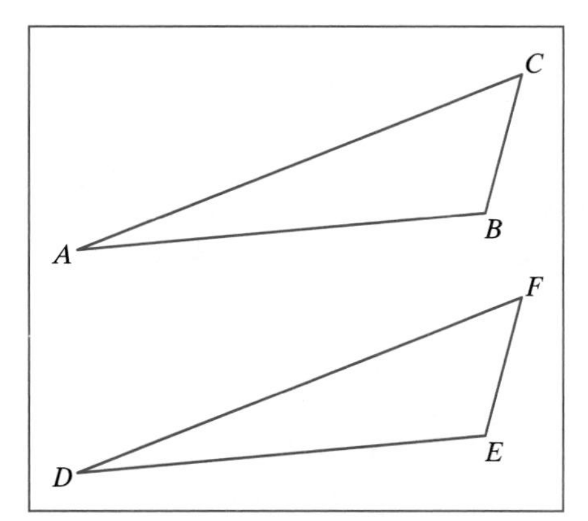

Figure 9.69

Related Topics

Triangle-Congruence Properties

TOPIC 9.21 Triangle-Congruence Properties

cornerstone **Triangles**
Congruence

Introduction

When two triangles are congruent to each other, their corresponding angles and sides are congruent. However, to show that two triangles are congruent, we do not always need to show that all six corresponding parts are congruent. There are certain cases when, if three corresponding parts are congruent, the other corresponding parts are forced to be congruent. In this section, we discuss the side-side-side, side-angle-side, angle-side-angle, angle-angle-side, and isosceles-triangle-congruence properties.

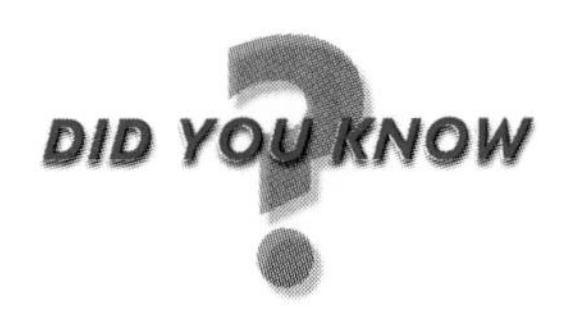

A phrase known to almost every student who has completed a course in high school geometry is "Corresponding parts of congruent triangles are congruent."

9.21.1 Side-Side-Side Triangle-Congruence Property

If three sides of one triangle are congruent to the three corresponding sides of a second triangle, then the triangles are congruent. This property is called the **Side-Side-Side (SSS) Triangle-Congruence Property** and is illustrated in Figure 9.70.

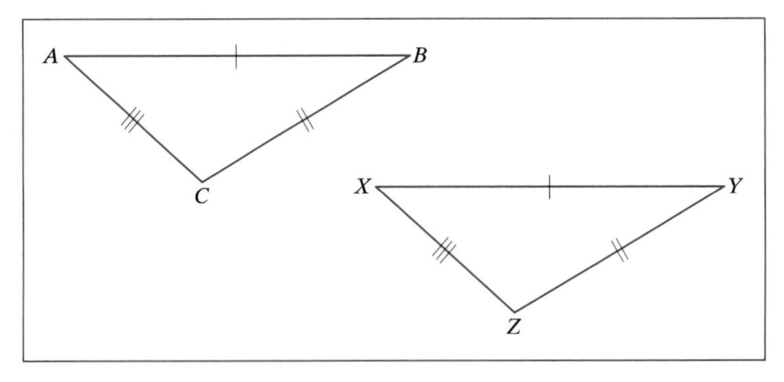

Figure 9.70

Because $\overline{AB} \cong \overline{XY}$, $\overline{BC} \cong \overline{YZ}$, and $\overline{AC} \cong \overline{XZ}$, we know $\Delta ABC \cong \Delta XYZ$.

9.21.2 Side-Angle-Side Triangle-Congruence Property

If two sides and the included angle (the angle formed by these two sides) of one triangle are congruent to the corresponding two sides and included angle of a second triangle, then the triangles are congruent. This property is called the **Side-Angle-Side (SAS) Triangle-Congruence Property** and is illustrated in Figure 9.71.

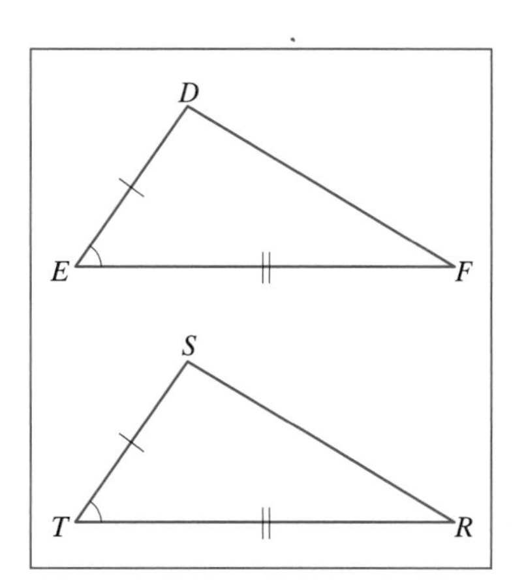

Figure 9.71

Because $\overline{DE} \cong \overline{ST}$, $\overline{EF} \cong \overline{TR}$, and $\angle E \cong \angle T$, we know $\Delta DEF \cong \Delta STR$.

9.21.3 Angle-Side-Angle Triangle-Congruence Property

If two angles and the included side (the side that is common to the two angles) of one triangle are congruent to the corresponding two angles and included side of a second triangle, then the triangles are congruent. This property is called the **Angle-Side-Angle (ASA) Triangle-Congruence Property** and is illustrated in Figure 9.72.

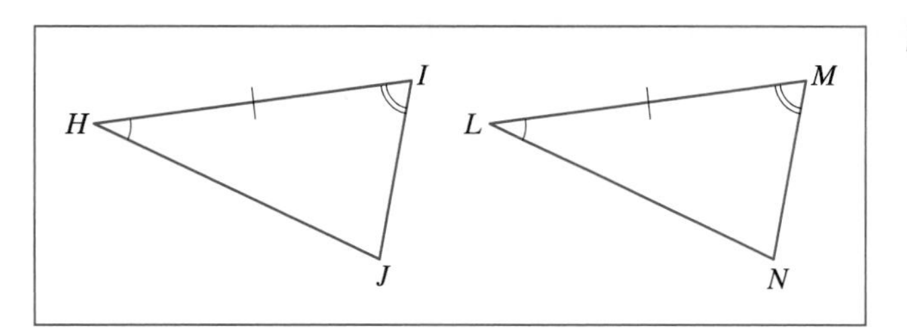

Figure 9.72

Since $\angle H \cong \angle L$, $\overline{HI} \cong \overline{LM}$, and $\angle I \cong \angle M$, we know $\Delta HIJ \cong \Delta LMN$.

9.21.4 Angle-Angle-Side Triangle-Congruence Property

If two angles and a nonincluded side of one triangle are congruent to the corresponding two angles and side of a second triangle, then the triangles are congruent. This property is called the **Angle-Angle-Side (AAS) Triangle-Congruence Property** and is illustrated in Figure 9.73.

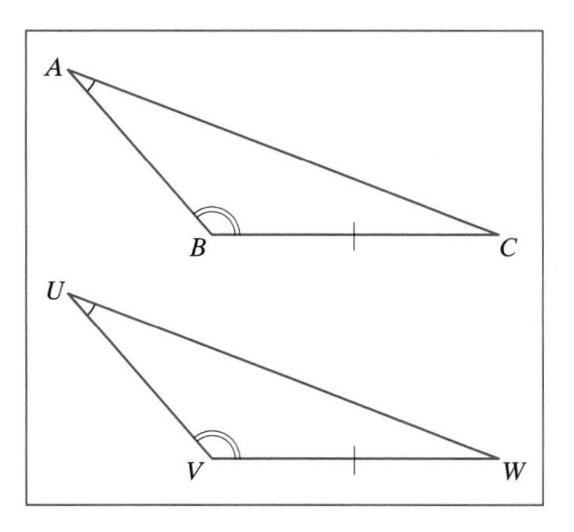

Figure 9.73

Because $\angle A \cong \angle U$, $\angle B \cong \angle V$, and $\overline{BC} \cong \overline{VW}$, we know $\Delta ABC \cong \Delta UVW$.

9.21.5 Isosceles-Triangle-Congruence Properties

Isosceles triangles have some properties that are true for every isosceles triangle. (Recall that an isosceles triangle is a triangle with at least two sides congruent.) These are: (a) Angles that are opposite congruent sides are congruent, and (b) the altitude drawn from the vertex common to the congruent sides bisects this vertex angle and is the perpendicular bisector of the opposite side. These properties are illustrated in Figure 9.74.

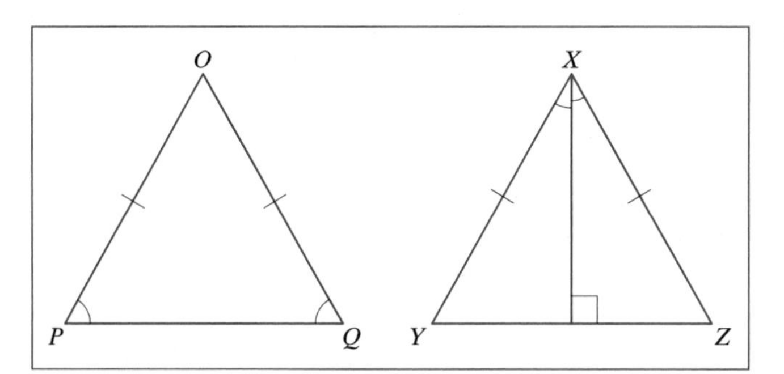

Figure 9.74

Related Topics Quadrilateral Properties

Quadrilateral Properties

cornerstone

Regular Polygons
Triangle-Congruence Properties

Introduction

Some special types of quadrilaterals have particular congruence properties, along with other relationships. In this section, we discuss some properties for trapezoids, parallelograms, rectangles, kites, rhombi, and squares.

The series of numbers, 1, 1, 2, 3, 5, 8, 13, …, is known as a Fibonacci sequence. The ratio of adjacent numbers in the sequence gets closer and closer to a fixed number known as the Golden Ratio. This number is approximately 1.618. A rectangle that has sides in this ratio is called a Golden Rectangle; many examples of the Golden Rectangle can be found in ancient Greek architecture.

For More Information Garland, T. H. (1987). *Fascinating Fibonaccis: Mystery and Magic in Numbers.* Palo Alto, CA: Dale Seymour.

Runion, G. E. (1990). *The Golden Section.* Palo Alto, CA: Dale Seymour.

9.22.1 Properties of Trapezoids

Recall that the sum of the measures of the interior angles in any convex quadrilateral is 360°. Because a trapezoid is a quadrilateral with at least one pair of parallel sides, it must be convex. A property of trapezoids is that consecutive angles contained between parallel lines are supplementary. We will prove this relationship.

Given trapezoid $ABCD$ with $\overline{AB} \parallel \overline{DC}$, draw a line through $\overline{AB}$ and $\overline{DC}$ that is perpendicular to these sides (as shown in Figure 9.75). We can draw this line because if a line m is perpendicular to a line n in a plane, line m is perpendicular to all other lines that are parallel to line n. Label the points of intersection E and F (as shown in Figure 9.75). We know that $\angle AEF$ and $\angle EFD$ are right angles because they are formed by perpendicular lines. Thus, $m\angle AEF = 90°$ and $m\angle EFD = 90°$. Because the sum of the measures of the interior angles in $AEFD$ is 360°, then $m\angle FDA + m\angle DAE = 180°$, and thus these angles are supplementary. The same argument can be used to show that $\angle EBC$ and $\angle BCF$ are supplementary.

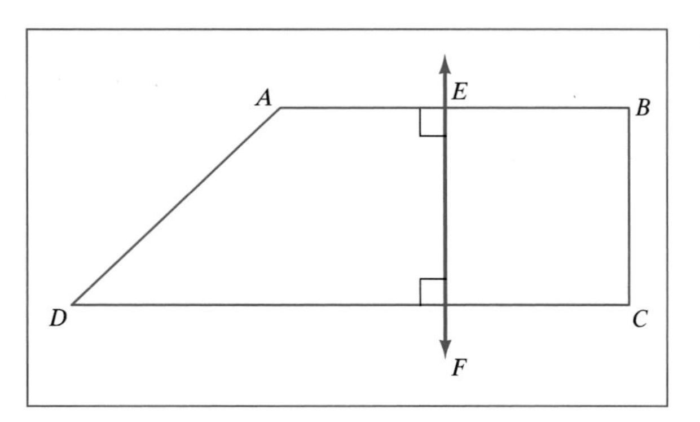

Figure 9.75

9.22.2 Properties of Parallelograms

Certain properties of parallelograms are especially noteworthy in the study of geometry:

- Opposite sides and angles of parallelograms are congruent.
- Diagonals in a parallelogram bisect each other.
- If a quadrilateral has both pairs of opposite sides congruent, it is a parallelogram.
- If a quadrilateral has two sides that are parallel and congruent, it is a parallelogram.
- If the diagonals of a quadrilateral bisect each other, it is a parallelogram.
- Each diagonal of a parallelogram separates it into two congruent triangles.

Because a parallelogram is a trapezoid, properties of a trapezoid are also valid for a parallelogram. We will prove two of these properties.

To prove that opposite sides and angles are congruent in a parallelogram, we will use a triangle-congruence property. Given parallelogram $QRST$, draw diagonals $\overline{TR}$ and $\overline{QS}$, and label their intersection point P, as shown in Figure 9.76. Because opposite sides of a parallelogram are parallel, and because alternate interior angles formed by parallel lines cut by a transversal are congruent, then $\angle TRS \cong \angle RTQ$ and $\angle RTS \cong \angle TRQ$. Because $\overline{TR} \cong \overline{TR}$ (every segment is congruent to itself), then $\Delta TRS \cong \Delta RTQ$ by the Angle-Side-Angle Triangle-Congruence Property. Then, because corresponding parts of congruent triangles are congruent, $\overline{TS} \cong \overline{RQ}$, $\overline{QT} \cong \overline{SR}$, and $\angle S \cong \angle Q$. We can use a similar argument to show that $\Delta QTS \cong \Delta SRQ$, and thus $\angle T \cong \angle R$.

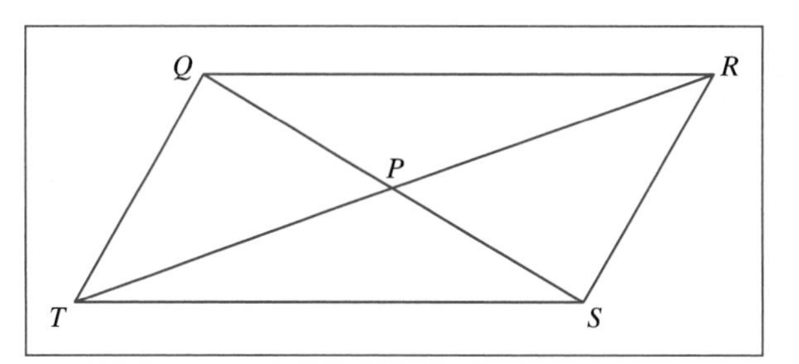

Figure 9.76

We will also use parallelogram $QRST$ in Figure 9.76 to prove that the diagonals of a parallelogram bisect each other. Consider ΔTPS and ΔRPQ. We know that $\angle PTS \cong \angle PRQ$ and $\angle PST \cong \angle PQR$ because they are alternate interior angles formed by a transversal cutting parallel lines. Because opposite sides in a parallelogram are congruent, $\overline{TS} \cong \overline{RQ}$. Then, $\Delta TPS \cong \Delta RPQ$ by the Angle-Side-Angle Triangle-Congruence Property. Because corresponding parts of congruent triangles are congruent, $\overline{PS} \cong \overline{PQ}$ and $\overline{PR} \cong \overline{PT}$. Thus, P must be the midpoint of $\overline{QS}$ and $\overline{TR}$, and so we have shown that the diagonals of a parallelogram bisect each other.

9.22.3 Properties of Rectangles

Because rectangles are parallelograms, all the properties that are true for parallelograms are also valid for rectangles. Additionally, rectangles have the following properties:

- A rectangle has four right angles.
- Any quadrilateral with four right angles must be a rectangle.

- The diagonals in a rectangle are congruent.
- Any quadrilateral with diagonals that bisect each other and are congruent is a rectangle.

We will prove the property that the diagonals of a rectangle are congruent.

Given rectangle $MNOP$ with diagonals $\overline{MO}$ and $\overline{NP}$, as shown in Figure 9.77, we know that $\overline{MP} \cong \overline{NO}$ and $\overline{MN} \cong \overline{OP}$ because opposite sides of a rectangle are congruent. We also know that all four angles in a rectangle are right angles, and because all right angles are congruent, $\angle MNO \cong \angle NOP$. Thus, $\Delta MNO \cong \Delta NOP$ by the Side-Angle-Side Triangle-Congruence Property. Because corresponding parts of congruent triangles are congruent, $\overline{MO} \cong \overline{NP}$. This shows that the diagonals in a rectangle are congruent.

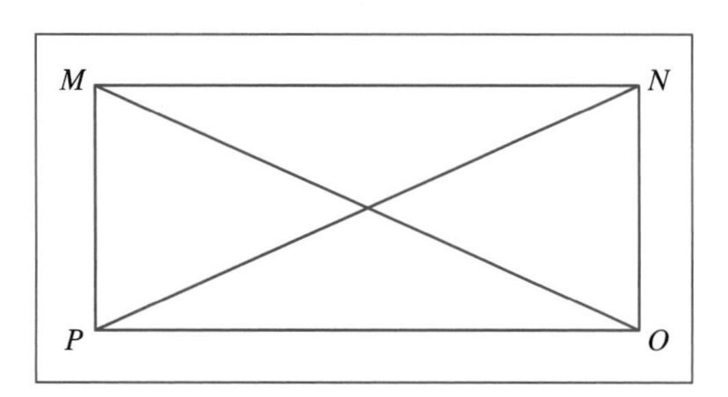

Figure 9.77

9.22.4 Properties of Kites

Some properties of a kite are:

- Diagonals are perpendicular to each other.
- Exactly one diagonal is the perpendicular bisector of the other diagonal.

We will prove that the diagonals are perpendicular to each other using kite $GHIJ$ with diagonals $\overline{IG}$ and $\overline{JH}$, as shown in Figure 9.78. Label the point of intersection point K. By definition of a kite, $\overline{JG} \cong \overline{GH}$ and $\overline{JI} \cong \overline{IH}$. Because $\overline{IG} \cong \overline{IG}$, then $\Delta GJI \cong \Delta GHI$ by the Side-Side-Side Triangle-Congruence Property. Because corresponding parts of congruent triangles are congruent, $\angle JGK \cong \angle HGK$. Using the fact that $\overline{GK} \cong \overline{GK}$, then $\Delta GJK \cong \Delta GHK$ by the Side-Angle-Side Triangle-Congruence Property. Because corresponding parts of congruent triangles are congruent, $\angle GKJ \cong \angle GKH$, and because these angles are linear pair angles, they each must be right angles. Thus, the diagonals are perpendicular to each other.

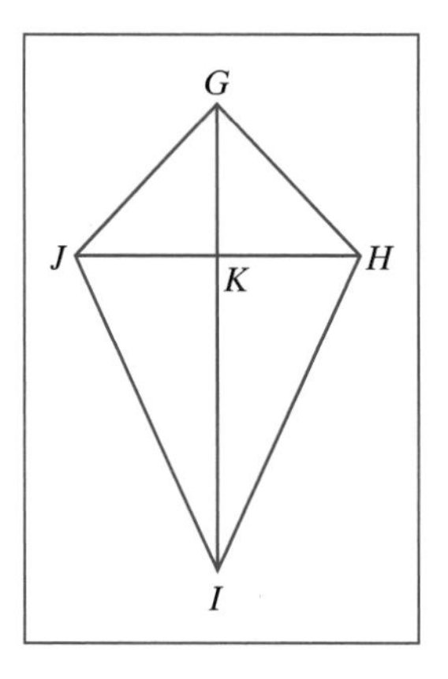

Figure 9.78

9.22.5 Properties of Rhombi

Because a rhombus is both a parallelogram and a kite, all the properties of a parallelogram and a kite also hold for a rhombus. Additionally, a rhombus has the following properties:

- If a quadrilateral has all sides congruent, it is a rhombus.
- Diagonals bisect opposite angles.

We will prove that in a rhombus, the diagonals bisect opposite angles. Given rhombus $ABCD$ with diagonals $\overline{AC}$ and $\overline{BD}$, as shown in Figure 9.79, we know from the proof showing that the diagonals in a kite are perpendicular to each other that $\Delta ABC \cong \Delta ADC$. Because corresponding parts of congruent triangles are congruent, we know that $\angle DAC \cong \angle BAC$ and $\angle DCA \cong \angle BCA$. Thus, we know that $\angle DAB$ and $\angle BCD$ are bisected by $\overline{AC}$. A similar argument can be used to show that $\angle ABC$ and $\angle CDA$ are bisected by $\overline{DB}$.

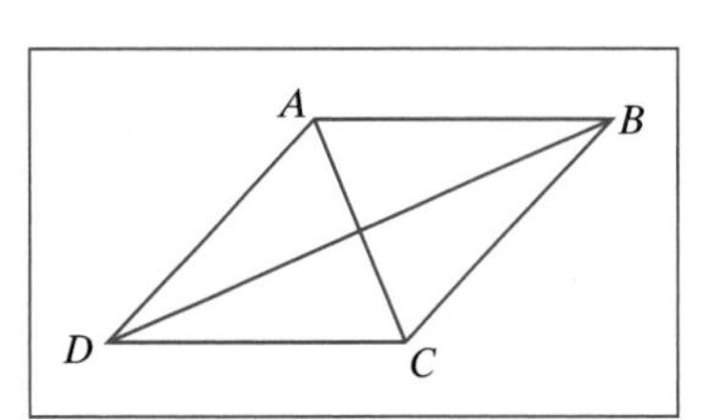

Figure 9.79

9.22.6 Properties of Squares

Because a square is both a rectangle and a rhombus, it has all the properties of a rectangle and a rhombus.

Related Topics Similarity

Similarity

cornerstone **Triangles**
Triangle-Congruence Properties

Introduction

The mathematical ideas of similarity and scale have been studied for thousands of years because they are so useful in solving problems. In this section, we discuss similarity and similar figures, similar triangles, and indirect measurement.

A Greek merchant named Thales, who lived about 600 B.C., is said to have determined the height of a pyramid in Egypt by using the length of its shadow.

For More Information Bunt, L. N. H., Jones, P. S., & Bedient, J. D. (1988). *The Historical Roots of Elementary Mathematics*. New York: Dover Publications.

9.23.1 Similarity and Similar Figures

Informally speaking, two geometric figures are **similar** if they have exactly the same shape but not necessarily the same size. More precisely, two geometric figures are said to be **similar** if their corresponding angles are congruent and their corresponding sides are proportional. The figures shown in Figure 9.80 are similar figures.

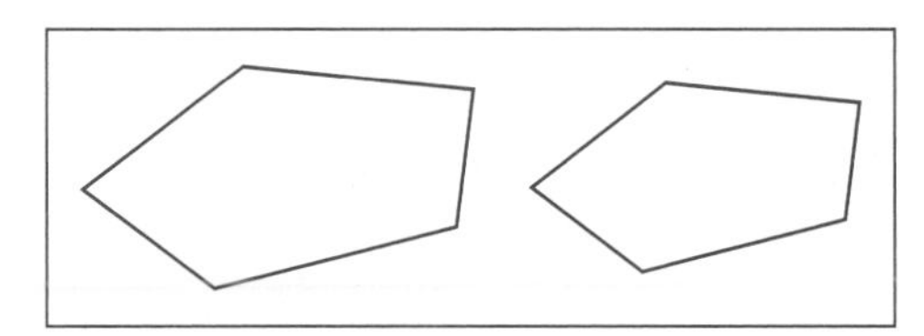

Figure 9.80

9.23.2 Similar Triangles

ΔJKL is **similar** to ΔMNO (written $\Delta JKL \sim \Delta MNO$) if and only if $\angle J \cong \angle M$, $\angle K \cong \angle N$, $\angle L \cong \angle O$, and $\frac{JK}{MN} = \frac{KL}{NO} = \frac{JL}{MO}$.In other words, two triangles are congruent if and only if their corresponding angles are congruent and their corresponding sides are proportional.

In the section called Triangle-Congruence Properties, we discuss how two triangles can be shown to be congruent if minimal congruence relationships are established. Likewise, two triangles can be shown to be similar if minimal conditions are satisfied. There are three similarity properties for triangles:

- If three sides of one triangle are proportional to the three corresponding sides of a second triangle, then the triangles are similar. This property is called the **Side-Side-Side (SSS) Triangle-Similarity Property** and is illustrated in the first drawing in Figure 9.81.
- If two sides of one triangle are proportional to two sides of a second triangle and the included angles of these triangles are congruent, then the triangles are

Figure 9.81

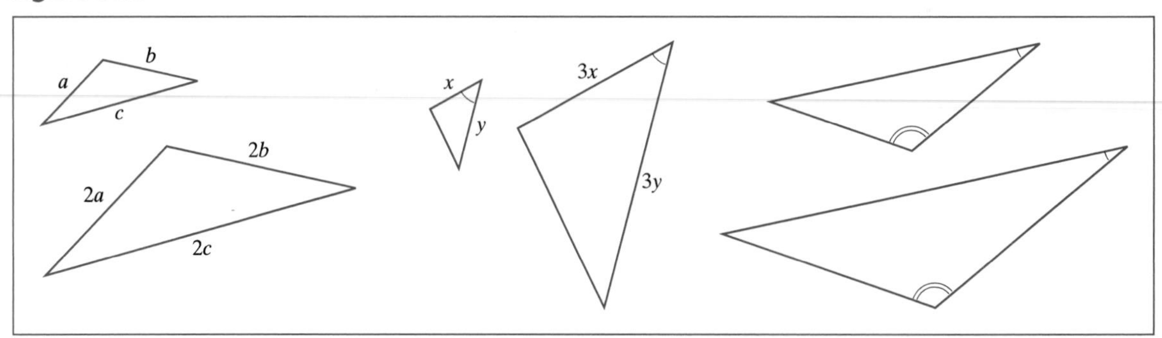

similar. This property is called the **Side-Angle-Side (SAS) Triangle-Similarity Property** and is illustrated in the second drawing in Figure 9.81.

- If two angles of one triangle are congruent to two angles of a second triangle (which means the third pair of angles is also congruent), then the triangles are similar. This property is called the **Angle-Angle (AA) Triangle-Similarity Property** and is illustrated in the third drawing in Figure 9.81.

9.23.3 Indirect Measurement

Similar triangles can be useful in measuring lengths of objects that cannot be measured directly. We call this indirect measurement. For example, if you wanted to measure the height of a building on a sunny day, you could use your shadow and the building's shadow to determine two triangles, as shown in Figure 9.82.

Figure 9.82

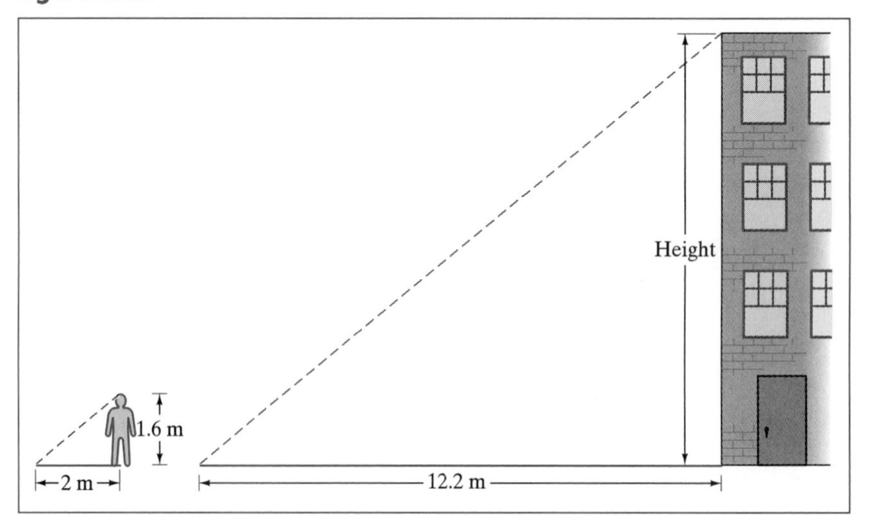

Because both you and the building are (we are assuming) forming right angles with the ground and the sun is shining down on you and the building at the same angle, the two triangles formed are similar. Thus, we can use the fact that similar triangles have proportional sides to find the height of the building. Suppose you are 1.6 m tall, your shadow is 2.0 m long, and the building's shadow is 12.2 m long. To find the height of the building, we have the proportion:

$$\frac{1.6\text{ m}}{2.0\text{ m}} = \frac{\text{height of building}}{12.2\text{ m}}$$

$$1.6\text{ m}(12.2\text{ m}) = 2.0\text{ m (height of the building)}$$

$$\frac{1.6\text{ m}(12.2\text{ m})}{2.0\text{ m}} = \text{height of the building}$$

$$9.76\text{ m} = \text{height of the building}$$

Related Topics

Triangle Proportion Properties
Fractals and Rep-Tiles
Size Transformations

TOPIC 9.24 Triangle Proportion Properties

cornerstone **Similarity**

Introduction

There are a number of proportion properties that are relationships within a triangle. In this section, we discuss one of these properties and its converse.

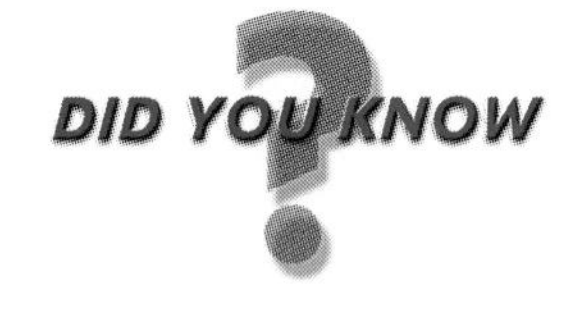

Ideas of proportionality are used in many fields, such as medicine, film, architecture, photography, and integrated circuits.

Two proportion properties for a triangle arise from a line intersecting two sides of a triangle. One property is the converse of the other; this means that one property has the hypothesis and the implication reversed from the other property. To illustrate a statement and its converse, consider the following statement about a square: If a quadrilateral is a square, then it must also be a rectangle. The converse of this statement is: If a quadrilateral is a rectangle, then it must also be a square. In this case, the statement is true, but the converse is not true.

One of the properties states: If a line parallel to one side of a triangle intersects the other sides, then it divides the intersected sides into proportional segments. We will prove this relationship using ΔDEF and line l parallel to $\overline{EF}$ and intersecting $\overline{DE}$ and $\overline{DF}$ in points G and H, as shown in Figure 9.83. Because $\overline{GH} \parallel \overline{FE}$, then $\angle DGH \cong \angle DFE$ and $\angle DHG \cong \angle DEF$ because they are corresponding angles, and $\Delta DGH \sim \Delta DFE$ by the Angle-Angle Triangle-Similarity Property. Thus, $\frac{DF}{DG} = \frac{DE}{DH}$, which can be written as

$$\frac{DG + GF}{DG} = \frac{DH + HE}{DH}$$

$$\frac{DG}{DG} + \frac{GF}{DG} = \frac{DH}{DH} + \frac{HE}{DH}$$

$$1 + \frac{GF}{DG} = 1 + \frac{HE}{DH}$$

$$\frac{GF}{DG} = \frac{HE}{DH}$$

which is the same as $\frac{DG}{GF} = \frac{DH}{HE}$,

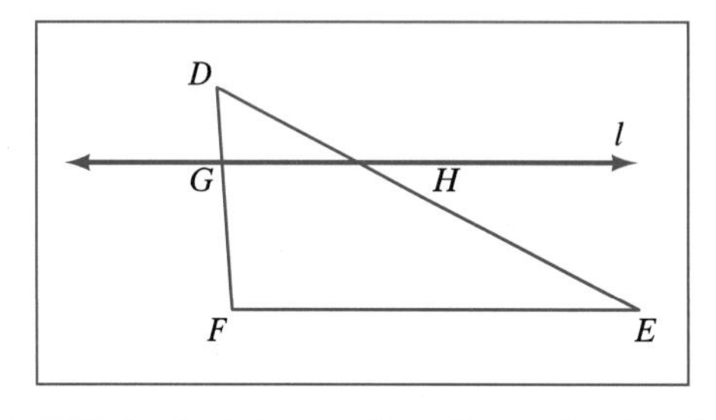

Figure 9.83

and thus the line divides the intersected sides into proportional segments.

The converse of this property is also true: If a line divides two sides of a triangle into proportional segments, then that line is parallel to the third side. If points G and H in Figure 9.83 are midpoints of $\overline{DF}$ and $\overline{DE}$, respectively, then the segment, $\overline{GH}$, that connects the midpoints of two sides of a triangle is called a **midsegment** of the triangle.

Related Topics

Size Transformations

TOPIC 9.25 Fractals and Rep-Tiles

cornerstone **Similarity**

Introduction

Although the geometry studied in the K–12 curriculum is usually limited to fairly simple, ideal shapes (e.g., circles, polygons, polyhedron), geometric shapes in nature come in a wide variety, often unrelated to the shapes studied in school. In this section, we discuss two especially interesting shapes: fractals and rep-tiles.

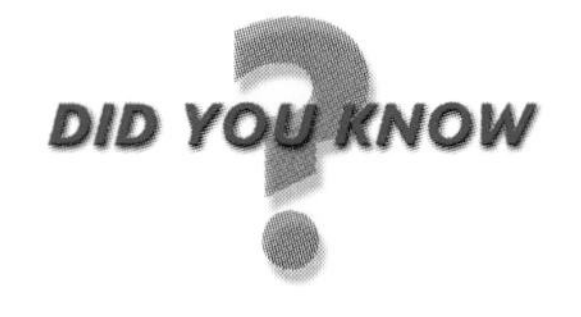

Although Benoit Mandelbrot coined the word *fractal* in 1975, mathematicians had begun to study the curves related to fractals before 1900.

For More Information Sibley, T. Q. (1998). *The Geometric Viewpoint: A Survey of Geometries.* Reading, MA: Addison-Wesley.

9.25.1 Fractals

Benoit Mandelbrot used the word **fractal** to describe chaotic curves and surfaces that can be used to represent shapes that occur in nature, such as mountains, coastlines, and even human lungs. These shapes are generally self-similar; in other words, smaller subsets of the shape's outline have the same outline as larger subsets of the shape's outline. A well-known example of a self-similar shape is called the Sierpinski triangle. This triangle is formed by starting with an equilateral triangle and removing equilateral triangles that are one-fourth the size of the original from the centers of the triangles formed in the previous stage. This process is illustrated in Figure 9.84.

Figure 9.84

Another well-known fractal is called the Koch curve. The construction begins with a segment that we separate into thirds. In the next step, we construct an equilateral - triangle "bump" in the middle third of the segment. In the next step, we construct equilateral-triangle bumps on the middle third of each side of the big triangle. We now continue to repeat this pattern infinitely many times. (See Figure 9.85.)

Figure 9.85

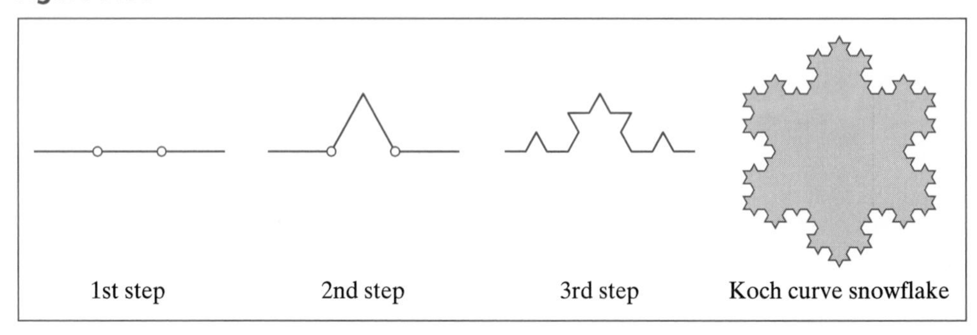

There are two important aspects to this curve: First, each new triangle that is constructed is similar to the original triangle—it will look exactly the same no matter how many times it is magnified; second, the curve has a finite area—the entire picture can be placed inside a circle. The area of the entire diagram cannot be more than the area of the circle. However, the Koch curve has an infinite perimeter! We thus have an amazing geometric figure here—physically, you may think of it as being a fixed amount of land that needs an infinite amount of wire fencing to cover its boundaries!

9.25.2 Rep-tiles

A **rep-tile** is a figure that serves as a building block for making a larger figure similar to the building block. For example, an equilateral triangle and a square are rep-tiles in the figures shown in Figure 9.86, which are similar to the rep-tiles, respectively.

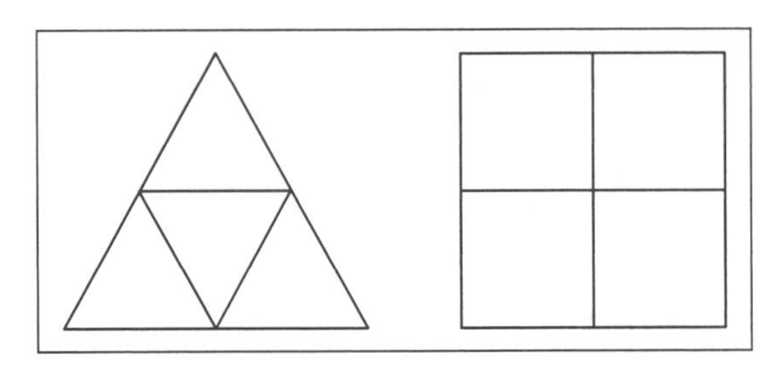

Figure 9.86

Related Topics

Size Transformations

TOPIC 9.26 Three-Dimensional Geometry Basics

cornerstone

Planes
Line Relationships
Rays and Angles

Introduction

An important part of the study of geometry deals with three-dimensional geometry. After all, the world in which we live is three dimensional. In this section, we discuss plane relationships and dihedral angles.

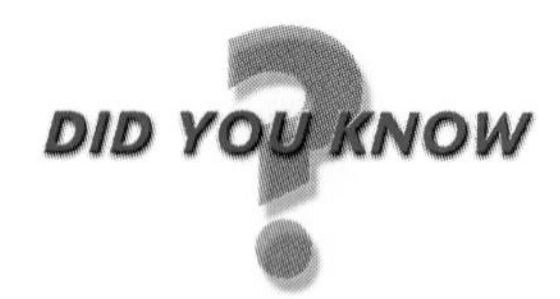

▸ The Greek philosopher Plato (427–347 B.C.) is said to have had a sign over the entrance to his academy that read: "Let no one ignorant of geometry enter here."

9.26.1 Plane Relationships

When planes intersect each other, they have various possible points of intersection. Two planes may not have any points in common and would be parallel planes. Think of the plane of a door intersecting the plane of a wall; those two planes intersect and have many points in common. In fact, if two distinct planes intersect, they intersect in a line, as in Figure 9.87.

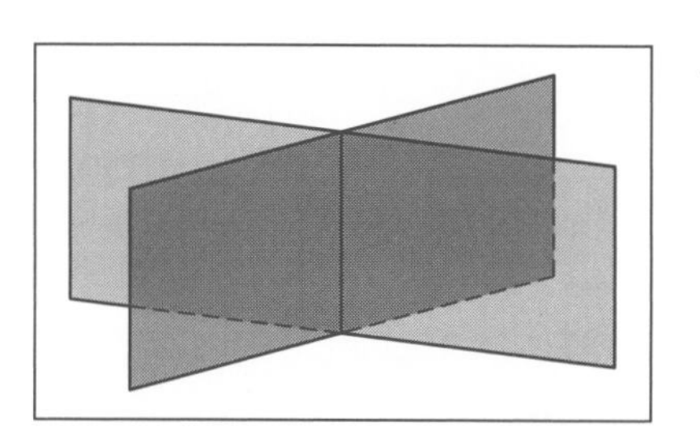

Figure 9.87

There are four possible ways that three distinct planes can relate to each other. They could have no points in common and thus be parallel to each other. They could have exactly one point in common among all three planes, as shown in Figure 9.88 in the illustration of the box. They could have a line in common, as shown in the illustration of a booklet where each page represents a plane. Two planes could also be parallel to each other and the third plane intersects each of the other two planes.

Figure 9.88

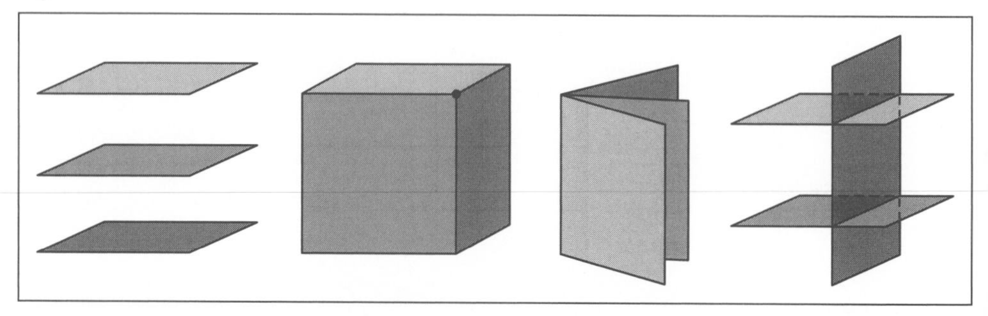

9.26.2 Dihedral Angles

When two planes intersect, they form an angle. This angle is called a **dihedral angle** and is formed by the line of intersection of the two planes and the union of the two half-planes. Whenever two planes intersect, there are four dihedral angles formed, as shown in Figure 9.89.

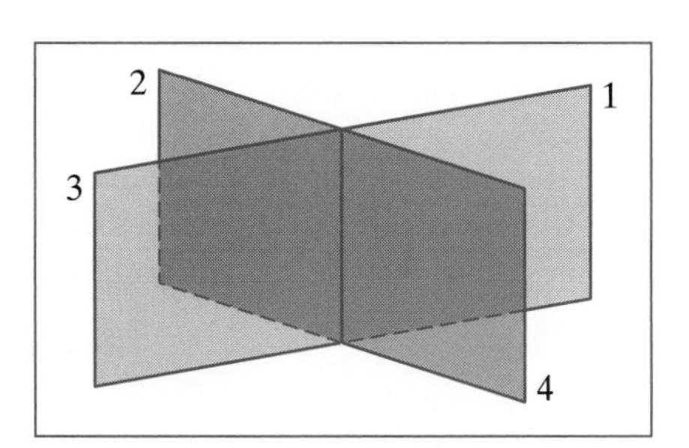

Figure 9.89

Related Topics

Angle Measurement
Polyhedra, Prisms, and Pyramids
Regular Polyhedra
Cylinders, Cones, and Spheres

TOPIC 9.27 Polyhedra, Prisms, and Pyramids

cornerstone **Three-dimensional Geometry Basics**

Introduction

All around us, in our everyday world, are types of three-dimensional figures. Many of these figures are polyhedra. In this section, we discuss a simple closed surface, a polyhedron and its features, convex and concave polyhedra, prisms, and pyramids.

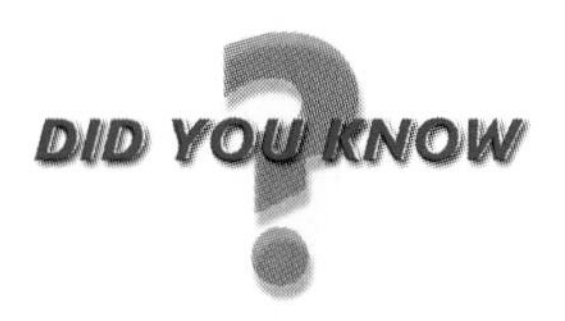

Leonard Euler (1707–1783), a Swiss mathematician, developed a formula that shows the relationship among the number of vertices, edges, and faces for many polyhedra, including all convex ones. This relationship has a number of applications in geometry, graph theory, and topology.

For More Information Sibley, T. Q. (1998). *The Geometric Viewpoint: A Survey of Geometries.* Reading, MA: Addison-Wesley.

9.27.1 A Simple Closed Surface

A **simple closed surface** is a three-dimensional surface that is closed and has exactly one interior, has no holes, and is hollow. In the same way that a polygon separates the plane into three disjoint sets, a simple closed surface separates space into three disjoint sets. These are the interior of the surface, the surface, and the exterior of the surface. A simple closed surface and its interior together make up the set of points known as a **solid.** (See Figure 9.90.)

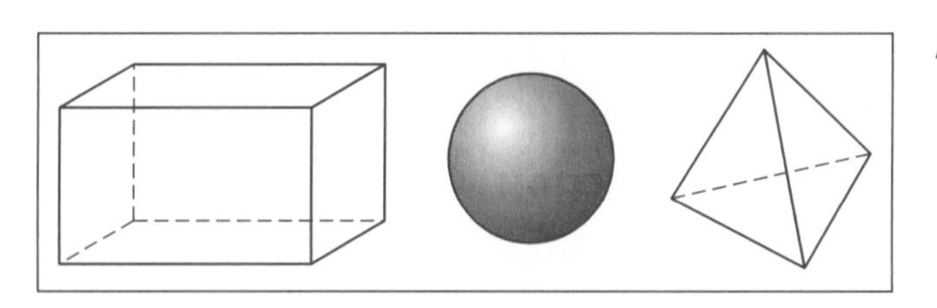

Figure 9.90

9.27.2 A Polyhedron and its Features

A **polyhedron** is a simple closed surface composed entirely of polygonal regions. The word *polyhedron* comes from *poly* meaning "many" and *hedron* meaning "flat surfaces." Figure 9.91 shows some examples of polyhedra and nonpolyhedra. The nonpolyhedra are not polyhedra because they either have holes, are not closed, or are not composed entirely of polygonal regions.

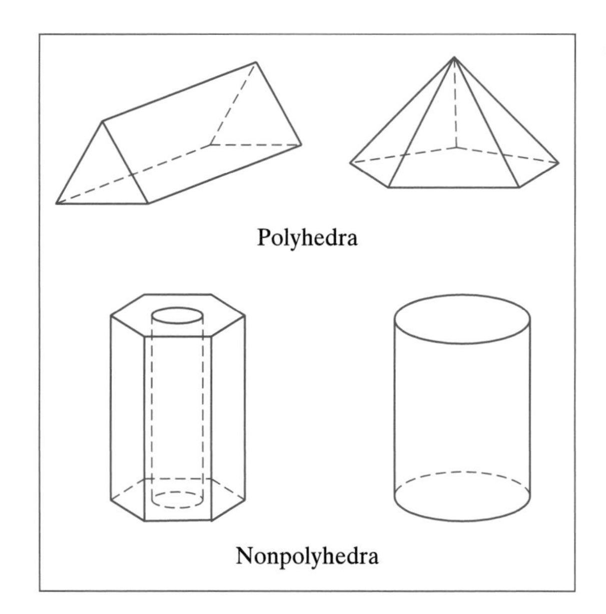

Figure 9.91

The polygonal regions that comprise a polyhedron are called the **faces** and the vertices of the faces are called the **vertices** of the polyhedron. The line segment where two faces meet is called an **edge** (see Figure 9.92).

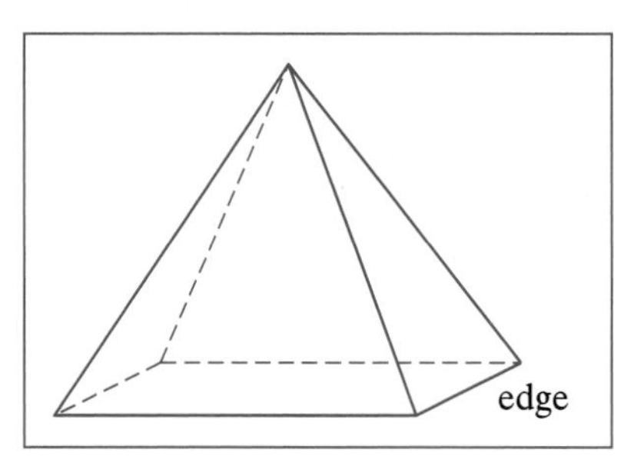

Figure 9.92

9.27.3 Convex and Concave Polyhedra

A polyhedron is **convex** if and only if, for two distinct points in the interior of the polyhedron, the segment that joins those points, is contained entirely in the interior of the polyhedron. If the segment that joins these two points does not lie in the interior of the polyhedron, then the polyhedron is **concave.** Figure 9.93 shows convex and concave polyhedra.

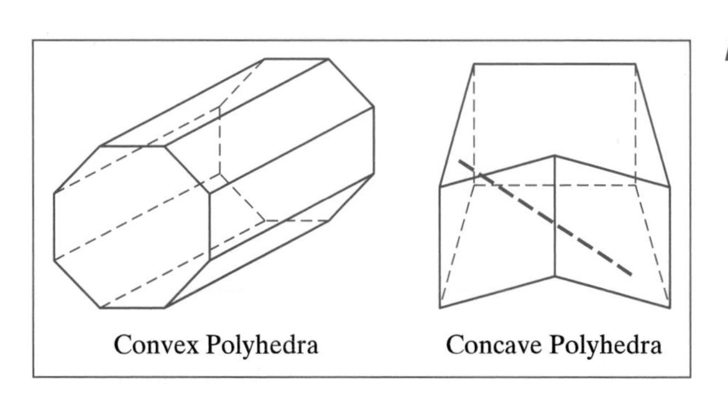

Figure 9.93

9.27.4 Prisms

A **prism** is a polyhedron comprised of two congruent faces that are in parallel planes and parallelograms for the remaining faces. The two congruent faces that are parallel to each other are called the **bases** of the prism, and the remaining parallelogram faces are called the **lateral faces** of the prism, as shown in Figure 9.94.

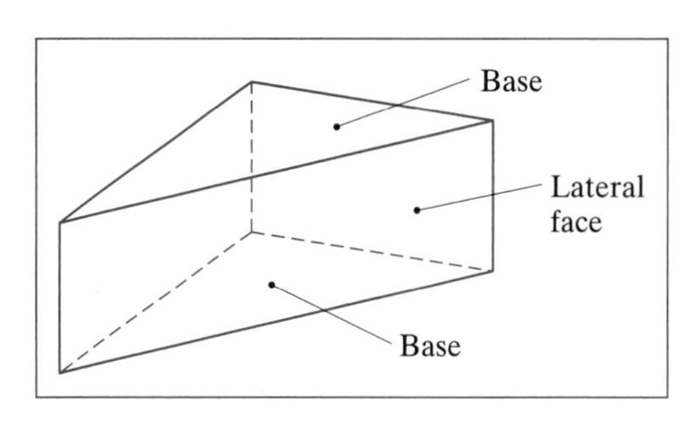

Figure 9.94

If the lateral faces are all rectangles, then the prism is called a **right** prism. If all the lateral faces are not rectangles, the prism is called an **oblique prism**. A prism is also named by its bases, as shown in Figure 9.95.

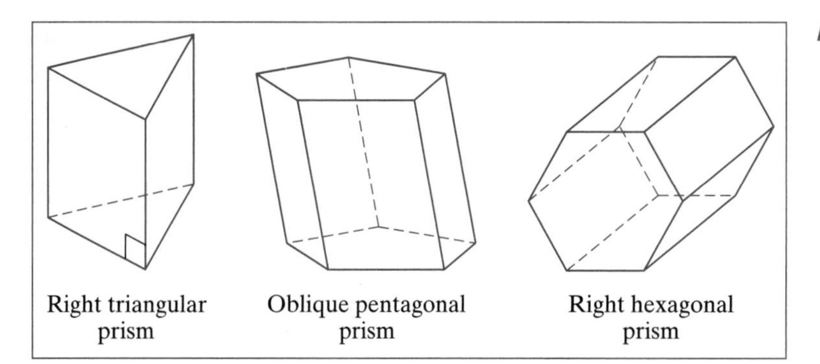

Figure 9.95

9.27.5 Pyramids

A **pyramid** is a polyhedron that is comprised of a polygon with n edges for its **base,** a point not in the plane of the base called the **apex,** and n triangular faces that are determined by the apex and the n edges of the base. The triangular faces are called the **lateral faces.** If the base of a pyramid is a regular polygon and the lateral faces are congruent isosceles triangles, the pyramid is called a **right regular pyramid.** If the pyramid's base is regular but the lateral faces are not isosceles triangles, the pyramid is called an **oblique regular pyramid.** A pyramid is also named by its base, as shown in Figure 9.96.

Figure 9.96

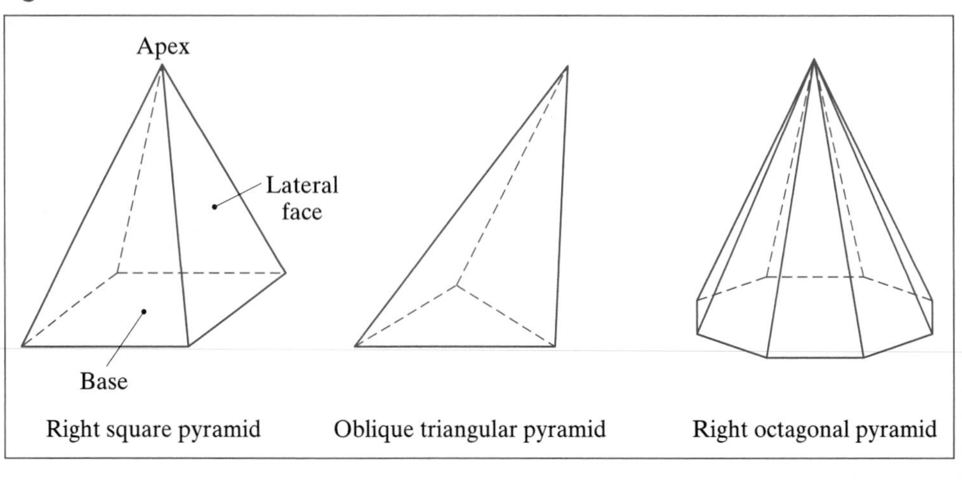

Related Topics

Regular Polyhedra
Cylinders, Cones, and Spheres
Cavalieri's Principle
Surface Area of Prisms and Cylinders
Surface Area of Pyramids, Cones, and Spheres
Volume of Prisms and Cylinders
Volume of Pyramids, Cones, and Spheres

TOPIC 9.28 Regular Polyhedra

cornerstone **Three-dimensional Geometry Basics**
Polyhedra, Prisms, and Pyramids

Introduction

Regular polyhedra are particular kinds of polyhedra, and as we will see, there are a limited number of regular polyhedra. In this section, we discuss regular and semiregular polyhedra.

DID YOU KNOW

One of the marks of the Greek philosopher Plato's influence on the development of mathematics is the fact that the regular polyhedra are often called the Platonic solids, due to his fascination with these solids. Plato's writings about the regular polyhedra are among the oldest writings about polyhedra that survive.

For More Information Bunt, L. N. H., Jones, P. S., & Bedient, J. D. (1988). *The Historical Roots of Elementary Mathematics*. New York: Dover Publications.

9.28.1 Regular Polyhedra

A **regular polyhedron** is a convex polyhedron comprised of congruent regular polygons where at each vertex of the polyhedron there are the same number of edges meeting. There are exactly five regular polyhedra, as shown in the following figure.

Figure 9.97

Polyhedron	Polygon Used	Number of faces, F	Number of vertices, V	Number of edges, E	Figure
Tetrahedron	Triangle	4	4	6	
Octahedron	Triangle	8	6	12	
Icosahedron	Triangle	20	12	30	
Cube	Square	6	8	12	
Dodecahedron	Pentagon	12	20	30	

9.28.2 Semiregular Polyhedra

A **semiregular polyhedron** is a convex polyhedron comprised of regular polygons but not necessarily with the same number of sides. Some semiregular polyhedra are shown in Figure 9.98.

Figure 9.98

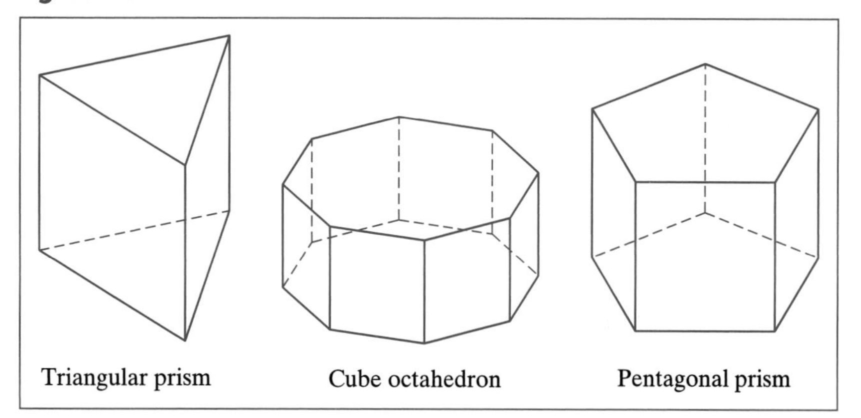

Related Topics

Cylinders, Cones, and Spheres
Cavalieri's Principle
Surface Area of Prisms and Cylinders
Surface Area of Pyramids, Cones, and Spheres
Volume of Prisms and Cylinders
Volume of Pyramids, Cones, and Spheres

TOPIC 9.29 Cylinders, Cones, and Spheres

cornerstone **Three-dimensional Geometry Basics**

Introduction

Not all simple closed surfaces are polyhedra. In this section, we discuss cylinders, cones, and spheres.

Astronomy and navigational needs have pushed the study of the geometry of the sphere. The inside surface of a sphere is used by astronomers to locate stars, planets, and other objects. Navigators use a sphere to represent Earth in seeking the most direct shipping or flying routes, where following the curved surface of Earth is more direct than a straight line on a planar map of Earth.

For More Information Sibley, T. Q. (1998). *The Geometric Viewpoint: A Survey of Geometries*. Reading, MA: Addison-Wesley.

9.29.1 Cylinders

A **cylinder** is a simple closed surface comprised of two congruent simple closed nonpolygonal curves that are in parallel planes, their interiors, and the line segments that connect corresponding points on the curves. The two nonpolygonal curves are called the **bases** of the cylinder. The remaining points of the cylinder compose the **lateral surface** of the cylinder, as shown in Figure 9.99.

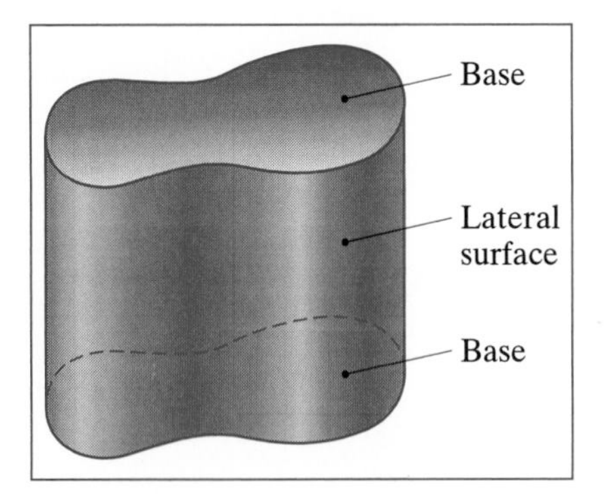

Figure 9.99

A particular kind of cylinder is a **circular cylinder,** which has two congruent circles for bases. If the line segments joining corresponding points on the bases are perpendicular to the bases, the cylinder is a **right cylinder.** An **oblique cylinder** is a cylinder that is not a right cylinder. Some examples of cylinders are shown in Figure 9.100.

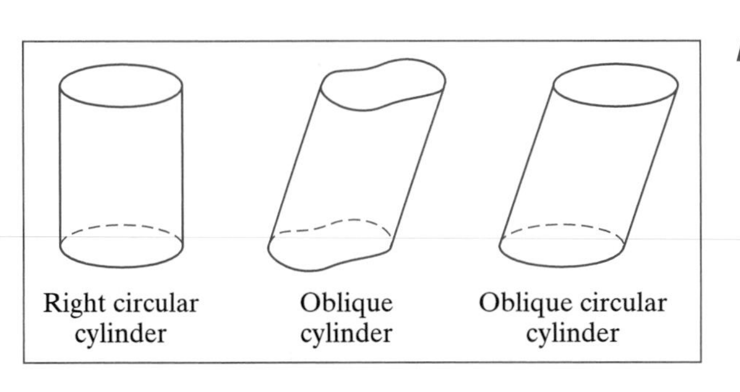

Figure 9.100

9.29.2 Cones

A **cone** is a simple closed surface comprised of a simple closed nonpolygonal curve, its interior, and the line segments that connect each point of the curve to a point not in the plane of the curve. The nonpolygonal curve is called the **base** of the cone, and the point not in the plane of the curve is called the **apex.** The **altitude** of the cone is the line segment from the apex that is perpendicular to the base, as shown in Figure 9.101.

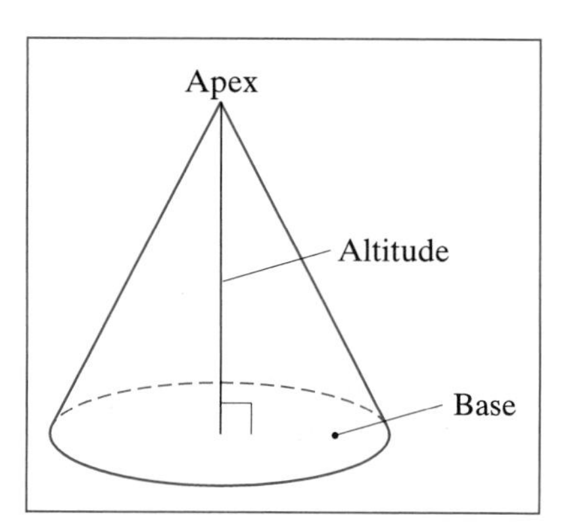

Figure 9.101

A cone with a circular base is called a **circular cone.** A circular cone whose altitude intersects the base at the center of the circular base is called a **right circular cone.** An **oblique circular cone** is one that is not a right circular cone, as shown in Figure 9.102.

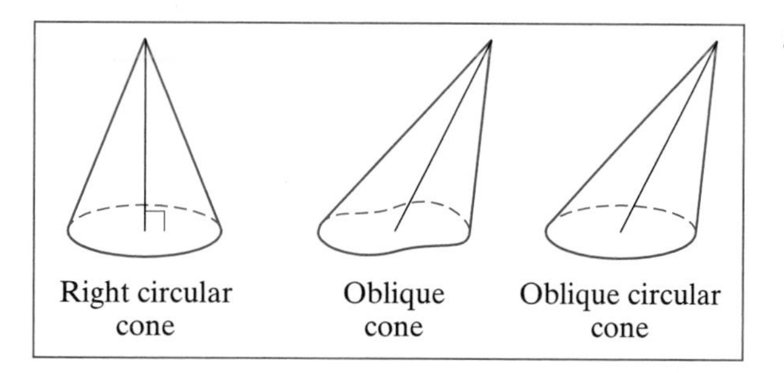

Figure 9.102

9.29.3 Spheres

A **sphere** is a simple closed surface comprised of all the points that are a given distance from a given point called the **center,** as shown in Figure 9.103. The **radius** of a sphere is the length of a segment with an endpoint on the sphere and having the center as the other endpoint. The segment is also called a radius. The **diameter** of a sphere is the length of a segment that has its endpoints on the sphere and passes through the center. The segment is also called a diameter.

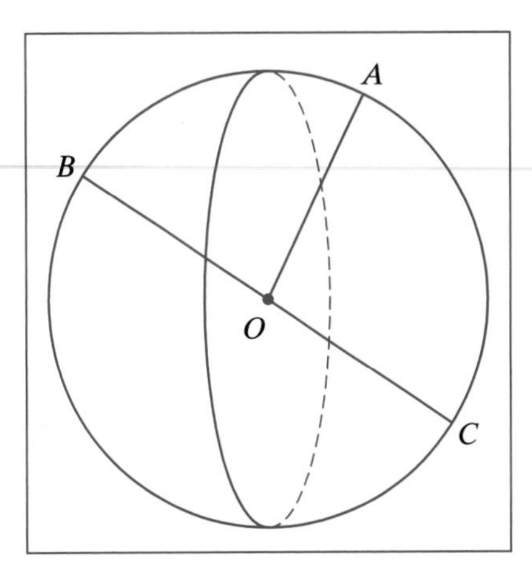

Figure 9.103

Polyhedra, Prisms, and Pyramids
Regular Polyhedra
Cavalieri's Principle
Surface Area of Prisms and Cylinders
Surface Area of Pyramids, Cones, and Spheres
Volume of Prisms and Cylinders
Volume of Pyramids, Cones, and Spheres

TOPIC 9.30 *van Hiele Levels*

cornerstone **Two-dimensional Geometry Basics**

Introduction

The study of geometry has been an important part of the school mathematics curriculum for a long time, but it has suffered from being poorly defined. Work done by Pierre van Hiele and Dina van Hiele-Geldof of the Netherlands has helped to better determine what geometry ideas should be included in the school curriculum and when they should be included. The van Hieles' work, conducted primarily in the 1950s, suggests that children move through developmental stages as they grow in their understanding of geometry ideas.

Although the van Hieles' work was given much attention in the 1960s and 1970s in what was then the Soviet Union, their work was largely ignored in the United States for nearly two decades. Now, however, their work has become very influential in determining the precollege geometry curriculum.

For More Information Fuys, D., Geddes, D., & Tischler, R. (1988). The van Hiele Model of Thinking in Geometry Among Adolescents. *Journal for Research in Mathematics Education Monograph No. 3*.

A central part of the van Hieles' model is a five-level developmental hierarchy of geometric thinking. The five levels describe the thinking processes that the van Hieles propose people use in geometric situations. Learners, according to the van Hieles, pass through the five levels assisted by appropriate instructional experiences and cannot reach a particular level of thinking without having gone through the previous levels. The descriptions provided here are only a brief overview of the van Hiele levels of geometric thinking and are taken from a monograph written by Fuys, Geddes, and Tischler (1988).

Level 0: The student identifies, names, compares, and operates on geometric figures (e.g., triangles, angles, intersecting or parallel lines) according to their appearance.

Level 1: The student analyzes figures in terms of their components and relationships among components and discovers properties/rules by giving or following informal arguments.

Level 2: The student logically interrelates previously discovered properties/rules by giving or following informal arguments.

Level 3: The student proves theorems deductively and establishes interrelationships among networks of theorems.

Level 4: The student establishes theorems in different postulational systems and analyzes/compares these systems. (p. 5)

Related Topics

Angle Relationships
Line Relationships
Triangle-Congruence Properties
Quadrilateral Properties
Proportion Properties

In the Classroom

While this chapter focuses on the *Geometry Content Standard*, it also emphasizes the *Communication Process Standard* (NCTM, 2000). All students should be enabled to "analyze characteristics and properties of two- and three-dimensional shapes and develop mathematical arguments about geometric relationships" (p. 41) and "use the language of mathematics to express mathematical ideas precisely" (p. 60). To enable your students in this manner, we believe you should explore the following in the context of geometry: (a) the importance of language and representation in mathematics (Activities 9.1–9.6), (b) the logic of classification and inherited properties and results (Activities 9.7–9.11), and (c) mathematical connections, proof, and changing assumptions (Activities 9.12–9.17).

Some Expectations of Your Future Students:

Pre-K–2 (NCTM, 2000, p. 96)

- recognize, name, build, draw, compare, and sort two- and three-dimensional shapes
- describe attributes and parts of two- and three-dimensional shapes
- describe, name, and interpret direction and distance in navigating space, and apply ideas about direction and distance
- relate ideas in geometry to ideas in number and measurement

Grades 3-5 (NCTM, 2000, p. 164)

- classify two- and three-dimensional shapes according to their properties, and develop definitions of classes of shapes such as triangles and pyramids
- explore congruence and similarity
- make and test conjectures about geometric properties and relationships, and develop logical arguments to justify conclusions
- build and draw geometric objects

How Activities 9.1–9.17 Help You Develop an Adult-level Perspective on the Above Expectations:

- By working with two-dimensional shapes, you discuss the importance of precise language and communication while exploring properties of shapes. You discuss the concept of necessary and sufficient attributes and analyze various ways of defining a single shape.
- By making geometric constructions, you continue to explore definitions and also make connections with ideas in number and measurement.
- By working with quadrilaterals, you further classify shapes according to their attributes.
- By working with triangles, you continue to make connections with measurement and number ideas. In particular, you focus on the measure of line segments and angles and on the ratios of sides of triangles in discussing the concepts of congruence and similarity.
- By initially making and testing conjectures about triangles and quadrilaterals, you develop logical arguments to justify your conclusions. Later, you build on these conclusions to prove simple results about more complex geometric shapes while learning to write your justifications in precise mathematical language.
- By exploring spherical geometry, you begin to describe and interpret shapes as the space and distance measures change from the familiar Euclidean and extend your thinking about definitions, geometry, and measurement.

Bibliography

Arithmetic Teacher. (1990, February). Focus issue on spatial sense, *37*.

Battista, M. T. (1998). *Shape makers: Developing geometric reasoning with the Geometer's Sketchpad*. Emeryville, CA: Key Curriculum Press.

Bennett, D. (1999). *Exploring geometry with the Geometer's Sketchpad* (2nd ed.) Emeryville, CA: Key Curriculum Press.

Brahier, D. J., & Speer, W. R. (1997). Worthwhile tasks: Exploring mathematical connections through geometric solids. *Mathematics Teaching in the Middle School, 3*, 20–28.

Bright, G. W., & Harvey, J. G. (1998). Learning and fun with geometry games. *Arithmetic Teacher, 35*, 22–26.

Bruni, V., & Seidenstein, R. B. (1990). Geometric concepts and spatial sense. In J. N. Payne (Ed.), *Mathematics for the young child* (pp. 203–207). Reston, VA: National Council of Teachers of Mathematics.

Cangelosi, J. S. (1985). A "fair" way to discover circles. *Arithmetic Teacher, 33*, 11–13.

Carroll, W. M. (1988). Cross sections of clay solids. *Arithmetic Teacher, 35*, 6–11.

Carroll, W. M. (1998). Middle school students' reasoning about geometric situations. *Mathematics Teaching in the Middle School, 3*, 398–403.

Chazan, D. (1990). Implementing the *Standards*: Students' microcomputer-aided exploration in geometry. *Mathematics Teacher, 83*, 628–635.

Clements, D. H. (1999). Geometric and spatial thinking in young children. In J. V. Copley (Ed.), *Mathematics in the early years* (pp. 66–79). Reston, VA: National Council of Teachers of Mathematics.

Clements, D. H., & Battista. M. (1986). Geometry and geometric measurement. *Arithmetic Teacher, 33*, 29–32.

Confer, C. (1994). *Math by all means: Geometry, grade 2*. Sausalito, CA: Math Solutions Publications.

Craine, T. V., & N. Rubenstein, R. N. (1993). A quadrilateral hierarchy to facilitate learning in geometry. *Mathematics Teacher, 86*, 30–36.

Dana, M. E. (1987). Geometry: A square deal for elementary teachers. In M. M Lindquist (Ed.), *Learning and teaching geometry K–12* (pp. 113–125). Reston, VA: National Council of Teachers of Mathematics.

de Villiers, M. D. (1999). *Rethinking proof with the Geometer's Sketchpad*. Emeryville, CA: Key Curriculum Press.

Del Grande, J. (1993). *Geometry and spatial sense: Addenda series, grades K–6*. Reston, VA: National Council of Teachers of Mathematics.

Ellington, B. (1983). Star Trek: A construction problem using compass and straightedge. *Mathematics Teacher, 76*, 329–332.

Enderson, M. C., & Manouchehri, A. (1998). Technology-based geometric explorations for the middle grades. In L. P. Leutzinger (Ed.), *Mathematics in the middle* (pp. 193–200). Reston, VA: National Council of Teachers of Mathematics.

Evered, L. J. (1992). Folded fashions: Symmetry in clothing design. *Arithmetic Teacher, 40*, 204–206.

Flores, A. (1993). Pythagoras meets van Hiele. *School* Science and Mathematics, 93, 152–157.

Fuys, D. J., & Liebov, A. K. (1993). Geometry and spatial sense In R. J. Jensen (Ed.), *Research ideas for the classroom: Early childhood mathematics* (pp. 195–222) Old Tappan, NJ: Macmillan.

Geddes, D. (1992). *Geometry in the middle grades. Addenda series, grades 5–8*. Reston. VA: National Council of Teachers of Mathematics.

Hill, J. M. (Ed.) (1987). *Geometry for grades K–6: Readings from The Arithmetic Teacher*. Reston, VA. National Council of Teachers of Mathematics.

Hoffer, A. R., & Hoffer, S. A. K. (1992). Geometry and visual thinking. In T.R. Post (Ed.), *Teaching mathematics in grades K–8: Research-based methods* (2nd ed.) (pp. 249–277). Needham Heights, MA: Allyn & Bacon.

Johnson, A., & Boswell, L. (1992). Geographic constructions. *Mathematics Teacher, 85*, 184–187.

Krause, E. F. (1986). *Taxicab geometry: An adventure in non-Euclidean geometry*. New York: Dover Publications, Inc.

Kriegler, S. (1991). The Tangram: It's more than an ancient puzzle. *Arithmetic Teacher, 38*, 38–43.

Lappan, G., & Even, R. (1988). Similarity in the middle grades. *Arithmetic Teacher, 35*, 32–35.

Lappan, G., Phillips, E. A., & Winter, M. J. (1984). Spatial visualization. *Mathematics Teacher, 77*, 618–625.

Lehrer, R., & Curtis, C. L. (2000). Why are some solids perfect? Conjectures and experiments by third graders. *Teaching Children Mathematics, 6*, 324–329.

Malloy, C. (1999). Perimeter and area through the van Hiele model. *Mathematics Teaching in the Middle School, 5*, 87–90.

Manouchehri, A., Enderson, M. C., & Pugnucco, L. A. (1998). Exploring geometry with technology. *Mathematics Teaching in the Middle School, 3*, 436–442.

Masingila, J. O. (1993). Secondary geometry: A lack of evolution. *School Science and Mathematics, 93* (1), 38–44.

Mercer, S., & Henningsen, M.A. (1998). The Pentomino project: Moving students from manipulatives to reasoning and thinking about mathematical ideas. In L. P. Leutzinger (Ed.), *Mathematics in the middle* (pp. 184–192). Reston, VA: National Council of Teachers of Mathematics.

Morrow, L. J. (1991). Implementing the *Standards*: Geometry through the *Standards. Arithmetic Teacher, 38*, 21–25.

Nowlin, D. (1993). Practical geometry problems: The case of the Ritzville pyramids. *Mathematics Teacher, 86*, 198–200.

Rowan, T. E. (1990). Implementing the *Standards*: The geometry standards in K–8 mathematics. *Arithmetic Teacher, 37*, 24–28.

Rubenstein, R. N., Lappan, G., Phillips, E., & Fitzgerald, W. (1993). Angle sense: A valuable connector. *Arithmetic Teacher, 40*, 352–358.

Rubenstein, R. N., & Thompson, D R. (1995). Making connections with transformations in grades K–8. In P. A. House (Ed.). Connecting mathematics across the curriculum (pp. 65–78). Reston, VA: National Council of Teachers of Mathematics.

Sandberg, S. E. (1998). A plethora of polyhedra. *Mathematics Teaching in the Middle School, 3*, 388–391.

Thiessen, D., & Matthias, M. (1989). Selected children's books for geometry. *Arithmetic Teacher, 37*, 47–51.

Wilson, P. S. (1990). Understanding angles: Wedges to degrees. *Mathematics Teacher, 83*, 294–300.

Winter, J J., Lappan, G., Phillips, E., & Fitzgerald, W. (1986). *Middle grades mathematics project: Spatial visualization*. Menlo Park, CA: Addison-Wesley.

Woodward, E., Gibbs, V., & Shoulders, M. (1992). A fifth-grade similarity unit. *Arithmetic Teacher, 39*, 22–26.

Yeshurun, S. (1985). An improvement of the congruent angles theorem. *Mathematics Teacher, 78*, 53–54.

CHAPTER 10 TEN

Measurement

TOPIC 10.1 Angle Measurement
10.2 Linear Measurement
10.3 Perimeter
10.4 Circumference
10.5 Measuring Area
10.6 Area of Quadrilaterals
10.7 Area of Triangles
10.8 Area of Regular Polygons and Circles
10.9 Surface Area of Prisms and Cylinders
10.10 Surface Area of Pyramids, Cones, and Spheres
10.11 Measuring Volume
10.12 Volume of Prisms and Cylinders
10.13 Volume of Pyramids, Cones, and Spheres
10.14 Cavalieri's Principle
10.15 Mass and Temperature
10.16 Pythagorean Theorem and Other Triangle Relationships
10.17 Distance Formula
10.18 Tessellations
10.19 Tessellations of Regular Polygons
10.20 Transformations
10.21 Translations
10.22 Rotations
10.23 Reflections and Glide Reflections
10.24 Size Transformations
In the Classroom
Bibliography

CHAPTER OVERVIEW

Perhaps no part of mathematics is more clearly applicable to everyday life than measurement. As a consequence of the practicality of measurement in the real world, it is a very important strand in the elementary school mathematics curriculum. In particular, students learn extremely useful measurement skills (e.g., how to use a ruler), concepts (e.g., the concepts of area and perimeter), and key formulas (e.g., $A = l \cdot w$). Just as important is the fact that making measurements can be a source of many, very interesting problems. For example, did you know that two shapes can have the same perimeter but different areas? One natural question that follows from this is "Can two shapes have the same area, but different perimeters?" In this chapter, you can learn some interesting facts about measurement and will examine some concepts and procedures involving measurement ideas.

BIG MATHEMATICAL IDEAS

Problem-solving strategies, conjecturing; verifying; generalizing

NCTM PRINCIPLES & STANDARDS LINKS

Measurement; Problem Solving; Reasoning; Communication; Connections; Representation

TOPIC 10.1 Angle Measurement

cornerstone

Planes
Rays and Angles
Three-dimensional Geometry Basics

Introduction

Angles are essential parts of many geometric figures, and often it is important for us to be able to measure angles. In this section, we discuss angle measurement and notation, how to use a protractor, angle classification, and dihedral angle measurement.

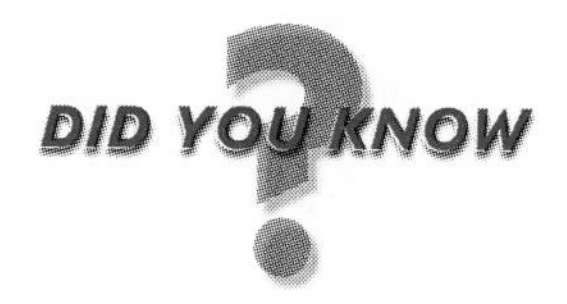

The Babylonians determined that there were approximately 360 days in one year and so divided a circle into 360 equal parts. This is how we have come to measure angles, in terms of degrees, where each degree is $\frac{1}{360}$ of a circle.

For More Information Bunt, L. N. H., Jones, P. S., & Bedient, J. D. (1988). *The Historical Roots of Elementary Mathematics.* New York: Dover Publications.

10.1.1 Angle Measurement and Notation

An angle is **measured** by considering the rotation needed to move one ray to fit on the other ray. A common way to measure this rotation is to use **degree units.** There are 360 degree units (or **degrees**) in a full circular rotation. We use the symbol ° to stand for degrees. The angles shown in Figure 10.1 have measures of 50° and 135°, respectively.

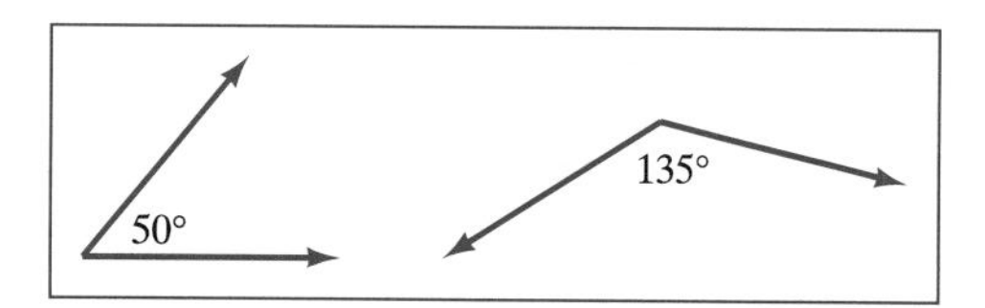

Figure 10.1

10.1.2 How to Use a Protractor

A geometric tool we use to measure angles is called a **protractor,** as shown in Figure 10.2. The unit of measurement on the protractor is the **degree.** Each degree is divided into 60 equal parts called **minutes,** and each minute is divided into 60 equal parts called **seconds.** If an angle has a measurement of 65 degrees, 30 minutes, 15 seconds, we write that as 60°30'15".

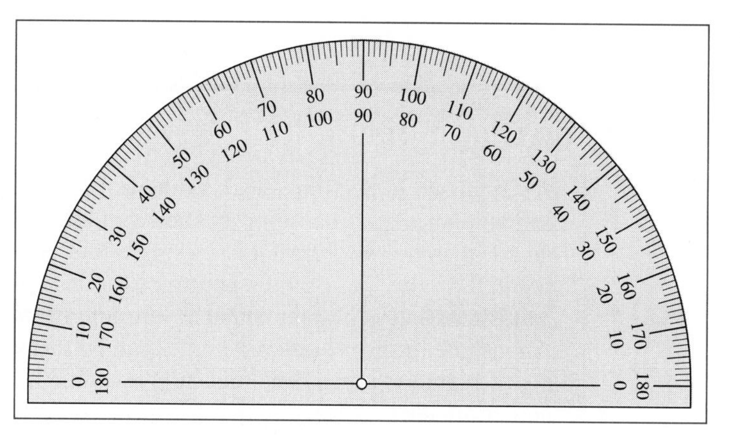

Figure 10.2

To use a protractor to measure an angle, first place the center mark of the protractor on the vertex of the angle so that one ray of the angle lies on the zero degree (0°) line on the protractor (see the first drawing in Figure 10.3). Then see where the other ray of the angle crosses the protractor's scale. Make sure that the number corresponds to the rotation needed to move the one ray (at 0°) to fit on top of the second ray, as shown in the second drawing in Figure 10.3.

Figure 10.3

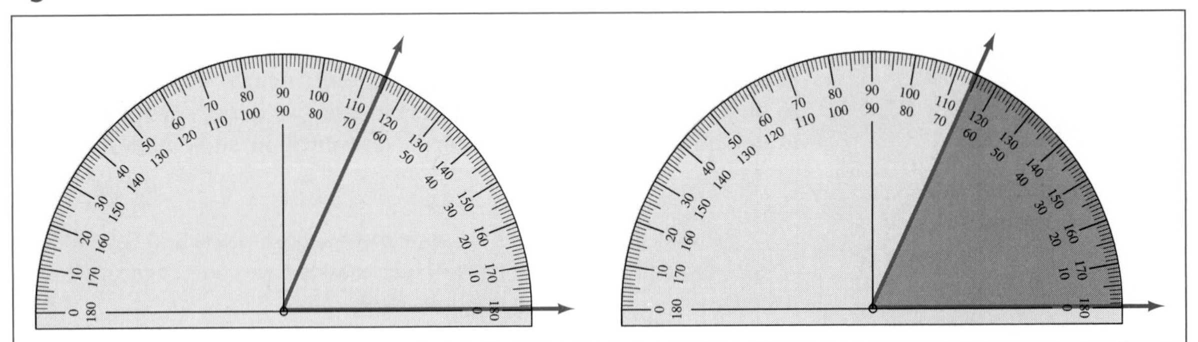

10.1.3 Angle Classification

Angles are usually classified as one of five type of angles: acute, right, obtuse, straight, or reflex. An **acute** angle is an angle that has a measure greater than 0° and less than 90°. An angle that measures 90° is called a **right** angle. An **obtuse** angle is one that has a measure greater than 90° and less than 180°. An angle that measures 180° is called a **straight** angle. An angle that measures more than 180° is called a **reflex** angle. In the drawings in Figure 10.4, x represents the measure of the angle.

Figure 10.4

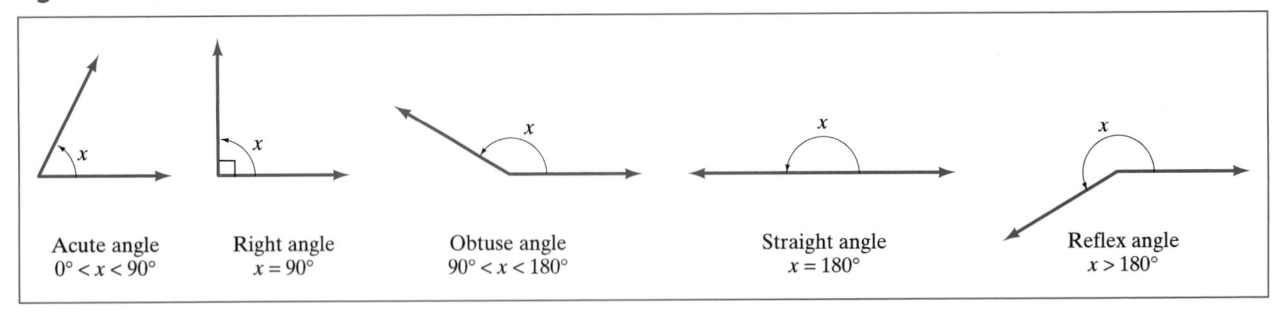

10.1.4 Dihedral Angle Measurement

As discussed in Section 9.26.2, a dihedral angle is formed by the line of intersection of two planes and the union of the two half-planes. To determine the measure of a dihedral angle, we form a planar angle that is on the dihedral angle by choosing a point B on the line of intersection of the two planes and by drawing a ray in each half-plane that is perpendicular to the line of intersection, as shown in Figure 10.5. An infinite number of planar angles can be formed for any dihedral angle, each having the same measure. The measure of a dihedral angle is the measure of any of its planar angles.

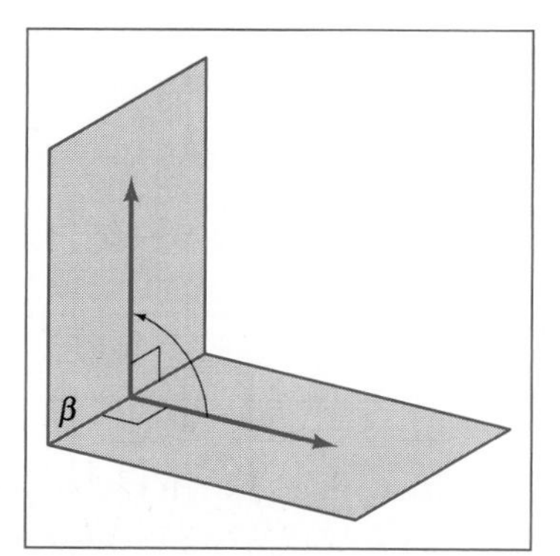

Figure 10.5

Related Topics

Angle Relationships
Congruence
Solid Geometry Basics

TOPIC 10.2 Linear Measurement

cornerstone **Line Segments**

Introduction

Throughout history, humans have chosen various units to measure length or distance. These nonstandard units evolved into standard units of measure. In this section, we discuss length and how to measure it, the English system of linear measurement, and the metric system of linear measurement.

The earliest recorded linear unit of measurement is the *cubit*, the distance from the point of the elbow to the outstretched tip of the middle finger. *Foot* was another early unit of measure, and its length was likely based on the length of the foot of Charlemagne, emperor of the Holy Roman Empire. In the tenth century, King Edgar I of England decreed that the *yard* would be the distance from the tip of *his* nose to the top of the middle finger of *his* outstretched arm. This decree was to regulate trade in textiles, so that one yard of cloth would be the same throughout England.

For More Information Bunt, L. N. H., Jones, P. S., & Bedient, J. D. (1988). *The Historical Roots of Eementary Mathematics*. New York: Dover Publications.

10.2.1 Length and How to Measure It

The **length** of a segment is the number of units that fit into the segment. To measure a segment, we must choose a unit of measure. Historically, groups of people developed various units of measure to be able to trade goods and record quantities. However, these measurements were not standardized, and eventually two main systems of measurement emerged. The English system is quite complicated, whereas the metric system is based on powers of 10 and is easier to remember and to convert among measures within the system.

10.2.2 The English System of Linear Measurement

Table 10.1

Unit	Equivalent
1 inch (in.)	1/12 ft
1 foot (ft)	12 in.; 1/3 yd
1 yard (yd)	3 ft; 36 in.
1 mile (mi)	5280 ft; 1760 yd

The basic units of linear measure in the **English system** (also called the U.S. customary system) are inch, foot, yard, and mile. The length of a pencil might be measured in inches, the length of a room might be measured in feet, a playing field might be measured in yards, and the distance between two cities might be measured in miles. Table 10.1 shows the equivalent values of one unit in terms of other units.

10.2.3 The Metric System of Linear Measurement

Almost all countries in the world, except the United States, use the **metric system** of measurement. The metric system is based on powers of 10. The basic unit of linear measure in the metric system is the **meter,** and the most commonly used linear measures are millimeter, centimeter, meter, and kilometer. The thickness of a pencil point might be measured in millimeters, the length of a piece of paper might be measured in centimeters, the distance a ball is thrown might be measured in meters, and the distance traveled in a vehicle might be measured in kilometers. Table 10.2 shows the equivalent values of one unit in terms of other units.

Table 10.2

Unit	Equivalent
1 millimeter (mm)	$\frac{1}{10}$ cm
1 centimeter (cm)	10 mm; $\frac{1}{100}$ m
1 meter (m)	100 cm; 1000 mm
1 kilometer (km)	1000 m

The prefixes used are shown in the chart in Table 10.3. Different units are obtained by multiplying the basic unit by a power of 10.

Table 10.3

Prefix	Factor
milli	0.001 (10^{-3})
centi	0.01 (10^{-2})
deci	0.1 (10^{-1})
basic unit	1 (10^{0})
deka	10 (10^{1})
hecto	100 (10^{2})
kilo	1000 (10^{3})

Related Topics

Measuring Area
Distance Formula
Perimeter
Circumference

TOPIC 10.3 Perimeter

cornerstone

Linear Measurement
Distance Formula

Introduction

Knowing the distance around a figure is important information when constructing a building, putting up a fence, or laying out a garden. In this section, we discuss the perimeters of polygons, that is, the distance around polygons.

The Romans divided 1 foot into 12 equal units, called *unciae,* from which we get the word *inch.*

For More Information Bunt, L. N. H., Jones, P. S., & Bedient, J. D. (1988). *The Historical Roots of Elementary Mathematics.* New York: Dover Publications.

10.3.1 Perimeter

The length of a simple closed curve, or the distance around the curve, is called the **perimeter.** Perimeter is a type of linear measure. A figure could have a perimeter of, for example, 8 cm or 10 ft or 135 km.

10.3.2 Perimeters of Polygons

The perimeter of a polygon is the sum of the lengths of the sides of the polygon. For example, for a pentagon with sides of length s_1, s_2, s_3, s_4, and s_5, the perimeter is $P = s_1 + s_2 + s_3 + s_4 + s_5$. The perimeters of the polygons in Figure 10.6 can be found by adding up the lengths of the sides.

Figure 10.6

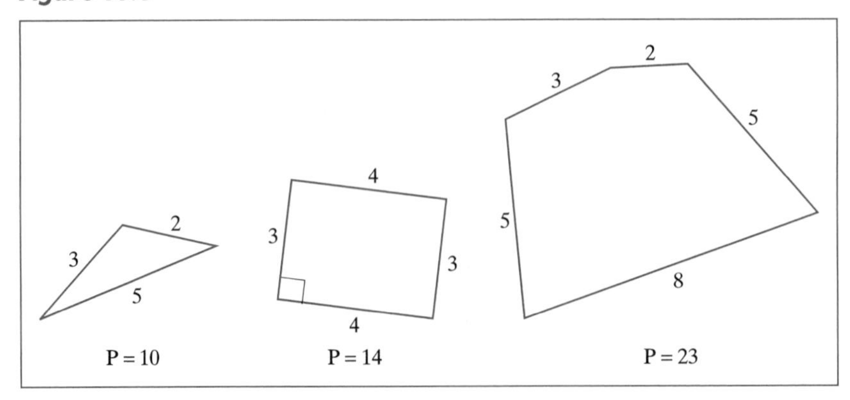

For some special polygons, we can simplify the perimeter formula. For an **equilateral triangle, $P = 3s$,** where s is the length of a side. For a **square, $P = 4s$;** for a **rectangle, $P = 2l + 2w$,** where l is the length of one side and w is the length of a consecutive side, as shown in Figure 10.7.

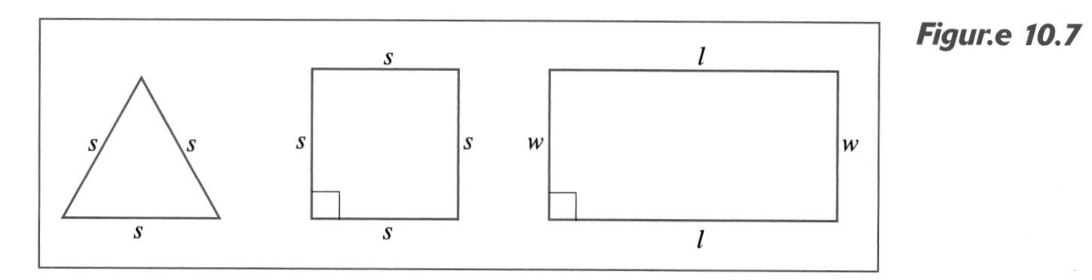

Figur.e 10.7

Related Topics

Circumference

TOPIC 10.4 Circumference

cornerstone

Angle Measurement
Linear Measurement
Circles

Introduction

While the distance around a circle is its perimeter, we have a special name for the perimeter of a circle. In this section, we discuss circumference, π and arc length.

The quest for more-accurate estimates of π began as early as with the ancient Egyptians. A problem from the Ahmes papyrus asks: A circular field has a diameter 9 khet. What is its area? In solving this problem, the Egyptians used an implicit estimate of π of approximately 3.1605.

For More Information Joseph, G. G. (1991). *The Crest of the Peacock: Non-European Roots of Mathematics.* London, England: Penguin Books.

10.4.1 Circumference and π

The distance around a circle is called the **circumference.** In other words, for a circle the perimeter is known as the circumference. The ratio of the circumference (C) of a circle to its diameter (d) has been found to be approximately 3.14 or $\frac{22}{7}$. This ratio is the irrational number pi (π). In other words, the circumference of a circle divided by its diameter will always equal π. Thus, the formula for the circumference of any circle is: $\frac{C}{d} = \pi$ or $C = \pi d$. Because the length of the diameter in a circle is twice the length of a radius in the same circle, this can also be written: $C = 2\pi r$, where r is the radius. The circumferences of the circles shown in Figure 10.8 can be found by using these formulas.

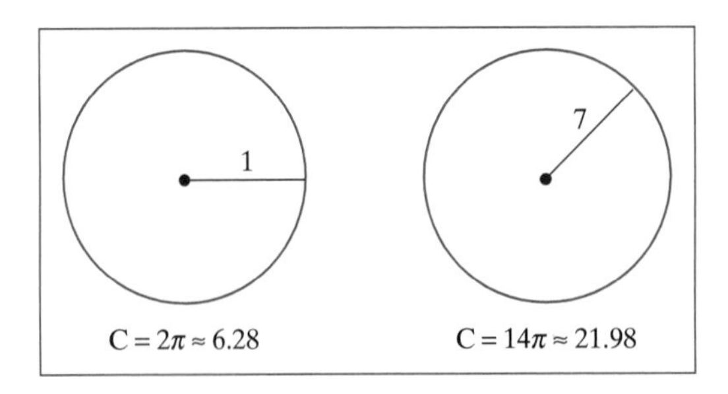

Figure 10.8

10.4.2 Arc Length

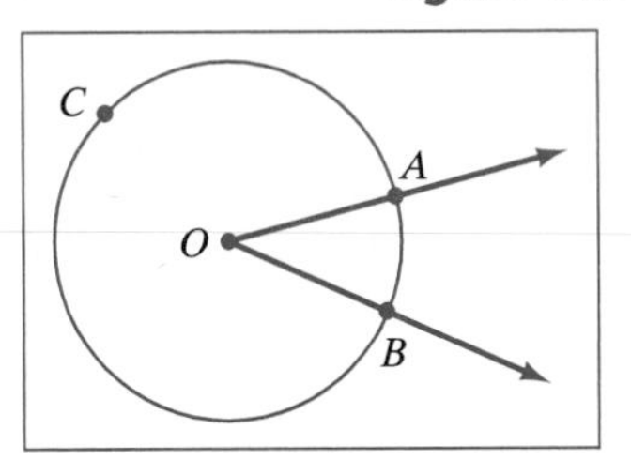

Figure 10.9

The length of an arc on a circle is determined by the radius of the circle and the central angle forming the arc. A **central angle** is an angle that has the center of a circle as its vertex and has sides intersecting the circle, as shown in Figure 10.9. When a central angle intersects the circle, two arcs are formed. For example, in Figure 10.9, $\overset{\frown}{AB}$ and $\overset{\frown}{ACB}$ are formed by $\angle AOB$ intersecting circle O.

A central angle with a measure of 180° forms two arcs that are semicircles. Because the circumference of a circle is $2\pi r$, where r is the radius of the circle, the arc length of a semicircle is half of the circumference, or πr, as shown in Figure 10.10.

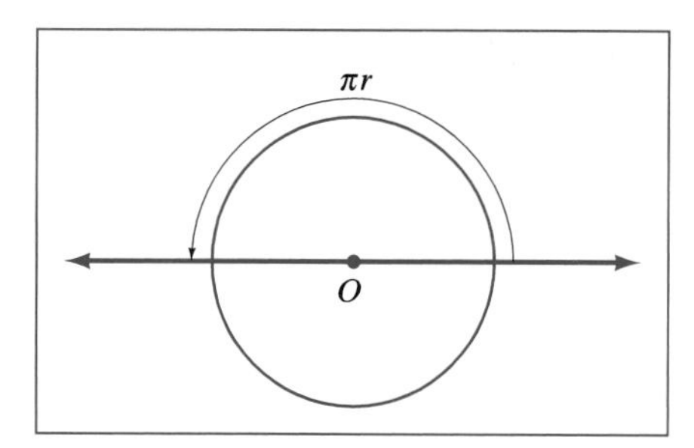

Figure 10.10

Any arc length can be determined by using the fact that a circle has 360° and an arc length is formed by a central angle having a particular measure, say $\theta°$. Because an arc is a portion of a circle, we can set up a proportion to find the arc length of an arc formed by a central angle with a measure of $\theta°$: $\frac{\text{arc length}}{2\pi r} = \frac{\theta°}{360°}$ or arc length $= \frac{\pi r \theta°}{180°}$ (see Figure 10.11).

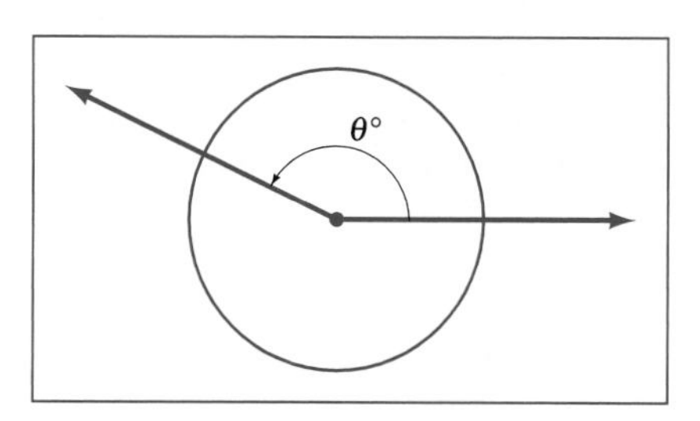

Figure 10.11

Related Topics

Perimeter

Measuring Area

cornerstone **Linear Measurement**

Introduction

Measuring the area of a region is important in many situations, such as laying carpet, sewing, and designing a brochure. In this section, we discuss area and how to measure it, the English system of area measurement, and the metric system of area measurement.

Historically, an *acre* was the amount of land that a farmer and two oxen could plow in one day. This varied from place to place. Eventually, King Henry VIII of England standardized the area of an acre as 40 *poles* long by 4 *poles* wide. A *pole* equals 16.5 feet.

For More Information Bunt, L. N. H., Jones, P. S., & Bedient, J. D. (1988). *The Historical Roots of Elementary Mathematics.* New York: Dover Publications.

10.5.1 Area and How to Measure It

The **area** of a planar shape is the number of square units enclosed by the shape. If we use a square as the basic unit, the area of a figure is the number of squares that is required to cover the region enclosed by the shape with no squares overlapping and no gaps (this is called *tessellating the plane* and is discussed in the section called Tessellations). Area is measured in square units, such as square inches, square meters, or square miles.

To measure the area of a planar shape, we can place a grid comprised of squares over the shape and estimate the number of square units needed to cover the shape. If we use a finer grid, we can arrive at a better estimate of the area, as illustrated in Figure 10.12.

Figure 10.12

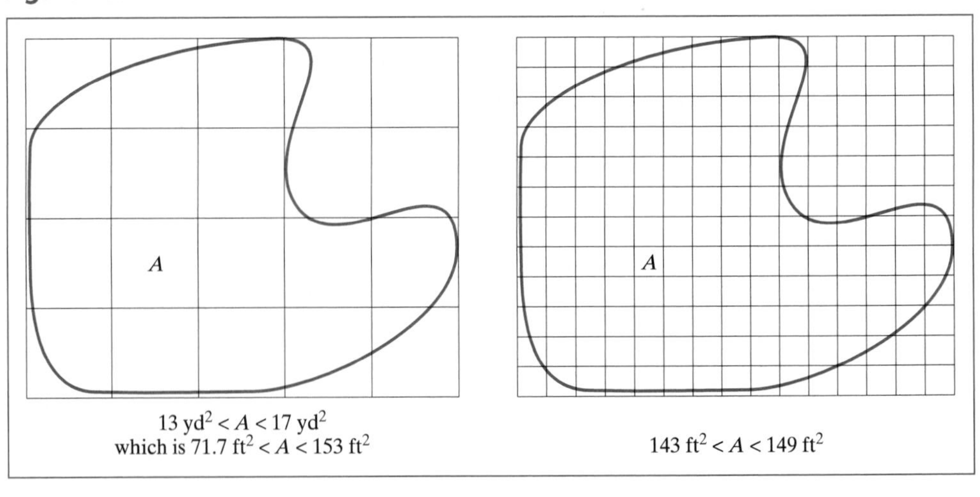

10.5.2 The English System of Area Measurement

In the English system of measurement, area is usually measured in square inches, square feet, square yards, acres, or square miles. Table 10.4 shows these units and their equivalents in other units in the English system.

Table 10.4

Unit	Equivalent
1 square inch (1 in.^{2})	$\frac{1}{144}$ ft^2
1 square foot (1 ft^2)	144 in.^{2}; $\frac{1}{9}$ yd^2
1 square yard (1 yd^2)	9 ft^2
1 acre (1 A)	43,560 ft^2
1 square mile (1 mi^2)	27,878,400 ft^2; 3,097,600 yd^2

Figure 10.13

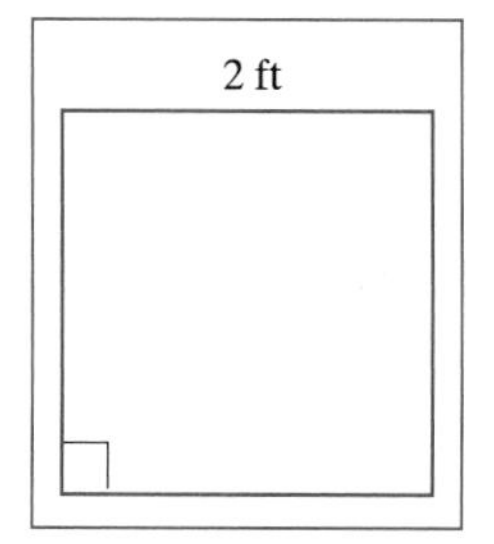

To convert from one area measure to another within the system, we need to remember that we are using square units. For example, a square with each side having a length of 2 ft would have an area of 4 ft^2, as shown in Figure 10.13. To determine what the area would be in terms of square inches, we substitute 24 inches for the length of each side because 2 ft = 24 in. Then the number of square inches that it would take to cover the region enclosed by the square would be $24 \cdot 24$ or 576. Thus, 4 ft^2 = 576 in.^{2}.

10.5.3 The Metric System of Area Measurement

In the metric system of measurement, area is usually measured by square millimeters, square centimeters, square decimeters, square meters, square dekameters (called ares), square hectometers (called hectares), and square kilometers. Table 10.5 shows these units and their equivalents in other units in the metric system.

Table 10.5

Unit	Equivalent
1 square millimeter (1 mm^2)	0.000001 m^2
1 square centimeter (1 cm^2)	0.0001 m^2
1 square decimeter (1 dm^2)	0.01 m^2
1 square meter (1 m^2)	10,000 cm^2
1 are (1 a)	100 m^2
1 hectare (1 ha)	10,000 m^2
1 square kilometer (1 km^2)	1,000,000 m^2

We can convert from one area measure to another within the system by substituting equivalent quantities. For example, to determine how many 1-mm squares are in a square centimeter, we replace the length of each side (1 cm) with 10 millimeters, as shown in Figure 10.14, because 1 cm = 10 mm. Then we can see that it takes 100 1-mm squares to equal 1 cm^2, and thus, 100 mm^2 = 1 cm^2.

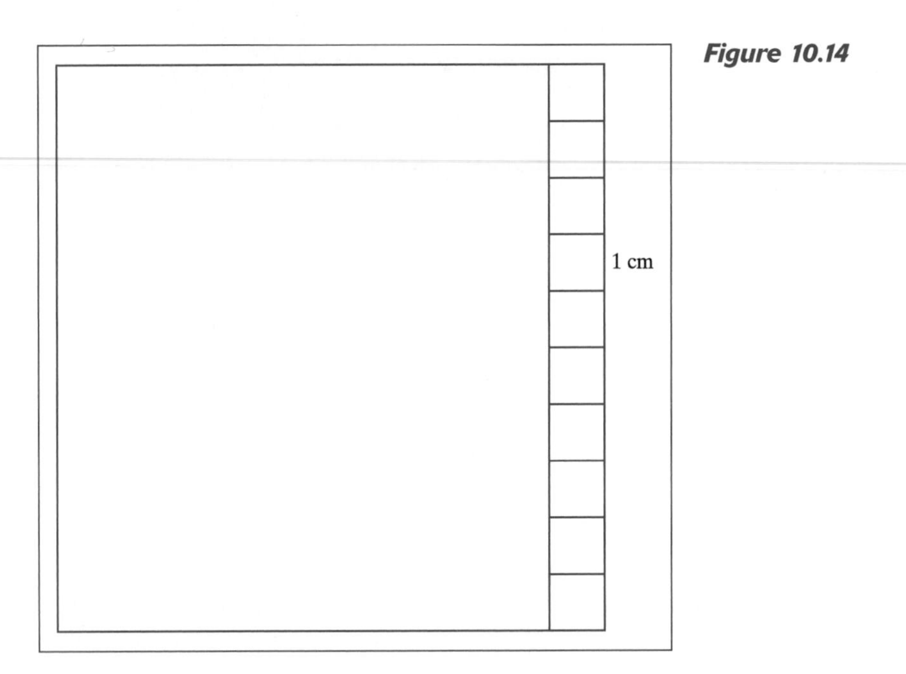

Figure 10.14

Related Topics

Area of Triangles
Area of Quadrilaterals
Area of Regular Polygons and Circles
Tessellations

TOPIC 10.6 *Area of Quadrilaterals*

cornerstone **Measuring Area**
Quadrilaterals

Introduction

Because area is measured in square units, the square is the foundational building block we can use to determine area formulas for many polygonal figures. In this section, we discuss the areas of various quadrilaterals—in particular, square, rectangle, parallelogram and trapezoid.

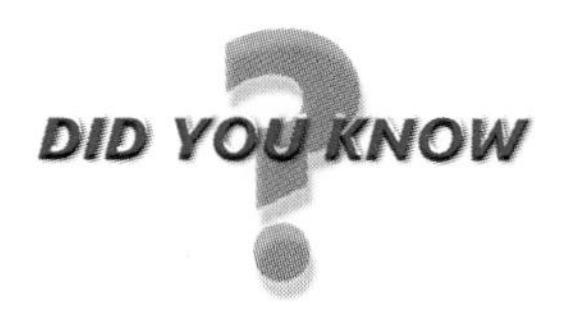

The early Egyptians had ways to calculate the areas of certain planar figures. Their procedures for calculating the area of a rectangle, a triangle, and a trapezoid were correct. However, they calculated the area of an arbitrary quadrilateral as the product of half the sum of two opposite sides and half the sum of the other two, which is valid for only some quadrilaterals.

For More Information Bunt, L. N. H., Jones, P. S., & Bedient, J. D. (1988). *The Historical Roots of Elementary Mathematics*. New York: Dover Publications.

10.6.1 Area of a Square

The area of a square is the number of square units required to cover the region enclosed by the square. Consider the square in Figure 10.15 with each side of length 4 units. If we place 1-unit squares inside the 4-unit square, we can see that there are four 1-unit squares along each dimension of the 4-unit square, and there are a total of 16 1-unit squares inside the 4-unit square. Thus, the area of the 4-unit square is 16 square units, or 16 units^2. This is the same as the product of the length of the square's sides, or the square of the length of one side. The formula for the area of a square is: $A = s^2$.

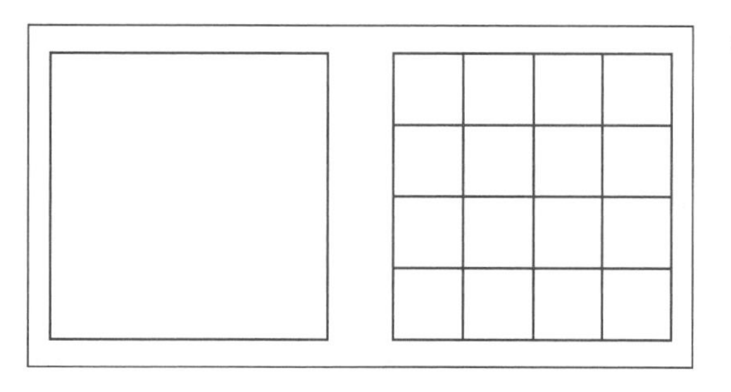

Figure 10.15

10.6.2 Area of a Rectangle

Using what we derived for the area of a square, we can see that if we place 1-unit squares inside a rectangle, we find that the area of the rectangle is equal to the product of the number of unit squares along the base and the number of unit squares along the height, as illustrated in Figure 10.16. This is the same as the product of the rectangle's base and height (also called the length and width).

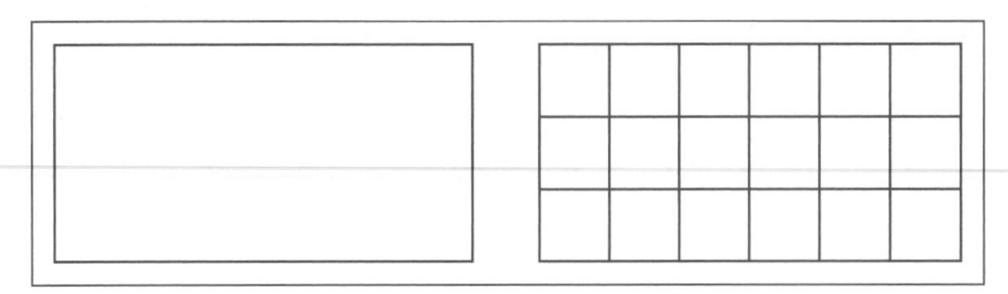
Figure 10.16

The formula for the area of a rectangle is: A = bh (or A = lw).

10.6.3 Area of a Parallelogram

We can use what we know about finding the area of a rectangle to find the area of a parallelogram. Consider the parallelogram shown in A in Figure 10.17. If we draw the altitude from one vertex to the opposite side, as shown in the first drawing in Figure 10.17, and then take the right triangle formed by drawing this altitude and move it to the opposite side of the parallelogram, we see that we now have, B, a rectangle with a base of b and a height of h as shown in the second drawing in Figure 10.17 .

Figure 10.17

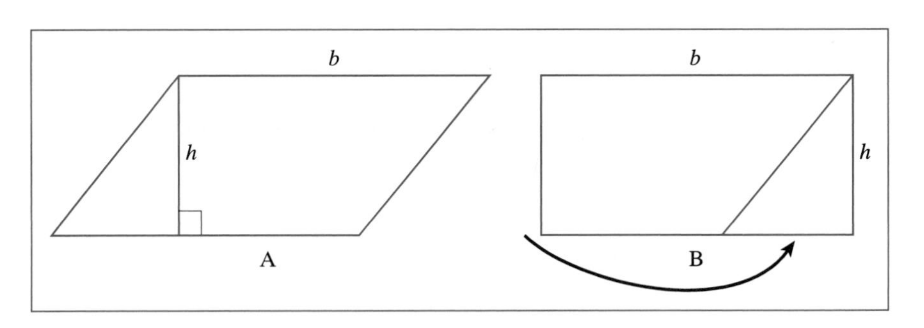

Thus, the formula for the area of a parallelogram is: A = bh.

10.6.4 Area of a Trapezoid

To find the area of a trapezoid, we can use what we know about finding the area of a parallelogram. Consider the trapezoid in A in Figure 10.18 with bases of length b_1 and b_2. Take a trapezoid that is congruent to A and position it as shown in B in Figure 10.18. We can see that a parallelogram is formed with a base of $b_1 + b_2$, and a height of h. Because this parallelogram is comprised of two congruent trapezoids, the area of the trapezoid must be half the area of the parallelogram.

Figure 10.18

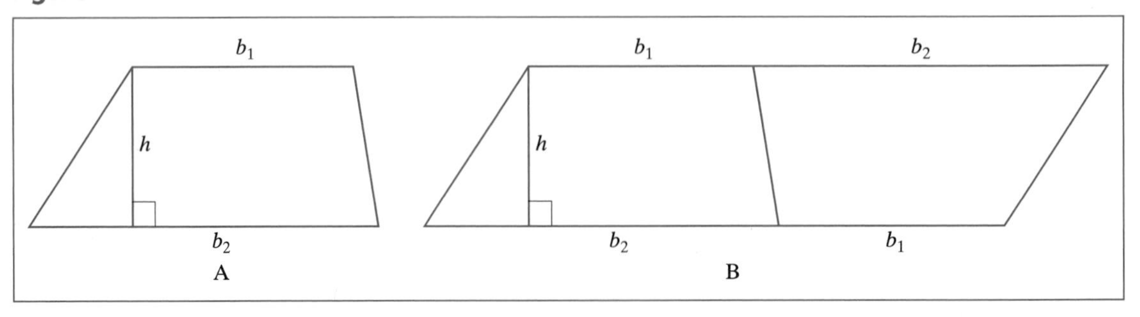

Because the area of the parallelogram is bh, which in this case is $(b_1 + b_2)h$, then the formula for the area of a trapezoid is: A = $\frac{1}{2}(b_1 + b_2)h$.

Related Topics
Area of Triangles
Area of Regular Polygons and Circles

TOPIC 10.7 *Area of Triangles*

cornerstone **Measuring Area**
Area of Quadrilaterals
Triangles

Introduction

The formula for the area of a triangle can be derived from the formula for the area of a parallelogram. In this section, we discuss the area of a triangle.

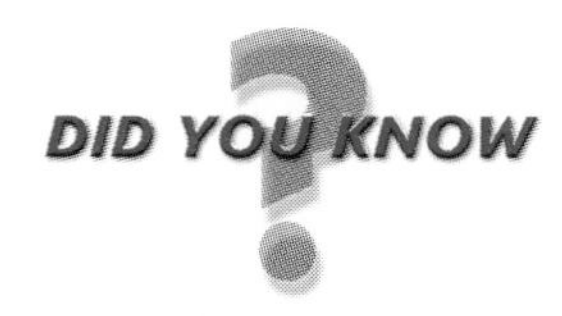

DID YOU KNOW

The name of Heron, a Greek mathematician, is connected with a formula for finding the area of a triangle using its side lengths (a, b, c): $A = \sqrt{s(s - a)(s - b)(s - c)}$, where $s = \frac{1}{2}(a + b + c)$. Although this algorithm may actually be the work of Archimedes, Heron's proof of the relationship is the oldest one that exists.

For More Information Bunt, L. N. H., Jones, P. S., & Bedient, J. D. (1988). *The Historical Roots of Elementary Mathematics.* New York: Dover Publications.

Consider ΔABC in Figure 10.19. Place a congruent triangle, $\Delta A'B'C'$ so that B and C' are the same vertex and C and B' are the same vertex, as shown in the second drawing in Figure 10.19. We will now prove that the quadrilateral formed by these two triangles is a parallelogram.

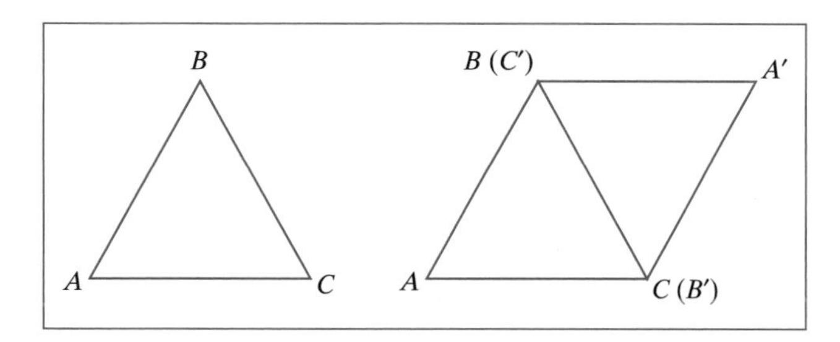

Figure 10.19

We know that $\Delta ABC \cong \Delta A'CB$. Because corresponding parts of congruent triangles are congruent, $\angle ABC \cong \angle BCA'$ and $\angle ACB \cong \angle CBA'$. Because these angles are alternate interior angles, then $\overline{AB} \parallel \overline{CA'}$ and $\overline{AC} \parallel \overline{BA'}$. Because both pairs of opposite sides are parallel, $ABA'C$ is a parallelogram.

This proof is valid for any triangle. Because any two congruent triangles can form a parallelogram, the area of a triangle is half the area of a parallelogram. Thus, the formula for the area of a triangle is: $A = \frac{1}{2}bh$.

Related Topics **Area of Regular Polygons and Circles**

TOPIC 10.8 Area of Regular Polygons and Circles

cornerstone

Measuring Area	**Circles**
Regular Polygons	**Circumscribed Circle Constructions**
Area of Triangles	**Circumference**

Introduction

Using what we know about finding the area of a triangle, we can find the area of any regular polygon and the area of a circle. In this section, we discuss the areas of regular polygons, circles, and sectors of circles.

To find the area of a circular region, the ancient Egyptians used a procedure that was equivalent to $A = (\frac{8}{9}d)^2$, where d is the length of the diameter. The Babylonians used a procedure equivalent to $A = \frac{1}{12}C^2$, where C is the circumference of the circle. Both of these procedures produced results that are very good approximations to the results produced by the algorithm we use today.

For More Information Bunt, L. N. H., Jones, P. S., & Bedient, J. D. (1988). *The Historical Roots of Elementary Mathematics.* New York: Dover Publications.

10.8.1 Area of a Regular Polygon

The area of a regular polygon can be found by dividing the polygon into congruent triangles and finding the sum of the areas of those triangles. Consider the regular pentagon with sides of length s shown in the first drawing in Figure 10.20. Divide the pentagon into five congruent triangles as shown in the second drawing in Figure 10.20. The point that is a common vertex for the five triangles is the center of the circle that could circumscribe the pentagon.

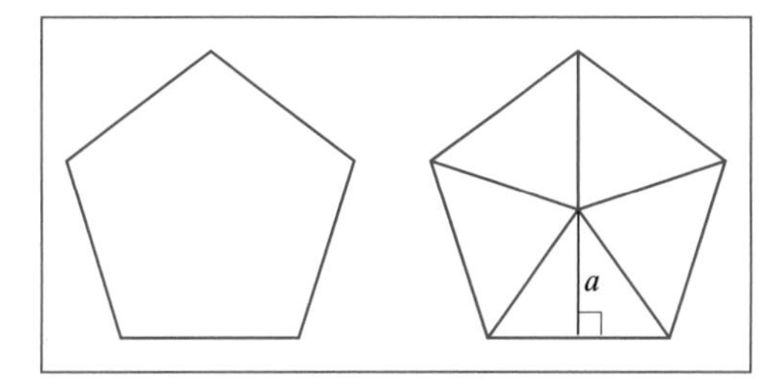

Figure 10.20

The height of the congruent triangles formed in a regular polygon, each with a vertex at the center, is called the *apothem* and is denoted by a. Thus, the area of each of these triangles is $\frac{1}{2}\,as$. Because there are five triangles, the area of the regular pentagon is $5(\frac{1}{2})\,as$. But $5s$ is the perimeter of the pentagon, so the area of the regular pentagon is $\frac{1}{2}\,ap$, where p is the perimeter. This same procedure can be used to derive the area of any regular polygon. Thus, the formula for the area of a regular polygon is: $A = \frac{1}{2}\,ap$, where a is the height of one of the congruent triangles and p is the perimeter of the polygon.

10.8.2 Area of a Circle

We can use what we know about finding the area of a regular polygon to derive the formula for the area of a circle. As the number of sides of a regular polygon increases, the perimeter and the area of the regular polygon get closer and closer to the circumference and area of a circle circumscribed about the polygon, as shown in Figure 10.21.

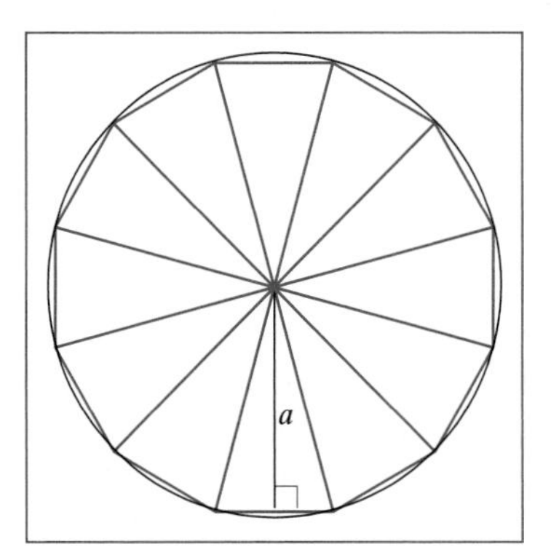

Figure 10.21

If we form the congruent triangles, each with a vertex at the center, we can see that the apothem in each triangle is approximately equal to the radius of the circle and that the perimeter of the polygon is approximately equal to the circumference of the circle. Thus, the formula for the area of a regular polygon can be used to derive the formula for the area of a circle: $A = \frac{1}{2}r(2\pi r) = \pi r^2$. The justification for this derivation is beyond the scope of this book. The interested reader is referred to the references noted in this topic.

10.8.3 Area of a Sector

A **sector** of a circle is a region of the circle formed by a central angle, as shown in Figure 10.22. If the central angle has a measure of 180°, the sector is half of the circle, and so the area of the sector is half the area of the circle. If the central angle has a measure of 90°, the area of the sector is one-fourth the area of the circle. In general, the formula for the area of a sector with a central angle of $\theta°$ is: $A = \frac{\theta}{360}(\pi r^2)$.

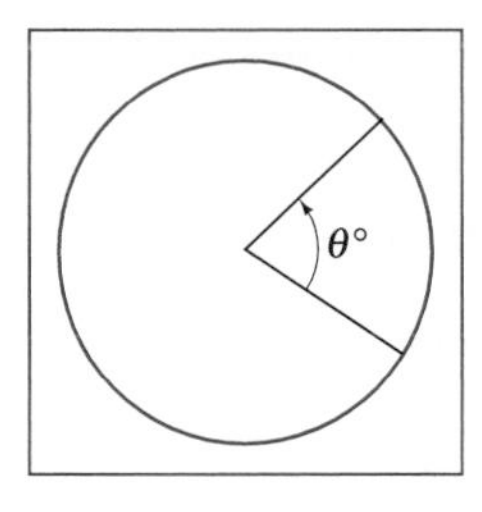

Figure 10.22

Related Topics
Surface Area of Prisms and Cylinders
Surface Area of Pyramids, Cones, and Spheres

TOPIC 10.9 Surface Area of Prisms and Cylinders

cornerstone
Measuring Area
Polyhedra, Prisms, and Pyramids
Cones, Cylinders, and Spheres
Area of Regular Polygons and Circles

Introduction

Package designers need to know the dimensions and shapes of package surfaces. The surface area of a package is important in designing the package. In this section, we discuss surface area, surface area of a right prism, and surface area of a right circular cylinder.

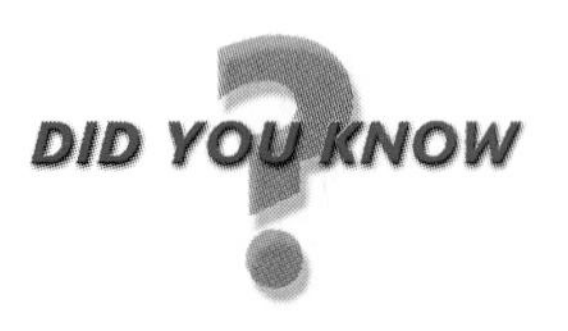

The world's largest oil tanks are located in Ju'ayman, Saudi Arabia. The five cylindrical tanks are each 72 feet tall and have a diameter of 386 feet.

For More Information Guinness Book of Records.

10.9.1 Surface Area

The **surface area** of an object is literally the area of its surface. We can find the surface area of three-dimensional geometric figures by separating the surface into planar figures, for which we can find the area, and then summing the areas of all the regions. We can represent the surface of a three-dimensional figure by using a two-dimensional pattern called a **net.** Recall that prisms, cylinders, pyramids, and cones all have at least one base and that their remaining surfaces are called the lateral surfaces. Thus, for these figures, their nets are comprised of their bases and their lateral surfaces.

10.9.2 Surface Area of a Right Prism

Recall from Section 9.27.4, that a prism is a polyhedron comprised of two congruent faces that are in parallel planes and parallelograms for the remaining faces. If the lateral faces are all rectangles, then the prism is called a right prism. We will discuss how to find the surface area for right prisms only.

To find the surface area of a right prism, we sum the areas of the bases and the lateral faces (called the **lateral surface area**). Figure 10.23 shows two right prisms, one with a triangular base and the other with a rectangular base. Figure 10.24 shows nets for these two prisms.

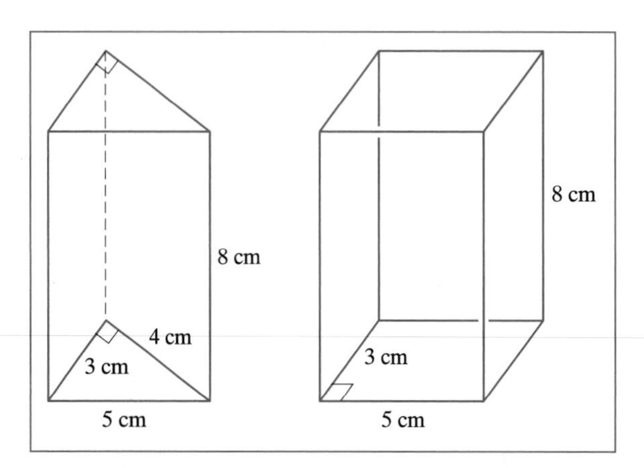

Figure 10.23

Figure 10.24

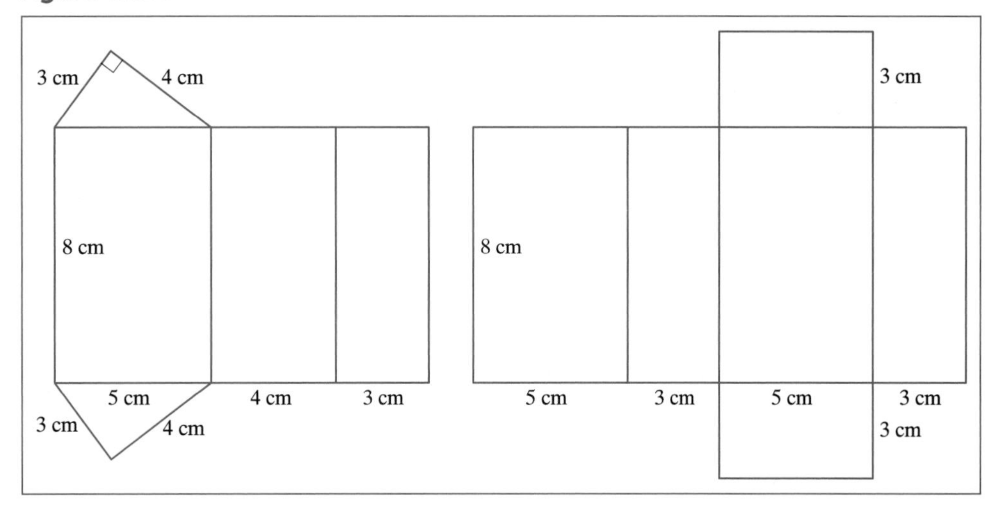

To find the surface area of the right triangular prism, we find the area of the bases [$2(\frac{1}{2} \cdot 3 \cdot 4)$, or 12 cm^2] and the lateral surface area [($3 \cdot 8 + 4 \cdot 8 + 5 \cdot 8$), or 96 cm^2]. Thus, the surface area of the right triangular prism is 12 cm^2 + 96 cm^2, or 108 cm^2.

To find the surface area of the right rectangular prism, we find the area of the bases [$2(3 \cdot 5)$, or 30 cm^2] and the lateral surface area [$2(3 \cdot 8) + 2(5 \cdot 8)$, or 128 cm^2]. Thus, the surface area of the right rectangular prism is 30 cm^2 + 128 cm^2, or 158 cm^2. The formula for the surface area of a prism is: S.A. = 2B + L.A., where S.A. is the surface area, B is the base area, and L.A. is the lateral surface area.

10.9.3 Surface Area of a Right Circular Cylinder

We can use what we know about finding the surface area of a prism to find the surface area of a cylinder because as the number of edges of a prism's bases increases, the prism moves closer to approximating a cylinder. We find the surface area of a right circular cylinder by summing the area of its bases and its lateral surface area. Figure 10.25 shows a right circular cylinder and a net. This net can be obtained by cutting off the bases and cutting the lateral surface along a line perpendicular to the bases. When flattened, we can see that the lateral surface is a rectangle.

Figure 10.25

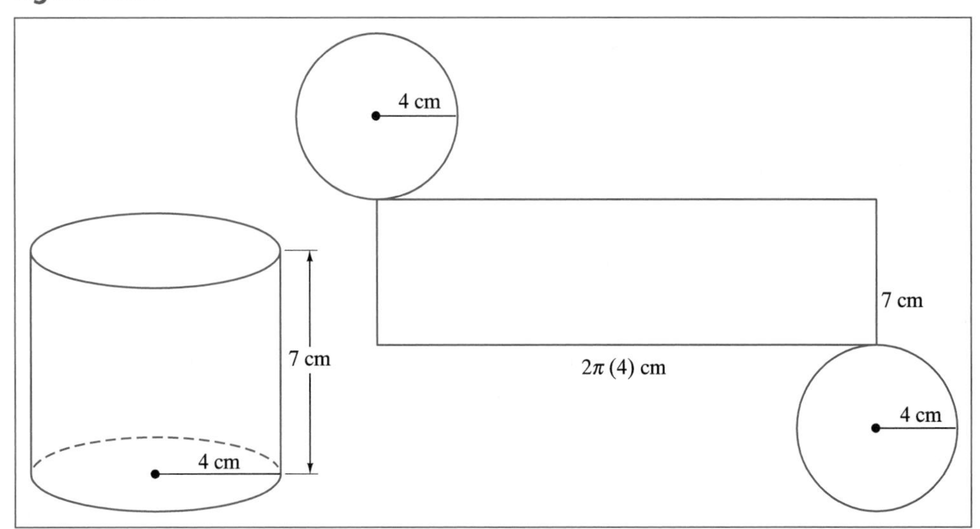

To find the surface area of this cylinder, we find the area of the bases $[2(\pi \cdot 4^2)$, or $32\pi\ cm^2]$ and the lateral surface area $[2 \cdot \pi \cdot 4 \cdot 7$, or $56\pi\ cm^2]$. Thus, the surface of this right circular cylinder is $88\pi\ cm^2$. In general, the formula for the surface area of a right circular cylinder is: $S.A. = 2\pi r^2 + 2\pi rh = 2\pi(r^2 + h)$, where each base has area πr^2, the length of the rectangular lateral surface is the circumference of a base—$2\pi r$—and h is the height of the rectangular lateral surface.

Related Topics

Surface Area of Pyramids, Cones, and Spheres

TOPIC 10.10 Surface Area of Pyramids, Cones, and Spheres

cornerstone

Measuring Area
Polyhedra, Prisms, and Pyramids
Cones, Cylinders, and Spheres
Area of Regular Polygons and Circles
Area of Triangles

Introduction

Some packages are in the form of pyramids, cones, and spheres. Thus, manufacturers need to be able to find the surface area of these three-dimensional figures in designing and producing these containers. In this section, we discuss surface area of a right regular pyramid, surface area of a right circular cone, and surface area of a sphere.

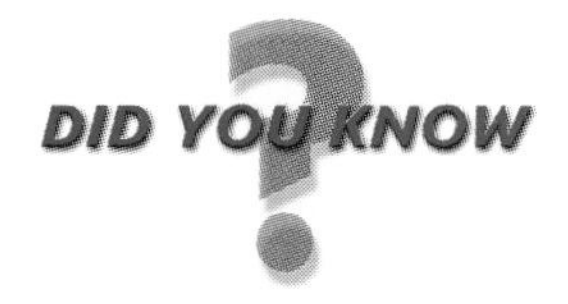

The early Egyptians may have derived an algorithm for finding the surface area of a hemisphere (half of a sphere) that is equivalent to the modern formula with a different value for π. However, some scholars have suggested that a term used in the Egyptian problem should be interpreted differently and that what the Egyptians actually derived was a counterpart for the modern formula for the area of the curved surface of a semicylinder.

For More Information Joseph, G. G. (1991). *The Crest of the Peacock: Non-European Roots of Mathematics*. London, England: Penguin Books.

10.10.1 Surface Area of a Right Regular Pyramid

Recall from Section 9.27.5, that a pyramid is a polyhedron comprised of a polygon with n edges for its base, a point not in the plane of the base called the apex, and n triangular faces that are determined by the apex and the n edges of the base. The triangular faces are called the lateral faces. If the base of a pyramid is a regular polygon and the lateral faces are congruent isosceles triangles, the pyramid is called a right regular pyramid. We will discuss how to find the surface for right regular pyramids only.

Because a pyramid has one base, the surface area of a pyramid is the sum of the base area and the lateral surface area. Figure 10.26 shows a right square pyramid and its net. The area of its base is 6^2 cm^2, or 36 cm^2. To find the lateral surface area, we need to find the area of one of the isosceles triangles that is a lateral face and then multiply that area by the number of lateral faces (which is the number of edges of the base). Each of these congruent isosceles triangles has an altitude of length 8 cm. The altitude on a lateral face of a pyramid is called the **slant height.** The lateral surface area for this pyramid is $4 \cdot \frac{1}{2} \cdot 6 \cdot 8$, or 96 cm^2. Thus, the surface area of the right triangular prism is 36 cm^2 + 96 cm^2, or 132 cm^2. Thus, the formula for the surface area of a pyramid is: $\text{S.A.} = \text{B} + \text{n} \cdot \frac{1}{2} \cdot \text{b} \cdot \ell$ where B is the area of the base, n is the number of edges of the base, b is the length of each side of the base, and ℓ is the slant height.

Figure 10.26

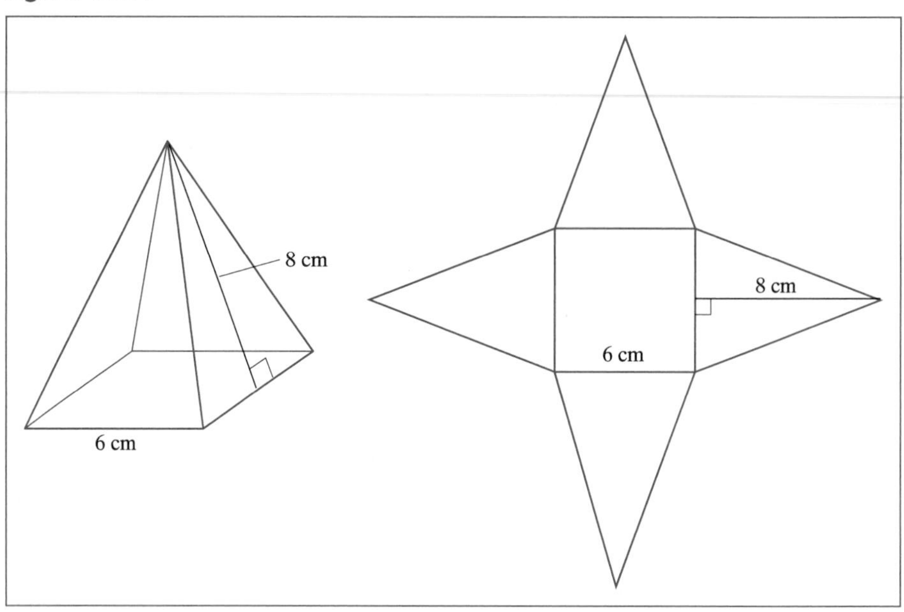

10.10.2 Surface Area of a Right Circular Cone

As the number of edges of a pyramid's base increases, the pyramid moves closer to approximating a cone. Thus, we can use what we know about finding the surface area of a pyramid to find the surface area of a cone. Figure 10.27 shows a pyramid, having a base with many edges, and its net, as a way of approximating a cone.

Figure 10.27

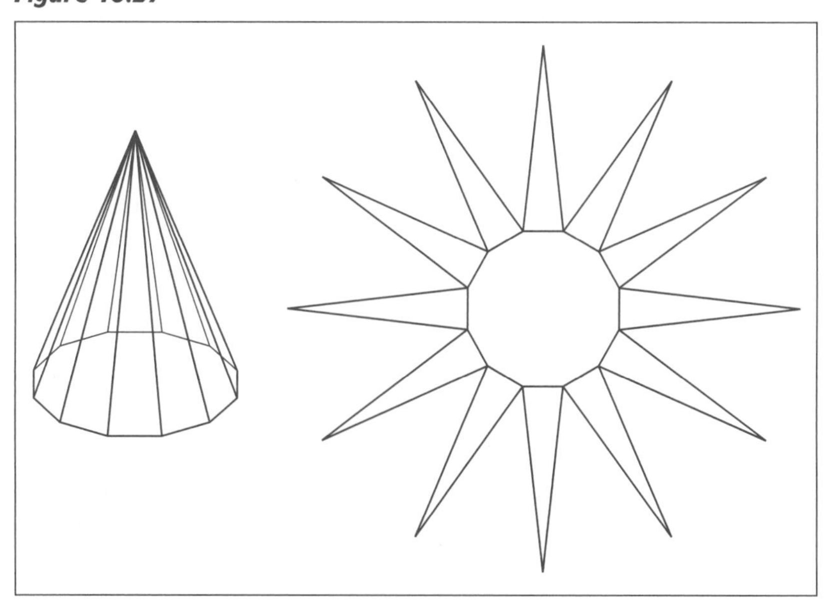

The surface area of a right circular cone is the sum of the base area (πr^2) and the lateral surface area. In a pyramid, the lateral surface area is $n \cdot \frac{1}{2} \cdot b \cdot \ell$ (as discussed in section 10.10.1). However, $n \cdot b$, the number of edges times the length of each side of the base, is equal to the perimeter of the base; in a cone, this would be the circumference of the circular base. Thus, the formula for the surface area of a right circular cone is: S.A. $= B + \frac{1}{2}(2\pi r)\ell = \pi r^2 + \pi r\ell = \pi r(r + \ell)$.

10.10.3 Surface Area of a Sphere

The formula for the surface area of a sphere can be derived using methods of calculus; however, that derivation is beyond our scope in this book. We will, however, discuss the formula, and to do so we need to define a great circle. A **great circle** is a circle on a sphere that is formed by the intersection of the sphere and a plane passing through the center of the sphere, as shown in Figure 10.28. The surface area of a sphere is four times the area of a great circle of the sphere. Thus, the formula for the surface area of a sphere is: S.A. $= 4\pi r^2$.

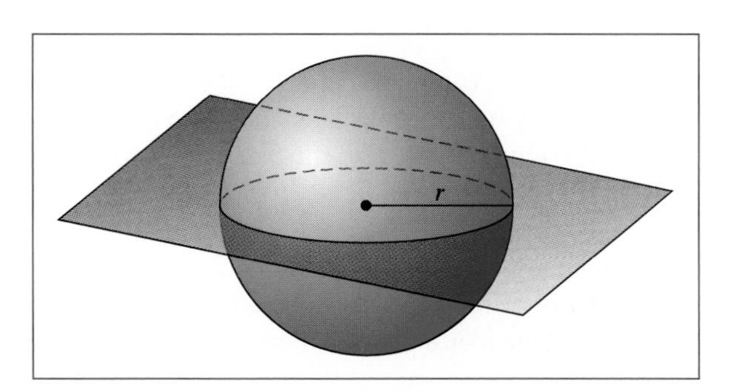

Figure 10.28

Related Topics **Surface Area of Prisms and Cylinders**

TOPIC 10.11 Measuring Volume

cornerstone

Measuring Area
Area of Triangles
Area of Quadrilaterals
Area of Regular Polygons and Circles

Introduction

Knowing the volume or capacity of a container is often very useful in real-world problem solving. In this section, we discuss volume and how to measure it, the English system of volume measurement, and the metric system of volume measurement.

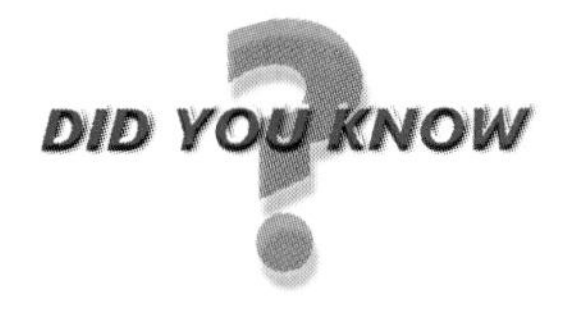

One of the major achievements of the ancient Egyptian mathematicians was deriving a formula for calculating the volume of a truncated pyramid (a pyramid with the top portion sliced off by a plane that is parallel to the base).

For More Information Joseph, G. G. (1991). *The Crest of the Peacock: Non-European Roots of Mathematics*. London, England: Penguin Books.

10.11.1 Volume and How to Measure It

The **volume** of a three-dimensional shape is the number of cubic units enclosed by the shape. A **cubic unit** is the amount of space enclosed by a cube that is 1 unit in length along each edge. If we use a cube as the basic unit, the volume of a figure is the number of cubes required to fill the space enclosed by the shape with no cubes overlapping and no gaps (this is called *tessellating the space* and is discussed in the section called Tessellations). Volume is measured in cubic units, such as cubic centimeters, cubic feet, or cubic meters.

10.11.2 The English System of Volume Measurement

In the English system of measurement, dry volume is usually measured in cubic inches, cubic feet, and cubic yards, whereas liquid volume is usually measured in cups, pints, quarts, and gallons. We will discuss the units for dry measure only. Table 10.6 shows these units and their equivalents in other units in the English system.

Table 10.6

Unit	Equivalent
1 cubic inch (1 in.^3)	$\frac{1}{1{,}728}\ \text{ft}^3$
1 cubic foot (1 ft^3)	$1{,}728\ \text{in.}^3;\ \frac{1}{27}\ \text{yd}^3$
1 cubic yard (1 yd^3)	$27\ \text{ft}^3$

To convert from one volume measure to another within the system, we need to remember that we are using cubic units. For example, a cube with each side having a length of 2 feet would have a volume of 8 ft^3, as shown in Figure 10.29. To determine what the volume would be in terms of cubic inches, we substitute 24 inches for the

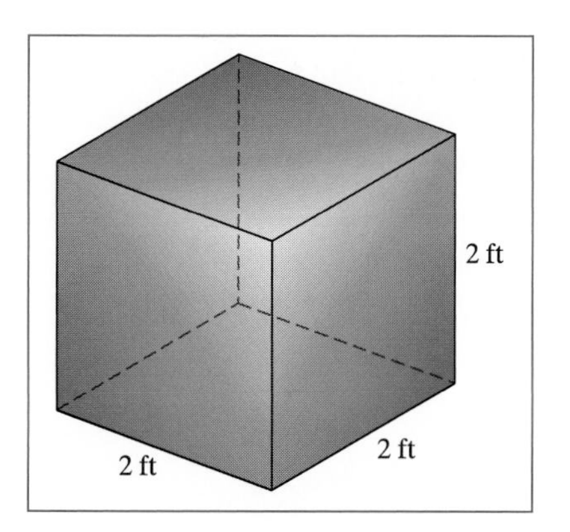

Figure 10.29

length of each side because 2 ft = 24 in. Then the number of cubic inches that it would take to fill the space enclosed by the cube would be $24 \cdot 24 \cdot 24$ or 13,824. Thus, $8\ \text{ft}^3 = 13,824\ \text{in.}^3$.

10.11.3 The Metric System of Volume Measurement

In the metric system of measurement, volume is usually measured by cubic centimeters and cubic meters, or milliliters and liters, depending on the type of measure. These units can be used for either dry or wet measure, but milliliters and liters are the metric units typically used for liquid measure. Table 10.7 shows these units and their equivalents in other units in the metric system.

Table 10.7

Unit	Equivalent
1 cubic centimeter (1 cm^3)	0.000001 m^3
1 cubic meter (1 m^3)	1,000,000 cm^3
1 milliliter (1 mL)	0.001 L; 1 cm^3
1 liter (1 L)	1 dm^3

We can convert from one volume measure to another within the system by substituting equivalent quantities. For example, to determine how many 1-cm cubes are in a cube measuring 2 m on a side, we replace the length of each side (2 m) with 200 cm, as shown in Figure 10.30 because 2 m = 200 cm. We can see that it takes 1,000,000 1-cm cubes to equal 1 m^3, and thus, 8,000,000 cm^3 = 8 m^3, which is the volume of the cube measuring 2 m on a side.

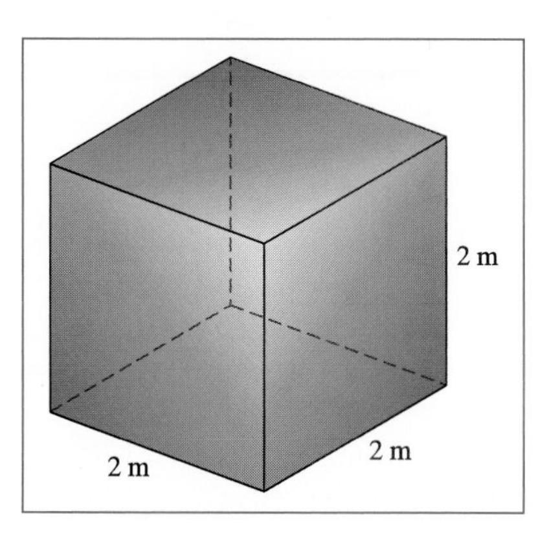

Figure 10.30

Related Topics

Volume of Prisms and Cylinders
Volume of Pyramids, Cones, and Spheres
Cavalieri's Principle

Volume of Prisms and Cylinders

Measuring Volume
Polyhedra, Prisms, and Pyramids
Cones, Cylinders, and Spheres

Introduction

If we were designing a container to hold as much of a product as possible for a certain amount of packaging material, we would need to be able to determine the volume of the container. In this section, we discuss the volume of a right prism and the volume of a right circular cylinder.

DID YOU KNOW

A typical drinking straw is 19.4 cm high with a diameter of 0.6 cm. Because 1 mL of water occupies 1 cm^3, then a typical straw can contain approximately 5.48 mL of water.

10.12.1 Volume of a Right Prism

The volume of a prism is the number of cubic units that are needed to fill the space enclosed by the prism. We will discuss the volume of a right prism in this section, while the volume of any prism is discussed in the section called Cavalieri's principle. Consider the right rectangular prism shown in Figure 10.31. We can see that one layer of cubes that is ℓ units long and w units wide will cover the base of the prism. If we have h layers, we can fill the space enclosed by the prism.

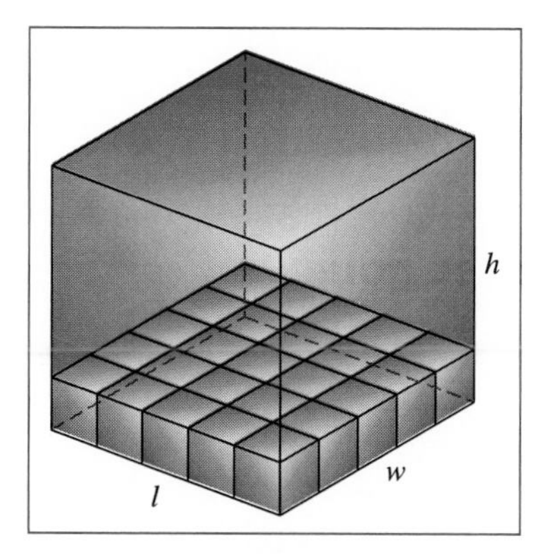

Figure 10.31

Thus, the formula for the volume of a right rectangular prism is: $V = \ell wh$. We can extend this to a right prism with any polygonal base, using the fact that ℓw is the area of the base of the right rectangular prism. Thus, the formula for the volume of a right prism is: $V = Bh$, where B is the area of the base. (Note: A right rectangular prism for which $\ell = w = h$ is commonly called a **cube.**)

10.12.2 Volume of a Right Circular Cylinder

We can use what we know about finding the volume of a right prism to find the volume of a right circular cylinder because as the number of edges of a prism's bases increases, the prism moves closer to approximating a cylinder. Because the base is a circle, its area is $B = \pi r^2$, where r is the radius. If the cylinder is h units tall, the formula for the volume of a right circular cylinder is: $V = Bh = \pi r^2 h$. Thus, the right circular cylinder shown in Figure 10.32 has a volume of 192π cm^3.

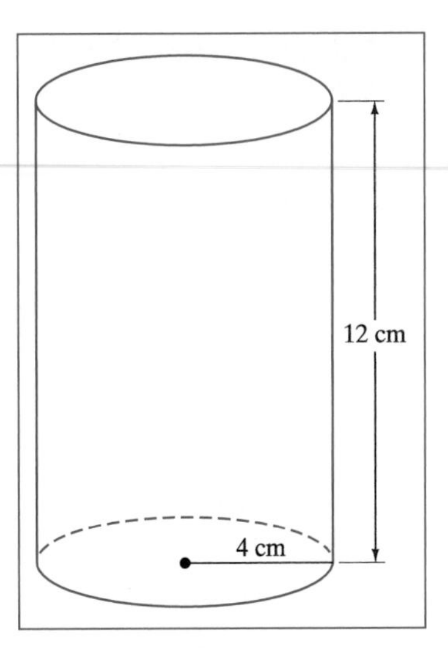

Figure 10.32

Related Topics

Volume of Pyramids, Cones, and Spheres
Cavalieri's Principle

TOPIC 10.13 Volume of Pyramids, Cones, and Spheres

Measuring Volume
Polyhedra, Prisms, and Pyramids
Cones, Cylinders, and Spheres
Volume of Prisms and Cylinders

Introduction

The relationships between the volumes of a right prism and a right pyramid with congruent bases and equal heights, and the volumes of a right circular cylinder and a right cone with congruent bases and equal heights are very interesting. In this section, we discuss volumes of right pyramids, right circular cones, and spheres.

The Greek mathematician Archimedes derived a formula for finding the volume of a sphere by demonstrating that the volume of a sphere is equal to the volume of a cylinder circumscribed about the sphere minus the volume of the cone inscribed in the cylinder. His proof involved slicing the three solids with planes and comparing the slices.

For More Information Bunt, L. N. H., Jones, P. S., & Bedient, J. D. (1988). *The Historical Roots of Elementary Mathematics.* New York: Dover Publications.

10.13.1 Volume of a Right Pyramid

The volume of a right pyramid is one-third the volume of a right prism with congruent bases and equal height; in other words, three right pyramids could fit into one right prism with congruent bases and equal height, as shown in Figure 10.33. Thus, the formula for the volume of a right pyramid is: $V = \frac{1}{3}Bh$, where B is the base area and h is the height.

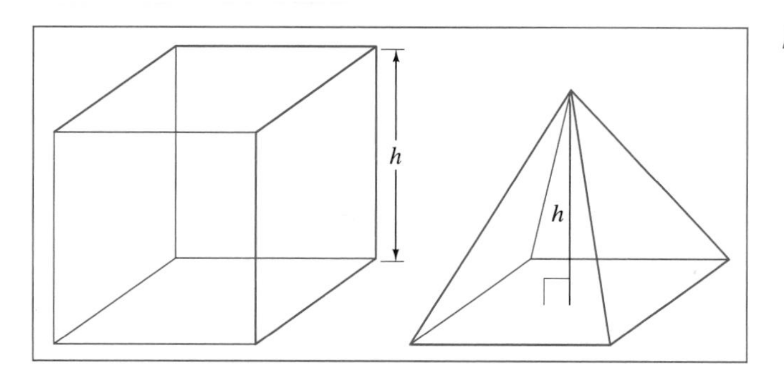

Figure 10.33

10.13.2 Volume of a Right Circular Cone

The volume of a right circular cone is one-third the volume of a right circular cylinder with congruent bases and equal height; in other words, three right circular cones could fit into one right circular cylinder with congruent bases and equal height, as shown in Figure 10.34. Thus, the formula for the volume of a right circular cone is: $V = \frac{1}{3}\pi r^2 h$ where h is the height.

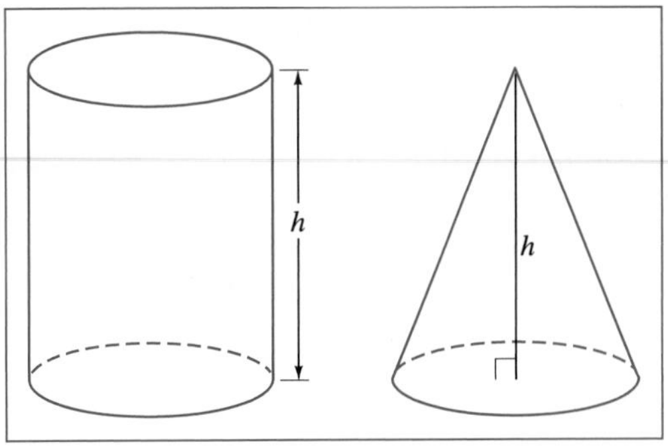

Figure 10.34

10.13.3 Volume of a Sphere

To understand better the volume formula for a sphere, think of a sphere as having a very large number of congruent pyramids with their apexes at the center of the sphere and with their vertices of the base touching the sphere, as shown in Figure 10.35. If the bases of the pyramids are small, then the radius, r, of the sphere is approximately equal to the height of each pyramid. Because we know that the volume of each pyramid is $\frac{1}{3}\mathrm{Bh}$ or $\frac{1}{3}\mathrm{Br}$, then the volume of all the pyramids is $n \cdot \frac{1}{3}\mathrm{Br}$, where n is the number of pyramids. Now, $n \cdot \mathrm{B}$ is the area of all the bases of the pyramids, and because this approximates the surface area of the sphere (which is $4\pi \mathrm{r}^2$), the formula for the volume of a sphere is: $\mathrm{V} = \frac{1}{3}(4\pi \mathrm{r}^2)\mathrm{r} = \frac{4}{3}\pi \mathrm{r}^3$.

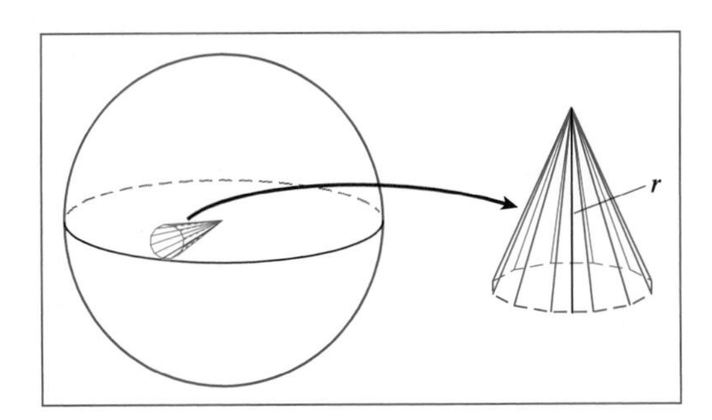

Figure 10.35

Related Topics **Cavalieri's Principle**

Cavalieri's Principle

cornerstone

Measuring Volume
Polyhedra, Prisms, and Pyramids
Cones, Cylinders, and Spheres
Volume of Prisms and Cylinders
Volume of Pyramids, Cones, and Spheres

Introduction

We can derive the formulas for the volumes of various three-dimensional figures using the formula for the volume of a right prism. In this section, we discuss Cavalieri's principle which details this.

Italian mathematician Bonaventura Cavalieri (1598–1647) was one of Galileo's students. His ideas were used to find volumes before calculus techniques were invented.

For More Information Sibley, T. Q. (1998). *The Geometric Viewpoint: A Survey of Geometries.* Reading, MA: Addison-Wesley.

Cavalieri developed a principle for finding the volume of many three-dimensional figures given a particular relationship. His principle states: Let A and B be two solids, included between two parallel planes (as shown in Figure 10.36). If every plane parallel to the given planes intersects A and B in sections with the same area, then A and B have the same volume.

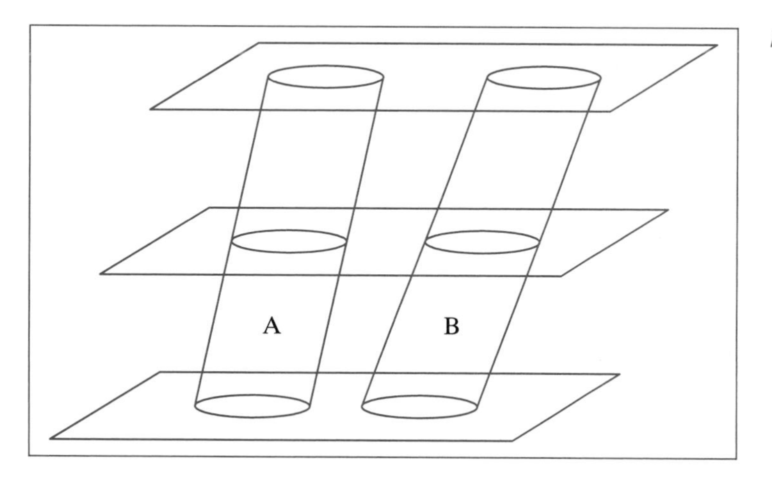

Figure 10.36

Figure 10.37 illustrates another example of solids with the same volume. If two pyramids have the same height and their bases have the same area, then the two pyramids have the same volume.

Figure 10.37

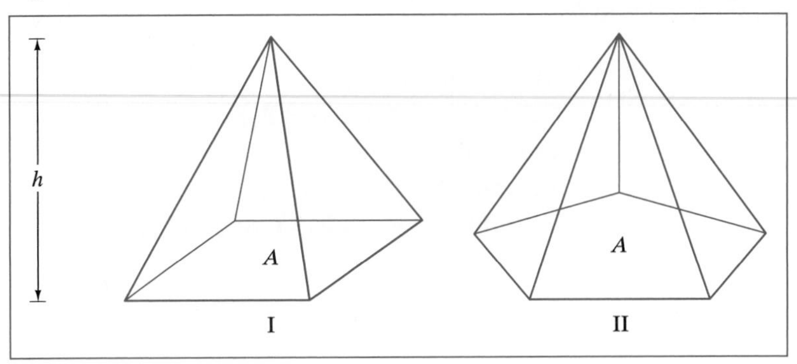

Pyramid I and pyramid II have the same volume because their bases have the same area.

TOPIC 10.15 *Mass and Temperature*

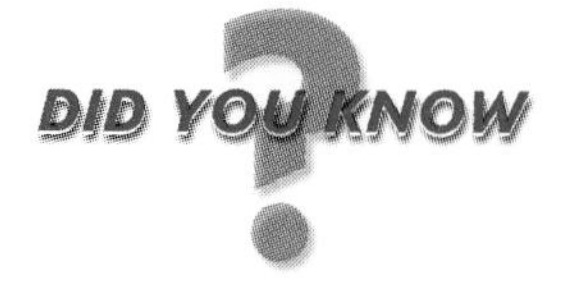

Linear Measure

Introduction

We measure a wide variety of things besides length, area, and volume. Two common characteristics that are measured are mass and temperature. In this section, we discuss mass and weight, and temperature.

DID YOU KNOW

The gram is the basic unit of mass in the metric system. One gram is officially equal to the weight of 1 cubic centimeter of distilled water at 4°C.

For More Information *Webster's New World Dictionary.*

10.15.1 Mass and Weight

Mass and weight are two quantities that are often confused. **Mass** is a quantity of matter; **weight** is a force exerted by gravitational pull. This is why the weight of astronauts changes when measured on Earth and when measured on the moon, but their masses remain the same.

In everyday usage, we interchange the quantities of mass and weight, but we use different units of measure to quantify them. We measure mass in metric units, with the **gram** being the basic unit. Table 10.8 shows metric units for measuring mass and their equivalents in other units in the metric system.

Conversions between different units of mass measure within the metric system are calculated in the same way that we did when working with metric units for linear measures in Section 10.2.3.

In the English system of measurement, weight is measured in avoirdupois units such as ounces, pounds, and tons. Table 10.9 shows these units and their equivalents.

Table 10.8

Unit	Equivalent
1 milligram (1 mg)	0.001 g
1 centigram (1 cg)	0.01 g
1 decigram (1 dg)	0.1 g
1 gram (1 g)	1,000 mg
1 dekagram (1 dg)	10 g
1 hectogram (1 hg)	100 g
1 kilogram (1 kg)	1,000 g
1 ton (metric) (1 t)	1,000,000 g

Table 10.9

Unit	Equivalent
1 ounce (1 oz)	$\frac{1}{16}$ lb
1 pound (1 lb)	16 oz; $\frac{1}{2,000}$ t
1 ton (English) (1 t)	2,000 lb

10.15.2 Temperature

Temperature, the characteristic that indicates how cold or how hot something is, can be measured using either the metric system or the English system. In the metric system, the **degree Kelvin** is used for scientific measurements; the temperature for the freezing point of water on the Kelvin scale is 273°.

The more common unit of measure in the metric system is the **degree Celsius.** This scale is named after the Swedish scientist Anders Celsius who invented it. The Celsius scale is marked off in 100 equal parts between 0 degrees Celsius (0°C), which is the freezing point of water, and 100 degrees Celsius (100°C), which is the boiling point of water.

In the English system of measurement, the Fahrenheit scale is used to measure temperature. This scale is named after Gabriel Fahrenheit, a German instrument maker who made the first mercury thermometer in 1714. He labeled the lowest temperature that he could create in his laboratory as 0° and the normal temperature of the body as 98°. The Fahrenheit scale is marked off in 180 equal parts between 32°F, which is the freezing point of water, and 212°F, which is the boiling point of water.

To convert from temperature measures using a Celsius scale to those using a Fahrenheit scale, or vice versa, we can derive a conversion formula. Because the Celsius scale has 100 equal parts between the freezing point and boiling point of water, and the Fahrenheit scale has 180 equal parts, the ratio is 100 to 180, or 5 to 9. Thus, for every 5 degree increase on the Celsius scale, there is a 9 degree increase on the Fahrenheit scale. Because the scales have the freezing point of water at different values, we must also adjust for that. The formulas for converting from the Celsius scale to the Fahrenheit scale, and the Fahrenheit scale to the Celsius scale are, respectively:

$$°F = \frac{9}{5}°C + 32° \text{ and } °C = \frac{5}{9}(°F - 32°).$$

Thus, if we have a temperature of 25°C that we want to convert to the Fahrenheit scale, we have: $F = \frac{9}{5}(25) + 32 = 45 + 32 = 77$. So, 25°C = 77°F. Likewise, if we have a temperature of 50°F that we want to convert to the Celsius scale, we have: $C = \frac{5}{9}(50 - 32) = \frac{5}{9}(18) = 10$. So, 50°F = 10°C.

TOPIC 10.16 Pythagorean Theorem and Other Triangle Relationships

cornerstone

Triangles
Measuring Area

Introduction

Quite possibly the most famous theorem in all of mathematics—known to school children and mathematicians alike—is what is known as the *Pythagorean theorem*. In this section, we discuss the Pythagorean theorem, its converse, special right triangles, and the triangle inequality.

Pythagoras was a Greek mathematician who lived about 570–500 B.C. He and his followers, called the Pythagoreans, divided mathematics into four parts: music, arithmetic (what is commonly known today as number theory), astronomy, and geometry. These four subjects were called the *quadrivium* and were later adopted by Plato and Aristotle and became the school curriculum until the Renaissance period.

For More Information Bunt, L. N. H., Jones, P. S., & Bedient, J. D. (1988). *The Historical Roots of Elementary Mathematics*. New York: Dover Publications.

10.16.1 The Pythagorean Theorem

Long before the time of the Pythagoreans, people knew about and used the relationship we call the Pythagorean theorem. Before discussing the theorem, we need to identify some parts of a right triangle. In a right triangle, the side opposite the right angle is called the **hypotenuse,** and the two sides that form the right angle are called the **legs,** as illustrated in Figure 10.38.

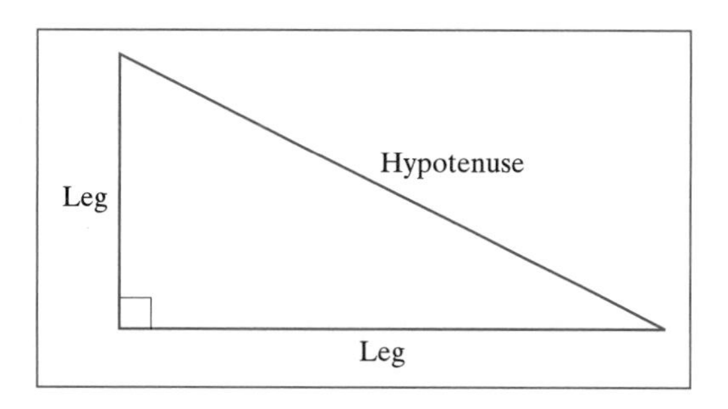

Figure 10.38

The Pythagorean theorem, as stated in Euclid's *Elements* is: In right-angled triangles, the square on the side subtending the right angle is equal to the squares on the sides containing the right angle. In textbooks today, we are more likely to find it stated as: If a right triangle has legs of length a and b and hypotenuse of length c, then $a^2 + b^2 = c^2$, as illustrated in Figure 10.39.

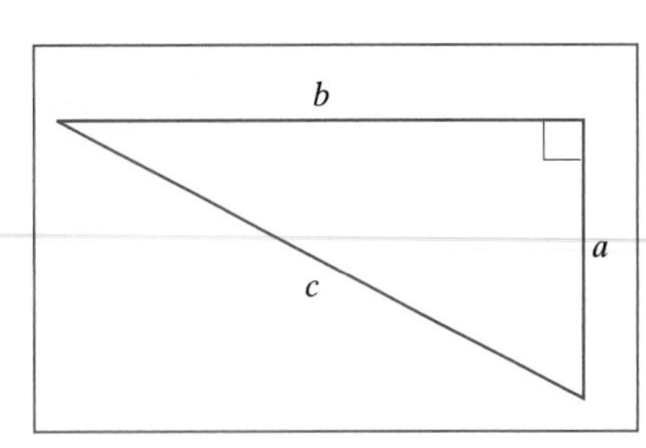

Figure 10.39

The illustration in Figure 10.39 shows the relationship among the sides of the right triangle, but the original statement of the Pythagorean theorem demonstrates the relationship among the area of the squares formed on the sides of a right triangle, as illustrated in Figure 10.40.

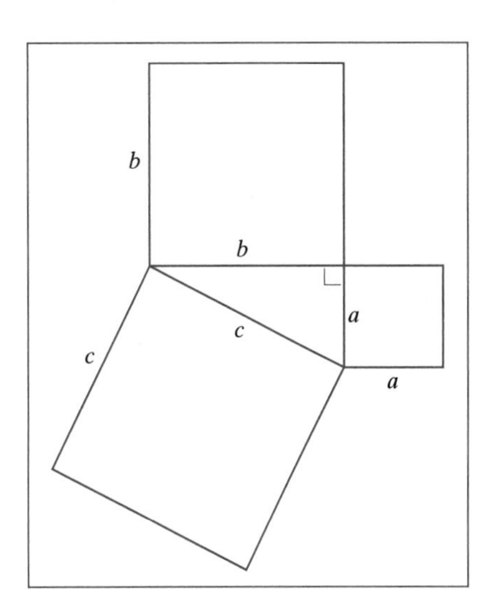

Figure 10.40

10.16.2 The Converse of the Pythagorean Theorem

The converse of the Pythagorean theorem is also true. Euclid stated the converse as: If in a triangle the square on one of the sides be equal to the squares on the remaining two sides of the triangle, the angle contained by the remaining two sides of the triangle is right. Textbooks today might state the converse as: If a triangle with sides of lengths *a, b,* and *c* has the relationship $a^2 + b^2 = c^2$, then the triangle is a right triangle with the right angle opposite the side of length *c*.

10.16.3 Special Right Triangles

Figure 10.41

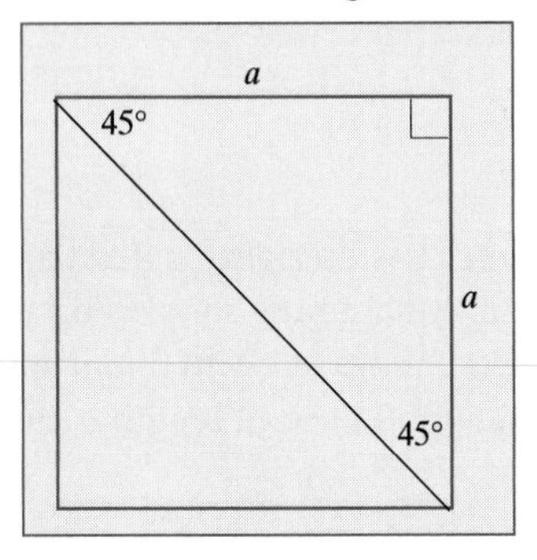

Besides the Pythagorean theorem, there are other relationships that exist in right triangles. We will discuss two special kinds of right triangles and the relationships that exist within these triangles.

The first triangle we will discuss is called a **45°–45°–90° right triangle.** This triangle is a right triangle with two 45° angles and two congruent legs and can be formed by drawing a diagonal of a square, as shown in Figure 10.41. Using the relationship discussed in the Pythagorean theorem, we can see that if each leg in this triangle has length a, then the hypotenuse must have length $a\sqrt{2}$ because $a^2 + a^2 = c^2$ gives us $2a^2 = c^2$, which gives us $c = a\sqrt{2}$. Thus, a property of a $45° - 45° - 90°$ triangle is that if the legs have length a, then the hypotenuse will have length $a\sqrt{2}$.

Another special right triangle is a **30°–60°–90° right triangle.** This triangle is a right triangle with a 30° angle and a 60° angle and can be formed by dividing an equilateral triangle in half, as shown in Figure 10.42.

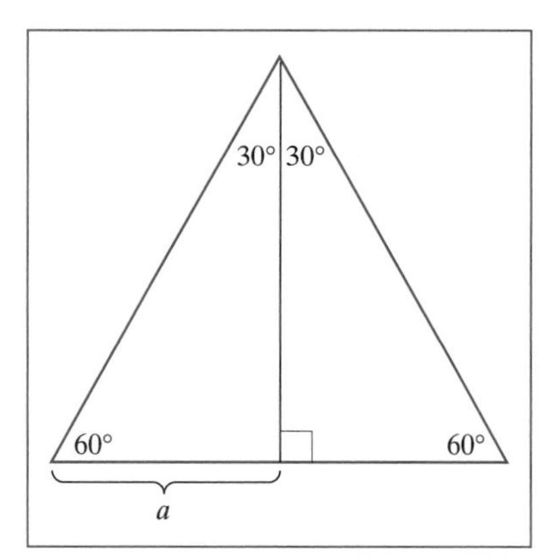

Figure 10.42

From the equilateral triangle in Figure 10.42, we can see that the side opposite the 30° angle will be half the length of the hypotenuse. Suppose the side opposite the 30° angle has length a, then the hypotenuse has length $2a$. Using the Pythagorean theorem relationship, we can see then that the side opposite the 60° angle must have length $a\sqrt{3}$ because $a^2 + b^2 = (2a)^2$ gives us $b^2 = 3a^2$ which gives us $b = a\sqrt{3}$. Thus, a property of a $30° - 60° - 90°$ triangle is that the length of the hypotenuse is twice as long as the leg opposite the 30° angle, and the leg opposite the 60° angle is $\sqrt{3}$ times the length of the leg opposite the 30° angle.

10.16.4 The Triangle Inequality

To construct a triangle using three segment lengths, there has to be a particular relationship among the lengths of the segments or sides. This relationship is known as the **Triangle Inequality.** The Triangle Inequality theorem states: The sum of the lengths of any two sides of a triangle is greater than the length of the third side. In other words, in any triangle, ΔXYZ; $XY + YZ > XZ$ and $XY + XZ > YZ$ and $YZ + XZ > XY$ (see Figure 10.43).

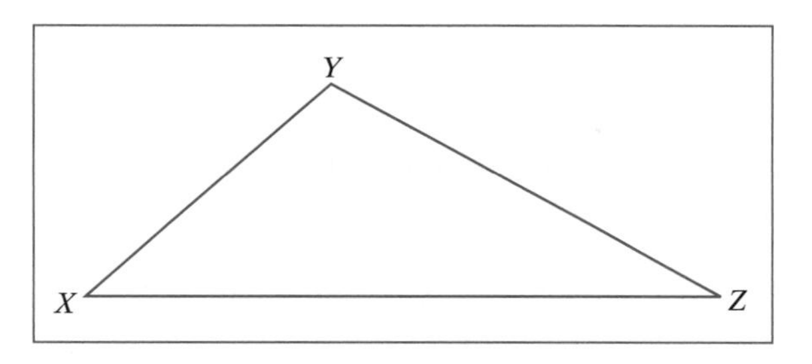

Figure 10.43

Related Topics **Distance Formula**

TOPIC 10.17 Distance Formula

cornerstone **Line Segments**
Cartesian Coordinate System

Introduction

There is a common saying that the shortest distance between any two points is a straight line. In this section, we discuss distance and how to find the distance between two points.

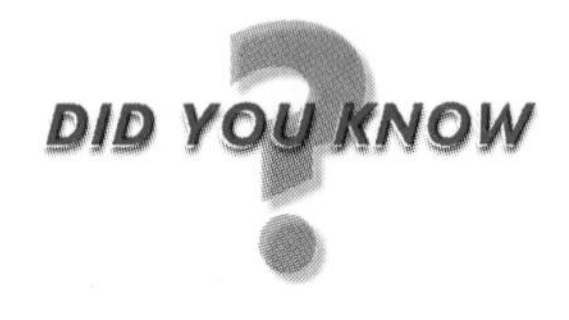

DID YOU KNOW

Before standardized units of measure were adopted, distance was measured in various ways by various peoples. In some African societies, long distances were measured by the number of days it would take to travel between two points or by time between meals.

For More Information Zaslavsky, C. (1979). *Africa Counts: Number and Pattern in African Culture*. Brooklyn, NY: Lawrence Hill Books.

The **distance postulate** states: To every pair of different points, there corresponds a unique positive number. In other words, for every two points that are different, we can assign a unique positive value that is the distance between those two points.

If two points are on the coordinate axes and lie on a horizontal or vertical line, then to find the distance between these two points, we simply count the number of units in the horizontal or vertical direction. In Figure 10.44 we can see that A and B are 5 units apart, so $AB = 5$, and C and D are 4 units apart, so $CD = 4$. We can generalize counting units to say that (a) the distance between two points that lie on a horizontal line is the absolute value of the difference of their x-coordinates ($|x_2 - x_1|$), and (b) the distance between two points that lie on a vertical line is the absolute value of the difference of their y-coordinates ($|y_2 - y_1|$). Furthermore, the distance between any two points, X and Y, is the same as the distance between Y and X. Thus, in Figure 10.44, $BA = 5$ and $DC = 4$.

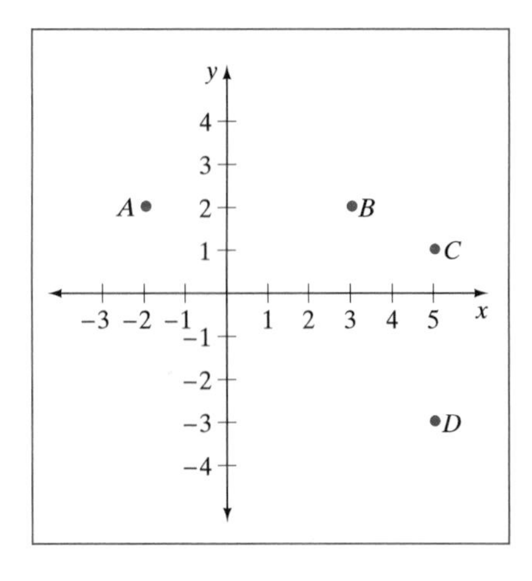

Figure 10.44

Using the Pythagorean theorem relationship, we can derive the formula to find the distance between any two points, regardless of where they are located on the coordinate axes. Consider point J with ordered pair (x_1, y_1) and point K with ordered pair (x_2, y_2) in Figure 10.45. We can draw a horizontal line through J and a vertical line through K, labeling the point of intersection of the two lines as L. Notice that L corresponds to the ordered pair (x_2, y_1), and we have formed a right triangle, ΔJKL.

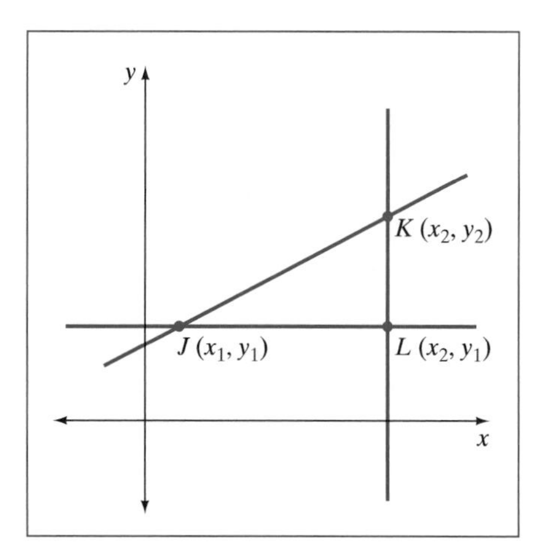

Figure 10.45

We know that $(JL)^2 + (KL)^2 = (JK)^2$, from the Pythagorean theorem. Substituting in the values of these lengths, we have $(x_2 - x_1)^2 + (y_2 - y_1)^2 = (JK)^2$. Thus, we find that $JK = \sqrt{(x_2 - x_1)^2 + (y_2 - y_1)^2}$. This formula is called the **distance formula** and we can state it as: The distance between the points (x_1, y_1) and (x_2, y_2) is $\sqrt{(x_2 - x_1)^2 + (y_2 - y_1)^2}$.

One other formula that is related to the distance between two points is the **midpoint formula.** We know that every segment has a unique midpoint and that this midpoint is equal distance from both endpoints of the segment. The midpoint formula gives us the coordinates of the midpoint of a segment. It states: given A with ordered pair (x_1, y_1) and B with ordered pair (x_2, y_2), the midpoint of $\overline{AB}$ is the point M with ordered pair $(\frac{x_1 + x_2}{2}, \frac{y_1 + y_2}{2})$.

Slope
Coordinate Geometry
Pythagorean Theorem and Other Triangle Relationships

TOPIC 10.18 Tessellations

cornerstone

Polygons
Regular Polygons
Planes
Angle Measurement

Introduction

There are tessellations all around us—from tile floors, to brick buildings, to patterns on fabric. In this section, we discuss what a tessellation is, how any triangle can tessellate the plane, and how any quadrilateral can tessellate the plane.

Some of the most celebrated examples of tessellations are found in Islamic designs. These designs often use geometric figures and floral designs.

For More Information Zaslavsky, C. (1996). *The Multicultural Math Classroom: Bringing in the World.* Portsmouth, NH: Heinemann.

10.18.1 What Is a Tessellation?

A **tessellation** has been defined in different ways. Here are the two most common definitions:

I. A tessellation is a filling of the plane with repetitions of polygonal regions (polygons and their interiors), without gaps and overlapping of regions, and where all corners meet at a vertex.

II. A tessellation is a filling of the plane with repetitions of polygonal regions, without gaps and overlapping of regions, and where corners do not necessarily meet at a vertex.

We can see that a *tiling of the plane* (filling the plane) that fits the first definition also fits the second definition, but a tiling of the plane that fits the second definition does not necessarily fit the first definition. The first drawing in Figure 10.46 shows a tessellation that fits both the first and second definition, whereas the second drawing shows a tessellation that fits only the second definition.

Figure 10.46

10.18.2 Any Triangle Can Tessellate the Plane

Any triangle can tessellate the plane; in other words,—any triangle can repeatedly be used to fill the plane without gaps and overlapping. Figure 10.47 illustrates the process of tessellating the plane with a scalene triangle whose angles have been labeled $\angle 1$, $\angle 2$, and $\angle 3$. To fill the plane, rotate the triangle so that congruent sides are placed against each other. Notice that at any vertex where the corners of the triangles meet, there are six angles that meet at the vertex ($\angle 1$, $\angle 2$, $\angle 3$, $\angle 1$, $\angle 2$, $\angle 3$). Because we know that the sum of the measures of the three interior angles of a triangle is 180°, we can see that the sum of the measures of the angles at any vertex in this tessellation is 360°. Because one complete revolution about a point is 360°, we can see that this triangle is filling the plane without gaps or overlapping. This is true for any type of triangle.

Figure 10.47

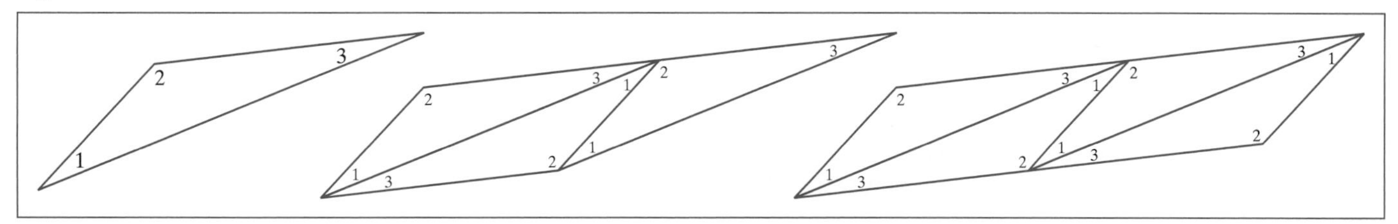

10.18.3 Any Quadrilateral Can Tessellate the Plane

Using what we learned about any triangle tessellating the plane, we can see that any quadrilateral can tessellate the plane. Figure 10.48 illustrates the process of tessellating the plane with a quadrilateral whose angles have been labeled $\angle 1$, $\angle 2$, $\angle 3$, and $\angle 4$. To fill the plane, rotate the quadrilateral so that congruent sides are placed against each other. Notice that at any vertex where the corners of the quadrilaterals meet, there are four angles that meet at the vertex ($\angle 1$, $\angle 2$, $\angle 3$, $\angle 4$). Because we know that the sum of the measures of the four interior angles of a quadrilateral is 360°, we can see that the sum of the measures of the angles at any vertex in this tessellation is 360°. Because one complete revolution about a point is 360°, we can see that this quadrilateral is filling the plane without gaps or overlapping. This is true for any type of quadrilateral.

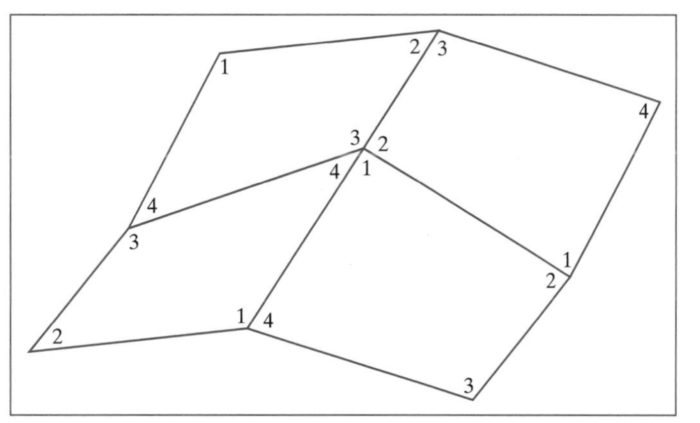

Figure 10.48

Related Topics — **Tessellations of Regular Polygons**

TOPIC 10.19 Tessellations of Regular Polygons

cornerstone Tessellations
Regular Polygons

Introduction

We can find tessellations of some regular polygons in nature. Two examples are the tessellation of a regular hexagon in any honeycomb and the tessellation of regular polygons in the molecular structures of some crystals. In this section, we discuss which regular polygons can tessellate the plane and semiregular tessellations. There are other types of tessellations that we will not address here.

The twentieth-century Dutch artist Maurits Cornelis Escher created tessellations using design elements such as fish, reptiles, birds, and humans. Some of his inspiration came from observing the Islamic designs in the Alhambra in Granada, Spain.

For More Information Schattschneider, D. (1990). *Visions of Symmetry: Notebooks, Periodic Drawings, and Related Work of M. C. Escher*. New York: W. H. Freeman and Company.

10.19.1 Which Regular Polygons Can Tessellate the Plane?

For any polygon to tessellate the plane, with repetitions of the same polygon, the measures of the angles that come together must sum to 360° because a complete revolution about a point is 360°. If the measures of the angles that come together did not sum to 360°, then there would be gaps (if the sum of the angle measures was less than 360°) or overlapping (if the sum of the angle measures was more than 360°).

We can determine which regular polygons can tessellate the plane because we know that in a regular polygon all the interior angles have the same measure, and thus, the measure of an interior angle must be a factor of 360 for the sum of the angle measures to equal 360°. Table 10.10 shows the regular polygons from $n = 3$ to 10 sides, and the measure of each interior angle in these polygons.

Table 10.10

Polygon	Measure of each Interior Angle
Equilateral Triangle	60°
Square	90°
Regular Pentagon	108°
Regular Hexagon	120°
Regular Heptagon	about 128.6°
Regular Octagon	135°
Regular Nonagon	140°
Regular Decagon	144°

We can see that the equilateral triangle, the square, and the regular hexagon all have interior angle measures that are factors of 360. These, in fact, are the only regular polygons that have this characteristic, and thus, an equilateral triangle, a square, and a regular hexagon are the only regular polygons that will tessellate the plane. Figure 10.49 shows tessellations made from each of these three regular polygons.

Figure 10.49

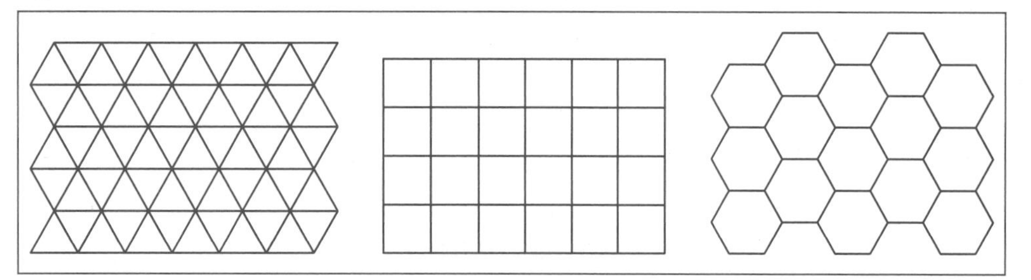

An interesting observation is that when we were determining which regular polygons will tessellate the plane, we could have stopped checking at a regular hexagon (where $n = 6$) because for that polygon, the measure of each interior angle is 120°. The only factor of 360 that is larger than 120 is 180. However, the measure of each angle in a regular polygon must be less than 180°. Thus, no regular polygon with more than six sides could tessellate the plane.

10.19.2 Semiregular Tessellations

A **semiregular tessellation** is a tessellation comprised of more than one regular polygon such that each vertex has the same polygonal arrangement. Combining squares and regular octagons, as shown in the first drawing in Figure 10.50, is an example of a semiregular tessellation. The second drawing in Figure 10.50 shows another semiregular tessellation, formed by equilateral triangles and squares.

Figure 10.50

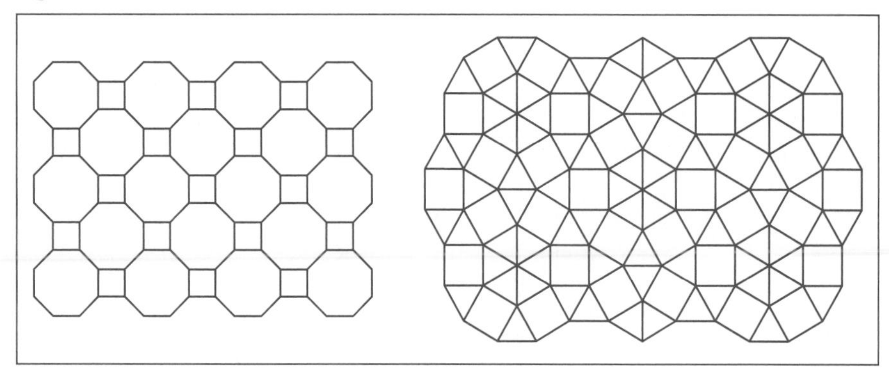

Related Topics **Transformations**

TOPIC 10.20 Transformations

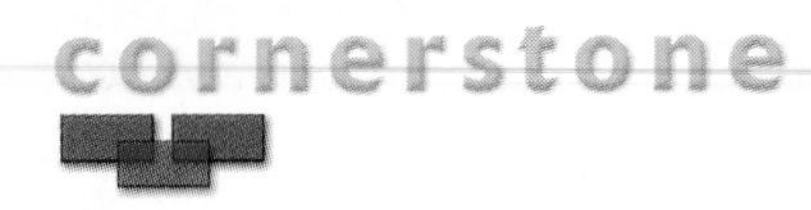

Congruence
Line Segments

Introduction

The study of geometry concerns not only geometric figures, but also the movement of geometric figures. In this section, we discuss the meaning of a transformation and the different types of transformations.

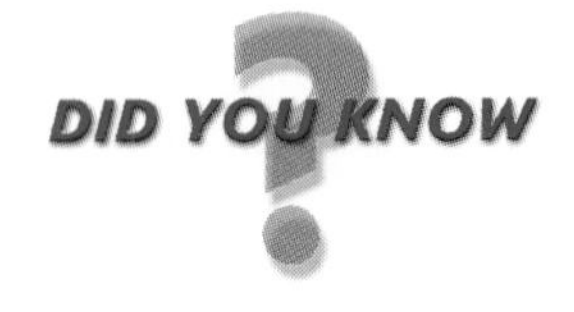

M. C. Escher (1889–1972), a Dutch artist, identified ten foundational systems (or types) of tessellations. The first nine involve quadrilaterals, and the tenth uses triangles. Escher described these ten systems in a chart that lists the transformations for each system.

For More Information Schattschneider, D. (1990). *Visions of Symmetry: Notebooks, Periodic Drawings, and Related work of M. C. Escher*. New York: W. H. Freeman and Company.

If we have a geometric figure in a plane and we move all the points of the figure in certain ways, we form a new geometric figure called the **image** of the original figure. A **transformation** is a correspondence between all the points of a geometric figure in a plane and all the points of its image, such that each point of the figure can be matched with exactly one point of its image, and each point of the image can be matched with exactly one point of the original figure. Figure 10.51 shows some figures and their images formed by transformations.

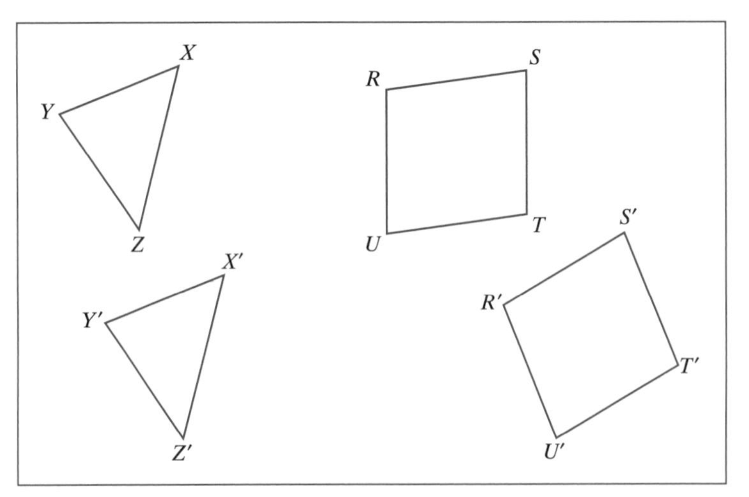

Figure 10.51

When we move all the points of a planar figure, we represent the points of the image using the corresponding letters with a special notation. Figure 10.52 shows figure $ABCD$ and its image $A'B'C'D'$.

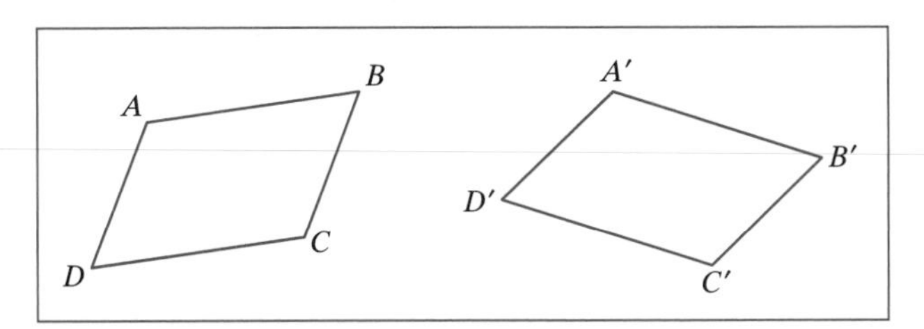

Figure 10.52

A transformation that preserves the size and shape of the original figure is called an **isometry,** or a **rigid transformation.** There are three types of transformations that are isometries: translation (slide), rotation (turn), and reflection (flip). Because an isometry does not change the size or shape from the original figure to the image, the figure and its image are congruent. Not all transformations preserve both size and shape. For example, size transformations preserve the shape of a figure but either reduce or enlarge the size.

Related Topics

Translations
Rotations
Reflections
Size Transformations

Translations

cornerstone

Line Segments
Congruence
Transformations
Coordinate Geometry

Introduction

A translation is often thought of as a slide, like the sliding action involved in playing checkers or chess. In this section, we discuss translations and their features.

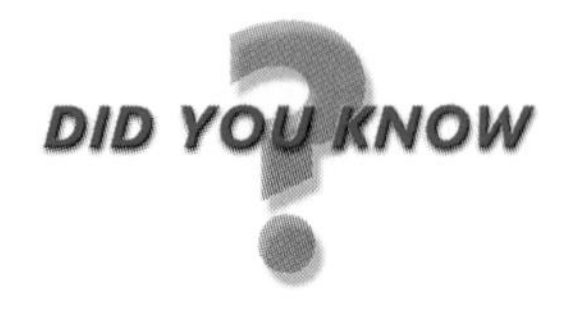

Felix Klein (1849–1925) gave his inaugural address as a professor of mathematics at the University of Erlangen in 1872 at the age of 23. The presentation is now known as Klein's Erlanger Programm. In his address, he defined geometry by using groups of transformations: Geometry is the study of those properties of a set that are preserved under a group of transformations on that set.

For More Information Sibley, T. Q. (1998). *The Geometric Viewpoint: A Survey of Geometries.* Reading, MA: Addison-Wesley.

One transformation that is an isometry is a **translation.** Formally, a translation is a transformation on a plane in which every point in a set is moved the same number of units in the same direction. Thus, a translation involves both distance and direction. Figure 10.53 shows a rectangle ($UVWX$) and its image ($U'V'W'X'$) under a translation. The translation is determined by the **directed line segment** from J to K (denoted by $\overrightarrow{JK}$). A directed line segment is a line segment that has a direction, in this case from J to K. In Figure 10.53, the image of a point U in the plane, is the point U' that results from moving U along a line that is parallel to $\overrightarrow{JK}$ a distance of JK in the direction from J to K. Because a translation is an isometry, the image is congruent to the original figure (i.e., congruence is preserved).

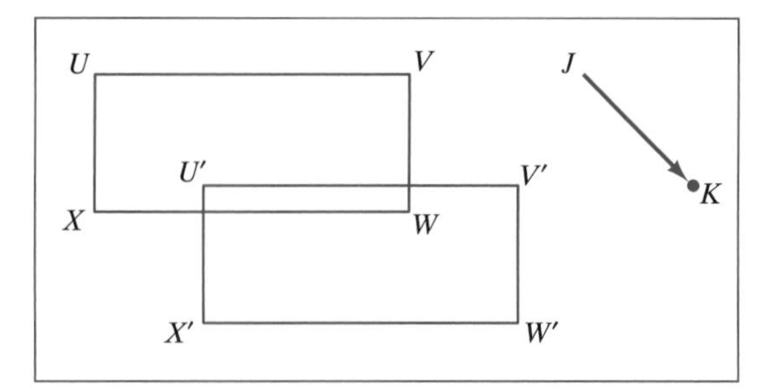

Figure 10.53

Thus, we obtain the image of $UVWX$ by moving each point in the set of points that comprise $UVWX$ along a line parallel to $\overrightarrow{JK}$ a distance of JK in the direction from J to K.

On a coordinate plane, we can specify a translation, using the horizontal and vertical movements in the plane. For example, $\overline{EF}$ is shown with its image $\overline{E'F'}$ in Figure 10.54. We can see that each point of $\overline{EF}$ has been moved four units to the right and two units down. We can denote this translation symbolically by writing $(x, y) \rightarrow (x + 4, y - 2)$.

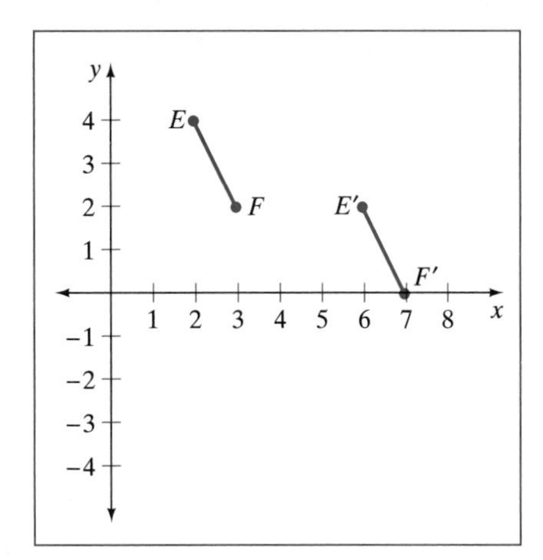

Figure 10.54

Related Topics

Rotations
Reflections

TOPIC 10.22 Rotations

cornerstone **Transformations**

Introduction

A rotation is often thought of as a turn, like the turning of a steering wheel or of a screw. In this section, we discuss rotations and their features.

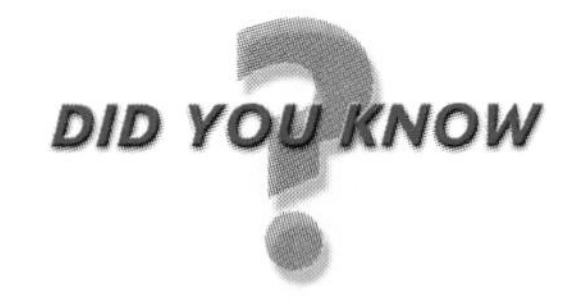

DID YOU KNOW

Because one complete revolution around a point is 360°, if, you rotate any figure 360°, the image will match the original figure. In basketball or skateboarding, this motion is called "doing a 360."

Like a translation, a **rotation** is a transformation that is an isometry. Formally, a rotation is a transformation on a plane in which every point in a set is moved the same number of units (e.g., degrees) in the same direction by rotating about a point. Thus, a rotation involves both the amount of the turn and direction. The point in the plane about which the set of points is rotated is called the **center of rotation.** The direction of the rotation can be either clockwise or counterclockwise. In any rotation, three key elements are the center of rotation, the amount of the turn, and the direction.

Figure 10.55 shows a triangle (ABC) and its image ($A'B'C'$) under a rotation about point A through 45° in a clockwise direction.

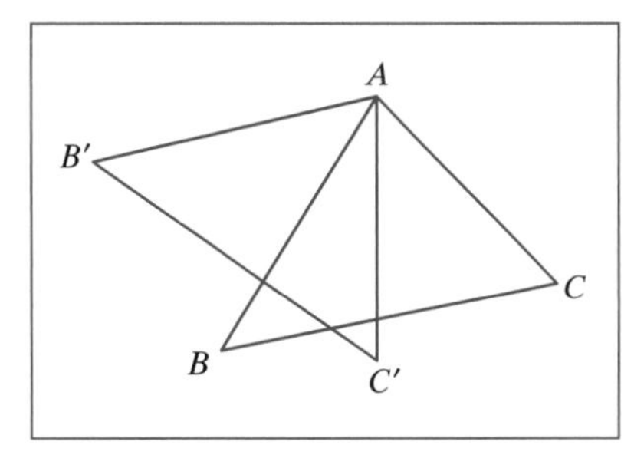

Figure 10.55

Figure 10.56 shows a triangle (ABC) and its image ($A'B'C'$) under a rotation about point K through 60° in a counterclockwise direction. Because rotation is an isometry, the image is congruent to the original figure (i.e., congruence is preserved).

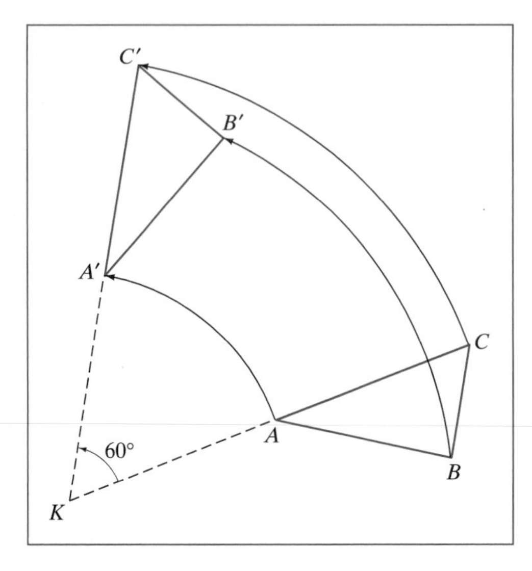

Figure 10.56

Related Topics

- Translations
- Reflections
- Symmetry

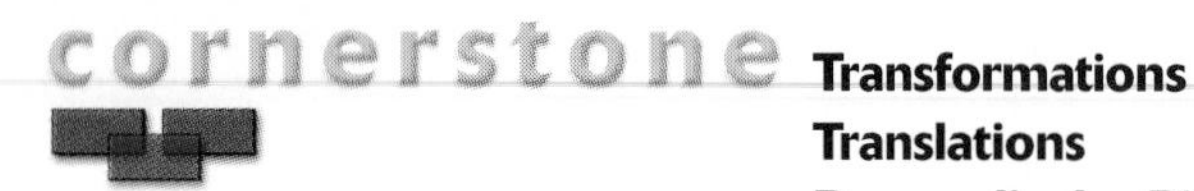

TOPIC 10.23 Reflections and Glide Reflections

cornerstone

Transformations
Translations
Perpendicular Bisector Construction

Introduction

A reflection is often thought of as a flip, like a tree's image in the water of a pond. In this section, we discuss reflections, including glide reflections, and their features.

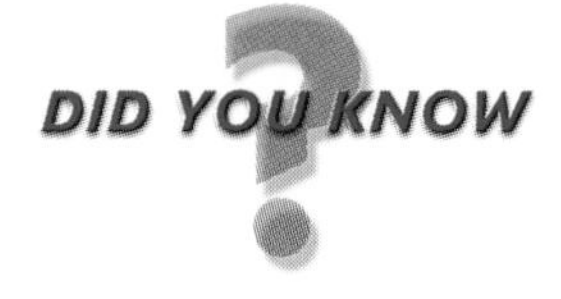

Although we may think that the human face is symmetrical—with one side being a reflection of the other side over a line dividing the face vertically—in reality, the human face is rarely symmetrical, and computer-generated photos of human faces that are symmetrical have been found to be less attractive than photos of the original human faces.

Another transformation that is an isometry is a **reflection.** Formally, a reflection is a transformation on a plane in which every point in a set is paired with a point in the image such that there is a line that is the perpendicular bisector of the line segment joining a point and its image. The line that is the perpendicular bisector of the line segments joining points and their images is called the **line of reflection.** If a point of the set being reflected happens to lie on the line of reflection, the point will be paired with itself under the reflection.

Figure 10.57 shows a pentagon $(DEFGH)$ and its image $(D'E'F'G'H')$ under a reflection in line m. Because a reflection is an isometry, the image is congruent to the original figure (i.e., congruence is preserved).

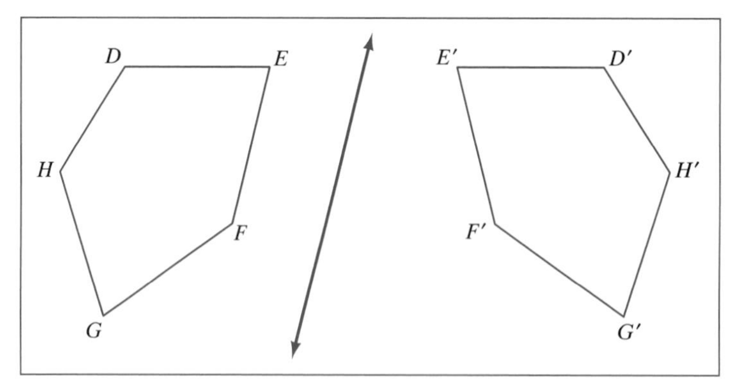

Figure 10.57

By combining two transformations, we can obtain another transformation. If we slide a figure and then flip it, we can perform a glide reflection. A **glide reflection** is a transformation on a plane in which every point in a set is translated and then reflected in a line parallel to the directed line segment. Figure 10.58 shows a parallelogram $STUV$ and its image $(S'T'U'V')$ under a glide reflection, where $STUV$ is first translated from A to B, and then reflected in line n, which is parallel to $\overrightarrow{AB}$. Because a glide reflection consists of a translation and a reflection, it is an isometry, and, thus, the image is congruent to the original figure (i.e., congruence is preserved).

Figure 10.58

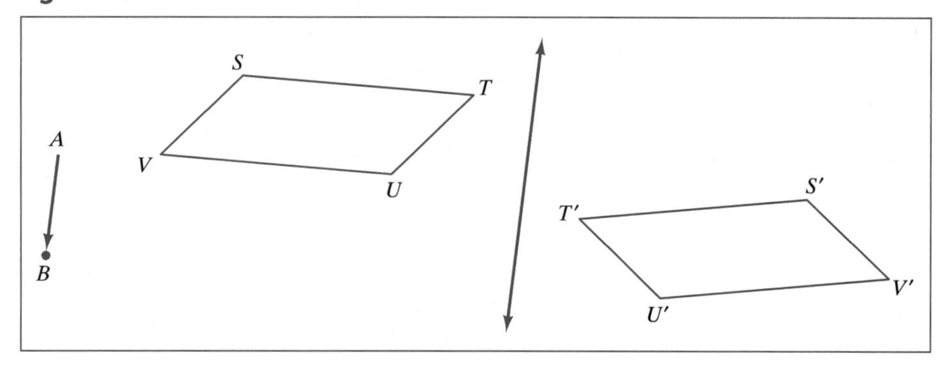

Related Topics Rotations
Symmetry

TOPIC 10.24 Size Transformations

cornerstone **Transformations**

Introduction

A size transformation is often thought of as shrinking or enlarging, such as enlarging a photo. In this section, we discuss size transformations and their features.

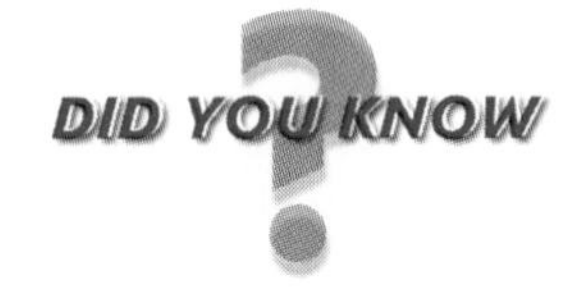

Overhead projectors and photocopy machines are common machines that make use of size transformations.

One type of transformation that is not an isometry is a **size transformation.** A size transformation can be thought of as a reduction or an enlargement; formally, a size transformation is a transformation on a plane to the plane with center O and scale factor $s(s > 0)$ in which every point P in a set is paired with a point P', such that $OP' = s \cdot OP$ and O, P and P' are collinear with O not between P and P'.

Figure 10.59 shows a triangle (XYZ) and its image ($X'Y'Z'$) under a size transformation with a scale factor of 2.5.

Figure 10.59

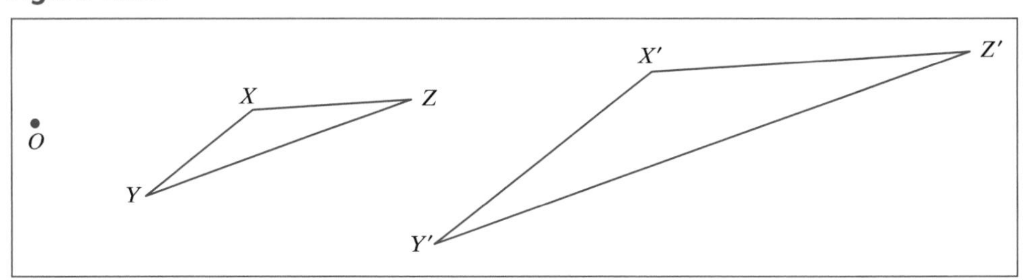

Figure 10.60 shows a rectangle ($JKLM$) and its image ($J'K'L'M'$) under a size transformation with a scale of 0.5. If the scale factor is greater than 1, the image will be an enlargement of the original figure. If the scale factor is less than 1, the image will be a reduction of the original figure. If the scale factor is equal to 1, the image will be congruent to the original figure. Unless the scale factor is 1, a size transformation is not an isometry and the image will not be congruent to the original figure (i.e., congruence is not preserved). However, the shape of the original figure is preserved.

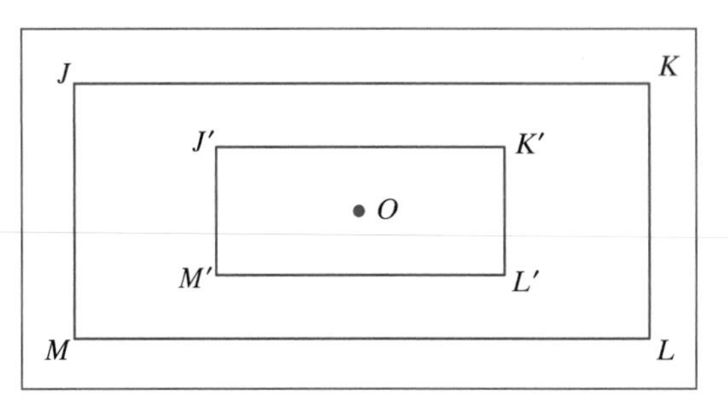

Figure 10.60

Related Topics

- **Translations**
- **Rotations**
- **Reflections**
- **Fractals and Rep-tiles**

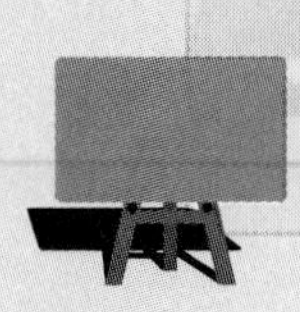

In the Classroom

This chapter emphasizes standards from both *Measurement* and *Geometry Content Standards* as well as the *Connections Process Standard.* To prepare instruction that enables your students to "understand measurable attributes of objects and the units, systems, and processes of measurement [and to] apply appropriate techniques, tools, and formulas to determine measurements" (NCTM, 2000, p. 44), we believe you should (a) work with measures of objects in ways that extend your current thinking by challenging your misconceptions and by rethinking familiar results in multiple ways (Activities 10.1–10.12), and (b) use measurement to explore new concepts and ways of reasoning (Activities 10.13–10.17).

Some Expectations of Your Future Students:

Pre-K–2 (NCTM, 2000, pp. 96, 102)

- recognize and apply slides, flips, and turns
- recognize and represent shapes from different perspectives
- select an appropriate unit and tool for the attribute being measured
- use repetition of a single unit to measure something larger than the unit

Grades 3–5 (NCTM, 2000, pp. 164, 170)

- make and use coordinate systems to specify locations and to describe paths
- predict and describe the results of sliding, flipping, and turning two-dimensional shapes
- explore what happens to measurements of a two-dimensional shape, such as its perimeter and its area when the shape is changed in some way
- develop strategies for estimating the perimeters, areas, and volumes of irregular shapes, and to determine the surface areas and volumes of rectangular solids
- develop, understand, and use formulas to find the area of rectangles and related triangles and parallelograms

How Activities 10.1–10.18 Help You Develop an Adult-level Perspective on the Above Expectations:

- By working with area and perimeter of polygons, you understand the importance of units of measurement and of using appropriate tools for measuring, as well as their limitations.
- By developing strategies for measuring irregular objects, you continue to work with the big idea of decomposition. You draw, construct, or model ways to make indirect measurements by building formulae that enable calculation of the required measurement.
- By exploring a number proofs of Pythagoras's theorem, you realize that length and area measurements are the theorem's conceptual basis while developing reasoning strategies.
- By investigating the circumference to diameter ratio of a circle, you work with the idea that physical measurement is an estimation and make important connections to concepts from number, geometry, data analysis, and algebra. You use your knowledge from these content areas to argue for the existence of the irrational number p and to develop a formula for calculating the circumference and area of a circle in terms of its radius.
- By working with *Geometer's Sketchpad*™, you continue to associate the attribute being measured with the analogous physical object(s) while exploring the geometric concepts of translations, rotations, and reflections, all which depend on well-defined measures.
- By investigating three-dimensional shapes and coordinate geometry, you explore the concepts of perspective, volume, surface area, and systems to specify locations.

Bibliography

Battista, M. T. (1999). Fifth graders' enumeration of cubes in 3D arrays: Conceptual progress in an inquiry-based classroom. *Journal for Research in Mathematics Education, 30,* 417–448.

Beaumont, V., Curtis, R., & Smart, J. (1986). *How to teach perimeter, area, and volume.* Reston, VA: National Council of Teachers of Mathematics.

Binswanger, R. (1988). Discovering perimeter and area with Logo. *Arithmetic Teacher, 36,* 18–24.

Bledsoe, G. (1987). Guessing geometric shapes. *Mathematics Teacher, 80,* 78–80.

Bright, G. W., & Hoeffner, K. (1993). Measurement, probability, statistics, and graphing. In D. T. Owens (Ed.), *Research ideas for the classroom: Middle grades mathematics* (pp. 7–98) Old Tappan, NJ: Macmillan.

Clemens, S. R. (1985). Applied measurement: Using problem solving. *Mathematics Teacher, 78,* 176–180.

Clopton, E. L. (1991). Area and perimeter are independent. *Mathematics Teacher, 84,* 33–35.

Clopton, E. L. (1991). Sharing teaching ideas: Area and perimeter are independent. *Mathematics Teacher, 84,* 33–35.

Coburn, T. G., & Shulte, A. P. (1986). Estimation in measurement. In H. L. Schoen (Ed.), *Estimation and mental computation* (pp. 195–203). Reston, VA: National Council of Teachers of Mathematics.

Corwin, R. B., & Russell, S. J. (1990). Measuring: From paces to feet. (A unit of study for grades 3–5 from *Used numbers: Real data in the classroom*) Palo Alto, CA: Dale Seymour.

Fay, N., & Tsairides, C. (1989). Metric mall. *Arithmetic Teacher, 37,* 6–11.

Fosnaugh, L. S., & Harrell, M. E. (1996). Covering the plane with reptiles. *Mathematics Teaching in the Middle School, 1,* 666–670.

Gerver, R. (1990). Discovering pi: Two approaches. *Arithmetic Teacher, 37,* 18–22.

Giganti. P., Jr., & Cittadino, M. J. (1990). The art of tessellation. *Arithmetic Teacher, 31*(7), 6–16.

Harrison, W. R. (1987). What lies behind measurement? *Arithmetic Teacher, 34,* 19–21.

Hart, K. (1984). Which comes first—length, area, or volume? *Arithmetic Teacher, 31,* 16–18, 26–27.

Hawkins, V. J. (1989). Applying Pick's theorem to randomized areas. *Arithmetic Teacher, 36,* 47–49.

Hawkins, V. J. (1984). The Pythagorean theorem revisited: Weighing the results. *Arithmetic Teacher, 32,* 36–37.

Hiebert, J. (1984). Why do some children have trouble learning measurement concepts? *Arithmetic Teacher, 31*(7), 19–24.

Jensen, R. (1984). Multilevel metric games. *Arithmetic Teacher, 32,* 36–39.

Kaiser, B. (1988). Explorations with tessellating polygons. *Arithmetic Teacher, 36,* 19–24.

Kastner, B. (1989). Number sense: The role of measurement applications. *Arithmetic Teacher, 36,* 40–46.

Kennedy, J. B. (1993). Activities: Area and perimeter connections. *Mathematics Teacher, 86,* 218–221, 231–232.

Liedtke, W. W. (1990). Measurement. In J. N. Payne (Ed.), *Mathematics for the young* child (pp. 229–249). Reston, VA: National Council of Teachers of Mathematics.

Lindquist, M. M. (1989). Implementing the *Standards:* The measurement standards. *Arithmetic Teacher, 37,* 22–26.

Marche, M. M. (1984). A Pythagorean curiosity. *Mathematics Teacher, 77,* 611–613.

Masingila, J. O. (1995). Carpet laying: An illustration of everyday mathematics. In P. A. House (Ed.), *Connecting mathematics across the curriculum* (pp. 163–169). Reston, VA: National Council of Teachers of Mathematics.

McClain, K., Cobb, P., Gravemeijer, K., & Estes, B. (1999). Developing mathematical reasoning within the context of measurement. In L. V. Stiff (Ed.), *Developing mathematical reasoning in grades K–12* (pp. 93–106). Reston, VA: National Council of Teachers of Mathematics.

McLaughlin, H. (1992). Activities: Determining area and calculating cost: A "model" application. *Mathematics Teacher, 85,* 360–361, 367–370.

Miller, L. D., & Miller, J. (1989). Metric week—The capitol way. *Mathematics Teacher, 82,* 454–458.

Miller, W. A., & Wagner, L. (1993). Activities: Pythagorean dissection puzzles. *Mathematics Teacher, 86,* 302–308, 313–314.

Neufeld, K. A. (1989). Body measurement. *Arithmetic Teacher, 36,* 12–15.

Parker, J., & Widmer, C. C. (1993). Teaching mathematics with technology: Patterns in measurement. *Arithmetic Teacher, 40,* 292–295.

Rectanus, C. (1998). *Math by all means: Area and perimeter, grades 5–6.* Sausalito, CA: Math Solutions Publications.

Renshaw, B. S. (1986). Symmetry the trademark way. *Arithmetic Teacher, 34,* 6–12.

Rhone, L. (1995). Measurement in a primary-grade integrated curriculum. In P. A. House (Ed.), *Connecting mathematics across the curriculum* (pp. 124–133). Reston, VA: National Council of Teachers of Mathematics.

Schultz, J. E. (1991). Area models: Spanning the mathematics of grades 3–9. *Arithmetic Teacher, 39,* 42–46.

Shaw, J. M. (1983). Student-made measuring tools. *Arithmetic Teacher, 31*(3), 12–15.

Shaw, J. M., & Cliatt, M. J. P. (1989). Developing measurement sense. In P. R. Trafton (Ed.), *New directions for elementary school mathematics* (pp. 149–155). Reston, VA: National Council of Teachers of Mathematics.

Smith, R. F. (1986). Let's do it: Coordinate geometry for third graders. *Arithmetic Teacher, 36,* 6–11.

Souza, R. (1988). Golfing with a protractor. *Arithmetic Teacher, 35,* 52–56.

Szetela, W., & Owens, D. T. (1986). Finding the area of a circle: Use a cake pan and leave out the pi. *Arithmetic Teacher, 33*(9), 12–18.

Terc, M. (1985). Coordinate geometry: Art and mathematics. *Arithmetic Teacher, 33,* 22–24.

Thompson, C. S., & Van de Walle, J. A. (1985). Learning about rulers and measuring. *Arithmetic Teacher, 32*(8), 8–12.

Vissa, J. M. (1987). Coordinate graphing: Shaping a sticky situation. *Arithmetic Teacher, 35,* 6–10.

Whitman, N. (1991). Activities: Line and rotational symmetry. *Mathematics Teacher, 84,* 296–302.

Wilson, P. S., & Adams, V. M. (1992). A dynamic way to teach angle and angle measure. *Arithmetic Teacher, 39,* 6–13.

Wilson. P. S., & Osbome, A. (1992). Foundational ideas in teaching about measure. In T. R. Post (Ed.), *Teaching mathematics in grades K–8: Research-based methods* (2nd ed.) (pp. 89–121). Needham Heights, MA: Allyn & Bacon.

Wilson, P. S., & Rowland, R. E. (1993). Teaching measurement In R. J. Jensen (Ed.), *Research ideas for the classroom: Early childhood mathematics* (pp. 171–194). Old Tappan, NJ: Macmillan.

Zaslavsky, C. (1990). Symmetry in American folk art. *Arithmetic Teacher, 38,* 6–12.

Zweng, M. J. (1986). Introducing angle measurement through estimation. In H. L. Schoen (Ed.), *Estimation and mental computation* (pp. 212–219). Reston, VA: National Council of Teachers of Mathematics.

Appendix A
Resources and Tools

Helpful Resources/Websites

Contact information for *Standards*-based Curricula

NCTM *Standards*

Black-line Masters

Organizations:

www.awm-math.org/	Association for Women in Mathematics
www.nctm.org	National Council of Teachers of Mathematics
www.nsf.gov/nsb/documents/	National Science Board
www.nsf.gov/home/ehr/start.htm	National Science Foundation: Education and Human Resources

Math Lessons and Resources:

www.aplusmath.co/	Aplus Math—Interactive activities for teaching and reinforcing math skills
www.coolmath.com	Cool Math—Collection of activities, games and lessons
www.enc.org/	Eisenhower National Clearinghouse (ENC)—Classroom activities and databases
www.forum.swarthmore.edu/	Math Forum—Mathematics Internet Library, discussion groups, and classroom activities
www.mathgoodies.com	Math Goodies—Collection of lessons, puzzles and professional networking
www.mcrel.org/resources/links/hotlinks.asp	McRel Educator Resources—Math activities and lessons
www.scienceu.com/geometry/	Science U—Interactive activities and classroom lessons

Math Competitions and Projects:

www.unl.edu/amc/	American Mathematical Competitions
www.figurethis.org/index40.htm	Figure This!—Math challenges for grades 5–8 and families
www.kidlink.org/KIDPROJ/math00/	Kidlink—Collaborative project designed to make students aware of the math around them
www.mathcounts.org/	MATHCOUNTS—Annual mathematics competition

Calculator Activities:

www.ti.com	Texas Instruments—Newsletter for grades K–8, calculator activities

Research and support:

www.showmecenter.missouri.edu/ShowMe/RelatedLinks.html	Links to National Science Foundation Curriculum Implementation Projects
www.middleweb.com/	MiddleWeb: Exploring Middle School Reform
www.terc.edu/index.cfm	TERC—Research and Development of Math, Science and Technology Teaching
www.ustimss.msu.edu/	Third International Mathematics and Science Study

Software:

www.clearinghouse.k12.ca.us/	Clearinghouse—Educators Guide to high quality instructional technology resources
www.handygraph.com	Handygraph—Create and edit grids, blank graphs and number lines with Microsoft Word
www.mathforum.com/mathed/math.software.reviews.html	Internet Math Library for catalog of software reviews
www.keypress.com	Key Curriculum Press—Dynamic Geometry software
www.mathgoodies.com/software/	Math goodies—Mathematics software referrals
www.mathtype.com/	Math Type—create mathematical notation
www.terc.edu/mathequity/gw/html/gwhome.html	NSF Funded research projects describes 50 games and criteria for evaluating mathematical games
www.superkids.com	Superkids Educational Software Review
www.math.ucalgary.ca/~laf/colorful/colorful.html	U of Calgary—Mathematics Computer Games

Buyer's Guide:

www.nctm.org/buyersguide/	Buyer's Guide and information on math education vendors, consultation and software producers.

Standards-based curricula:

Investigations in Number, Data, and Space-*Grades K–5*

Phone: (617) 547-0430
Email: tasha_morris@terc.edu **www.terc.edu/byterc/invest.html**

Publisher: Dale Seymour Publications
PO Box 5026
White Plains, NY 10602
(800) 872-1100 ext. 5237 **www.aw.com/dsp/**

University of Chicago School Mathematics Project (UCSMP)-*Grades K–6*

To receive more information, including a project brochure and recent newsletters, e-mail UCSMP at ucsmp@uchicago.edu

Publisher: Everyday Learning Corporation
(800) 382-7670 **www.Everydaylearning.com/index.html**

The Connected Mathematics Project-*Grades 6–8*

Michigan State University
8715 Wells Hall
East Lansing, MI 48824
(517) 353-3835
E-mail: *ephillips@math.msu.edu* **www.ns.msu.edu/CMP/cmp.html**

Publisher: Dale Seymour Publications
10 Bank Street
White Plains, NY 10602
Phone: (800) 872-1100

Seeing and Thinking Mathematically: MathScape-*Grades 6–8*

Phone: (617) 969-7100

Publisher: Creative Publications
5623 West 115th Street
Alsip, Illinois 60482
Phone: (800) 624-0822

Mathematics In Context: A Connected Curriculum-*Grades 5–8*

(608) 263-4285

Publisher: Encyclopedia Britannica Educational Corporation
310 South Michigan Ave.
Chicago, Illinois 60604
(800) 554-9862 Ext. 7007
E-mail: info@ebec.com

Number and Operations

Standard	Pre-K–2	Grades 3–5	Grades 6–8
Instructional programs from prekindergarten through grade 12 should enable all students to—	**Expectations** In prekindergarten through grade 2 all students should—	**Expectations** In grades 3–5 all students should—	**Expectations** In grades 6–8 all students should—
Understand numbers, ways of representing numbers, relationships among numbers, and number systems	• count with understanding and recognize "how many" in sets of objects; • use multiple models to develop initial understandings of place value and the base-ten number system; • develop understanding of the relative position and magnitude of whole numbers and of ordinal and cardinal numbers and their connections; • develop a sense of whole numbers and represent and use them in flexible ways, including relating, composing, and decomposing numbers; • connect number words and numerals to the quantities they represent, using various physical models and representations; • understand and represent commonly used fractions, such as 1/4, 1/3, and 1/2.	• understand the place-value structure of the base-ten number system and be able to represent and compare whole numbers and decimals; • recognize equivalent representations for the same number and generate them by decomposing and composing numbers; • develop understanding of fractions as parts of unit wholes, as parts of a collection, as locations on number lines, and as divisions of whole numbers; • use models, benchmarks, and equivalent forms to judge the size of fractions; • recognize and generate equivalent forms of commonly used fractions, decimals, and percents; • explore numbers less than 0 by extending the number line and through familiar applications; • describe classes of numbers according to characteristics such as the nature of their factors.	• work flexibly with fractions, decimals, and percents to solve problems; • compare and order fractions, decimals, and percents efficiently and find their approximate locations on a number line; • develop meaning for percents greater than 100 and less than 1; • understand and use ratios and proportions to represent quantitative relationships; • develop an understanding of large numbers and recognize and appropriately use exponential, scientific, and calculator notation; • use factors, multiples, prime factorization, and relatively prime numbers to solve problems; • develop meaning for integers and represent and compare quantities with them.
Understand meanings of operations and how they relate to one another	• understand various meanings of addition and subtraction of whole numbers and the relationship between the two operations; • understand the effects of adding and subtracting whole numbers; • understand situations that entail multiplication and division, such as equal grouping of objects and sharing equally.	• understand various meanings of multiplication and division; • understand the effects of multiplying and dividing whole numbers; • identify and use relationships between operations, such as division as the inverse of multiplication, to solve problems; • understand and use properties of operations, such as the distributivity of multiplication over addition.	• understand the meaning and effects of arithmetic operations with fractions, decimals, and integers; • use the associative and commutative properties of addition and multiplication and the distributive property of multiplication over addition to simplify computations with integers, fractions, and decimals; • understand and use the inverse relationships of addition and subtraction, multiplication and division, and squaring and finding square roots to simplify computations and solve problems.

Table of Standards and Expectations. Reprinted with permission from "Principles and Standards for School Mathematics,"

Algebra

Standard	Pre-K–2	Grades 3–5	Grades 6–8
Instructional programs from prekindergarten through grade 12 should enable all students to—	**Expectations** In prekindergarten through grade 2 all students should—	**Expectations** In grades 3–5 all students should—	**Expectations** In grades 6–8 all students should—
Understand patterns, relations, and functions	• sort, classify, and order objects by size, number, and other properties; • recognize, describe, and extend patterns such as sequences of sounds and shapes or simple numeric patterns and translate from one representation to another; • analyze how both repeating and growing patterns are generated.	• describe, extend, and make generalizations about geometric and numeric patterns; • represent and analyze patterns and functions, using words, tables, and graphs.	• represent, analyze, and generalize a variety of patterns with tables, graphs, words, and, when possible, symbolic rules; • relate and compare different forms of representation for a relationship; • identify functions as linear or nonlinear and contrast their properties from tables, graphs, or equations.
Represent and analyze mathematical situations and structures using algebraic symbols	• illustrate general principles and properties of operations, such as commutativity, using specific numbers; • use concrete, pictorial, and verbal representations to develop an understanding of invented and conventional symbolic notations.	• identify such properties as commutativity, associativity, and distributivity and use them to compute with whole numbers; • represent the idea of a variable as an unknown quantity using a letter or symbol; • express mathematical relationships using equations.	• develop an initial conceptual understanding of different uses of variables; • explore relationships between symbolic expressions and graphs of lines, paying particular attention to the meaning of intercept and slope; • use symbolic algebra to represent situations and to solve problems, especially those that involve linear relationships; • recognize and generate equivalent forms for simple algebraic expressions and solve linear equations.
Use mathematical models to represent and understand quantitative relationships	• model situations that involve the addition and subtraction of whole numbers, using objects, pictures, and symbols.	• model problem situations with objects and use representations such as graphs, tables, and equations to draw conclusions.	• model and solve contextualized problems using various representations, such as graphs, tables, and equations.
Analyze change in various contexts	• describe qualitative change, such as a student's growing taller; • describe quantitative change, such as a student's growing two inches in one year.	• investigate how a change in one variable relates to a change in a second variable; • identify and describe situations with constant or varying rates of change and compare them.	• use graphs to analyze the nature of changes in quantities in linear relationships.

Table of Standards and Expectations. Reprinted with permission from "Principles and Standards for School Mathematics,"

Number and Operations (continued)

Standard	Pre-K–2	Grades 3–5	Grades 6–8
Instructional programs from prekindergarten through grade 12 should enable all students to—	**Expectations** In prekindergarten through grade 2 all students should—	**Expectations** In grades 3–5 all students should—	**Expectations** In grades 6–8 all students should—
Compute fluently and make reasonable estimates	• develop and use strategies for whole-number computations; with a focus on addition and subtraction; • develop fluency with basic number combinations for addition and subtraction; • use a variety of methods and tools to compute, including objects, mental computation, estimation, paper and pencil, and calculators.	• develop fluency with basic number combinations for multiplication and division and use these combinations to mentally compute related problems, such as 30×50; • develop fluency in adding, subtracting, multiplying, and dividing whole numbers; • develop and use strategies to estimate the results of whole-number computations and to judge the reasonableness of such results; • develop and use strategies to estimate computations involving fractions and decimals in situations relevant to students' experience; • use visual models, benchmarks, and equivalent forms to add and subtract commonly used fractions and decimals; • select appropriate methods and tools for computing with whole numbers from among mental computation, estimation, calculators, and paper and pencil according to the context and nature of the computation and use the selected method or tool.	• select appropriate methods and tools for computing with fractions and decimals from among mental computation, estimation, calculators or computers, and paper and pencil, depending on the situation, and apply the selected methods; • develop and analyze algorithms for computing with fractions, decimals, and integers and develop fluency in their use; • develop and use strategies to estimate the results of rational-number computations and judge the reasonableness of the results; • develop, analyze, and explain methods for solving problems involving proportions, such as scaling and finding equivalent ratios.

Table of Standards and Expectations. Reprinted with permission from "Principles and Standards for School Mathematics,"

Geometry

Standard	Pre-K–2	Grades 3–5	Grades 6–8
Instructional programs from prekindergarien through grade 12 should enable all students to—	**Expectations** In prekindergarten through grade 2 all students should—	**Expectations** In grades 3–5 all students should—	**Expectations** In grades 6–8 all students should—
Analyze characteristics and properties of two- and three-dimensional geometric shapes and develop mathematical arguments about geometric relationships	• recognize, name, build, draw, compare, and sort two- and three-dimensional shapes; • describe attributes and parts of two- and three-dimensional shapes; • investigate and predict the results of putting together and taking apart two- and three-dimensional shapes.	• identify, compare, and analyze attributes of two- and three-dimensional shapes and develop vocabulary to describe the attributes; • classify two- and three-dimensional shapes according to their properties and develop definitions of classes of shapes such as triangles and pyramids; • investigate, describe, and reason about the results of subdividing, combining, and transforming shapes; • explore congruence and similarity; • make and test conjectures about geometric properties and relationships and develop logical arguments to justify conclusions.	• precisely describe, classify, and understand relationships among types of two- and three-dimensional objects using their delining properties; • understand relationships among the angles, side lengths, perimeters, areas, and volumes of similar objects; • create and critique inductive and deductive arguments concerning geometric ideas and relationships, such as congruence, similarity, and the Pythagorean relationship.
Specify locations and describe spatial relationships using coordinate geometry and other representational systems	• describe, name, and interpret relative positions in space and apply ideas about relative position; • describe, name, and interpret direction and distance in navigating space and apply ideas about direction and distance; • find and name locations with simple relationships such as "near to" and in coordinate systems such as maps.	• describe location and movement using common language and geometric vocabulary; • make and use coordinate systems to specify locations and to describe paths; • find the distance between points along horizontal and vertical lines of a coordinate system.	• use coordinate geometry to represent and examine the properties of geometric shapes; • use coordinate geometry to examine special geometric shapes, such as regular polygons or those with pairs or parallel or perpendicular sides.
Apply transformations and use symmetry to analyze mathematical situations	• recognize and apply slides, flips, and turns; • recognize and create shapes that have symmetry.	• predict and describe the results of sliding, flipping, and turning two-dimensional shapes; • describe a motion or a series of motions that will show that two shapes are congruent; • identify and describe line and rotational symmetry in two- and three-dimensional shapes and designs.	• describe sizes, positions, and orientations of shapes under informal transformations such as flips, turns, slides, and scaling; • examine the congruence, similarity, and line or rotational symmetry of objects using transformations.
Use visualization, spatial reasoning, and geometric modeling to solve problems	• create mental images of geometric shapes using spatial memory and spatial visualization; • recognize and represent shapes from different perspectives; • relate ideas in geometry to ideas in number and measurement; • recognize geometric shapes and structures in the environment and specify their location.	• build and draw geometric objects; • create and describe mental images of objects, patterns, and paths; • identify and build a three-dimensional object from two-dimensional representations of that object; • identify and build a two-dimensional representation of a three-dimensional object; • use geometric models to solve problems in other areas of mathematics, such as number and measurement; • recognize geometric ideas and relationships and apply them to other disciplines and to problems that arise in the classroom or in everyday life.	• draw geometric objects with specified properties, such as side lengths or angle measures; • use two-dimensional representations of three-dimensional objects to visualize and solve problems such as those involving surface area and volume; • use visual tools such as networks to represent and solve problems; • use geometric models to represent and explain numerical and algebraic relationships; • recognize and apply geometric ideas and relationships in areas outside the mathematical classroom, such as art, science, and everyday life.

Table of Standards and Expectations. Reprinted with permission from "Principles and Standards for School Mathematics," copyright by the National Council of Teachers of Mathematics.

Measurement

Standard	Pre-K–2	Grades 3–5	Grades 6–8
Instructional programs from prekindergarten through grade 12 should enable all students to—	**Expectations** In prekindergarten through grade 2 all students should—	**Expectations** In grades 3–5 all students should—	**Expectations** In grades 6–8 all students should—
Understand measurable attributes of objects and the units, systems, and processes of measurement	• recognize the attributes of length, volume, weight, area, and time; • compare and order objects according to these attributes; • understand how to measure using nonstandard and standard units; • select an appropriate unit and tool for the attribute being measured.	• understand such attributes as length, area, weight, volume, and size of angle and select the appropriate type of unit for measuring each attribute; • understand the need for measuring with standard units and become familiar with standard units in the customary and metric systems; • carry out simple unit conversions, such as from centimeters to meters, within a system of measurement; • understand that measurements are approximations and understand how differences in units affect precision; • explore what happens to measurements of a two-dimensional shape such as its perimeter and area when the shape is changed in some way.	• understand both metric and customary systems of measurement; • understand relationships among units and convert from one unit to another within the same system; • understand, select, and use units of appropriate size and type to measure angles, perimeter, area, surface area, and volume.
Apply appropriate techniques, tools, and formulas to determine measurements	• measure with multiple copies of units of the same size, such as paper clips laid end to end; • use repetition of a single unit to measure something larger than the unit, for instance, measuring the length of a room with a single meterstick; • use tools to measure; • develop common referents for measures to make comparisons and estimates.	• develop strategies for estimating the perimeters, areas, and volumes of irregular shapes; • select and apply appropriate standard units and tools to measure length, area, volume, weight, time, temperature, and the size of angles; • select and use benchmarks to estimate measurements; • develop, understand, and use formulas to find the area of rectangles and related triangles and parallelograms; • develop strategies to determine the surface areas and volumes of rectangular solids.	• use common benchmarks to select appropriate methods for estimating measurements; • select and apply techniques and tools to accurately find length, area, volume, and angle measures to appropriate levels of precision; • develop and use formulas to determine the circumference of circles and the area of triangles, parallelograms, trapezoids, and circles and develop strategies to find the areas of more-complex shapes; • develop strategies to determine the surface area and volume of selected prisms, pyramids, and cylinders; • solve problems involving scale factors, using ratio and proportion; • solve simple problems involving rates and derived measurements for such attributes as velocity and density.

Table of Standards and Expectations. Reprinted with permission from "Principles and Standards for School Mathematics,"

Data Analysis and Probability

Standard	Pre-K–2	Grades 3–5	Grades 6–8
Instructional programs from prekindergarten through grade 12 should enable all students to—	**Expectations** In prekindergarten through grade 2 all students should—	**Expectations** In grades 3–5 all students should—	**Expectations** In grades 6–8 all students should—
Formulate questions that can be addressed with data and collect, organize, and display relevant data to answer them	• pose questions and gather data about themselves and their surroundings; • sort and classify objects according to their attributes and organize data about the objects; • represent data using concrete objects, pictures, and graphs.	• design investigations to address a question and consider how data-collection methods affect the nature of the data set; • collect data using observations, surveys, and experiments; • represent data using tables and graphs such as line plots, bar graphs, and line graphs; • recognize the differences in representing categorical and numerical data.	• formulate questions, design studies, and collect data about a characteristic shared by two populations or different characteristics within one population; • select, create, and use appropriate graphical representations of data, including histograms, box plots, and scatterplots.
Select and use appropriate statistical methods to analyze data	• describe parts of the data and the set of data as a whole to determine what the data show.	• describe the shape and important features of a set of data and compare related data sets, with an emphasis on how the data are distributed; • use measures of center, focusing on the median, and understand what each does and does not indicate about the data set; • compare different representations of the same data and evaluate how well each representation shows important aspects of the data.	• find, use, and interpret measures of center and spread, including mean and interquartile range; • discuss and understand the correspondence between data sets and their graphical representations, especially histograms, stem-and-leaf plots, box plots, and scatterplots.
Develop and evaluate inferences and predictions that are based on data	• discuss events related to students' experiences as likely or unlikely.	• propose and justify conclusions and predictions that are based on data and design studies to further investigate the conclusions or predictions.	• use observations about differences between two or more samples to make conjectures about the populations from which the samples were taken; • make conjectures about possible relationships between two characteristics of a sample on the basis of scatterplots of the data and approximate lines of fit; • use conjectures to formulate new questions and plan new studies to answer them.
Understand and apply basic concepts of probability		• describe events as likely or unlikely and discuss the degree of likelihood using such words *as certain, equally likely,* and *impossible;* • predict the probability of outcomes of simple experiments and test the predictions; • understand that the measure of the likelihood of an event can be represented by a number from 0 to 1.	• understand and use appropriate terminology to describe complementary and mutually exclusive events; • use proportionality and a basic understanding of probability to make and test conjectures about the results of experiments and simulations; • compute probabilities for simple compound events, using such methods as organized lists, tree diagrams, and area methods.

Table of Standards and Expectations. Reprinted with permission from "Principles and Standards for School Mathematics," copyright by the National Council of Teachers of Mathematics.

Problem Solving

Standard
Instructional programs from prekindergarten through grade 12 should enable all students to—

- Build new mathematical knowledge through problem solving
- Solve problems that arise in mathematics and in other contexts
- Apply and adapt a variety of appropriate strategies to solve problems
- Monitor and reflect on the process of mathematical problem solving

Reasoning and Proof

Standard
Instructional programs from prekindergarten through grade 12 should enable all students to—

- Recognize reasoning and proof as fundamental aspects of mathematics
- Make and investigate mathematical conjectures
- Develop and evaluate mathematical arguments and proofs
- Select and use various types of reasoning and methods of proof

Communication

Standard
Instructional programs from prekindergarten through grade 12 should enable all students to—

- Organize and consolidate their mathematical thinking through communication
- Communicate their mathematical thinking coherently and clearly to peers, teachers, and others
- Analyze and evaluate the mathematical thinking and strategies of others
- Use the language of mathematics to express mathematical ideas precisely

Connections

Standard
Instructional programs from prekindergarten through grade 12 should enable all students to—

- Recognize and use connections among mathematical ideas
- Understand how mathematical ideas interconnect and build on one another to produce a coherent whole
- Recognize and apply mathematics in contexts outside of mathematics

Representation

Standard
Instructional programs from prekindergarten through grade 12 should enable all students to—

- Create and use representations to organize, record, and communicate mathematical ideas
- Select, apply, and translate among mathematical representations to solve problems
- Use representations to model and interpret physical, social, and mathematical phenomena

Centimeter Grid Paper

Inch Grid Paper

Half-Inch Grid Paper

Quarter-Inch Grid Paper

Isometric Dot Paper

Geoboard Dot Paper

Hundreds Charts

1	2	3	4	5	6	7	8	9	10
11	12	13	14	15	16	17	18	19	20
21	22	23	24	25	26	27	28	29	30
31	32	33	34	35	36	37	38	39	40
41	42	43	44	45	46	47	48	49	50
51	52	53	54	55	56	57	58	59	60
61	62	63	64	65	66	67	67	69	70
71	72	73	74	75	76	77	78	79	80
81	82	83	84	85	86	87	88	89	90
91	92	93	94	95	96	97	98	99	100

0	1	2	3	4	5	6	7	8	9
10	11	12	13	14	15	16	17	18	19
20	21	22	23	24	25	26	27	28	29
30	31	32	33	34	35	36	37	38	39
40	41	42	43	44	45	46	47	48	49
50	51	52	53	54	55	56	57	58	59
60	61	62	63	64	65	66	67	67	69
70	71	72	73	74	75	76	77	78	79
80	81	82	83	84	85	86	87	88	89
90	91	92	93	94	95	96	97	98	99

Variations of Hundreds Charts: A

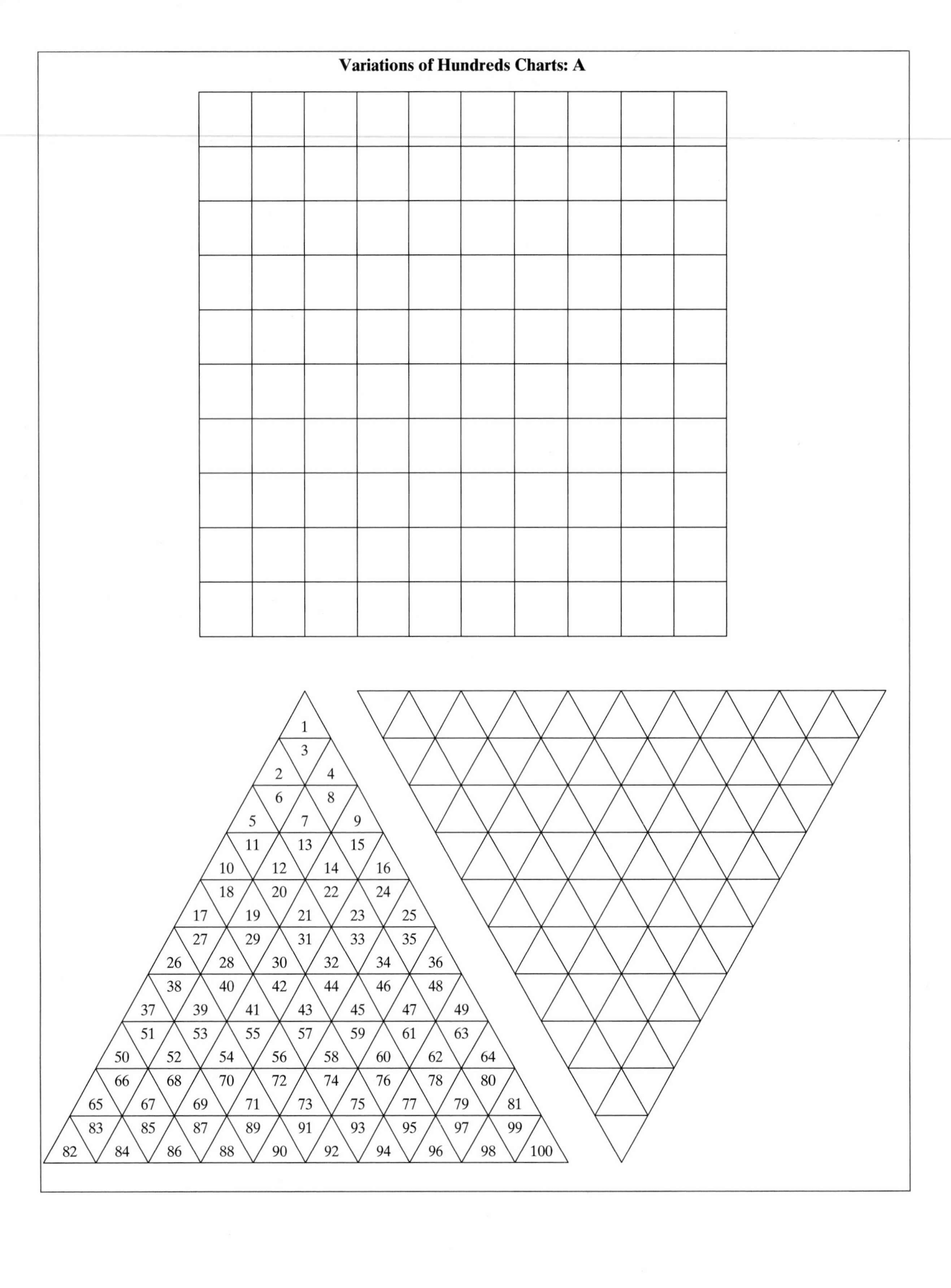

Variations of Hundred Charts: B

−9	−8	−7	−6	−5	−4	−3	−2	−1	0
1	2	3	4	5	6	7	8	9	10
11	12	13	14	15	16	17	18	19	20
21	22	23	24	25	26	27	28	29	30
31	32	33	34	35	36	37	38	39	40
41	42	43	44	45	46	47	48	49	50
51	52	53	54	55	56	57	58	59	60
61	62	63	64	65	66	67	67	69	70
71	72	73	74	75	76	77	78	79	80
81	82	83	84	85	86	87	88	89	90
91	92	93	94	95	96	97	98	99	100
101	102	103	104	105	106	107	108	109	110

101	102	103	104	105	106	107	108	109	110
111	112	113	114	115	116	117	118	119	120
121	122	123	124	125	26	127	128	129	130
131	132	133	134	135	136	137	138	139	140
141	142	143	144	145	146	147	148	149	150
151	152	153	154	155	156	157	158	159	160
161	162	163	164	165	166	167	167	169	170
171	172	173	174	175	176	177	178	179	180
181	182	183	184	185	186	187	188	189	190
191	192	193	194	195	196	197	198	199	200

Decimal or Percent Paper

Fraction Strips

1

$\frac{1}{2}$ $\frac{2}{2}$

$\frac{1}{3}$ $\frac{2}{3}$ $\frac{3}{3}$

$\frac{1}{4}$ $\frac{2}{4}$ $\frac{3}{4}$ $\frac{4}{4}$

$\frac{1}{5}$ $\frac{2}{5}$ $\frac{3}{5}$ $\frac{4}{5}$ $\frac{5}{5}$

$\frac{1}{6}$ $\frac{2}{6}$ $\frac{3}{6}$ $\frac{4}{6}$ $\frac{5}{6}$ $\frac{6}{6}$

$\frac{1}{8}$ $\frac{2}{8}$ $\frac{3}{8}$ $\frac{4}{8}$ $\frac{5}{8}$ $\frac{6}{8}$ $\frac{7}{8}$ $\frac{8}{8}$

$\frac{1}{9}$ $\frac{2}{9}$ $\frac{3}{9}$ $\frac{4}{9}$ $\frac{5}{9}$ $\frac{6}{9}$ $\frac{7}{9}$ $\frac{8}{9}$ $\frac{9}{9}$

$\frac{1}{10}$ $\frac{2}{10}$ $\frac{3}{10}$ $\frac{4}{10}$ $\frac{5}{10}$ $\frac{6}{10}$ $\frac{7}{10}$ $\frac{8}{10}$ $\frac{9}{10}$ $\frac{10}{10}$

$\frac{1}{12}$ $\frac{2}{12}$ $\frac{3}{12}$ $\frac{4}{12}$ $\frac{5}{12}$ $\frac{6}{12}$ $\frac{7}{12}$ $\frac{8}{12}$ $\frac{9}{12}$ $\frac{10}{12}$ $\frac{11}{12}$ $\frac{12}{12}$

$\frac{1}{20}$ $\frac{2}{20}$ $\frac{3}{20}$ $\frac{4}{20}$ $\frac{5}{20}$ $\frac{6}{20}$ $\frac{7}{20}$ $\frac{8}{20}$ $\frac{9}{20}$ $\frac{10}{20}$ $\frac{11}{20}$ $\frac{12}{20}$ $\frac{13}{20}$ $\frac{14}{20}$ $\frac{15}{20}$ $\frac{16}{20}$ $\frac{17}{20}$ $\frac{18}{20}$ $\frac{19}{20}$ $\frac{20}{20}$

$\frac{1}{24}$ $\frac{2}{24}$ $\frac{3}{24}$ $\frac{4}{24}$ $\frac{5}{24}$ $\frac{6}{24}$ $\frac{7}{24}$ $\frac{8}{24}$ $\frac{9}{24}$ $\frac{10}{24}$ $\frac{11}{24}$ $\frac{12}{24}$ $\frac{13}{24}$ $\frac{14}{24}$ $\frac{15}{24}$ $\frac{16}{24}$ $\frac{17}{24}$ $\frac{18}{24}$ $\frac{19}{24}$ $\frac{20}{24}$ $\frac{21}{24}$ $\frac{22}{24}$ $\frac{23}{24}$ $\frac{24}{24}$

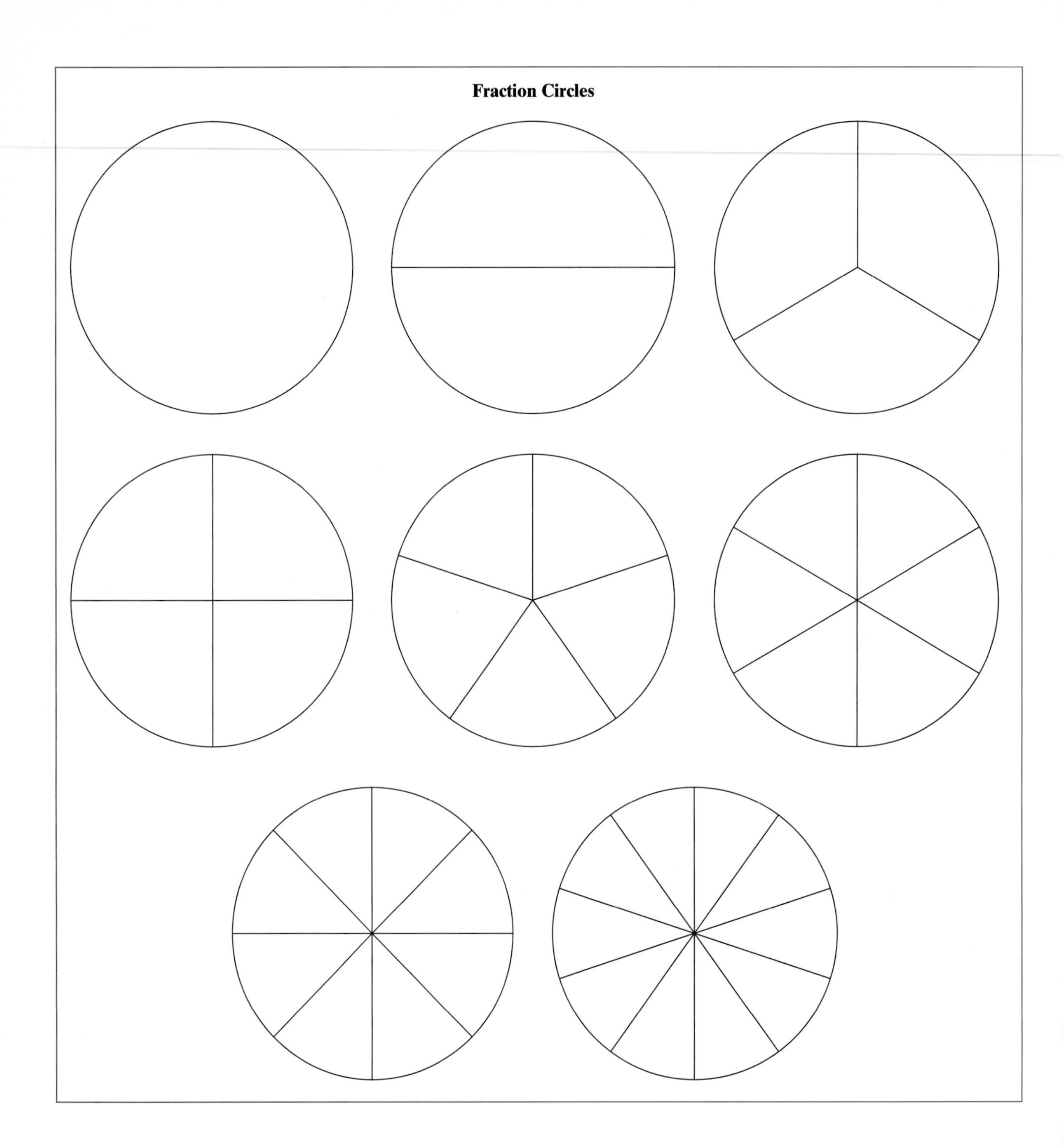
Fraction Circles

Basic Facts: Addition and Multiplication

+	0	1	2	3	4	5	6	7	8	9
0	0	1	2	3	4	5	6	7	8	9
1	1	2	3	4	5	6	7	8	9	10
2	2	3	4	5	6	7	8	9	10	11
3	3	4	5	6	7	8	9	10	11	12
4	4	5	6	7	8	9	10	11	12	13
5	5	6	7	8	9	10	11	12	13	14
6	6	7	8	9	10	11	12	13	14	15
7	7	8	9	10	11	12	13	14	15	16
8	8	9	10	11	12	13	14	15	16	17
9	9	10	11	12	13	14	15	16	17	18

×	0	1	2	3	4	5	6	7	8	9
0	0	0	0	0	0	0	0	0	0	0
1	0	1	2	3	4	5	6	7	8	9
2	0	2	4	6	8	10	12	14	16	18
3	0	3	6	9	12	15	18	21	24	27
4	0	4	8	12	16	20	24	28	32	36
5	0	5	10	15	20	25	30	35	40	45
6	0	6	12	18	24	30	36	42	48	54
7	0	7	14	21	28	35	42	49	56	63
8	0	8	16	24	32	40	48	56	64	72
9	0	9	18	27	36	45	54	63	72	81

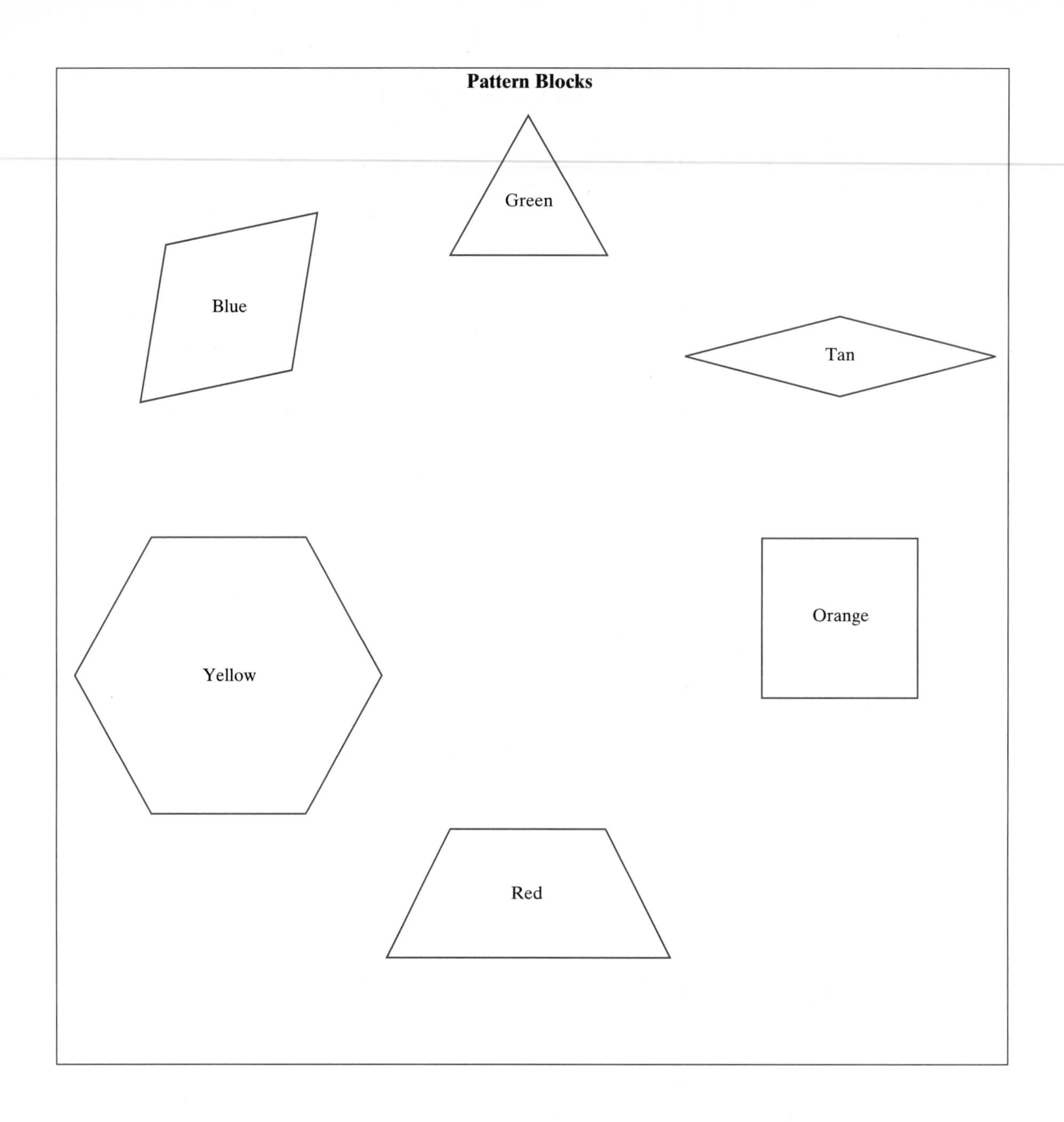
Pattern Blocks
Green
Blue
Tan
Yellow
Orange
Red

Latice Multilication Blanks

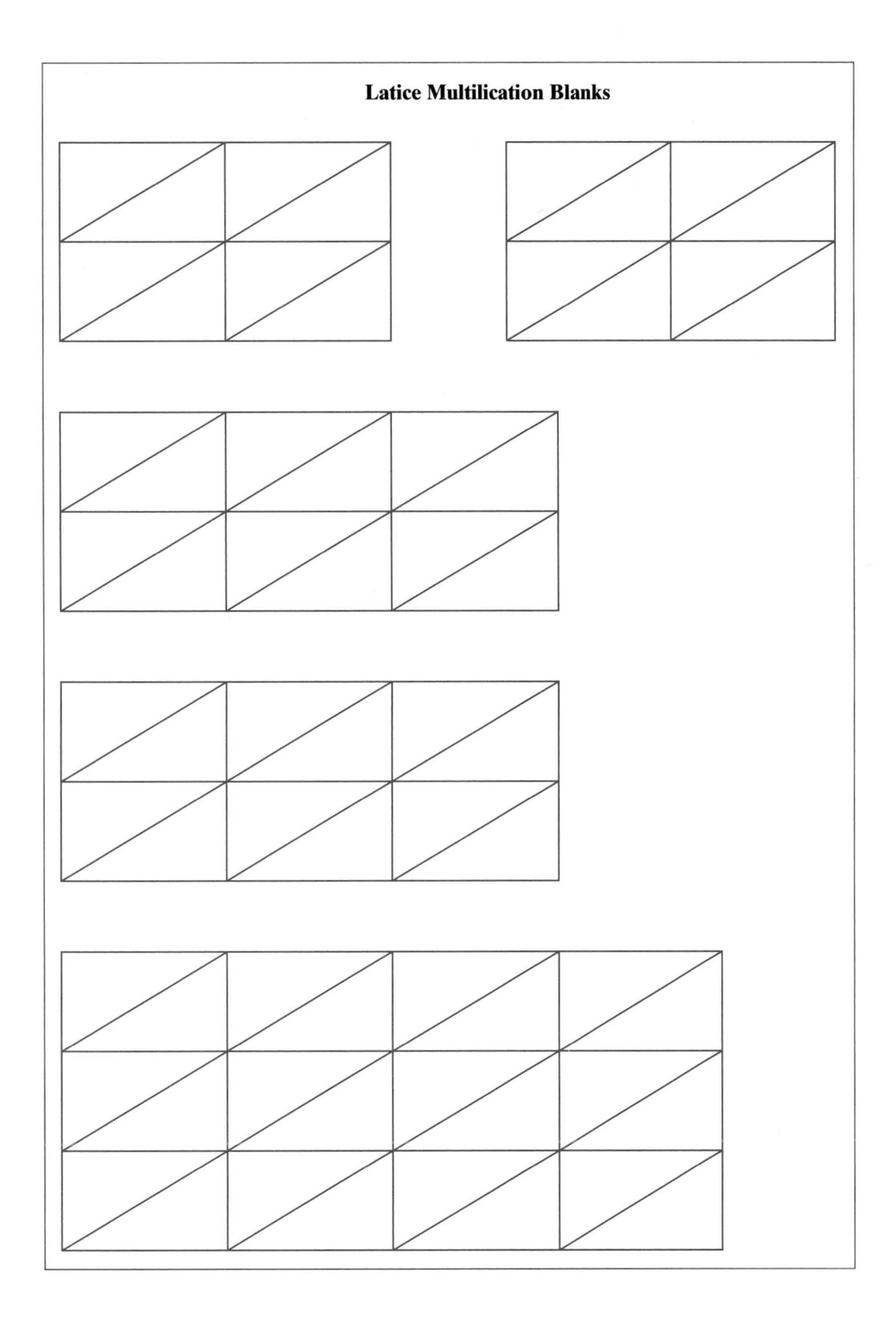

Tangram

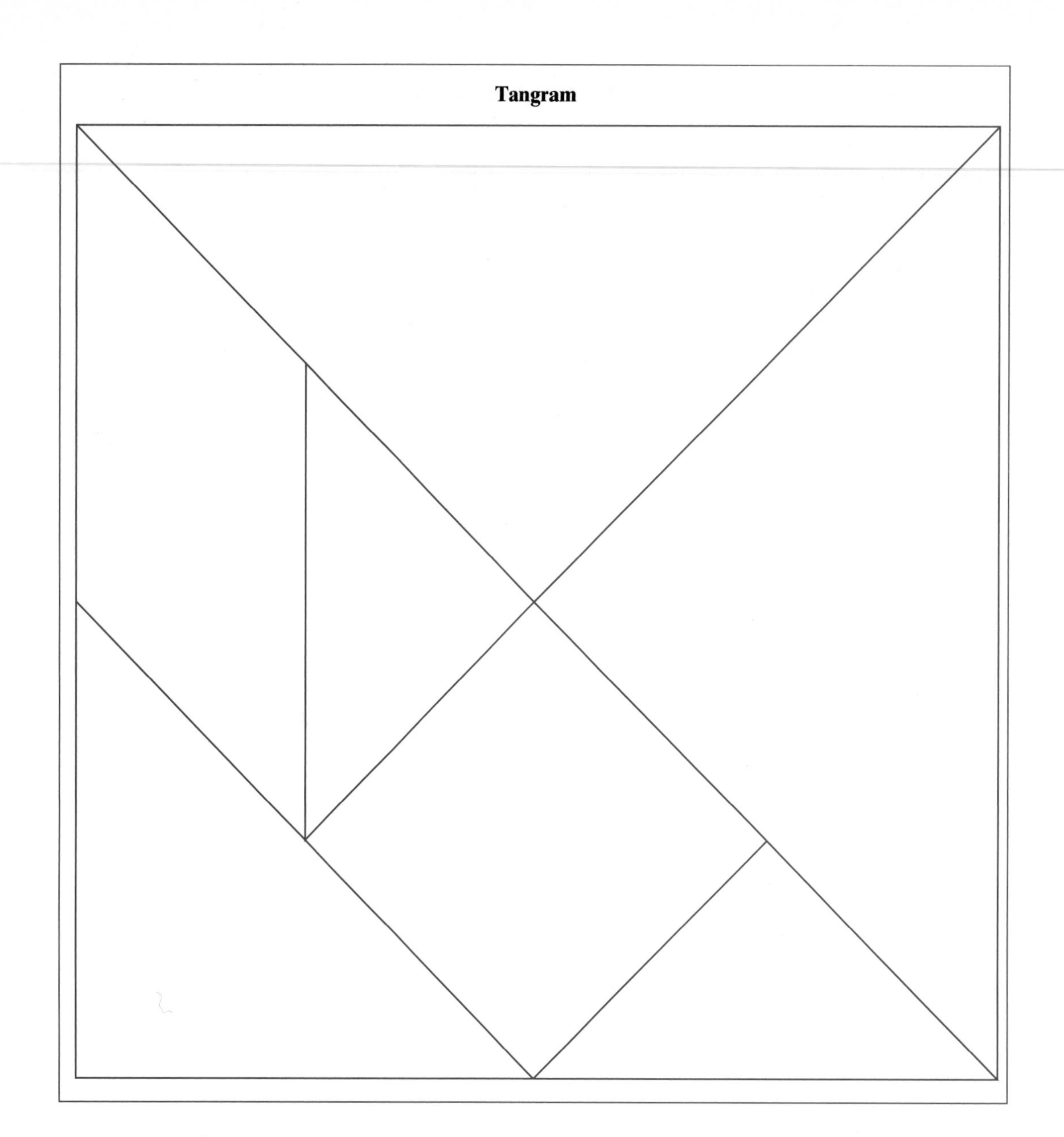

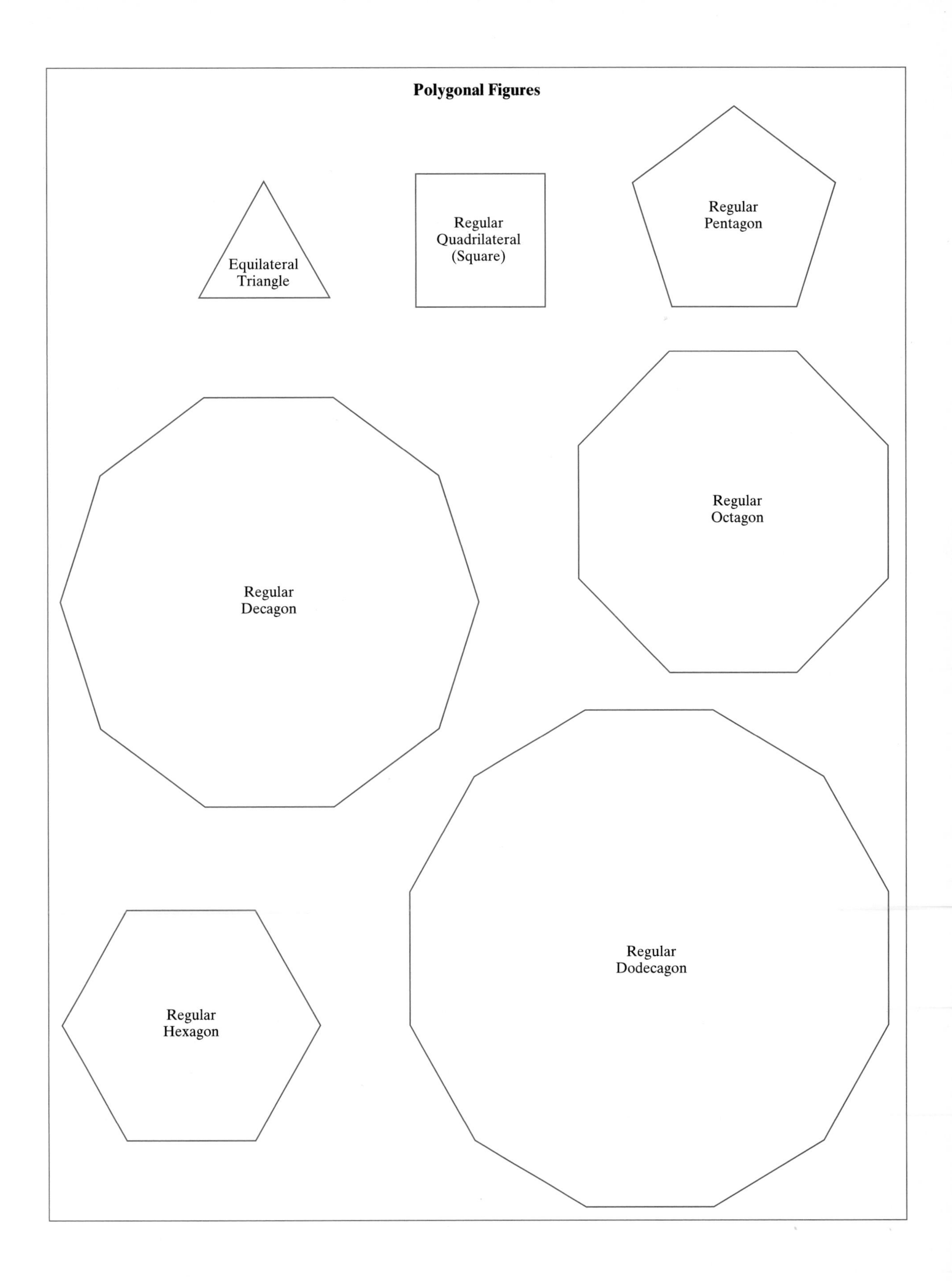
Polygonal Figures
Equilateral
Triangle
Regular
Quadrilateral
(Square)
Regular
Pentagon
Regular
Decagon
Regular
Octagon
Regular
Hexagon
Regular
Dodecagon

Glossary

Absolute value: The absolute value of a number is the number of units the number is from zero.

Acute angle: An angle with measure x, such that $0° < x < 90°$.

Acute triangle: A triangle with only acute angles.

Additive: A numeration system is said to be additive if the value of the set of symbols representing a number is the sum of the values of the individual symbols.

Adjacent angles: Angles in a plane that share a common ray without their interiors intersecting are called adjacent angles.

Algorithm: A step-by-step procedure for performing a mathematical task.

Angle: An angle is formed when two rays join together at their endpoint.

Angle bisector: A ray that separates an angle into two congruent angles.

Arc: A connected subset of points of a circle.

Area of a planar shape: The area of a planar shape is the total region enclosed by the shape.

Arithmetic mean: The arithmetic mean of n numbers is their sum divided by n.

Arithmetic sequence: A pattern of numbers where the difference between any two terms of the sequence is the same quantity.

Associativity: A set is said to have the Property of Associativity under an operation * if for any three elements a, b, and c in it, $a * (b * c) = (a * b) * c$.

Average: A term widely used to refer to the arithmetic mean.

Bar graph: A graph, used to present frequency distributions or time series, that consists of bars (rectangles) of equal width, whose lengths are proportional to the frequencies (or values) they represent. A sample bar graph is shown below.

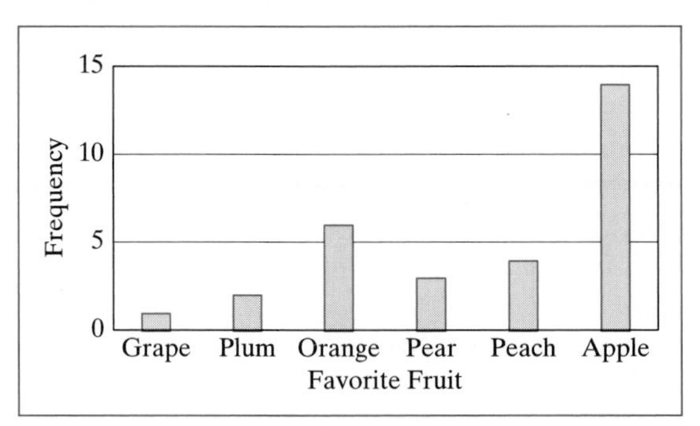

Base: A numeration system has a base if the numbers in that system reflect a process of repeated grouping by some number greater than 1. The number is then called the base.

Box-and-whisker plot (or box plot): A graph that displays the *5-number summary* of a set of data. A box-and-whisker plot for the following 12 quiz scores (10 points was the maximum score) is shown below: 8, 4, 9, 10, 10, 5, 7, 3, 8, 2, 9, 7.

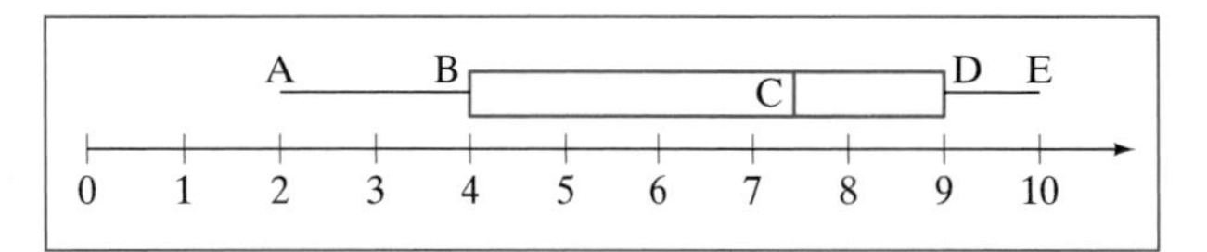

A—lowest score; B—1st quartile; C—2nd quartile
D—3rd quartile; E—highest score

Cardinal number: A number that refers to the number of objects in a set.

Cartesian coordinate system: The set of all points in a plane that are identified by a pair of x- and y-coordinates.

Cartesian product: The Cartesian product of sets A and B, denoted $A \times B$, is the set of ordered pairs (x, y) that can be formed where the first element in the ordered pair is an element of set A and the second element in the ordered pair is an element of set B.

Central angle: An angle that has the center of a circle as its vertex and has sides intersecting the circle.

Central tendency or measure of central tendency: This expression is sometimes used to refer to statistics—such as the mean, median, or mode—that indicate the average value of a data set.

Centroid: The point that is the intersection of the medians of a triangle.

Chord: Any line segment that has its endpoints on a circle.

Circle: A circle is formed by a set of points, all of which are at a constant distance from a fixed point.

Circular cone: A cone with a circular base.

Circular cylinder: A cylinder that has two congruent circles for bases.

Circumference: The distance around a circle.

Circumscribed circle: A circle that surrounds a polygon such that the vertices of the polygon lie on the circle.

Closure: A set is said to be closed with respect to an operation * if, for any two elements a and b in it, $a * b$ belongs to the set.

Collinear points: Points that lie on the same line are called collinear points.

Combination: A selection of one or more of a set of distinct objects without regard to order. The number of possible combinations, each containing k objects, that can be formed from a collection of n distinct objects is given by $(n!)/(k!)(n - k)!$

and is denoted by (n, k), ${}_nC_k$, $C(n, k)$, or C^n_k. For example, the number of combinations of the letters of the alphabet, taken four at a time, is $26!/4!(26 - 4)! = 14950$.

Commutativity: A set is said to have the Property of Commutativity under an operation * if for any two elements a and b in it, $a * b = b * a$.

Complement: The complement of a set A, denoted $\overline{A}$, is the set of all elements in the universal set that are not in A.

Complementary angles: Two angles whose sum is 90 degrees are called complementary angles.

Composite number: A positive integer that has more than two unique factors.

Compound statements: A statement formed by combining two statements.

Concave polygon: If every line joining any pair of two points inside a polygon does not lie completely within the polygon, we say it is concave.

Conclusion: A decision or judgment made regarding the truth of a hypothesis.

Conditional probability: The probability that A will occur given that B has occurred or will occur. If A and B are any two events and the probability of B is not equal to zero, then the conditional probability of A relative to B is denoted $P(A|B)$ and given by $P(A \cap B)/P(B)$, where $P(A \cap B)$ is the probability of the joint occurrence of A and B.

Conditional statement: A statement of the form "if p then q." The statement p is called the hypothesis, and the statement q is called the conclusion.

Cone: A simple closed surface comprised of a simple closed nonpolygonal curve, its interior, and the line segments that connect each point of the curve to a point not in the plane of the curve.

Congruence modulo m: A number a is said to be congruent to a number b modulo m if m divides the number $(a - b)$.

Congruent angles: Angles that have the same measure.

Congruent polygons: Two polygons are said to be congruent if, when we superimpose one upon the other, the two polygons coincide completely.

Congruent segments: Segments that have equal lengths.

Conjecture: A guess or opinion based on inconclusive evidence.

Convex polygon: If a line joining any two points inside a polygon lies completely within the polygon, we say it is convex.

Coplanar points: Points that lie in the same plane are called coplanar points.

Cube: A right rectangular prism for which the length, width, and height are equal.

Cylinder: A simple closed surface comprised of two congruent simple closed nonpolygonal curves that are in parallel planes, their interiors, and the line segments that connect corresponding points on the curves.

Data: The results of an experiment, census, survey, and any kind of process or operation.

Decimal: A decimal representation is a means of writing a fraction in the base ten place-value system.

Deductive reasoning: Reasoning from a set of axioms or given general statements to derive or prove other specific statements or conclusions.

Dependent events: *See* INDEPENDENT EVENTS.

Dependent variable: When considering the relationship between two variables, where the value of one *depends* on the value of the other, one of the variables is called a *dependent variable*, and the other is called an *independent variable*. For example, the number of apples you can purchase *depends* on the amount of money you have. In this situation, "number of apples" is a dependent variable, and "amount of money" is an independent variable.

Descriptive statistics: Any treatment of data that does not involve generalizations. (*See* INFERENTIAL STATISTICS.)

Diameter: The length of a segment that has its endpoints on a circle and passes through the center of the circle.

Dihedral angle: An angle formed by the line of intersection of two planes and the union of the two half-planes.

Disjoint sets: Sets that have no elements in common other than the empty set.

Distribution: This term is used to refer to the overall scattering of observed data. It is often used as a synonym for *frequency distribution*.

Distributivity: Given a set with two operations, * and Â, we say * distributes over Â if, for any three elements, a, b, and c in the set, $a * (b \hat{A} c) = a * b \hat{A} a * c$.

Divisible: A number, b, is said to be divisible by another number, a, if a is a factor of b. In other words, when b is divided by a, the remainder is 0.

Divisibility test for a number m: A quick and relatively easy method of determining if a given number is divisible by m without performing the actual division.

Divisor: A number, a, is said to be a divisor of another number, b, if a divides b without leaving a remainder.

Domain of a function: The set of all values that are allowed as input values of a function. For example, in the function

$$f(x) = \frac{1}{x - 5}$$

all real numbers *except* 5 could be in the domain 5 cannot be in the domain because

$$F(5) = \frac{1}{5 - 5} = \frac{1}{0}$$

is undefined.

Element: A member of a set.

Empirical probability: The probability determined from data collected by conducting multiple rounds of an experiment.

Empty set: A set with no elements; also called the null set.

Equilateral triangle: A triangle with all three sides congruent.

Equally likely outcomes: Outcomes that have the same probability of occurring.

Equivalent fractions: Two or more fractions in which if all common factors between the numerator and the denominator of each fraction are removed, the fractions are the same.

Even number: A number is said to be even if it is divisible by 2.

Event: An event, in probability theory, is a subset of a sample space. For example, the event of getting an odd number when a die is tossed is the subset $\{1, 3, 5\}$ of the sample space $\{1, 2, 3, 4, 5, 6\}$.

Expanded notation (or form): A form of writing a number so as to represent the value of each digit using multiples of powers of the base.

Expected value: The expected value of a random variable is the mean of its distribution.

Experiment: Any activity in which different results can occur.

Experimental probability: Probability of an event that is found by performing the experiment multiple times.

Exponent: A symbol placed above and after a number or expression, referred to as the base, that raises the base to the power designated by the symbol.

Face value: The numerical quantity represented by the symbol for a digit.

Factor: A number, a, is said to be a factor of another number, b, if a divides b without leaving a remainder.

Fair game: A game in which each participant has the same expectation; that is, a game that does not favor any player or group of players.

Fibonacci sequence: A sequence where each successive term is found by summing the two previous terms.

Figurate number: A number that can be represented by dots in some geometrical pattern.

Finite population: A well-defined set consisting of a finite number of elements, taken from a larger set. This larger set is called a population.

Five-number summary: The smallest value, the largest value, and the 3 quartile values—the first quartile, the second quartile (or median), and the third quartile—for a data set.

Fractal: A geometric figure that may result from an iterated function. Fractals typically have self-similarity, in which part of the figure is similar to larger parts of the figure. Examples include Koch's snowflake curve and Sierpinski's triangle.

Fraction: A fraction is a number of the form a/b, where a and b are any numbers, and b is 0.

Fractions in simplest terms: A fraction is said to be in its simplest terms if the numerator and the denominator do not share a common factor.

Frequency: The number of items, or cases, falling (or expected to fall) into a category or classification.

Frequency polygon: The graph of a frequency table obtain by drawing straight lines joining successive points representing the class frequencies, plotted at the corresponding class marks. (*See* FREQUENCY TABLE.)

Frequency table: A table (or other sort of arrangement) that displays the classes into which a set of data has been grouped, together with the corresponding frequencies, that is, the number of items falling into each class. Below is an example of a frequency table of the weights of 35 bags of beans taken from the production of a filling machine.

Weight (in pounds)	Number of Bags
12.60—12.79	3
12.80—12.99	12
13.00—13.19	5
13.20—13.39	9
13.40—13.59	4
13.60—13.79	2

Function: A correspondence between a set A (the *domain*) and a set B (the *range*) in which each element of A is paired with exactly one element of B. The table below on the left could be a function. The table on the right could not be a function because one element of the domain (1) is paired with *two* elements of the range (2 and 3).

a	b
1	2
2	5
3	2

a	b
1	2
2	5
1	3

Geometric sequence: A pattern of numbers where each term is a constant multiple of the preceding term.

Glide reflection: A transformation on a plane in which every point in a set is translated and then reflected in a line parallel to the directed line segment.

Graph: A picture of information. In other words, a graph is a visual representation of data.

Great circle: A circle on a sphere that is formed by the intersection of the sphere and a plane passing through the center of the sphere.

Greatest common divisor: The greatest common divisor of two (or more) integers is the largest number that divides both (or all) of the numbers; also called the greatest common factor.

Heptagon: A polygon that has seven sides.

Hexagon: A polygon that has six sides.

Histogram: A graph of a frequency table obtained by drawing rectangles whose bases coincide with the class intervals and whose areas are proportional to the class frequencies. The sample histogram below shows the amount of money (in dollars) spent on food by a family over a five-month period.

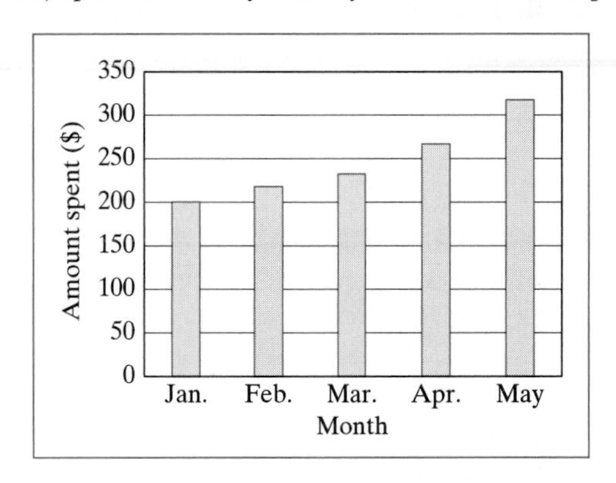

Hypotenuse: The side opposite the right angle in a right triangle.

Hypothesis: A statement that is assumed to be true.

Identity: We say that a set of numbers has an identity I, under the operation *, if, for any element x in the set, we have, $I * x = x * I = x$. We call I the *-identity of the set.

Incenter: The point that is the intersection of the angle bisectors of a triangle.

Independent events: Two events, A and B, are independent if the occurrence of either is not affected by the occurrence of the other (see mutually exclusive events). If the events are not independent, they are said to be *dependent.*

Independent variable: *See* DEPENDENT VARIABLE.

Inductive reasoning: Reasoning from the particular to the general.

Inferential statistics: Any decision, generalization, or estimate based on a sample.

Input variable: Same as INDEPENDENT VARIABLE. See DEPENDENT VARIABLE for a description.

Inscribed circle: A circle that is surrounded by a polygon such that the inscribed circle intersects each side of the polygon in exactly one point.

Integers: The set of numbers $\{\ldots, -3, -2, -1, 0, 1, 2, \ldots\}$, or numbers belonging to the set.

Interquartile range: A measure of variation given by the difference between the values of the third and first quartiles of a set of data; it represents the length of the interval that contains the middle 50% of the data.

Intersection: The intersection of two sets is the set of all elements that are in both sets.

Inverse: An element, say x, belonging to a set that has a *-identity, say I, is said to have a *-inverse in the set if there exists an element y in the set such that $x * y = y * x = I$.

Irrational number: A number that is not a rational number.

Isometry: A transformation that preserves the size and shape of the original figure.

Isosceles triangle: A triangle with at least two sides congruent.

Iterated function: A function that is "repeated" by taking the output and "feeding it back" to the function as the new input value. The first input of an iterated function is often referred to as the *seed.*

Kite: A quadrilateral with at least two distinct pairs of consecutive sides congruent.

Least common multiple: The least common multiple of two (or more) positive integers is the smallest nonzero multiple that is common to each number.

Legs of a right triangle: The two sides that form the right angle.

Line: Although *line* is an undefined term in geometry, we can describe a line as a one-dimensional set of an infinite number of points with direction.

Line of best fit: A line imposed on paired data in a scatterplot that best fits the trend of the data.

Line segment: A subset of a line that consists of two points on the line, called the endpoints of the line segment, and all the points on the line that are between the two endpoints.

Linear pair angles: Two adjacent angles in which their noncommon sides form a straight angle (an angle of 180°).

Mathematical expectation: Consider the events $E_1, E_2, \ldots,$ and E_k to be mutually exclusive events. Furthermore, suppose a player receives an amount of money $M_1, M_2, \ldots,$ and M_k when events $E_1, E_2, \ldots,$ and E_k occur respectively. If $P_1, P_2, \ldots, P_k$ are the respective probabilities of these events, then the player's *mathematical expectation*—the amount of money he/she is expected to win in the long run—is given by $E(M) = M_1P_1 + M_2P_2 + \cdots + M_kP_k$. In gambling game, for the game to be fair the expectation should equal the charge for playing the game.

Mean: The mean of n numbers is their sum divided by n. The mean is also know as the *arithmetic mean* or *average.*

Measures of variation: Statistical description, such as the standard deviation, the mean deviation, or the range, that are indicative of the spread, or dispersion, of a set of data.

Median: If a list of numbers are arranged in increasing order, the median is the numerical value that marks the middle of the list. Hence, when there is an odd number of values, the median is the middle value in the ordered set of data. When there is an even number of values, the median is the mean of the two middle values.

Mode: This term is used to refer to the value that occurs with the highest frequency (most often) in a set of data. Note that a set of data can have more than one mode, or no mode at all when no two values are alike.

Modular arithmetic: For integers a, b, and m, a and b are congruent modulo m, written $a \equiv b$ mod m, if a and b have the same remainders when divided by m.

Multiple: A number, b, is said to be a multiple of another number, a, if a divides b without leaving a remainder.

Multiplicative: A numeration system is said to be multiplicative if each symbol in a number in that system represents a different multiple of the face value of that symbol.

Mutually exclusive events: Two events A and B are mutually exclusive if both events cannot occur at the same time.

Natural numbers: The set of counting numbers, that is, the set $\{1, 2, 3, \ldots\}$, or numbers belonging to this set.

Negation: The negation of a statement p, which is denoted $\sim$p, is the statement "it is false that p."

Negative numbers: Numbers that are less than zero.

Normal distribution: A distribution of data in which the graphical representation of the data is a smooth, bell-shaped curve.

Null set: A set with no elements; also called the empty set.

Number: An idea that represents a numerical quantity.

Numeral: A written symbol used to represent a numerical quantity or number.

Numeration system: A system of writing numbers created to measure numerical quantities.

Oblique cylinder: A cylinder that is not a right cylinder.

Oblique prism: A prism with not all rectangles for the lateral faces.

Oblique regular pyramid: A pyramid with a regular polygon for a base but the lateral faces are not isosceles triangles.

Obtuse angle: An angle with measure x, such that $90° < x < 180°$.

Obtuse triangle: A triangle with one obtuse angle.

Octagon: A polygon that has eight sides.

Odd number: A number is said to be odd if it is not divisible by 2.

One-to-one correspondence: Two sets, A and B, have one-to-one correspondence if and only if each element of set A can be paired with exactly one element of set B and each element of set B can be paired with exactly one element of set A.

Operation: An arithmetic operation such as addition, subtraction, multiplication, or division, defined on the elements of a set.

Ordinal number: A number that designates a comparative position or an order of some set of objects.

Orthocenter: The point that is the intersection of the altitudes of a triangle.

Outcome: A result that can occur during an experiment.

Output variable: Same as DEPENDENT VARIABLE.

Palindrome: A number that reads the same backward and forward.

Parallel lines: Lines in a plane that do not intersect no matter how far they are extended.

Parallelogram: A quadrilateral that has both pairs of opposite sides parallel.

Pascal's triangle: A triangular pattern of numbers, written such that the first and last term in every row is 1, and every other term in a row is the sum of the two numbers directly above it.

Pattern: A design or relationship that is predictable in some manner or a schematic that can be followed.

Pentagon: A polygon that has five sides.

Percent: A ratio where one of the quantities being compared is 100.

Percentiles: The percentiles $P_1, P_2, \ldots$, and P_{99} are values at or below which lie, respectively, the lowest $1, 2, \ldots$, and 99% of a set of data.

Perfect number: A number is said to be perfect if the sum of all its proper factors is equal to the number.

Perimeter: The perimeter of a planar shape is the total length of its boundaries.

Permutation: Any ordered subset of a collection of n distinct objects. The number of possible permutations, each containing k objects, that can be formed from a collection of n distinct object is given by $n(n-1)\ldots(n-k+1) = (n!)/(n-k)!$, and denoted $_nP_k$, $P(n, k)$, $P_k{}^n$ or $(n)_k$. For example, the number of three-digit codes that can be formed using the number $0, 1, 2, \ldots, 9$ is $(10!)/(10-3)! = 720$.

Perpendicular bisector: A perpendicular bisector of a line segment is a line that is perpendicular to the line segment at its midpoint.

Perpendicular lines: Lines in a plane that intersect each other at an angle of 90 degrees.

Pictograph: A graph in which pictures are used to represent data.

Pie graph: A graph used to represent categorical distributions, especially categorical percentage distributions. A pie graph consists of a circle subdivided into a sectors whose sizes are proportional to the quantities or percentages they represent. An example is shown below. During Spring Break, the expenses of three friends went to food, drink, gas, and lodging. The pie graph shows how the money was divided among these four categories (in percent).

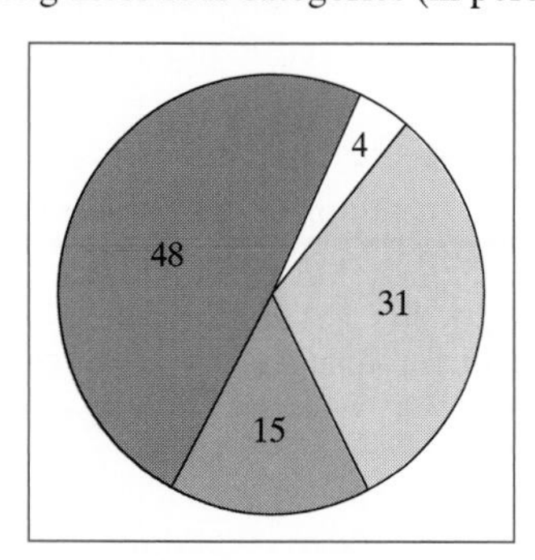

Place value: A numeration system is said to be a place-value system if the value of each digit in a number in the system is determined by its position in the number. The place value of a digit is a description of its position in a given number that determines the value of the digit.

Plane: Although *plane* is an undefined term in geometry, we can describe a plane as a two-dimensional set of an infinite number of points extending forever in all directions.

Point: Although *point* is an undefined term in geometry, we can describe a point as a location.

Polygon: A closed curve composed of line segments that starts and ends at the same point without crossing or retracing itself anywhere in its path is called a polygon.

Polyhedron: A simple closed surface composed entirely of polygonal regions.

Population: *See* FINITE POPULATION; SET.

Positive numbers: Numbers that are greater than zero.

Prime factorization: The decomposition of a number to express it as a product of its prime factors.

Prime number: A positive integer that has exactly two unique factors, 1 and itself.

Prism: A polyhedron comprised of two congruent faces that are in parallel planes and parallelograms for the remaining faces.

Probability of an event: If an event A occurs n times in t trials, the probability of event A, denoted by P(A), is

$$P(A) = \frac{[\text{total occurrences of event A}]}{[\text{total number of trials}]}.$$

Probability distribution: A list of all possible outcomes to an experiment along with the probabilities associated with each outcome.

Proof: A logical, irrefutable argument to demonstrate the truth of a mathematical result.

Proper subset: A is a proper subset of B if A is a subset of B, but there is at least one element of B that is not an element of A.

Proportion: We say two ratios are in proportion if they represent equivalent fractions.

Pyramid: A polyhedron comprised of a polygon with n edges for its base, a point not in the plane of the base called the apex, and n triangular faces that are determined by the apex and the n edges of the base.

Pythagorean Theorem: If a right triangle has legs of lengths a and b and hypotenuse of length c, then $c^2 = a^2 + b^2$.

Quadrilateral: A polygon that has four sides is called a quadrilateral.

Quartiles: The quartiles Q_1, Q_2, and Q_3 are values at or below which lie, respectively, the lowest 25, 50, and 75% of a set of data.

Random event: An unbiased event.

Randomness: An indiscriminate phenomenon.

Random Process: A random process, also called a stochastic process, is a process that is carried at least in part by some random mechanism.

Random sample: A sample of size n from a finite population of size N is said to be random if it is chosen so that each of the $_NC_n$ possible samples has the same probability of being selected.

Random variables: A real-valued function defined over a sample space. It is often referred to as a chance variable, a stochastic variable, or a variate. The number of tails obtained in 12 flips of a coin is an example of random variable. A capital letter, say X, is often used to denote the random variables, and a lowercase letter, say x, is used to denote their values.

Range: The set of all numbers that could occur as "output values" of a function. For example, in the function

$$f(x) = \frac{1}{x - 5}$$

all real numbers *except* 0 could be in the range. Note that 0 cannot be in the range because for the value of the function to be 0, the numerator of the fraction would have to be equal 0.

Range of data set: The range of a set of data is given by the difference between the largest value and the smallest. For instance, the range of the number 3, 1, 5, 2, and 9 is $(9 - 1) = 8$.

Rate of change: Change in one variable, divided by the corresponding change in another variable. For example, if the temperature increases 10° in two hours, then the rate of change can be expressed as 5° per hour (i.e., $10°/2\,\text{hr} = 5°/\text{hr}$).

Ratio: A ratio is a pair of numbers, written as $a : b$, with $b \neq 0$.

Rational number: A rational number is a number of the form a/b, where a and b are any integers, and $b \neq 0$.

Raw data: Data that has not been subjected to any sort of statistical treatment.

Ray: A ray is a collection of points, all of which lie on a line, that continues indefinitely in one direction.

Rectangle: A parallelogram with a right angle.

Reflection: A reflection is a transformation on a plane in which every point in a set is paired with a point in the image such that there is a line that is the perpendicular bisector of the line segment joining a point and its image.

Reflex angle: An angle that measures more than 180°.

Regular polygon: A polygon all of whose sides are of equal length.

Regular polyhedron: A convex polyhedron comprised of congruent regular polygons where at each vertex of the polyhedron there are the same number of edges meeting.

Relatively prime: Two integers are relatively prime if the only factor they have in common is 1.

Representative sample: A sample that contains the relevant characteristics of the population from which it came.

Rep-tile: A figure that serves as a building block for making a larger figure similar to the building block.

Rhombus: A parallelogram with all sides congruent.

Right angle: An angle measuring 90°.

Right circular cone: A circular cone is the set of all points joining the circumference of a circle in a plane to a point in space, such that the line joining the point to the center of the circle is perpendicular to the plane of the circle.

Right cylinder: A cylinder where the line segments joining corresponding points on the bases are perpendicular to the bases.

Right prism: A prism with all rectangles for the lateral faces.

Right regular pyramid: A pyramid with a regular polygon for its base and the lateral faces congruent isosceles triangles.

Right triangle: A triangle with one right angle.

Rotation: A rotation is a transformation on a plane in which every point in a set is moved the same number of units in the same direction by rotating about a point.

Sample: The subset of a population or sample space.

Sample space: The set of points representing the possible outcomes of an experiment is called the sample space for the experiment. The sample space is often denoted by the letter S. For example, when a die is tossed the sample space is $S = \{1, 2, 3, 4, 5, 6\}$.

Scalene triangle: A triangle with no sides congruent.

Scatter plot: A graph used to display the relationship between two variables. The scatter plot below displays the advertising expenditures and sales volume for a company for eight selected months (both are measured in thousands of dollars).

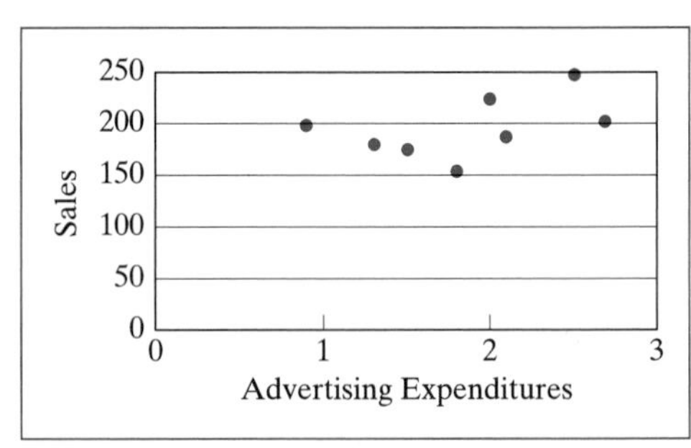

Sector of a circle: A region of a circle formed by a central angle.

Seed: The first input value of an iterated function. For example, if we begin with 7, square it, square the new result, and so on, 7 is referred to as the seed.

Semicircle: An are that consists of half of a circle.

Semiregular tessellation: A tessellation comprised of more than one regular polygon such that each vertex has the same polygonal arrangement.

Set: A collection of objects or ideas.

Sieve of Eratosthenes: A method of finding all the prime numbers less than a given number, where numbers that are not prime are systematically crossed off a list of the natural numbers from 1 to the given number.

Similar polygons: Two polygons are said to be similar if their corresponding sides are proportional and their corresponding angles are equal.

Simple closed surface: A three-dimensional surface that is closed and has exactly one interior, has no holes, and is hollow.

Size transformation: A transformation on a plane to the plane with center O and scale factor s ($s > 0$) in which every point P in a set is paired with a point P', such that $OP' = s \cdot OP$ and O, P and P' are collinear with O not between P and P'.

Skew lines: Any two lines that cannot lie in the same plane.

Skewed data: A distribution of data is skewed if it is not symmetric, and it extends to one side more than the other.

Slope: Graphically, the "steepness" of a line. Numerically, slope is found in the same way as rate of change. It is calculated as follows:

$$slope = \frac{change\ in\ dependent\ variable}{change\ in\ independent\ variable}$$

A positive slope would indicate a positive rate of change and would appear graphically as a line slanting upward from left to right. A negative slope would indicate a negative rate of change and would appear as a line slanting downward from left to right. A slope of zero would indicate no change in the dependent variable and would be seen as a horizontal line.

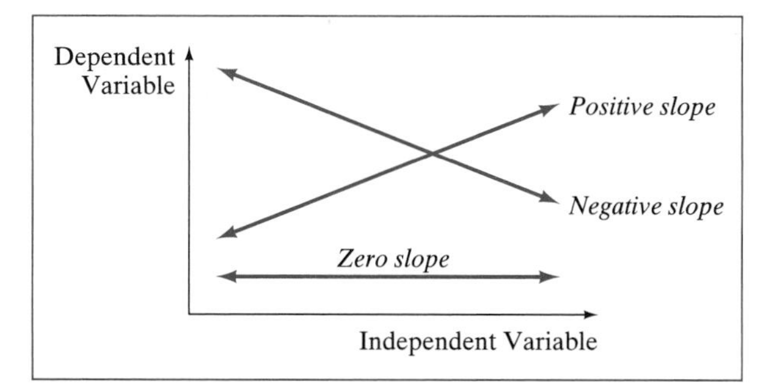

Solid: The set of points comprised of a simple closed surface and its interior.

Sphere: A sphere is the set of all points in space whose distance from a fixed point also in space is constant.

Square: A rectangle with all sides congruent.

Square root: The square root of a number is a number that when multiplied by itself produces the given number.

Standard Deviation: The standard of a sample of size n is given by the square root of the sum of the squared deviations from the mean divided by $(n - 1)$. Some mathematicians prefer to divide by n rather than by $(n - 1)$. If a set of data has x_j elements and $av(x)$ designates the mean of all the x_j's, for $j = 1, 2, \dots k$. then the standard deviation is given by

$$SD = \text{Sqrt}\left[\frac{(\hat{A}\,(x_j - av(x))^2)}{(n \text{ or } n - 1)}\right], \text{ for } j = 1, 2, \dots k.$$

Statistic: A number obtained on the basis of a sample. Standard deviation is an example of a statistic.

Statistics: The totality of methods employed in the collection and analysis of any kind of data and, more broadly, that branch of mathematics that deals with all aspects of the science of decision making in the face of uncertainty.

Stem-and-leaf plot: A two-column chart used to represent data in which the left column represents stems of data and the right column represents the leafs.

Straight angle: An angle that measures 180°.

Subset: A set A is a subset of another set B if every element in set A is also in set B.

Surface area: The area of an object's surface

Supplementary angles: Two angles whose sum is 90 degrees are called supplementary angles.

Surface area: The surface area of a solid figure is the total area of its external surfaces.

Table: Data that have been arranged in rows and columns.

Tally chart: A chart used to record the frequency (how often) a certain piece of data occurs. Each occurrence of the piece of data is recorded by a short vertical bar. (*See* FREQUENCY TABLE.) Below is an example of a tally chart.

Data	Tally	Frequency
apple	||| |	4
orange	||| ||| ||| |	10
prune	||| ||| ||	8

Tessellation: A tessellation of the plane is the filling of the plane with repetitions of polygonal regions so that all corners meet at a vertex.

Theoretical probability: The probability determined by analyzing an experiment or situation.

Tiling: A tiling of the plane is the filling of the plane with repetitions of polygonal regions where corners do not necessarily meet at a vertex.

Transformation: A correspondence between all the points of a geometric figure in a plane and all the points of its image such that each point of the figure can be matched with exactly one point of its image, and each point of the image can be matched with exactly one point of the original figure.

Translation: A transformation on a plane in which every point in a set is moved the same number of units in the same direction.

Trapezoid: A trapezoid is a quadrilateral with at least one pair of parallel sides.

Tree diagram: A diagram in which individual paths represent all possible outcomes of an experiment.

Trial: A series of repeated experiments, such as that of tossing a die until a 5 is obtained.

Triangle: A triangle is a polygon that has three sides.

Twin primes: Two prime numbers that differ from each other by 1.

Union: The union of two sets is the set of all elements that are in either set or in both sets.

Unique representation: A numeration system is a unique representation system if each numeral refers to one and only one number.

Universal set: The set of all elements in a particular category.

Variable: A quality or quantity that may change or has the potential to change. For example, the amount of money in a person's savings account is variable because it may change as a result of interest, deposits, withdrawals, and so on.

Variance: A statistic that indicates how data are dispersed around the mean. The variance is the mean of the squared deviation from the mean.

Vertical angles: Pairs of angles that are formed at the point of intersection of two intersecting lines.

Volume: The volume of a solid geometrical object is a measure of the amount of space that it encloses.

Weighted average: The average of a set of numbers obtained by multiplying each number by a weight expressing its relative importance and then dividing the sum of these products by the sum of the weights.

Consider the set of numbers $x_1, x_2, \dots, x_k$ with their respective weight $w_1, w_2, \dots, w_k$. The weighted average is given by $W_a = [x_1w_1 + x_2w_2 \cdots + x_kw_k]/[w_1 + w_2 + \cdots + w_k]$.

For example, if you work 3 hours at job A at \$18 per hour, and 5 hours at job B for \$24 per hour, then your (weighted) average earning per hour is $[(18)(3) + (5)(24)]/[8] = \$21.75$.

Whole numbers: The set of numbers $\{0, 1, 2, \dots\}$, or numbers belonging to the set.

Z-score: The number of standard deviations that a given data value is above or below the mean.

Index

A

absolute value, 81
absolute value notation, 81
abundant integer, 108
acute angle, 330
acute triangle, 277
addends, 46
addition, 45
addition algorithms, 52
Addition Property of Equations, 237
Addition Property of Inequalities, 238
addition table, 48
additive, 30
additive identity, 49, 212
Additive Inverse Property, 82, 213
adjacent angles, 269
Agnesi, Maria, 241
Ahmes Papyri, 170, 336
algorithm, 51
al-Khwàrizmì, 51
alternate exterior angles, 270
alternate interior angles, 270
altitude, 277, 279, 323
angle, 267
angle bisector, 289
Angle-Angle-Side Triangle-Congruence Property, 301
Angle-Angle Triangle-Similarity Property, 308
Angle-Side-Angle Triangle-Congruence Property, 301
annual percentage rate, 205
apex of a cone, 323
apex of a pyramid, 318
arc, 281
Archimedes, 201, 357
area of a planar shape, 338
area formulas, 341, 342, 343, 344
area model of multiplication, 64, 200
arithmetic sequence, 121
array, 62
array multiplication, 62
array division, 75
Associative Property of the Intersection of Sets, 9
Associative Property of the Union of Sets, 9
Associative Property of Addition, 49, 212
Associative Property of Multiplication, 67, 212

B

Babylonian mathematics, 34, 177, 281, 329
Babylonian numeration system, 34
bar graph, 158
base, 251
base of a cone, 323
base of a cylinder, 322
base of a polygon, 277, 279
base of a prism, 318
base of a pyramid, 318
base-60 system, 34
base-ten block model of addition, 52, 195
base-ten block model of division, 77
base-ten block model of multiplication, 69
base-ten block model of subtraction, 59, 197
base-ten system, 32
base-two system, 32
basic addition facts, 48
Bernoulli, Jakob, 130
between, 259
Bhàskara, 51
bilateral symmetry, 285
bimodal, 148
binary system, 32
Boethius, Anicius Manlius, 66
Borghi, Piero, 186
box-and-whisker plot, 156
Brahmagupta, 170, 177

C

Cantor, Georg, 2, 227
cardinal number, 42
Carroll, Lewis, 7
Cartesian Coordinate System, 239
Cartesian product, 9, 62
Cavalieri, Bonaventura, 359
Cavalieri's Principle, 359
Celsius, Anders, 362
center of a circle, 281
center of a sphere, 323
center of rotation, 376
central angle, 336
centroid, 297
certain event, 130
Charlemagne, 332
chord, 281
circle, 281
circle graph, 163
circular cone, 323
circular cylinder, 322
circumcenter, 294
circumference, 336
circumscribed circle, 294
class mark, 148
clock arithmetic, 118
Closure Property for Addition, 49, 212
Closure Property for Integer Subtraction, 86
Closure Property for Multiplication, 67, 212
cluster sampling, 146
collinear, 259
combination, 143
common denominator, 173
common fraction, 189
Commutative Property of Addition, 48, 49, 212
Commutative Property of the Intersection of Sets, 9
Commutative Property of Multiplication, 67, 212
Commutative Property of the Union of Sets, 9, 48
comparison subtraction, 57
compass, 282, 287
compensation, 46, 58
complement of a set, 8
complement of an event, 136
complementary angles, 270
composite number, 103

composition, 235
compound statement, 15
compounded interest, 206
concave polygon, 273
concave polyhedron, 317
conclusion, 14
concrete representation of an object, 42
concurrent lines, 263
conditional probability, 138
conditional statement, 16
cone, 323
congruent, 283, 298
conjecture, 14
conjunction, 15
constant of proportionality, 185
contrapositive statement, 16
convenience sampling, 146
converse statement, 16
convex polygon, 273
convex polyhedron, 317
coordinates, 239
coplanar, 260, 262
correlation, 163
corresponding angles, 270
counting back, 57
counting up, 46
counting numbers, 41, 80
cross multiplication, 174, 184
cube, 251, 355
cube of a number, 251
cubic unit, 352
cylinder, 322

D

Dantzig, George, 245
data, 145
data class, 159
de Meurs, Jean, 186
decagon, 273
decimal, 36, 186
decimal fraction, 186
decimal point, 186
decimal system, 36
deductive reasoning, 12
deficient integer, 108
degree Celsius, 362
degree Fahrenheit, 362
degree Kelvin, 361
degree unit, 329
denominator, 173
Density Property of Rational Numbers, 211
dependent events, 138
Descartes, René, 239
diagonal, 272
diameter of a circle, 281
diameter of a sphere, 323
difference, 56
dihedral angle, 315, 330
direct reasoning, 17
directed line segment, 374
directly proportional, 185
discrete models of division, 75
disjoint sets, 5, 45
disjunction, 15
distance formula, 367
distance postulate, 366
distribution curve, 152
Distributive Property of Intersection over Union of Sets, 10
Distributive Property of Multiplication over Addition, 68, 213
dividend, 73
divisibility tests, 109
divisibility theorems, 108
divisible, 108
division, 73
division algorithms, 77
division by zero, 76
Division Property of Equations, 237
Division Property of Inequalities, 238
divisor, 73
dodecahedron, 320
domain, 232
double bar graph, 158
double plus one, 47
double plus two, 47

E

edge of a polyhedron, 317
Egyptian mathematics, 31, 62, 170, 173, 214, 220, 341, 344, 349, 352
Egyptian numeration system, 31
element, 2
elimination method of solving systems of equations, 249
empirical probability, 131
empty set, 2
endpoints, 266
English System of Measurement, 332, 338, 352, 361, 362
equal additions method of subtraction, 59
equal sets, 6
equals sign, 237
equally likely outcomes, 130
equation, 233, 237
equiangular polygon, 274
equilateral polygon, 274
equilateral triangle, 278, 334
equivalent, 5
equivalent fractions, 173
equivalent ratios, 183
Escher, Maurits Cornelis, 370, 372
estimating products division algorithm, 79
Euclid, 113, 260
Euclidean algorithm, 114
Euler, 316
even number, 109
event, 128
expanded notation, 36
expected value, 132
experiment, 128
experimental probability, 131
exponent, 251
exterior angles, 270
exterior of an angle, 268

F

face of a polyhedron, 317
factor, 62, 109
factorial, 142
Fahrenheit, Gabriel, 362
Fermat, Pierre de, 128
Fermat primes, 274
Fermat's last theorem, 116
Fibonacci, 45, 121, 199
Fibonacci sequence, 122, 199, 303
figurate number, 123
finite, 2
finite set, 2, 211
45°-45°-90° right triangle, 364
fractal, 312
fraction, 170
fraction as part of a whole, 170
frequency, 159
frequency polygon, 161
frequency table, 159
function, 232
function machine, 234
Fundamental Counting Principle, 144
Fundamental Theorem of Arithmetic, 107

G

Gauss, Carl Friedrich, 82, 118, 151, 274
Gaussian curve, 151
geometric sequence, 122
geometry, 258

Germain, Sophie, 116
glide reflection, 378
Goldbach's conjecture, 103
Golden Ratio, 182, 303
Golden Rectangle, 182, 303
Golden Rule, 184
gram, 361
graph, 233
Graunt, John, 145
great circle, 351
greatest common divisor, 113
greatest common factor, 113, 175
Grouping Property, 49
growing pattern, 122
guess and check, 20

H

heptagon, 273
Heron, 343
hexagon, 273
hieroglyphics, 170
Hindu-Arabic numeration system, 30, 36
Hippasus, 225
histogram, 161
horizontal bar graph, 158
hypotenuse, 363
hypothesis, 14, 145

I

icosahedron, 320
Identity Property of Multiplication, 67
image, 372
impossible event, 131
improper fraction, 177
incenter, 297
independent events, 138
indirect reasoning, 17
indirectly proportional, 185
inductive reasoning, 11
inequality, 238
infinite, 2
infinite set, 2, 211
inscribed circle, 296
integer, 81
integer addition, 82
integer division, 94
integer multiplication, 90
integer subtraction, 86
interest, 205
interest rate, 205
interior angles, 270
interior of an angle, 268
intersecting lines, 263
inverse, 16
inverse statement, 16
irrational number, 225
Islamic mathematics, 368, 370
isometry, 373
isosceles trapezoid, 280
isosceles triangle, 278
Isosceles-Triangle-Congruence Properties, 302

K

Karmarkar, Narendra, 249
kite, 279, 305
Klein, Felix, 374
Klein's Erlanger Programm, 374
Koch curve, 312
Kovalevsky, Sonya, 232

L

lateral face of a prism, 318
lateral face of a pyramid, 318
lateral surface area, 346
lateral surface of a cylinder, 322
least common multiple, 116
legs of a right triangle, 363
length of a segment, 332
Leonardo of Pisa, 121, 199
Lindemann, Ferdinand, 294
line, 258
line graph, 161
line of best fit, 162
line of reflection, 378
line of symmetry, 285
line segment, 266
line symmetry, 285
linear model of a rational number, 208
linear pair angles, 269
Lobachevsky, Nicholai, 262
look for a pattern, 22

M

make a table, 22
make an organized list, 23
Mandelbrot, Benoit, 312
mapping, 234
mass, 361
Mayan numeration system, 31
mean, 147
median, 149
Mendel, Gregor, 147
mental computation, 46
Mesopotamia, 45
meter, 332
Metric System of Measurement, 332, 339, 353, 361, 362
midpoint, 266
midpoint formula, 367
midrange, 149
midsegment, 311
minuend, 56
missing-addend subtraction, 56
missing-factor division, 73
mixed decimal, 186
mixed number, 177
mode, 148
modular arithmetic, 119
modulus, 119
multimodal, 148
multiple, 109, 116
multiplication, 62
multiplication algorithms, 69
Multiplication Property of Equations, 237
Multiplication Property of Inequalities, 238
multiplication table, 66
multiplicative, 36
multiplicative identity, 67, 212
Multiplicative Inverse Property, 213
mutually exclusive events, 133

N

n-gon, 273
name, 42
natural number, 41, 80
negation, 15
negative correlation, 163
negative number, 80
negative sign, 80
negative slope, 241
negatively skewed data distribution, 153
net of the surface of a three-dimensional figure, 346
Nightingale, Florence, 147
nonagon, 273
noncollinear, 259
noncoplanar, 260, 262
non-mutually exclusive events, 133
nonterminating, nonrepeating decimal, 225
normal curve, 151
normal distribution, 153
nth term, 122
null set, 2
number, 42

number line, 46
number-line model of addition, 46, 84
number-line model of division, 75
number-line model of multiplication, 64, 92
number-line model of subtraction, 57, 88
numeral, 42

O

oblique circular cone, 323
oblique cylinder, 322
oblique prism, 318
oblique regular pyramid, 318
obtuse angle, 330
obtuse triangle, 277
octagon, 273
octahedron, 320
odd number, 109
odds against an event, 136
odds in favor of an event, 136
one-to-one correspondence, 5, 43
opposites, 81
order of a set, 6
order of operations, 97
ordered pair, 234, 239
Ordering Property, 49
ordinal number, 42
origin, 239
orthocenter, 297
outcome, 128
outlier, 163

P

Pacioli, Luca, 69, 73
paired data, 161
parallel lines, 263
Parallel Postulate, 262
parallelogram, 280, 304
Pascal, Blaise, 142
Pascal's triangle, 142
partial products algorithm, 70
partial sums algorithm, 53
pattern, 121
Pellos, Francesco, 186
pentagon, 273
pentagonal number, 123
percent, 203
percent change, 204
percentile, 155
perfect integer, 108
perimeter, 334
perimeter formulas, 334
permutation, 143
perpendicular bisector, 292
perpendicular lines, 264
Petty, William, 145
pi, 225
pictograph, 157
pictorial model, 45
pie graph, 163
place value, 32, 34
plane, 260
plane of symmetry, 286
Plato, 314, 320
point, 258
point-slope form, 244
point symmetry, 285
Polya, George, 19
polygon, 272
polygonal region, 272
polyhedron (polyhedra), 316
population, 146
positive correlation, 163
positive number, 80
positive sign, 80
positive slope, 241
positively skewed data distribution, 153
power, 251
powers of 10, 31
prime factorization, 105
prime number, 103
Primitive numeration system, 30
principal, 206
prism, 318
probability, 128
probability distribution, 129
problem solving, 19
product, 62
proper divisor, 108
proper subset, 4
proportion, 184
proportional, 185
protractor, 287, 329
pyramid, 318
Pythagoras, 66, 225, 363
Pythagorean Theorem, 363, 366
Pythagoreans, 225, 363

Q

quadrant, 239
quadrilateral, 273, 279
quartile, 153
Quételet, Lambert Adolph, 151
quotient, 73, 172

R

radius (radii) of a circle, 281
radius (radii) of a sphere, 323
random sampling, 146
range, 149, 159, 232
rate, 183
ratio, 171, 182
rational exponent, 226
rational number, 207, 210
ray, 267
real number, 227
reciprocal, 223
rectangle, 280, 304, 334
reflection, 378
reflection symmetry, 285
reflex angle, 330
region model of a rational number, 207
regular polygon, 274
regular polyhedron, 320
relative frequency, 164
relatively prime, 111, 113, 175
repeated addition multiplication, 62
repeated subtraction division, 73
repeatand, 192
repeating bar, 192
repeating decimal, 192
rep-tile, 313
rhombus, 280, 306
right angle, 330
right circular cone, 323
right cylinder, 322
right prism, 318
right regular pyramid, 318
right triangle, 277
rigid transformation, 373
Robinson, Julia Bowman, 80
Roman numeration system, 30, 173
rotation, 376
rotational symmetry, 285
Rudolff, Christoff, 186
Rule of Three, 184
ruler, 287

S

sample, 146
sample space, 128
scaffolding division algorithm, 78
scalene triangle, 278
scatterplot, 162
scientific notation, 253
scratch algorithm, 54
sector of a circle, 345
semicircle, 281

semiregular polyhedron, 320
semiregular tessellation, 371
set, 2
set model of a rational number, 208
set-builder notation, 2
set model of integer addition, 83
set model of integer division, 95
set model of integer multiplication, 91
set model of integer subtraction, 87
sharing division, 74
Side-Angle-Side Triangle-Congruence Property, 300
Side-Angle-Side Triangle Similarity Property, 308
Side-Side-Side Triangle-Congruence Property, 300
Side-Side-Side Triangle-Similarity Property, 307
sides of a polygon, 272
sides of an angle, 267
Sierpinski triangle, 312
Sieve of Eratosthenes, 103
significant digits, 253
similar, 307
simple closed surface, 316
simple interest, 205
simplest form, 175
simplex method, 245
simulation, 129
size transformation, 380
skew lines, 263
skewed data distribution, 153
slant height, 349
slope, 241
slope-intercept form, 244
solid, 316
solve a simpler problem, 24
sphere, 323
square, 251, 280, 306, 334
square number, 123
square of a number, 251
square root, 225
standard addition algorithm, 52
standard deviation, 151
standard multiplication algorithm, 71
statement, 15
statistical data, 145
statistics, 145
stem-and-leaf plot, 164
Stevin, Simon, 195
straight angle, 330
straightedge, 287
stratified sampling, 146
subset, 4
substitution method of solving systems of equations, 249
subtraction, 56
subtraction algorithms, 60
Subtraction Property of Equations, 237
Subtraction Property of Inequalities, 238
subtractive, 32
subtrahend, 56
sum, 46
supplementary angles, 270
surface area, 346
symbol for zero, 35
system of equations, 245
systematic sampling, 146

T

table, 234
take-away subtraction, 56
temperature, 361
terminating decimal, 192
tessellation, 368
tetrahedron, 320
Thales, 11, 307
theoretical probability, 131
30°-60°-90° right triangle, 365
transformation, 372
transitional algorithm for addition, 54
transitional algorithm for division, 79
translation, 374
transversal, 270
trapezoid, 279, 303
tree diagram, 63, 105, 140
triangle, 277
Triangle Inequality, 365
triangular number, 123
triple bar graph, 158
truth tables, 15

U

undefined terms, 258
union of sets, 7, 45
unique representation, 30
Universe set, 5
use a visual aid, 20
use algebra, 24

V

van Hiele, Pierre, 325
van Hiele-Geldof, Dina, 325
variance, 151
Venerabilis, Beda, 66
Venn diagram, 4, 134
Venn, John, 4
vertex (vertices), 267, 272, 317
vertical angles, 270
vertical bar graph, 158
vertical-line test, 235
volume, 352

W

Wallis, John, 251
Wantzel, Pierre, 296
Weierstrass, Karl, 94
weight, 361
Weyl, Hermann, 285
whole number, 44, 80
whole-number addition, 45
Wiles, Andrew, 116
work backward, 23

X

x-axis, 239
x-intercept, 243

Y

y-axis, 239
y-intercept, 244
Yoruba numeration system, 31

Z

z-score, 152
Zero Property of Multiplication, 67